CLASSICAL
ABSTRACT
ALGEBRA

CLASSICAL ABSTRACT ALGEBRA

Richard A. Dean
California Institute of Technology

HARPER & ROW, PUBLISHERS, New York
Grand Rapids, Philadelphia, St. Louis, San Francisco,
1817 London, Singapore, Sydney, Tokyo

Sponsoring Editor: Peter Coveney
Project Editor: Ellen MacElree
Art Direction/Cover Coordinator: Heather A. Ziegler
Cover Design: Wanda Lubelska Design
Production: Kewal Sharma

Classical Abstract Algebra

Library of Congress Cataloging-in-Publication Data

Dean, Richard A.
　Classical abstract algebra/Richard A. Dean
　　p.　cm.
　ISBN　0-06-041601-7
　1. Algebra, Abstract.　I. Title.
QA162.D43　1990
512′.02--dc20　　　　　　　　　　　89-48329
　　　　　　　　　　　　　　　　　　　CIP

89　90　91　92　9　8　7　6　5　4　3　2　1

Contents

Chapter 2   RELATIONS AND LATTICES   40

Chapter 3 INTRODUCTION TO GROUPS   66

Chapter 6 PERMUTATION GROUPS   158

Chapter 7 INTRODUCTION TO RINGS   178

Chapter 9 POLYNOMIALS 216

PART TWO   SELECTED TOPICS   308

Chapter 11   TOPICS IN GROUPS   309

Chapter 12   TOPICS IN FIELDS   343

Chapter 13   GALOIS THEORY FOR FIELDS OF CHARACTERISTIC ZERO

Preface

To the Instructor

This text is designed in two parts. The first part introduces the main algebraic concepts and constructs usually implied by the term "abstract algebra." This is the subject area called "modern algebra" by the Dutch mathematician B. L. van der Waerden and by two American mathematicians, Garrett Birkhoff and Saunders MacLane, who wrote excellent texts in the 1940s. Part One of this text provides first-time students of abstract algebra with a leisurely paced introduction, filled with many examples. For students with more mathematical experience Part One can be pruned to give a direct development of the tools needed to confront the classical problems of abstract algebra. Part Two then discusses topics of intermediate sophistication, which are often contained in first-year graduate courses in algebra. Here the proofs are given an elementary treatment. As often as possible proofs are chosen that are both constructive and algorithmic, and these important ideas and results are accompanied by many examples. Thus, advanced results are made accessible to students in their very first year of study.

The formal prerequisites for Part One are only the facility with algebraic manipulations and mathematical proofs that might come from a good course in Algebra II taken in high school or from a standard course in calculus. It is true that some mathematical sophistication is expected. Students fresh from high school will have to proceed slowly to gain experience with reading and writing proofs. A peek at the table of contents for Part One shows that it deals with the basics of integer arithmetic, mathematical induction, sets, groups, rings, and field extensions.

Chapters 1–10: A Basic Course Chapter 1 begins with some naive set theory to remind students that we need to use this terminology and notation. Our major focus is the treatment of mathematical induction, some elementary number theory dealing with congruences and greatest common divisors, and the all-important euclidean algorithm. In this way interesting examples can be constructed, and elementary computations can be carried out immediately. I have included quite a bit about mathematical induction because it is used everywhere and in different forms. It is introduced as both a weak and a strong induction principle, and as a well-ordering principle. Mathematical induction is used throughout this text not only to prove theorems but to describe algorithms for computation, to construct examples, and, occasionally, to give definitions. If your students are familiar with these concepts you should hurry along to Chapter 2. Also in Chapter 1 we pause to introduce the notion of an ideal in the integers. This is not a crucial point of the development, but it is very easy ground to cover here and provides a good preview of the ring theory concept to come. It also permits some economy in the proofs about cyclic groups in Chapter 3.

In Chapter 2 we give a brief introduction to partially ordered sets and lattices. These topics are, of course, the setting for a great deal of mathematical research, but this is not the emphasis here. What is important for us is their ability to give a concrete picture of the containment relations between subsystems in groups, rings, and fields. This material may be new to some students, but I have found that it is helpful. By the way, one of the most difficult hurdles for beginning students in algebra, including linear algebra, is that subsets are treated as individual elements and manipulated very much the way individual numbers or functions are treated in calculus. Dealing with sets is not a factor in learning calculus! And it is at this point that lattice diagrams help. By using lattices a subset can be pictured as a single point in the lattice diagram.

The next four chapters are devoted to group theory. This account begins with axiomatics to give some experience with group computations and then proceeds through subgroups, homomorphisms, and factor groups to end with direct products and permutation groups. The classes of groups that can be described in almost complete detail are cyclic groups, dihedral groups, multiplicative groups of integers modulo n, symmetric groups, and alternating groups. It pays to proceed slowly with students in this first part of the course. Beginning students should see lots of examples and do lots of exercises. As students become comfortable with a higher level of abstraction, and learn that they too can construct clarifying examples, the pace can be efficiently increased.

Chapter 3 begins with the briefest of axioms systems and proves the two-sidedness and uniqueness of the identity and the companion properties for the inverse of an element through a series of easy lemmas. Chapter 3 ends with a combinatorial proof of Cauchy's theorem. Chapter 4 covers group homomorphisms and normal subgroups including their lattice properties. It begins with the example of the group of rotations of a regular tetrahedron, complete with directions for making a model from an envelope. Chapter 5 begins with a listing of groups of small order and proceeds with motivation from the groups of order 8 and 9 to discuss direct products. A section on abelian groups states the structure theorem for finite abelian groups, but no proof is given at this stage. Chapter 6 concludes this introduction to group theory with a brief account of permutation groups. Included is an interesting inductive proof of the simplicity of the alternating group for $n \geq 5$.

If you wish, you may proceed directly to the presentation of the Sylow theorems in Sections 11.1 and 11.2. The reason they are not included earlier is that often students grow tired of the group setting and seem eager to push on to new ground. Another good reason is that while the Sylow theorems help to determine the structure of possible groups of a given order, they do not provide methods for construction of these groups. For that reason I have chosen to put this material with the discussion of semidirect products. All groups of reasonable orders (for example, order 24) may then be determined together.

Chapters 7 and 8 deal with rings and integral domains. Here, of course, the algebraic constructs are the same as those for groups. Many of the theorems have the same form. We can also pick up the pace of the presentation. There are plenty of examples to keep ideas concrete and to use for motivation. The definition of a prime element is based on the divisibility property, and some time is spent on examples to make clear the distinction between irreducible and prime elements. Fields are introduced by citing familiar ones and constructing the finite field of four elements. Wedderburn's theorem on the nonexistence of finite division rings that are not fields is mentioned, but naturally the proof has to be postponed to Part Two (Section 12.2).

Chapter 9 on polynomials is one of the longer chapters in this text. There is a bit of a to-do over the nature of the indeterminate x; rules for substitution for x are covered in a special theorem. The distinction between polynomials and polynomial functions is made precise with appropriate examples. The so-called "division algorithm," Theorem 9.1.11, is stated and proved in a general setting so that it may be used in rings of matrices rather than simply in the ring of polynomials over a field. Perhaps the most important of all theorems is the Kronecker existence theorem, which, given a field F and a polynomial, provides for the construction of a field containing F and a zero of the given polynomial. The ring of polynomials over a field is used to motivate the definition of euclidean domains. Gaussian integers and their application to the determination of those primes that can be expressed as the sum of two squares has been postponed to Section 12.3. You may want to include that material now. The chapter concludes with the lemma of Gauss and Eisenstein's irreducibility criterion.

The final chapter in Part One discusses field extensions. For this we require a little vector space theory over arbitrary fields. In Section 10.1 there is a short section to introduce this concept and to obtain the basic properties including independence, basis, and dimension. We give a neat proof that the row rank and column rank of a matrix are equal. However, no matrix operations are introduced here. We determine all quadratic extensions of degree 2 (for fields whose characteristic is not 2) and prove the multiplicative property of the degrees of a sequence of algebraic extensions. We end with a discussion of splitting fields and finite fields. It would not be wrong to stop an introductory course in algebra at this point. The chief concepts have been introduced and enough theory has been presented to be useful. The information on finite fields finds many applications outside abstract algebra. A student who later needs or is interested in the topics of Part Two is now ready to study and read independently.

Chapters 11–16: Special Topics   Part Two contains a number of independent topics in classical algebra. The background needed from Part One for each topic varies. Each chapter of Part Two contains a list of its prerequisites. In my course at Caltech I select several of these topics, but never all of them! Often I suggest to the class that they browse through the pages of Part Two and make a proposal. This often initiates a lively discussion of what interests or seems important to them. And we do what they select.

I recommend doing the Sylow theory and the Fundamental Theorem of Galois Theory (FTGT). The latter gives the Galois correspondence between subgroups of the group of automorphisms and subfields of an extension. The major simplifying assumption I have made here is to discuss only fields of characteristic zero. The FTGT does a beautiful job of pulling together ideas from groups and from fields in a mutually supporting role. Examples make these ideas concrete. And one can stop there; you do not have to do the whole criterion for "solution in radicals." But all the details are there if you want them. They are surprisingly complex and only so because one has to cope with a bunch of roots of unity which may not be the original field.

I like to include the last chapter on finitely generated modules. Modules abound in algebra, analysis, and topology and play an increasingly important role in contemporary mathematics. Chapter 16 is designed to feature the application of the central structure theorem to obtain canonical forms for finitely generated abelian groups (not just finite ones) and also for matrices over a field. Again, this material provides an opportunity to see the beauty and power of abstract thinking.

The topics chosen for inclusion in Part Two arise from perfectly natural questions but ones whose answers are not readily available with elementary arguments. For example:

- When is the multiplicative group of integers modulo n cyclic?
- Give an example of a finite non-abelian simple group other than an alternating group.
- Are there finite systems that satisfy all field axioms except commutativity of multiplication?
- For what primes p does the congruence $x^2 + 1 = 0 \pmod{p}$ have a solution?
- What is algebraic about the Fundamental Theorem of Algebra?

- So you have used the Sylow theorems to determine the possibilities for all groups of order n—but how do you show they exist?
- How do you find the cyclic summands in the direct product representation of a finite abelian group?
- Can a finite extension of the rationals have an infinite number of subfields?
- Is there an algorithm for determining Jordan Canonical Form?

The material in Part Two is not intrinsically more difficult than that of Part One. But it is more complex, and by this time proofs call on many different results from Part One. Again, patience in the classroom, together with lots of examples and concrete exercises, makes it possible for first-year students to assimilate this material.

Organization Every numbered definition, example, and theorem has been given a name and is listed in the table of contents. Thus, you can see at a glance what is involved. Students can also use this feature of the table of contents for review. Exercises appear at the end of every section. The ones with an asterisk are more difficult or time consuming, and should be tackled under those caveats.

To the Student

The term "modern" abstract algebra was introduced sometime in the 1930s. The most influential text was written by the Dutch mathematician B. L. van der Waerden. The problems it treated were old, as are the problems treated in this text. For example, Euclid must have wondered whether or not it was possible to trisect every angle using straightedge and compass alone. You can read here why this is impossible.

What is remarkable about mathematics is that the ideas invented to handle one class of problems turn out to have wide repercussions and applications to other problems. This is very much the case with algebra. And, of course, because mathematics has so many concrete applications in the "real world," it follows that these concepts have played an important role in the development of modern-day science and technology.

So it is not because of these ancient problems that you are embarking on a study of abstract algebra. It is because the ideas generated to answer them form a cornerstone of modern mathematics. The ideas have names: groups, rings, fields, vector spaces, and modules. There is a rhythm to their development. Groups are basic; they appear in each construct that follows. Rings can be thought of as a generalization of the integers. Fields generalize the rational numbers and have two groups inside them. Vector spaces and modules describe the algebraic properties associated with actions on points in space. The problems discussed in connection with groups and the methods developed to solve them are repeated again in the theory of rings, fields, vector spaces, and modules.

Each algebraic system is introduced by its axioms. A number of examples follow these axioms. Then come a number of definitions. You probably will need to refer back to these definitions often! Then come theorems and lemmas. With each lemma or theorem there is a proof. Before you begin a proof be sure you understand the statement of the theorem: What are the hypotheses and what are the conclusions? If at all possible, take one of the examples and fit it to the theorem. When you feel comfortable with the statement of the theorem you are ready for the proof.

You will learn mathematics in at least two ways. One is by understanding the proofs. The other is by doing the exercises. (Each theorem can be converted into an exercise by simply ignoring the written proof and giving your own.) Another interesting activity is to compose your own theorems! Suppose you are on page N. Make a guess as to what theorem will appear on page $N + 1$. At first this may seem to be a formidable task, but as you practice this technique you will find that your insight deepens and your creativity expands.

Exercises Each section has its own set of exercises. In general, easier exercises are placed at the beginning of each set. Problems that seem to require more than the usual sophistication or long computations are denoted by an asterisk (*). If you are interested in one of these problems, by all means give it a try, but don't stay up nights worrying over it. That is good advice for all problems. Many problems yield to the "on again, off again" approach. Once you have a problem firmly in mind and have made some initial attempts on a solution, but find yourself stuck, put it aside. A rest period is a good idea. When you return afresh, new strategies may come to the fore and calculations will proceed more easily. In this way even the most recalcitrant of problems may be solved. A word to the wise, then: Don't wait to begin a homework assignment on the night before it is due!

Problem-Solving Strategies Books have been written on the subject of problem solving. Every one of you must build your own personal library of methodology. Here are some things you might try if you are stuck. Not all make sense for all problems.

- *Do a simpler case of the exercise.* For example, if the problem concerns a pair of integers (a, b), where a and b have no common divisors except 1, try the case that a and/or b is a prime.
- *Do a specific case.* For example, if the problem is about groups try it out on your favorite group.
- *Add a simplifing hypothesis.* For example, if the problem is about all groups, try it for Abelian groups.
- *Try to find a counterexample.* This may show where you have misinterpreted the statement. On an exam, who knows, maybe your instructor erred. Perhaps even your author erred! I can assure you that there are no deliberate traps, but never bet against typographical and pilot error!

- *Draw some kind of a picture*. For example, problems often involve sets of elements. A schematic drawing of the inclusion relations between the sets is often helpful.

I hope you enjoy abstract algebra as much as I do. But don't feel badly if you don't. Some of my best friends prefer analysis or topology. One thing I do know is this: Whatever field of mathematics you study, you'll speak about groups, rings, fields, vector spaces, and modules.

Acknowledgments

This text would not exist if I had not had wonderful students and colleagues who posed good questions and gave encouragement to me to write the notes from which this text emerged. I thank them for many years of mathematical engagement!

A number of reviewers have provided useful suggestions for improving the quality of this text. In particular I would like to thank the following:

Herman Gollwitzer	*Drexel University*
Curtis Greene	*Haverford College*
John Hogan	*Marshall University*
Peter Jones	*Marquette University*
Philip Montgomery	*University of Kansas*
Donald Passman	*University of Wisconsin–Madison*
Robert Snider	*Virginia Polytechnic Institute*
Guido Weiss	*Washington University*

Finally, I would like to thank the editorial and production staffs at Harper & Row.

Richard A. Dean

PART ONE

BASIC CONCEPTS

Chapter 1

Integers and Sets

In this book we study several different algebraic systems. Each system is defined by a set of axioms. These are properties that are to be accepted without proof. In that sense they define the algebraic system. We give lots of examples of systems that satisfy these axioms. Then we develop many of the properties of systems satisfying these axioms by proving a series of theorems. Many of these examples as well as the proofs of the theorems entail properties of the integers, probably the algebraic system best known to you. We begin our work by developing some further properties of the integers with which you may not be familiar. In Section 1.2 the most important concept is mathematical induction, for that is a major tool in the proofs in this text. In Section 1.3 we discuss the greatest common divisor of two integers, the euclidean algorithm, and prime integers. In Section 1.4 we review operations on sets, functions, and binary operations.

Since we want to capitalize on your familiarity with the integers we shall not treat them with the formality we do the systems of groups, rings, fields, and vector spaces that occupy most of this text. However, to be sure we have a common ground we begin with a list of the basic axioms that the integers obey, properties well known to you, and then prove some others. These arguments will give you a chance to get the flavor of how proofs are presented in this text. The exercises will give you a chance to practice making proofs.

Just to state some of the things we need to know about the integers we find it useful to use a little set notation and a few properties about sets. So we begin with these things. You will probably be familiar with these concepts, and the main purpose of the next section is to show the notations and conventions we use in this text.

1.1 SETS

An informal understanding of the elements of set theory is helpful. Most of what is needed has long ago found its way into the language of mathematics taught in elementary and secondary school curricula.

The undefined, or primitive, terms that underlie informal set theory are

set, element, and membership

A set either has no elements or there are elements that are members of the set. The set with no elements is called the empty set. In this text we use the symbol $\varnothing$ to denote the empty set. Further, we usually denote *elements* by lowercase letters, *sets* by uppercase letters, *membership* by $\in$. Thus we might let T be the set of even integers and then write

$$28 \in T$$

to say that 28 belongs to this set.

We also have a symbol to show that an element is not a member of a set. Simply draw a line through the membership symbol. Thus to say that 3 is not an even integer we write

$$3 \notin T$$

We constantly refer to four sets of numbers, the integers, the rational numbers, the real numbers, and the complex numbers. We have special symbols for these sets, and in these cases our notation is pretty standard among mathematicians.

$\mathbb{Z}$ denotes the set of integers.

$\mathbb{Q}$ denotes the set of rational numbers.

$\mathbb{R}$ denotes the set of real numbers.

$\mathbb{C}$ denotes the set of complex numbers.

1.1.1 Definition.   Equality of Two Sets

Two sets, A and B, are equal if and only if they have the same elements. We write

$$A = B$$

to mean that for all elements x,

$$x \in A \quad \text{if and only if} \quad x \in B$$

Notation.   Sets may be described in several ways. One way is to give a list of all the members of the set. The list is traditionally enclosed in braces $\{\dots\}$.

A second way to describe a set is to give a rule by which the elements of the set have been chosen. If this rule is clear, it is sometimes enough to give a few examples that suggest how the elements have been selected.

For example, here are four ways to describe the same set:

1. $\{2, 4, 6, 8, 10\}$
2. $\{a: a \text{ is a positive even integer less than } 12\}$
3. $\{a \in \mathbb{Z}: 0 < a < 11 \text{ and } 2 \text{ divides } a\}$
4. $\{2, 4, \ldots, 10\}$

Doubtless you will be able to think of still other ways to record this set.

If the number of elements in a set is an integer, we write

$$|A| = \text{number of elements in } A$$

If

$$|A| = 1 \quad \text{that is, if} \quad A = \{a\}$$

we say that A is a *singleton* set. Sometimes we say that A is a *doubleton* if $|A| = 2$.

1.1.2 Definition.   Subset

The set B is a subset of a set A if, whenever $x \in B$, then $x \in A$.

We write

$$B \subseteq A$$

if B is a subset of A. Note that empty set is a subset of every set and a set is always a subset of itself.

We also say that "B is contained in A" or that "B is included in A." We also write

$$A \supseteq B \quad \text{for} \quad B \subseteq A$$

and say that "A contains B" or that "A includes B."

If $B \subseteq A$ and if B is not equal to A, we say that B is properly contained in A and write

$$B \subset A$$

for emphasis.

Example.   The subsets of $\{1, 2, 3\}$ are

1. The set A itself: $A = \{1, 2, 3\}$
2. The three subsets having two elements: $\{1, 2\}$, $\{1, 3\}$, $\{2, 3\}$
3. The three singleton sets: $\{1\}$, $\{2\}$, $\{3\}$
4. The empty set: $\emptyset$

Here are the basic properties that set inclusion ($\subseteq$) satisfies.

1.1.3 Theorem.   Properties of the (⊆) Relation

For all sets A, B, and C:

1. Inclusion is reflexive: $A \subseteq A$ for all sets A
2. Inclusion is antisymmetric: If $A \subseteq B$ and $B \subseteq A$ then $A = B$
3. Inclusion is transitive: If $A \subseteq B$ and $B \subseteq C$ then $A \subseteq C$

We leave the proofs of these properties which are consequences of the definitions of the symbols to the exercises.

Notation. The set whose elements are the subsets of a set A is called the *power* set of A and is denoted $P(A)$. Many also use the notation

$$2^A$$

for the power set because if A has a finite number of elements then the number of elements in the power set of A is $2^{|A|}$. We shall prove this fact in Theorem 1.4.1.

As an example of a power set, let

$$A = \{1, 2, 3\}$$

The preceding example lists all eight elements of the power set $P(A)$. This is also an example of a set whose elements are themselves sets.

1.2 BASIC PROPERTIES OF THE INTEGERS

The integers are the familiar set of counting numbers,

$$1, 2, 3, \ldots$$

together with their negatives and zero. As we have stated, we denote this set with the letter Z in open-face font, rendered $\mathbb{Z}$. Z is used as a reminder of the German word "Zahlen," which means "numbers."

The integers are in many ways typical of the more general algebraic systems we consider in this text. Two binary operations, addition and multiplication, are defined on the integers. By "binary" we simply mean that each of these operations combines two elements to produce a third. In Section 1.4 we make the concept of a binary operation more precise, but here the idea is so familiar it hardly needs an introduction. Addition of two numbers a and b is denoted

$$a + b$$

and multiplication is denoted either by a dot, times, or by juxtaposition:

$$a \cdot b \quad a \times b \quad \text{or} \quad ab$$

1.2.1 Axioms of Addition and Multiplication for the Integers

Addition and multiplication obey the associativity axiom: For any three integers a, b, and c,

$$(a + b) + c = a + (b + c)$$

and

$$(a \cdot b) \cdot c = a \cdot (b \cdot c)$$

Addition and multiplication obey the commutativity axiom: For any two integers a and b,

$$a + b = b + a \quad \text{and} \quad a \cdot b = b \cdot a$$

Addition and multiplication have unique identity elements, 0 and 1, respectively:

$$a + 0 = a \quad \text{and} \quad a \cdot 1 = a$$

Multiplication and addition interact through the distributive axiom; multiplication is said to distribute over addition: For all elements a, b, and c,

$$a \cdot (b + c) = (a \cdot b) + (a \cdot c)$$

For the addition operation there is, for each element, an inverse element. The term "inverse" anticipates a more general setting for this concept. Here this axiom may be stated: For each element $a \in \mathbb{Z}$ there is a unique integer b such that $a + b = 0$. The element is, as usual, denoted $-a$. Its defining property is that

$$a + (-a) = 0$$

Using the commutative axiom this same equation shows that the additive inverse of $-a$ is a:

$$-(-a) = a$$

As we know, not every nonzero integer has an inverse with respect to the multiplication operation in the set of integers. Indeed that is what sets apart the integers from the rational numbers, $\mathbb{Z}$. Pun intended!

From these axioms we could prove a large number of important results that are so well known as to seem trivial, facts like

$$0 \cdot 0 = 0 \qquad a \cdot 0 = 0 \quad \text{and} \quad (-a) \cdot (-b) = a \cdot b$$

These results will be proved in later chapters for more general systems called rings which satisfy these same axioms of arithmetic.

There is an additional important axiom for integer arithmetic, a fact that is not a consequence of any of the axioms or properties we have already mentioned. It is simply that the product of two nonzero integers is never equal to zero. Or to state it in the form of an implication, a form that is most useful: For any two integers a and b,

$$\text{If} \quad a \cdot b = 0 \qquad \text{then either} \quad a = 0 \quad \text{or} \quad b = 0$$

With the help of the distributive law this gives the cancellation axiom,

$$\text{If } a \cdot b = a \cdot c \quad \text{and} \quad a \neq 0 \qquad \text{then } b = c$$

Next we need to include axioms about the order relation imposed upon $\mathbb{Z}$ by inequality.

1.2.2 Axioms of the Order Relation on the Integers

There is an inequality relation, "less than or equal to," defined on $\mathbb{Z}$. The axioms that this relation is assumed to satisfy are described in this section.

We write

$$b \leq a$$

to mean that "b is less than or equal to a." The notation

$$a \geq b$$

is read "a is greater than or equal to b" and it means the same as $b \leq a$. We also define

$$b < a \text{ to mean "} b \leq a \text{ and } a \neq b \text{"}$$

and, of course $a > b$ means the same as $b < a$.

Here are the basic axioms that the order relation enjoys. Compare them with the properties of set inclusion ($\subseteq$).

The order relation satisfies the axiom of reflexivity:

$$\text{For all integers } a, \qquad a \leq a$$

The order relation satisfies the axiom of transitivity: For integers a, b, and c,

$$\text{If } a \leq b \quad \text{and} \quad b \leq c \quad \text{then} \quad a \leq c$$

The order relation satisfies the axiom of trichotomy: For each pair of integers, a and b, exactly one of the following holds:

$$a > b \qquad a = b \quad \text{or} \quad b > a$$

The order relation satisfies the axiom of preservation under addition and multiplication by a positive integer:

$$\text{If } a \leq b \quad \text{then} \quad a + c \leq b + c \qquad \text{for all } c \in \mathbb{Z}$$

and

$$\text{If } a \leq b \quad \text{and} \quad c > 0 \quad \text{then} \quad a \cdot c \leq b \cdot c$$

A number of further properties now follow from these axioms:

$$\text{If } a > 0 \quad \text{then} \quad 0 > -a$$

For all integers $a \neq 0$, $a \cdot a = a^2 > 0$; in particular, $1 > 0$, and

$$\text{If } a \neq 0 \quad \text{or} \quad b \neq 0 \quad \text{then} \quad a^2 + b^2 > 0$$

The absolute value of an integer a is denoted by $|a|$ and is defined by

$$|a| = a \text{ if } a \geq 0 \quad \text{and} \quad |a| = -a \text{ if } a < 0$$

It is often convenient to introduce the maximum and minimum of two integers through these definitions:

$$\max(a, b) = \begin{cases} a \text{ if } a \geq b \\ b \text{ if } b > a \end{cases}$$

and

$$\min(a, b) = \begin{cases} b \text{ if } a \geq b \\ a \text{ if } b > a \end{cases}$$

In these terms we may express

$$|a| = \max(a, -a)$$

Perhaps the most famous and widely applicable inequality is the so-called triangle inequality: For any two integers a and b,

$$|a + b| \le |a| + |b|$$

And from this we may obtain an equally useful inequality: For any two integers a and b,

$$|a - b| \ge |(|a| - |b|)|$$

Remark. The order relation on $\mathbb{Z}$ can of course be extended to the rational numbers $\mathbb{Q}$ and then to the real numbers $\mathbb{R}$. However, the order properties do not apply to the complex numbers. For example, in $\mathbb{C}$, the field of complex numbers

$$1^2 + i^2 = 1 + -1 = 0$$

in contrast to the property that the sum of two squares of nonzero integers (or real numbers) is never zero. On the other hand, it is possible to define an absolute value; if z is a complex number, say,

$$z = x + iy$$

then

$$\|z\| = x^2 + y^2$$

and the two triangle inequalities are true for complex numbers as well. The real number $\|z\|$ defined in this way is called the norm of z. Its nonnegative square root $|z|$ is the distance of the number z from the origin in the complex plane.

We often refer to the set of integers that are greater than zero and we have a notation for this set.

$$\mathbb{Z}^+ = \{n \in \mathbb{Z} : n > 0\}$$

The set $\mathbb{Z}^+$, under the ordering ($\le$), satisfies one more important axiom: The set of positive integers is "well-ordered" by ($\le$). By this we mean that every *nonempty* subset $S \subseteq \mathbb{Z}^+$ has a minimal element; that is, there is an element $m \in S$ such that $m \le x$ for every element $x \in S$.

We have used the integers so long that this will seem as natural as the preceding properties of the integers that we have just recalled. On the other hand, the well-ordering of the set $\mathbb{Z}^+$ cannot be proved from the other axioms we have stated. And please note too that the set of positive *rational* numbers is not well-ordered under ($\le$). For example, consider the set of positive rational numbers itself. This set has no minimal element, for if q is a positive rational number then so is $q/2$, yet $q/2$ is less than q.

It is remarkable that the axioms we have just cited, the axioms governing addition and multiplication together with the axiom of inequality, and the well-orderedness of the positive integers completely characterize this algebraic system. For all intents and purposes there is only one system of integers! A proof of this can be found in the author's earlier text, *Elements of Abstract*

Algebra. This uniqueness will not be the case for other systems we study; there are a great many different examples of groups, rings, fields, and vector spaces.

Our first use of the well-orderedness of the positive integers is the following lemma. Although this lemma comes as no surprise, it is a good opportunity to see how the well-ordered property can be used in a mathematical argument.

1.2.3 Lemma. Discrete Nature of $\mathbb{Z}$

There is no integer a such that $1 > a > 0$.

For any integer a there is no integer t such that $a + 1 > t > a$.

Proof. The second statement of the lemma follows from the first since subtracting a from each term of the second inequality yields $1 > t - a > 0$, which is a violation of the first statement. To prove the first statement we actually prove an equivalent statement:

$$\text{If } k \text{ is an integer} \quad \text{and} \quad k > 0 \quad \text{then} \quad k \geq 1$$

Consider the set $\mathbb{Z}^+$ itself. It is nonempty since $1 \in \mathbb{Z}^+$ as we have seen. Thus there is a least element, call it k, in this set. We want to show that $k \geq 1$. If this is not so, the trichotomy property implies $1 > k$ and, of course, $k > 0$ since k is in $\mathbb{Z}^+$. But then by preservation of $(>)$ under positive multiplication

$$k > k^2 \quad \text{and} \quad k^2 > 0$$

Thus k^2 is in $\mathbb{Z}^+$ and k^2 is less than k, a contradiction of the choice of k as the least element in $\mathbb{Z}^+$.

Thus $k \geq 1$. In other words, we have just shown that 1 is the least positive integer.

Now let us use this property to prove another familiar result but one that is used frequently in this text.

1.2.4 Theorem. Division Algorithm for $\mathbb{Z}$

Let a be a nonzero integer. For every integer b there an integer d and an integer r such that

$$b = a \cdot d + r \quad \text{and} \quad 0 \leq r < |a|$$

Proof. If $b = 0$, then we may choose $d = r = 0$, so hereafter we may suppose that $b \neq 0$. It is also easier to treat first the case that $a > 0$. So suppose that $a > 0$.

Consider the subset T of nonnegative integers that can be written in the form $b - n \cdot a$ for some $n \in \mathbb{Z}$:

$$T = \{ b - n \cdot a \ : n \in \mathbb{Z} \text{ and } b - n \cdot a > 0 \}$$

First we must show that T is nonempty. If b is positive, then by setting $n = 0$ we see that $b \in T$. If b is negative, then $-b > 0$ and since $2a > 1$ it follows that $-b \cdot 2a > -b$ so that $b > b \cdot 2a$. From this, $b - 2b \cdot a > 0$ and so $(b - 2b \cdot a) \in T$.

Thus by the well-ordering property T has a minimal element; call it r and let d be the integer such that $r = b - d \cdot a$. Since $r \in T$, we see that $r > 0$. We claim that $r \le a$. Were this not so, that is, if $r > a$, then

$$b - (d + 1) \cdot a = (b - d \cdot a) - a = r - a > 0$$

hence $(r - a) \in T$. But $r > r - a$. This is a contradiction of the minimal property of r.

If $r < a$, then there is nothing more to prove. But if $r = a$, then we have $a = b - d \cdot a$ so that $b = (d + 1) \cdot a + 0$, as required.

Finally to complete the proof we must consider what happens if a is negative. But that is easy, for $-a$ is positive in that case and so we can conclude from the first case that

$$b = d \cdot (-a) + r = (-d) \cdot a + r$$

and now the whole proof is complete.

Remark. This theorem is called the division algorithm, for it states the result of a process used to divide one integer by another until zero or a remainder less than the divisor is obtained. Yet when the well-ordered property of the integers is applied, the algorithmic nature of the result has vanished. Next we discuss a method of proof, mathematical induction, which often yields an algorithm to obtain the final result.

1.2.5 Mathematical Induction

One of the most important tools for making proofs of theorems that depend upon the integers $\mathbb{Z}$ is the principle of mathematical induction. We give two equivalent forms of the induction principle. At any given time we use whichever seems the easier to use. Each principle gives a method for proving that a subset of $\mathbb{Z}$ contains all integers greater than or equal to a particular one.

Weak Induction (WI) Suppose that S is a subset of $\mathbb{Z}$ such that these two properties hold:

WI. 1. There is an integer b in S.

WI. 2. If $k \ge b$ and k is an integer in S, then $(k + 1)$ is in S.

Then S contains all the integers greater than or equal to b.

In many applications of WI the set S is a set of integers for which a particular *statement* is true. The induction principle WI provides a way of showing that S contains all the integers greater than or equal to b, that is, of showing that the *statement* is true for all integers greater than or equal to b. To do this there are two steps that must be carried out.

Step 1. Property WI. 1 must be verified; that is, it must be shown that the *statement* is true for the integer b.

Step 2. Property WI. 2 must be verified. Notice that WI. 2 is really a lemma; the lemma must be proved. The hypothesis of the lemma is that "k is in S." The hypothesis that k is in S means that the *statement* holds for k. This is often called the inductive hypothesis. The conclusion of the lemma is that "$(k + 1)$ is in S," that is, that the *statement* is true for $(k + 1)$.

Example.   Prove that $2^n > n^2$ if $n \geq 5$.

We begin step 1 by verifying that 2^5 is greater than 5^2. Indeed $32 > 25$. Now we prove the lemma in step 2. Suppose that $k \geq 5$ and that

$$2^k > k^2 \tag{1}$$

We are to show that

$$2^{k+1} > (k + 1)^2 = k^2 + 2k + 1$$

Multiply both sides of the inequality (1) by 2 to get

$$2^{k+1} > 2k^2 = k^2 + k^2 \tag{2}$$

By comparing the inequalities (1) and (2) we see that all we need to do is to show that if $k \geq 5$ then $k^2 > 2k + 1$. This is so because if, in fact, $k \geq 4$, then

$$(k - 1)^2 \geq 9 > 2$$

and so $k^2 - 2k + 1 > 2$, and hence $k^2 > 2k + 1$.

This completes the proof of step 2. The principle WI assures that $2^k > k^2$ for all $k \geq 5$.

Remark.   Sometimes the interplay between the choice of the integer b in step 1 where the induction begins and what can be proved in the lemma of step 2 is a delicate balance. In this example the lemma of step 2 holds if $k \geq 3$ but we cannot replace 5 by 3 in step 1, for 2^3 is not greater than 3^2. Nor can we begin the induction at $k = 0$, although step 1 is true for $k = 0$, for then the proof of step 2 will not hold.

A second formulation of the induction principle is often used. It is referred to as "strong induction," SI.

Strong Induction Principle (SI)   Suppose S is a subset of the integers $\mathbb{Z}$ such that these two properties hold:

SI. 1. There is an integer b in S.

SI. 2. If, for each $k > b$, all integers h such that $k > h \geq b$ are in S, then k is in S.

Then S contains all integers greater than or equal to b.

Again SI. 2 is called the inductive step and constitutes a lemma to be proved. The induction hypothesis is that all integers h for which $b \leq h < k$ are in S. The conclusion is that k is in S.

As an example we give a second proof of the division algorithm using SI. We do only the case when a and b are positive. Specifically, we prove that if $a > 0$ and $b \geq 0$ then there exist integers d and r such that $b = d \cdot a + r$ where $0 \leq r < a$. We fix a and induct on b. Let S be the set of positive integers b for which there exist integers d and r such that $b = d \cdot a + r$ and $0 \leq r < a$. The first step is to show that $0 \in S$. That is easy, for we may choose $d = r = 0$ and find $0 = 0 \cdot a + 0$.

Now for the inductive lemma. Suppose that if $0 < h < k$ then $h \in S$. We are to show that $k \in S$. If $k < a$ then we may choose $d = 0$ and $r = h$ and so find

$$k = 0 \cdot a + k$$

On the other hand, if $k \geq a$, we consider $h = k - a \geq 0$. Now $k > h$ and so by the induction hypothesis, $h \in S$ and so there exist integers d and r such that

$$k - a = h = d \cdot a + r \quad \text{and} \quad 0 \leq r < a$$

so

$$k = (d + 1) \cdot a + r$$

Thus the inductive lemma is proved. And so by SI it follows that the division algorithm holds for any $b \geq 0$ and the fixed positive integer a. But since a was any positive integer, this argument shows that the division algorithm must hold for any pair of positive integers a and b.

Now in what sense does this proof set up an algorithm? Let's be specific. Suppose $a = 13$ and $b = 29$. Well, the lemma shows that if we knew what to do for $a = 13$ and $b = 29 - 13 = 16$ then we could construct the necessary d and r for $a = 13$ and $b = 29$. Surely we are now left with a simpler problem because the number to consider for b has been reduced from 29 to 16. That is the virtue of the proof of the inductive lemma. So we again subtract off 13 from 16 and find $16 - 13 = 3$ and notice that $3 < 13$, so we are done. To find the d and r for 29 we simply work our way back from that last step:

$$16 = 1 \cdot 13 + 3 \quad \text{and} \quad 29 = 16 + 13$$

so therefore

$$29 = (1 \cdot 13 + 3) + 13 = 2 \cdot 13 + 3$$

Thus the inductive lemma presents division in the integers as repeated subtraction! And we could write a little program to count how many times we must subtract a from b until we reach a remainder less than a. What SI has done is put the informal process of "counting...until" into careful language. Often in the course of this text you will find situations in which your intuition about an algorithm will lead to a proof of a desired result by induction; other times you will find that the formal proof suggests a constructive way to compute what is desired.

The difference between WI and SI is that more is assumed in the inductive hypothesis in SI. Thus WI can be proved from SI. It turns out that SI can be proved from WI. This assertion is left for the exercises if you are interested. The way is through a strengthened form of the well-ordering principle.

Here is the general well-ordering principle for $\mathbb{Z}$.

Well-Ordering Principle (WO)   Let $b \in \mathbb{Z}$ and let $\mathbb{Z}^{b+} = \{n \in \mathbb{Z}: n \geq b\}$. If S is a nonempty subset of $\mathbb{Z}^{b+}$ then S has a *least* element k; that is, there exists an element $k \in S$ such that if $h \in S$ then $h \geq k$.

EXERCISE SET 1.2

In the proofs for these exercises it is fair to use anything you know about the integers, the rational numbers, the real numbers. However, solutions can be obtained by referring to the properties given in this section. Try to do this!

1.2.1   Prove the second triangle inequality from the first. Give geometric interpretations for both in the case of complex numbers.

1.2.2   Prove that if a and b are integers and $ab = 1$ then $a = b = 1$ or $a = b = -1$.

1.2.3   Use order properties to prove that if $a \neq 0$ and $b \neq 0$ then $ab \neq 0$. Prove that if $ab = ac$ and $a \neq 0$ then $b = c$.

1.2.4   Prove the $\max(a, \min(b, c)) = \min((\max(a, b), \max(a, c))$.

1.2.5   Prove that if $0 \neq a \in \mathbb{Z}$ and $b \in \mathbb{Z}$ then there are integers d and r such that $b = a \cdot d + r$ and $0 \leq |r| \leq |a|/2$.

1.2.6   Prove that the sum of the first n integers is $n(n + 1)/2$ using WI.

1.2.7   Prove that the sum of the squares of the first positive n integers is

$$\frac{n^3}{3} + \frac{n^2}{2} + \frac{n}{6}$$

1.2.8   Prove the generalized distributive law:

$$(a_1 + b_1)(a_2 + b_2) \cdots (a_n + b_n) = \sum e_1 e_2 \cdots e_n$$

where each $e_i = a_i$ or b_i and the sum extends over all 2^n choices for $e_1, e_2, \ldots, e_n$.

1.2.9   Prove that for any integer a the set $\mathbb{Z}^{a+} = \{n \in \mathbb{Z}: n > a\}$ is well-ordered under $(>)$.

1.2.10   Prove that induction principle SI implies induction principle WI for the integers.

1.2.11   Prove that the induction principle WI implies the well-ordering principle WO for the integers. *Hint:* Given a nonempty set $A \subseteq \mathbb{Z}$. Let $S(k)$ be the statement: "If for some integer h, $1 \leq h \leq k$, $h \in A$ then A has a least element." Prove $S(k)$ holds for all $k \in \mathbb{Z}$ using WI.

*1.2.12   Prove the induction principle SI from the well-ordering principle WO.

1.3 INTEGER ARITHMETIC

One reason that the integers are so useful as a source of examples of general algebraic systems is that some integers are divisible by others and some are not! Out of this situation comes the concept of a prime number and the important theorem that every integer can be written as a product of primes. Moreover, except for a plus or a minus sign, the primes and the number of each occurring in the product is unique. This section presents a development of this theorem based on the properties of the integers given in Section 1.2. In addition, the elementary number theory of this section is indispensable in the development of abstract algebra.

1.3.1 Definition. Division in the Integers

If d is a nonzero integer and a is an integer, we say that d divides a if there is an integer b such that $a = db$. We also say that a is a multiple of d in this case. In symbols we write $d \mid a$ if d divides a and $d \nmid a$ if d does not divide a.

Note that any nonzero integer divides 0 since $0 = a \cdot 0$. However, by definition, zero divides no integer.

1.3.2 Theorem. Division Properties

For all integers a, b, c, and d ($d \neq 0$):

   i. $d \mid d$ and if $d \mid a$ then $d \mid (-a)$ and $(-d) \mid a$.
   ii. Division is transitive; that is, if $d \mid a$ and $a \mid b$ then $d \mid b$.
   iii. If $d \mid a$ then $d \mid (ab)$.
   iv. If $d \mid a$ and $d \mid b$ then $d \mid (a + b)$.
   v. If $d \mid a$ and $d \mid b$ then $d \mid (au + bv)$ for any integers u and v.
   vi. If $a \neq 0$ and $d \mid a$ then $|a| \geq |d|$. In particular if $d \mid 1$ then $|d| = 1$.

We leave the proofs of these familiar results to the exercises at the end of this section. Note, however, that Property vi follows from the preservation of order under positive multiplication. And Property v results from a combination of Properties iii and iv. Property v is often summarized by saying the set of integers divisible by d is "closed under addition and multiplication."

We can use Property vi to prove that the integers obtained in the Division Algorithm 1.2.4 are unique.

Suppose that for some v and u ($u \neq 0$) it is true that

$$v = ud_1 + r_1$$

and

$$v = ud_2 + r_2$$

where $0 \leq r_1 < |u|$ and $0 \leq r_2 < |u|$.

Subtracting these equations yields

$$0 = u(d_1 - d_2) + (r_1 - r_2) \tag{1}$$

and hence we see that

$$|u| \mid |r_1 - r_2|$$

But $|u| > |r_1 - r_2|$ and so from Division Property vi it follows that

$$r_1 - r_2 = 0$$

But then (1) yields

$$d_1 - d_2 = 0$$

Hence $r_1 = r_2$ and $d_1 = d_2$.

The division algorithm is used to find the greatest common divisor of two integers.

1.3.3 Definition.   Greatest Common Divisor (gcd)

An integer that divides two nonzero integers a and b is said to be a common divisor of a and b.

The integer d is said to be a greatest common divisor of a and b if it is a common divisor and if whenever $c \mid a$ and $c \mid b$ then $c \mid d$.

Example.   The common divisors of 12 and 30 include 2, 3, and -6. In fact -6 is a greatest common divisor of 12 and 30. To see this note that

$$-6 = 12 \cdot 2 - 30$$

so that if $c \mid 12$ and $c \mid 30$ then c also divides -6 from Division Property v.

This definition does not use directly the notion of the ordering of the integers. An exercise asks you to prove that a greatest common divisor has the greatest absolute value among all divisors. The definition given here will be used in other algebraic systems where there is no "absolute value" to rely on. Notice that there is no immediate guarantee that every pair of numbers has a greatest common divisor. If we had used the absolute value of the divisor to determine the "greatest," then it would be clear that one exists but it would not be clear that it had the divisibility properties we require.

It follows from Division Property i that if d is a greatest common divisor of a and b then so is $-d$. The converse is true as well for if d and d' are both greatest common divisors of a and b, then each divides the other and so $d = d'$ or $d = -d'$ from Division Property vi. To avoid this slight ambiguity we have the following notation. Let $\gcd(a, b)$ denote the positive greatest common divisor of a and b.

Terminology.   If $\gcd(a, b) = 1$ we say that a and b are relatively prime; some authors use the word *coprime*.

Note that if $a \mid b$ then the common divisors of a and b are just the divisors of a. Thus, if $a \mid b$, $\gcd(a, b) = |a|$.

The next major result is to show that any two nonzero integers have a greatest common divisor, to find an algorithm for computing it, and also to find integers u and v such that

$$\gcd(a, b) = au + bv$$

While you may believe that 78 and 21 have a gcd and can readily verify that it is 3 it may be more difficult to find integers u and v such that

$$3 = 78 \cdot u + 21 \cdot v \tag{2}$$

The key to all of this is the next lemma.

1.3.4 Lemma.   The Division Algorithm and a gcd

If $a \neq 0$ and $b = a \cdot d + r$ then any common divisor of a and b is a common divisor of a and r and conversely.

Proof.   Any common divisor of a and r is a divisor of $a \cdot d + r$ by Division Property v; hence any common divisor of a and r is a divisor of b. Conversely, since $r = b - ad$, any common divisor of a and b is a divisor of r. Thus the integers a and b and the integers a and r have the same set of common divisors.

For our little teaser above, this means that since

$$78 = 21 \cdot 3 + 15$$

it follows that $\gcd(78, 21) = \gcd(21, 15)$ if either exists. And while we have still to determine the u and v we now can work with smaller numbers. Indeed, suppose we knew x and y such that $3 = 21 \cdot x + 15 \cdot y$, then we could use the fact that $15 = 78 - 21 \cdot 3$ and substitute this for 15 in the equation above to obtain $3 = 21 \cdot (x - 13 \cdot y) + 78 \cdot y$. So after an application of the division algorithm, the lemma enables us to replace the original pair of numbers by a pair of smaller ones. And then this process can be repeated until we find actually the common divisor. In our example here is how this goes.

Example.   Find the gcd of 21 and 78.

$$78 = 21 \cdot 3 + 15 \tag{1}$$

By Lemma 1.3.4 the common divisors of $(21, 78) = $ the common divisors of $(15, 21)$.

$$21 = 15 \cdot 1 + 6 \tag{2}$$

By Lemma 1.3.4 the common divisors of $(15, 21) = $ the common divisors of $(6, 15)$.

$$15 = 6 \cdot 2 + 3 \tag{3}$$

By Lemma 1.3.4 the common divisors of $(6, 15) = $ the common divisors of $(3, 6)$.

$$6 = 3 \cdot 2 + 0$$

So $3 \mid 6$ and so $\gcd(3, 6) = 3$. Thus $\gcd(21, 78) = 3$.

To find integers u and v so that

$$3 = 21u + 78v$$

we may proceed up the chain of equalities by substituting in one equation the result of the next above it. Begin with equation (3) and solve for 3. We write this as equation (4). Solve equation (2) for 6 and substitute in (4). After collecting coefficients of 15 and 21 we get equation (5).

$$3 = 15 - 6 \cdot 2 \tag{4}$$

$$= 15 - (21 - 15) \cdot 2 = 15 \cdot (1 + 1 \cdot 2) + 21 \cdot (-2)$$

$$3 = 15 \cdot 3 + 21 \cdot (-2) \tag{5}$$

Solve equation (1) for 15 and substitute in equation (5):

$$3 = (78 - 21 \cdot 3) \cdot 3 + 21 \cdot (-2)$$

$$= 78 \cdot 3 + 21 \cdot (-11)$$

Now we want to prove in general what we have just done in this example.

1.3.5 Theorem.   Existence of the gcd

If a and b are nonzero integers, then $\gcd(a, b)$ exists and there are integers u and v such that

$$\gcd(a, b) = au + bv$$

Remark.   The preceding example gives us a good clue about how the proof should go. Because of an application of the division algorithm and the lemma, we know that the numbers involved are replaced by smaller ones. This tells us that some sort of induction will work. Suppose $a < b$ and that $a \nmid b$. We go from consideration of the pair (a, b) to another pair (r, a) where $0 < r < a < b$. Hence $\min\{r, a\} < \min\{a, b\}$. This suggests an application of the strong induction principle. The statement we will prove will involve the minimum of the two integers considered. Here are the details.

Proof.   Without loss of generality we may suppose that a and b are positive since divisibility is not affected by multiplication by 1 or -1.

Our first observation is that if $a \mid b$ then $\gcd(a, b) = a$, and in this case

$$a = a \cdot 1 + b \cdot 0$$

The proof now proceeds by induction. We shall use SI. Let S be the set of positive integers m for which the following statement is true:

For any pair of positive integers $\{a, b\}$, if $\min\{a, b\} < m$ then $\gcd(a, b)$ exists and there exist integers u and v such that

$$\gcd(a, b) = au + bv$$

We use SI to show that S contains all positive integers, thus proving our theorem. For the first step we are to show that $1 \in S$. If $m = 1$, then either a

or b is 1 and so divides the other element of the pair. By our first observation it follows that $1 \in S$.

For the second step in establishing the induction, we state the induction hypothesis:

Suppose that for $1 \le k < m$ we know that $k \in S$. That means that if for any pair of positive integers (p, q), $\min\{p, q\} \le k$ then $\gcd(p, q)$ exists and there are integers x and y such that

$$\gcd(p, q) = px + qy$$

We are to show that $m \in S$. Thus suppose that we are given a pair of positive integers $\{a, b\}$ and that $\min\{a, b\} = m$. Without loss of generality we suppose that $a = m = \min\{a, b\}$. Thus $a \le b$. If $a \mid b$ our first observation tells us that $\gcd(a, b) = a$ and $a = a + 0 \cdot b$. If $a \nmid b$ then the division algorithm yields integers d and r such that

$$b = a \cdot d + r \quad \text{and} \quad 0 < r < a$$

Since $\min\{r, a\} = r < a = m = \min\{a, b\}$ it follows from the induction hypothesis that $\gcd(r, a)$ exists and that there exist integers u and v such that $\gcd(r, a) = u \cdot a + v \cdot r$. From Lemma 1.3.4 it follows that $\gcd(a, b) = \gcd(r, a)$ and substituting for r we obtain

$$\gcd(a, b) = u \cdot a + v \cdot (b - a \cdot d) = (u - v \cdot d) \cdot a + v \cdot b$$

This completes the proof by SI that gcds exist and are integral combinations of the given pair.

1.3.6 The Euclidean Algorithm to Find gcd(a, b)

Without loss of generality assume that $1 \le a \le b$. Perform the division algorithm

$$b = aq + r$$

Now if $r = 0$, $\gcd(a, b) = a$. If not $\gcd(a, b) = \gcd(a, r)$ and $1 \le r < a \le b$. Now simply repeat the division on the smaller pair r and a.

The sequence of remainders generated by this process decreases until a remainder is reached that divides the preceding one in the sequence. At this point the process stops. The "worst" case is that the last remainder is one. This happens of course when the original integers were relatively prime.

Our example shows how this works and how we may work our way back up the chain to express $\gcd(a, b)$ as an integral combination of a and b.

For computing purposes it is more convenient to proceed down the chain of equalities because this would avoid keeping all these equations in the computor's memory. And there is a nice algorithm to do this, one that is suitable for most programmable calculators. The trick is to keep the remainders of the sequence always expressed as an integral combination of a and b.

In the beginning we have

$$b = 0 \cdot a + 1 \cdot b \tag{1}$$

$$a = 1 \cdot a + 0 \cdot b \tag{2}$$

Now $b = d_1 \cdot a + r_1$ so

$$r_1 = -d_1 \cdot a + 1 \cdot b \tag{3}$$

Next $a = d_2 \cdot r_1 + r_2$ so $r_2 = a - d_2 \cdot r_1$. Now subtract $d_2 \cdot$ [equation (3)] from equation (2):

$$r_2 = (1 + d_1 d_2) \cdot a + (0 - d_2) \cdot b \tag{4}$$

Here is the general step: Suppose we start with the pair (a, b) and have reached consecutive remainders (c, d) with $c < d$ so that $\gcd(a, b) = \gcd(c, d)$. Suppose we have

$$c = ax + by \tag{t}$$

$$d = aw + bz \tag{$t+1$}$$

Now divide c by d: $c = d \cdot q + r$ so that $r = c + d \cdot (-q)$. Subtract $q \cdot$ [equation $(t + 1)$] from equation (t):

$$r = a(x - wq) + b(y - zq) \tag{$t+2$}$$

Thus $\gcd(a, b) = \gcd(c, d) = \gcd(d, r)$. The first two equations in the algorithm are

$$b = a \cdot 0 + b \cdot 1 \quad \text{and} \quad a = a \cdot 1 + b \cdot 0$$

Carry out the above process until a zero remainder is obtained. Here is another example. Find $\gcd(83, 121)$ and integers u, v such that

$$\gcd(83, 121) = 83u + 121v$$

Initialize:

$$121 = 83 \cdot 0 + 121 \cdot 1 \tag{1}$$

$$83 = 83 \cdot 1 + 121 \cdot 0 \tag{2}$$

Divide 121 by 83. $121 = 83 \cdot 1 + 38$, so $38 = 83 \cdot (-1) + 121$. Subtract $1 \cdot$ [equation (2)] from equation (1):

$$38 = -83 + 121. \tag{3}$$

Next divide 83 by 38: $83 = 38 \cdot 2 + 7$. To express the new remainder, 7, subtract $2 \cdot$ [equation (3)] from equation (2):

$$7 = 83(1 + 2) + 121 \cdot (0 - 2)$$

$$7 = 83(3) + 121 \cdot (-2) \tag{4}$$

Divide 38 by 7: $38 = 7 \cdot 5 + 3$. To express the new remainder, 3, subtract $5 \cdot$ [equation (4)] from equation (3):

$$3 = 83(-1 - 5 \cdot 3) + 121(1 - 5 \cdot (-2))$$

$$3 = 83 \cdot (-16) + 121 \cdot (11) \tag{5}$$

Divide 7 by 3: $7 = 3 \cdot 2 + 1$. To express the new remainder, 1, subtract $2 \cdot$ [equation (5)] from equation (4):

$$1 = 83 \cdot (3 - 2 \cdot (-16)) + 121 \cdot (-2 - 2 \cdot 11)$$

$$1 = 83 \cdot 35 + 121 \cdot (-24) \tag{6}$$

Since 1 divides the preceding remainder, 3, the process stops. We conclude that

$$1 = \gcd(23, 121)$$

Equation (6) expresses 1 as an integral combination of 83 and 121.

Many useful results follow from the greatest common divisor theorem and the euclidean algorithm. The following properties of gcds are often useful.

1.3.7 Theorem. Greatest Common Divisor Properties

For nonzero integers a and b:

   i. $\gcd(a, b) = 1$ if and only if there are integers u and v such that $1 = au + bv$.
   ii. If $d = \gcd(a, b)$ then $1 = \gcd(a/d, b/d)$.
   iii. If $a \mid bc$ and if $\gcd(a, b) = 1$ then $a \mid c$.
   iv. If $a \mid c$ and $b \mid c$ and $\gcd(a, b) = 1$ then $ab \mid c$.

Proof of i.   If $\gcd(a, b) = 1$ then there are integers u and v such that

$$1 = au + bv$$

Conversely if such an equation holds then any divisor of a and b divides 1 and hence is 1 or -1. Thus

$$\gcd(a, b) = 1$$

Proof of ii.   More generally than in i, if $d = \gcd(a, b)$ then the greatest common divisor theorem gives integers u, v such that

$$d = au + bv$$

Divide through by d:

$$1 = \frac{a}{d}u + \frac{b}{d}v$$

and now ii follows from i

Proof of iii.   From $\gcd(a, b) = 1$ it follows that

$$1 = au + bd$$

and hence

$$c = acu + bcv$$

so if $a \mid bc$ it follows that $a \mid c$ from Division Property v.

Proof of iv.   Let $c = ad$ and write $1 = au + bv$ as before. Then

$$d = adu + bvd$$

and from this conclude that

$$d = cu + bdv$$

and hence, since $b \mid c$, that $b \mid d$. Thus $d = be$ and substituting for d,

$$c = ad = abe$$

shows that c is a multiple of ab.

Comment.   In a fashion analogous to that of "common divisor" a common multiple and a least common multiple can be defined. The standard notation is to write

$$[a, b] = \text{least positive common multiple of } a \text{ and } b$$

Then one can prove (see the exercises at the end of this section) that

$$a \cdot b = [a, b]\gcd(a, b)$$

Now we are ready to define a prime number and prove the unique factorization theorem for integers.

1.3.8 Definition.   Prime Integer

An integer p not 0, 1 or -1, is called a prime if whenever $p \mid ab$ then $p \mid a$ or $p \mid b$.

1.3.9 Theorem.   Alternative Definition of a Prime

For integers p not equal to 0, 1, or -1 the following statements are equivalent:

   i. Whenever $p \mid ab$ then $p \mid a$ or $p \mid b$.
   ii. The only divisors of p are p, $-p$, 1, or -1.

If p satisfies either (and hence both) of these conditions, p is called a *prime integer*.

Proof That i Implies ii.   Suppose that c is a divisor of p; $c \mid p$. Then $p = cd$ and so $p \mid cd$. From i, $p \mid c$ or $p \mid d$. First suppose that $p \mid c$. Thus $c = pf$ for some f. Substituting

$$p = cd = pfd$$

By cancellation $1 = fd$ and thus $f = 1$ or -1; thus $c = pf = p$ or $-p$.
   Next, suppose that $p \mid d$ then $d = pe$ and so

$$p = cd = cpe$$

and again $1 = ce$ so that $c = 1$ or -1. Thus ii holds.

Proof That i Implies i.   Suppose that $p \mid ab$ and suppose $p \nmid a$. Consider $\gcd(a, p)$. It must be a divisor of p. By condition ii the only possibilities are 1, -1, p, or $-p$. But the gcd must also divide a and since $p \nmid a$, the only possibility is

$$\gcd(a, p) = 1$$

Now it follows from the gcd Property iii that $p \mid b$.

Remark.   Property ii is usually taken as the definition of a prime integer. In more general algebraic systems Property i will be used as the definition of a prime. Notice that the proof that i implies ii uses the definitions and cancellation law, while the proof that ii implies i uses the euclidean algorithm and so may not be valid in some number systems.

Note too that if p is a prime then so is $-p$. However, neither 1 nor -1 is considered a prime. These numbers are called *units* because they divide every integer. Integers that are neither primes nor units are called *composite*. Composite integers have proper factorization as the product of integers neither of which is a unit.

The next theorem has central importance for the integers. Generalizations of it will be equally important in some of the algebraic systems we consider. We say an integer other than 0, 1, or -1 can be factored into a product of primes or has a prime factorization when it can be expressed as the product of primes.

1.3.10 Theorem.   Unique Factorization

Every nonzero integer not a unit can be factored into a product of primes. This factorization is unique in the following sense: If

$$n = \pm p_1 p_2 \cdots p_r = \pm q_1 q_2 \cdots q_s$$

where each p_i and q_j is a positive prime, then $r = s$ and upon proper numbering, $p_i = q_i$ for all i.

Proof.   Two things must be proved, the existence of a factorization and the essential uniqueness of that factorization as described in the theorem. It is apparent that we may assume that we may deal only with positive numbers, and we shall make that assumption from now on.

Proof of Existence.   The proof is made by strong induction SI. Let S be the set of integers greater than 1 having a prime factorization. First notice that any positive prime p is in S; indeed, $p = p$ is its prime factorization. Now 2 is a prime, as we now prove. If d is a positive divisor of 2 then by Division Property vi, $d \leq 2$ and hence $d = 1$ or 2, being the only positive integers less than or equal to 2 (Exercises Set 1.2). Thus 2 is in S.

Now suppose that for $2 \le h < k$ it is true that h is in S. We are to show that k is in S. If k is prime there is nothing to prove. If not, say

$$k = h_1 h_2$$

where neither factor is $\pm k$. Now we may assume that both h's are positive. By Division Property iv,

$$0 < h_i < k \qquad \text{for } i = 1 \text{ and } 2$$

Thus by the induction hypothesis each h_i has a factorization into a product of primes. Concatenating these gives a factorization for k into a product of primes. Thus using SI we find that S is the set of all positive integers greater than 1; that is, all integers greater than 1 have factorizations into a product of primes.

Proof of Uniqueness. This proof also uses strong induction. Again let S be the set of integers greater than 1 for which the factorization into primes is unique in the sense of the theorem. The number 2 must belong to this set, as indeed must any prime p. By definition a positive prime cannot be divided by any other prime so if $p = q_1 \cdots q_s$ then it must be that $s = 1$ and $p = q_1$.

Now suppose that for $2 \le h < k$ the factorization of h into a product of primes is unique. We're to show that the factorization of k into a product of primes is unique. Suppose that

$$k = p_1 \cdots p_r = q_1 \cdots q_s$$

are two factorizations of k into a product of positive primes. If k were a prime then $r = s = 1$ and we are done. Otherwise, since p_1 is a prime and it divides $k = q_1 \cdots q_s$ it must divide one of the factors q_i. By reordering the factors if necessary we may suppose that $p_1 \mid q_1$. Since q_1 is a prime it follows that $p_1 = q_1$. Now cancel this prime from both sides and find

$$h = k/p_1 = p_2 \cdots p_r = q_2 \cdots q_s$$

Now by induction, since $h < k$, it follows that $r - 1 = s - 1$ and, reordering if necessary, each $p_i = q_i$. Thus we have shown that k has a unique factorization as a product of primes.

There are many ways to determine whether a given number is a prime or to compile lists of prime numbers. There is increasing interest to test very large numbers as being prime, and some very fast computer algorithms have been discovered.

Here are two methods that we can carry out. First is the brute force method: Given n, simply divide n by all numbers less than n. In view of Theorem 1.3.10 we realize that as trial divisors only primes need be tested. Then, since both a factor and its quotient cannot be greater than the square root of n, only primes less than or equal to the square root of n need be tested as possible prime factors of n.

A more systematic method of compiling a list of all primes less than a given number is attributed to the early Greek mathematician Eratosthenes.

1.3.11 The Sieve of Eratosthenes

Suppose you want to make a list of the primes less than 100. First make a list of all the numbers from 2 to 100 in order. (In this example just to complete the table with 12 numbers on each row we have listed numbers from 2 to 109, inclusive.)

2	3	4	5	6	7	8	9	10	11	12	13
14	15	16	17	18	19	20	21	22	23	24	25
26	27	28	29	30	31	32	33	34	35	36	37
38	39	40	41	42	43	44	45	46	47	48	49
50	51	52	53	54	55	56	57	58	59	60	61
62	63	64	65	66	67	68	69	70	71	72	73
74	75	76	77	78	79	80	81	82	83	84	85
86	87	88	89	90	91	92	93	94	95	96	97
98	99	100	101	102	103	104	105	106	107	108	109

Now 2 is a prime. This means that every second number after 2 is a multiple of 2 and so is not a prime. Cross them out; they "fall through the sieve." We indicate this below, showing 2 with a caret to denote a prime; and replacing the even numbers by a slash. They are not primes. Thus we get

$\hat{2}$	3	/	5	/	7	/	9	/	11	/	13
/	15	/	17	/	19	/	21	/	23	/	25
/	27	/	29	/	31	/	33	/	35	/	37
/	39	/	41	/	43	/	45	/	47	/	49
/	51	/	53	/	55	/	57	/	59	/	61
/	63	/	65	/	67	/	69	/	71	/	73
/	75	/	77	/	79	/	81	/	83	/	85
/	87	/	89	/	91	/	93	/	95	/	97
/	99	/	101	/	103	/	105	/	107	/	109

Now the first number after 2, namely, 3, must be a prime, and hence every third number after that is not a prime. These are of course the multiples of 3. So place a caret over 3 and cross out every third number (or slash) after 3 to obtain

$\hat{2}$	$\hat{3}$	/	5	/	7	/	/	/	11	/	13
/	/	/	17	/	19	/	/	/	23	/	25
/	/	/	29	/	31	/	/	/	35	/	37
/	/	/	41	/	43	/	/	/	47	/	49
/	/	/	53	/	55	/	/	/	59	/	61
/	/	/	65	/	67	/	/	/	71	/	73
/	/	/	77	/	79	/	/	/	83	/	85
/	/	/	89	/	91	/	/	/	95	/	97
/	/	/	101	/	103	/	/	/	107	/	109

The reason the pattern looks so regular at this point is that there are 12 numbers on each line of the display so that the numbers divisible by 2 and by 3 lie in columns.

To find the next prime, simply pick the first number without a hat, namely, 5. It cannot be a multiple of the previous primes since it remained after we had crossed out the multiples of 2 and 3. Every fifth number after 5, the multiples of 5, will not be primes. Put a caret on 5 to show it is a prime and cross out every fifth number or slash after 5. Obtain

2̂	3̂	/	5̂	/	7	/	/	/	11	/	13
/	/	/	17	/	19	/	/	/	23	/	/
/	/	/	29	/	31	/	/	/	/	/	37
/	/	/	41	/	43	/	/	/	47	/	49
/	/	/	53	/	/	/	/	/	59	/	61
/	/	/	/	/	67	/	/	/	71	/	73
/	/	/	77	/	79	/	/	/	83	/	/
/	/	/	89	/	91	/	/	/	/	/	97
/	/	/	101	/	103	/	/	/	107	/	109

The next number not crossed out is 7. Place a caret over 7 to show it is a prime and cross out every seventh number after 7 to obtain

2̂	3̂	/	5̂	/	7̂	/	/	/	11	/	13
/	/	/	17	/	19	/	/	/	23	/	/
/	/	/	29	/	31	/	/	/	/	/	37
/	/	/	41	/	43	/	/	/	47	/	/
/	/	/	53	/	/	/	/	/	59	/	61
/	/	/	/	/	67	/	/	/	71	/	73
/	/	/	/	/	79	/	/	/	83	/	/
/	/	/	89	/	/	/	/	/	/	/	97
/	/	/	101	/	103	/	/	/	107	/	109

The next number not crossed out is 11, so 11 is a prime. Now we claim that in this short table all multiples of 11 have already been crossed out! All multiples, $11k$, where $2 \le k \le 10$ are multiples of the earlier primes, 2, 3, 5, or 7 and have been crossed out. Thus the first possibility is $11 \cdot 11 = 121$. But 121 exceeds the limit of the table. Thus it follows that all the remaining numbers in the table are primes. The primes less than 110 are

$$2, 3, 5, 7, 11, 13, 17, 19, 23, 29, 31, 37, 41, 43, 47, 53,$$
$$59, 61, 67, 71, 73, 79, 83, 89, 97, 101, 103, 107, 109$$

1.3.12 Theorem.   Infinitude of Primes

There are an infinite number of prime integers.

Proof.   The proof is a classic example of a "proof by contradiction."
Suppose that there were only a finite number of primes, say, $p_1, p_2, \ldots, p_k$. Consider the number

$$N = 1 + (p_1 p_2 \cdots p_k)$$

Now any number dividing the product of the listed primes and N must also divide their difference, 1. Thus

$$(N, p_1 p_2 \cdots p_k) = 1$$

In particular N is not divisible by any one of the listed primes, $p_1, \ldots, p_k$. But from Theorem 1.3.11 the number N is the product of one or more primes. Such a prime cannot have been listed above. Hence the alleged list of all primes is not complete. Tilt!

A "freebie" from the proof is that if you have a list of primes and wish to generate a new prime, all you have to do is add 1 to the product of all primes in the list. Then factor the number into a product of primes. One of these will be a "new" prime. This works well in theory but it involves a lot of computation in practice!

The famous "prime number" theorem states that the number of primes less than or equal to n is approximately $n/\log_e(n)$.

Now we introduce an idea that may be less familiar but plays a frequent role in arguments involving integers, and its analog in more general algebraic systems is of key importance.

1.3.13 Definition. Ideal

A nonempty subset A of $\mathbb{Z}$ is called an ideal if these two conditions hold:

    i. Whenever a and b are in A then $a - b$ is also in A.
    ii. Whenever a is in A then na is A for all integers n in $\mathbb{Z}$.

There are always two trivial ideals:

The set consisting of 0 alone is an ideal.

The set consisting of all integers is an ideal.

1.3.14 Lemma. Elementary Ideal Properties

    iii. Every ideal contains 0.
    iv. If an ideal contains a then it also contains $-a$.

Proof.   An ideal is nonempty. Let x be an element of it. Then, by Property i the ideal contains $x - x$ and

$$x - x = 0$$

To prove iv, suppose that a is in an ideal. Then, since we now know that 0 is in the ideal, Property i implies that $0 - a$ is in the ideal and thus

$$0 - a = -a$$

is in the ideal.

Many times we shall be able to verify easily that a subset of $\mathbb{Z}$ is an ideal. The next theorem characterizes ideals in $\mathbb{Z}$, and it is this that is helpful.

Here is another example of an ideal.

The ideal $\langle m \rangle$ Let m be any integer. Let A be the set of multiples of m. We claim that A is an ideal. Surely the difference of two multiples of m is a multiple of m so i holds. And a multiple of a multiple of m is again a multiple of m. Thus conditions i and ii are met.

The next theorem shows that this is the only kind of ideal in $\mathbb{Z}$! So what's the fuss? Well, it is often easy to verify i and ii. Then the next theorem does the work.

1.3.15 Theorem. Characterization of the Ideals in $\mathbb{Z}$

If A is an ideal in $\mathbb{Z}$ then there is an integer m such that

$$A = \langle m \rangle$$

Proof.   If A contains only 0 then we choose $m = 0$; indeed

$$\langle 0 \rangle = \{0\}$$

Otherwise let m be chosen so that among the positive elements of A, m is minimal. Use the well-ordering principle WO to do this. We claim that

$$A = \langle m \rangle$$

To see this, let b be any other element in A and apply the division algorithm:

$$b = mq + r \qquad \text{where } r = 0 \text{ or } 0 < r < m$$

Now mq is in A and so is b and hence

$$r = b - mq$$

is also in A. If $r \neq 0$ then r is a positive integer in A less than m, contrary to the choice of m. Hence $r = 0$ and so b is a multiple of m.

By the way,

$$\{0\} = \langle 0 \rangle \quad \text{and} \quad \mathbb{Z} = \langle 1 \rangle$$

EXERCISE SET 1.3

1.3.1   Prove Division Properties i to vi of Theorem 1.3.2.

1.3.2   Prove that if $a \mid x$ and $a \mid (x + y)$ then $a \mid y$.

1.3.3   Show that 6 divides the product of every three consecutive integers. Prove that 24 divides the product of any four consecutive integers. Make a conjecture suggested by these two examples and settle it!

1.3.4   Show that if $a \neq 0$ and $b \neq 0$ then $c \leq \gcd(a, b)$ if c is a positive common divisor of a and b.

1.3.5   Make up a definition for common multiple and least common multiple and prove that

$$[a, b] \gcd(a, b) = ab$$

1.3.6   Show by an example that it is not always true that $a \mid x$ and $b \mid x$ imply $ab \mid x$.

1.3.7 For positive integers a ($a > 1$), n, and m, show that

$$(a^n - 1) \mid (a^m - 1) \qquad \text{if and only if} \quad n \mid m$$

1.3.8 Let a and b be relatively prime integers greater than 1. Show that integers u and v may be chosen so that

$$1 = au + bv$$

and $|u| \le b$ and $|v| \le a$.

 What can be said more generally about the possible size of x and y if

$$d = ax + by$$

1.3.9 If a and b are integers greater than 1 and relatively prime, for what d are there positive integers x and y such that

$$d = ax + by$$

(Good partial results are not hard to find; a definitive solution may be a little tricky.)

*1.3.10 Given three jugs having integer capacities: $a > b > c$. Suppose that the first is full and the second two are empty. Suppose further that $\gcd(b, c) = 1$ and that a is even. Show how to divide the liquid in the first jug into equal parts, one in the second jug and one in the third, if

$$a \le 2(b + c) \quad \text{and} \quad a \ge b + c - 2$$

The classic case is $a = 8$, $b = 5$, $c = 3$ and the liquid is wine. Assume of course that no liquid is lost in the pouring process!

1.3.11 Find necessary and sufficient conditions on three nonzero integers a, b, and c such that the equation

$$1 = ax + by + cs$$

has an integer solution.

1.3.12 Is 2003 a prime?

1.3.13 By hand (OK to use a nonprogrammable calculator) find 91,195 and integers such that

$$\gcd(91,195) = 91u + 195v$$

1.3.14 Write a computer program to implement the algorithm for determining the gcd of two numbers a and b and find u and v such that

$$\gcd(a, b) = au + bv$$

Use your program on $a = 5497$ and $b = 973541$.

1.3.15 Use the well-ordering principle to show that each rational number $a \mid b$ may be written in lowest terms.

1.3.16 Let A be a nonempty subset of $\mathbb{Z}$. Suppose that
   i. $(-a)$ is in A whenever a is in A.
   ii. $a + b$ is in A whenever a and b are in A.
   Show that A is an ideal in $\mathbb{Z}$.

1.3.17 Suppose that A is a nonempty subset of the integers such that if a is in A and if n is any integer then na is in A. Show that A is an ideal of $\mathbb{Z}$.

1.3.18 Suppose that A is a nonempty subset of the integers such that if a is in A and b is in A then $a - b$ is in A. Show that A is an ideal of $\mathbb{Z}$.

Remark. In view of Problems 1.3.17 and 1.3.18 it follows that an ideal in $\mathbb{Z}$ can be characterized by either condition i or condition ii of Definition 1.3.13. This will not be the case when ideals are introduced in more general situations.

1.3.19 Let A and B be ideals in $\mathbb{Z}$. Show that their set intersection, $A \cap B$, is an ideal. If $a = \langle m \rangle$, $B = \langle n \rangle$, and $A \cap B = \langle t \rangle$ describe t in terms of m and n.

1.3.20 Let a and b be integers. For what integer t is $\langle t \rangle$ the smallest ideal containing a and b?

1.3.21 We say that an integer n has the form $4k + 1$ if $4 \mid (n - 1)$; equivalently that for some m, $n = 4m + 1$. Show that a product of positive integers each of the form $4k + 1$ is again of the form $4k + 1$. Show that there are an infinite number of primes of the form $4k - 1$.

1.3.22 Use the prime number theorem to estimate the number of primes less than 110.

1.4 MORE ON SETS

In this section we present a few additional things we use about sets.

1.4.1 Theorem.   Number of Elements in a Power Set

If $|A|$ is finite, the number of elements in the power set of A is $2^{|A|}$.

Proof. The proof is by induction on the number of elements in A. The proof uses the weak induction principle WI.

Let S be the set of nonnegative integers n such that for all sets A if $|A| = n$ then $|P(A)| = 2^n = 2^{|A|}$. First verify that $0 \in S$. If $|A| = 0$ then $A = \varnothing$ and because $\varnothing$ is the empty set, $P(\varnothing) = \{\varnothing\}$, and so has just one subset, itself. Thus

$$1 = |P(\varnothing)| = 2^0$$

Now the inductive hypothesis is that

$$\text{If } |A| = k \quad \text{then} \quad |P(A)| = 2^k$$

Let A be a set with $k + 1$ members. We want to show that $|P(A)| = 2^{k+1}$.

Since $k + 1 \geq 0$, A is not empty. Let $x \in A$. We are going to count the subsets of A by first counting those that do not contain x and then those that do. Since every subset either contains x or does not, if we can count these two kinds of subsets, the total number will just be the sum of the two counts.

Now if we form the subset A^* of A consisting of those elements not equal to x, we see that

$$|A^*| = k$$

and that every subset of A not containing x is a subset of A^*. Thus, by the induction hypothesis, there are 2^k subsets of A^* and hence 2^k subsets of A that do not contain x.

On the other hand, every subset of A that contains x can be thought of as being constructed by adjoining the element x to a subset of A^*. In this way, different subsets of A^* lead, by adjoining x, to different subsets of A. Thus there are the same number of subsets of A containing x as there are subsets of A^*. Thus

$$|A| = 2^k + 2^k = 2^{k+1}$$

This proves the inductive step and thus completes the proof of the theorem.

There are two binary operations on sets that are very useful. Their properties resemble those for multiplication and addition.

1.4.2 Definition.   Set Operations

Let A and B be sets. Define

$$A \cap B = \{ x : x \in A \text{ and } x \in B \}$$
$$A \cup B = \{ x : x \in A \text{ or } x \in B \}$$

The set $A \cap B$ consists of the elements that A and B have in common. This operation is called *set intersection* or *meet*.

The set $A \cup B$ consists of the elements that are members of either A or B or *both*. This operation is called *set union* or *join*. For example, if

$$A = \{1, 2\} \quad \text{and} \quad B = \{2, 3\}$$

then

$$A \cap B = \{2\} \quad \text{and} \quad A \cup B = \{1, 2, 3\}$$

1.4.3 Definition.   Disjoint Sets

Two sets A and B are said to be *disjoint* if $A \cap B = \varnothing$. A family of sets is said to be disjoint if every pair of sets in the family is disjoint.

A set S is said to be the disjoint union of subsets A and B if

$$S = A \cup B \quad \text{and} \quad A \cap B = \varnothing$$

In this case, as a result of the definition of addition we find

$$|S| = |A| + |B|$$

More generally, if $S = A \cup B$ but is not necessarily the disjoint union of these sets, then

$$|S| = |A| + |B| - |A \cap B|$$

since the elements of $A \cap B$ have been counted twice, once as members of A and once again as members of B, and so $|A \cap B|$ must be subtracted off.

Here are the algebraic properties of the set operations.

1.4.4 Theorem. Properties of Set Operations

The set operations $\cup$ and $\cap$ satisfy these properties. Let A, B, and C be three sets.

i. Idempotence: $A \cup A = A$ and $A \cap A = A$

ii. Associativity: $A \cup (B \cup C) = (A \cup B) \cup C$
and $A \cap (B \cap C) = (A \cap B) \cap C$

iii. Commutativity: $A \cup B = B \cup A$ and $A \cap B = B \cap A$

iv. Absorptivity: $A = A \cup (A \cap B)$ and $A = A \cap (A \cup B)$

v. Distributivity: $A \cup (B \cap C) = (A \cup B) \cap (A \cup C)$,
$A \cap (B \cup C) = (A \cap B) \cup (A \cap C)$, and
$(A \cup B) \cap (A \cup C) \cap (B \cup C) = (A \cap B) \cup (A \cap C) \cup (B \cap C)$

Proofs of these properties are asked for in the exercises.

It is important to note that the laws are dual with respect to $\cup$ and $\cap$; that is, interchanging these operations does not change the list of properties.

It is interesting, and not entirely trivial, to show that each of the distributive laws follows from any other distributive law together with Properties i to iv.

Notation. The operations of union and intersection can be extended to any set of sets. If S is a set whose elements are subsets A of a set E we write

$$\bigcup_{A \in S} A = \{x : x \in A \text{ for some } A \in S\}$$

or more briefly

$$\cup S = \bigcup_{A \in S} A$$

and dually,

$$\bigcap_{A \in S} = \{x : x \in A \text{ for all } A \in S\} = \cup S$$

Here is an example. Let $E = \mathbb{Z}$ and let S be the set of all ideals of $\mathbb{Z}$. Then

$$\cup S = \mathbb{Z}$$

since every integer is in some ideal and

$$\cap S = \{0\}$$

since only zero belongs to every ideal.

Another useful set construction is that of *complementation*. If A is a subset of S then we write

$$A^c = \{x \in S \text{ and } x \notin A\}$$

More generally for any two sets A and B we write

$$A - B = \{x : x \in A \text{ and } x \notin B\}$$

Thus $\{1, 2\} - \{2, 3\} = \{1\}$.

1.4.5 Definition.   Partition

Let S be a set and let E be a family of disjoint subsets T of S. The family E is said to *partition* or to be a partition of S if

$$S = \cup E = \bigcup_{T \in E} T$$

A set of representatives for the partition is a subset R of S such that R contains one and only one element from each set $T \in E$.

It is easy to construct partitions. For example, Let $S = \{1, 2, 3, 4, 5, 6, 7, 8, 9, 10\}$. Then the four sets $\{1, 2\}$, $\{3, 4, 5\}$, $\{6, 7, 8, 9\}$, and $\{10\}$ partition S. A set of representatives for this partition is $\{1, 4, 8, 10\}$.

It is hard to determine the number of different ways a set can be partitioned as a function of the number of elements in the set. Only asymptotic results are known. An allied concept of considerable interest in number theory and combinatorics is the so-called partition function, the number of ways of writing a positive integer n as the sum of positive integers.

1.4.6 Diagrams

Relationships between subsets of a set will be pictured in two ways.

Venn Diagrams   In a Venn diagram each set is pictured as a subset of the plane. The subsets of interest are often shaded. This gives a schematic picture of the relationships and is often helpful in suggesting proofs. See Figure 1.1.

Hasse Diagrams   In a Hasse diagram each set is designated by a dot on the page. If $A \subset B$ the dot for B is located above the dot for A and a line or arc is drawn from A to B.

Redundant lines arising from transitivity of $\subset$ are erased. See Figures 1.2 and 1.3.

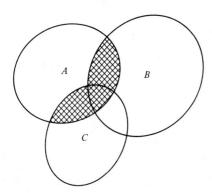

Figure 1.1 A Venn diagram showing $A \cap (B \cup C)$ as the shaded region. Note that this region is also the shaded region for $(A \cap B) \cup (A \cap C)$. Thus the Venn diagram shows that $A \cap (B \cup C) = (A \cap B) \cup (A \cap C)$.

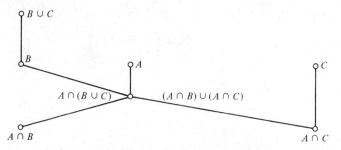

Figure 1.2 The Hasse diagram illustrating that $A \cap (B \cup C) = (A \cap B) \cup (A \cap C)$. The diagram must show a dot for each of the sets $A, B, C, (B \cup C), (A \cap B)$, and $(A \cap C)$. Note that $A \cap B \subseteq A \cap (B \cup C) \subseteq B \cup C$. However, in general $A \cap (B \cup C)$ is not a subset of B so no line is drawn from $A \cap (B \cup C)$ to B.

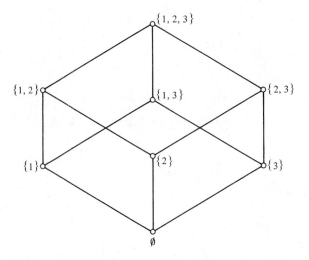

Figure 1.3 The Hasse diagram of all the subsets of $\{1, 2, 3\}$. A line goes up from the dot labeled $\{1\}$ to the dots labeled $\{1, 2\}$, and $\{1, 3\}$ since $\{1\}$ is a subset of these sets. No line is drawn directly from $\{1\}$ to $\{1, 2, 3\}$ because no extra line is needed. No line is drawn from $\{1\}$ to $\{2, 3\}$ since $\{1\}$ is not a subset of $\{2, 3\}$.

1.4.7 Definition. Ordered Pair

Let a and b be elements. The symbol (a, b) often denotes an ordered pair. The notation emphasizes the fact that the *order* in which the elements appear is important. Thus $(a, b) \neq (b, a)$ if $a \neq b$.

More formally we may invoke the definition given by Norbert Wiener and Kasimir Kuratowski:

$$(a, b) = \{\{a\}, \{a, b\}\}$$

is the set consisting of the singleton $\{a\}$ and the doubleton $\{a, b\}$. It is clear now that if $a \neq b$ then $(a, b) \neq (b, a)$ since (a, b) contains the singleton $\{a\}$ but not the singleton $\{b\}$, while (b, a) contains the singleton $\{b\}$.

1.4.8 Definition. Cartesian Product

The cartesian product of sets A and B is the set

$$A \times B = \{(a, b): a \in A \text{ and } b \in B\}$$

of all ordered pairs formed from A and B.

The use of the word cartesian is in honor of René Descartes, who is credited with coordinatizing the plane with ordered pairs of real numbers. The extension here to arbitrary sets is a natural one. Note that

$$A \times B \neq B \times A \qquad \text{if } A \neq B$$

If $|A|$ and $|B|$ are integers then

$$|A \times B| = |A| \cdot |B|$$

This rule also holds for infinite sets; in fact it can be made the basis for a definition of multiplication of infinite cardinal numbers. See, for example, *Naive Set Theory*, a text by Paul Halmos.

The notion and notation of a cartesian product of two sets can be extended easily to a cartesian product of several sets. Define

$$A_1 \times \cdots \times A_n = \{(a_1, \ldots, a_n): a_i \in A_i \text{ for } i = 1, 2, \ldots, n\}$$

as the set of all ordered n-tuples (a sequence) in which the ith element comes from the ith set.

Perhaps the most important activity with sets is the construction and study of functions. Ordered pairs and the cartesian product of sets A and B are often used to give a definition of function.

1.4.9 Definition. Function

Given sets A and B, a function f from A into B is a subset of $A \times B$ such that

  i. If $a \in A$ then there exist $b \in B$ such that $(a, b) \in f$
  ii. If (a, b) and (a, c) are in f then $b = c$

The *domain* of f is the set A. The *range* of f is

$$\{b \in B: \text{there exists } a \in A \text{ such that } (a, b) \in f\}$$

We shall often write more graphically:

$$f: A \to B$$

and for elements

$$f: a \to b \text{ or } f(a) = b \qquad \text{if } (a, b) \in f$$

We often say that f maps A into B or maps the element a to b.

Terminology. A function $f: A \to B$ is said to be *one-to-one* if

$$f(a) = f(b) \text{ implies } a = b$$

A function with this property is called "injective."

A function $f: A \to B$ is said to be *onto* if the range of f is all of B. A function with this property is called "surjective." Functions that are both surjective and injective are called "bijective." Thus a function $f: A \to B$ is bijective if it is one-to-one and onto A.

1.4.10 Definition.   Binary Operations and Tables

Let S be a nonempty set. A binary operation on S, denoted here by $\circ$, is a function from $S \times S$ into S. The image of the ordered pair (a, b) under this function is denoted

$$a \circ b$$

Informally, a binary operation combines pairs of elements in S to give an element of S. Examples abound; indeed that is about all we study in arithmetic and algebra.

Addition $(+)$ is a binary operation on $\mathbb{Z}$. The function $+$ maps an ordered pair of integers (a, b) to the sum, the integer $a + b$.

Multiplication is also a binary operation on $\mathbb{Z}$.

Division is not a binary operation on $\mathbb{Z}$ because the quotient of two integers is not always an integer. It is not a binary operation even on the rational number $\mathbb{Q}$ because division by 0 is not defined.

Esoteric examples may be constructed on a finite set by means of a table. Here is one example.

Let $S = \{a, b, c\}$. Let the binary operation # be defined by means of this table:

#	a	b	c
a	a	a	a
b	b	b	b
c	a	b	a

where the table is read in the following manner: The entry in the row labeled x and column headed y appears in that square. From the table we see, for example, that

$$b\#c = b \quad \text{and} \quad c\#a = a$$

The concept of the table is itself useful, and we shall refer to it when we begin our study of group theory in Chapter 3.

Here is another example of a binary function; this time the elements being combined are functions.

Example.   Composition of Functions.   Let T be a set and let S be the set of all functions from T into T,

$$S = \{ f: T \to T \}$$

A binary operation on S is created by composing two functions. Let $f \circ g$ be the composition of the function g followed by f. Thus if

$$g: a \to b \quad \text{and} \quad f: b \to c \quad \text{then} \quad f \circ g: a \to c$$

and so

$$f \circ g(a) = f(g(a))$$

We often represent this with a schematic diagram:

$$a \xrightarrow{g} b \xrightarrow{f} c$$
$$\underbrace{\qquad}_{f \circ g}$$

We need the fact that composition is associative. We prove this more generally in the next lemma.

1.4.11 Lemma.   Composition of Functions Is Associative

If R, S, T, and U are sets and f, g, and h are functions

$$h: R \to S \qquad g: S \to T \quad \text{and} \quad f: T \to U$$

then

$$(f \circ g) \circ h = f \circ (g \circ h)$$

Proof.   Compute, sprinkling parentheses liberally as you go. For each $r \in R$,

$$(f \circ g) \circ h(r) = f \circ g(h(r)) = f(g(h(r)))$$

and

$$f \circ (g \circ h)(r) = f(g \circ h(r)) = f(g(h(r)))$$

EXERCISE SET 1.4

1.4.1 Prove the properties of idempotence, associativity, commutativity, and absorption for the set operation $\cup$.

1.4.2 Prove the distributive properties for set operations.

1.4.3 Given an example to show that $A \cap (B \cup C)$ is not a subset of B in general.

*1.4.4 Assume that $\wedge$ and $\vee$ are operations on a set S that satisfy the properties of idempotence, associativity, commutativity, and absorption. Suppose also that for all A, B, and C in S

$$A \wedge (B \vee C) = (A \wedge B) \vee (A \wedge B)$$

Prove that the other two distributive laws hold.

1.4.5 Let $S = \{1, 2, 3, 4\}$. List all subsets $A \subseteq S$ such that

$$4 \in A \quad \text{or} \quad |A| = 2$$

Draw the Hasse diagram for this collection of subsets.

1.4.6 Let $S = \{1, 2, 3, 4\}$. List all subsets $A \subset S$ such that A has a nonempty intersection with $\{3, 4\}$. Draw a Hasse diagram for this collection of sets.

1.4.7 Draw a diagram, Venn or Hasse (your choice), for all 16 subsets of $\{1, 2, 3, 4\}$.

1.4.8 Let S be a set and let A be a subset. Show that

$$A \cap A^c = \varnothing \quad \text{and} \quad A \cup A^c = S$$

Show that $(A \cup B)^c = A^c \cap B^c$. Find a similar formula for $(A \cap B)^c$.

1.4.9 Define a set operation

$$A \oplus B = (A - B) \cup (B - A)$$

(In the literature this is called the symmetric difference of the sets A and B.) Which of Properties i to v of Theorem 1.4.4 hold when $\cup$ is replaced by $\oplus$ and $\cap$ retains its meaning as set intersection?

1.4.10 For each $n \in \mathbf{Z}^+$ show that if $|S| = n$ then S has 2^{n-1} subsets with an even number of elements. How many subsets of S have an odd number of elements?

1.4.11 Show that if A, B, and C are finite sets then

$$|A \cup B \cup C| = |A| + |B| + |C| - |A \cap B| - |A \cap C| - |B \cap C| + |A \cap B \cap C|$$

Exercises 1.4.12 to 1.4.17 develop the essential properties of the combinatorial symbol

$$\binom{n}{k}$$

without reference to $n!$.

1.4.12 Let n be a positive integer. If S is a set with n elements, let

$$\binom{n}{k} = \text{number of subsets of } S \text{ having } k \text{ elements}$$

Using this definition alone determine

$$\binom{n}{0} \quad \binom{n}{1} \quad \text{and} \quad \binom{n}{n}$$

Prove that $\binom{n}{k} = \binom{n}{n-k}$ if $0 \leq k \leq n$.

1.4.13 Use Theorem 1.4.1 to prove $2^n = \displaystyle\sum_{k=0}^{k=n} \binom{n}{k}$.

1.4.14 Prove, using Problem 1.2.8, that if a and b are integers then

$$(a + b)^n = \sum_{k=0}^{k=n} \binom{n}{k} a^k b^{n-k}$$

1.4.15 Using the definitions and results, if necessary, of Problems 1.4.12 to 1.4.14 prove

$$\binom{n}{k} = \binom{n-1}{k-1} + \binom{n-1}{k}$$

without using the formula of Exercise 1.4.16.

1.4.16 Use Problem 1.4.15 to show that $\binom{n}{k} = n!/k!(n-k)!$.

1.4.17 Use a set theoretic argument to prove

$$\binom{n}{k} = \sum_{h=0}^{h=\min(m,\,k)} \binom{m}{h}\binom{n-m}{k-h}$$

1.4.18 List all the functions f from $T = \{0,1,2\}$ into T such that $f(x) = 0$ or 1. Find a function such that $f \circ f = f$.

Is there a function g such that $g \circ g \circ g = g$ but $g \circ g \neq g$?

1.4.19 Let $\circ$ be composition of functions on the real numbers $\mathbb{R}$. Also define addition $(+)$ of functions as follows. If f and g are functions $(f + g)$ is the function such that

$$(f + g)(x) = f(x) + g(x)$$

Which of the distributive laws hold when $\cup$ is replaced by $+$ and $\cap$ is replaced by $\circ$?

1.4.20 Let A and B be ideals in $\mathbb{Z}$. Construct an example to show that the set union $A \cup B$ is not always an ideal. Find necessary and sufficient conditions for $A \cup B$ to be an ideal.

Chapter 2

Relations and Lattices

2.1 EQUIVALENCE RELATIONS AND CONGRUENCE IN $\mathbb{Z}$

In Chapter 1 we used the word *relation* in an informal way. Thus we spoke of an order relation on the integers in Section 1.2 and of an inclusion relation on the subsets of set in Section 1.1.

Formally we say that a relation R on a set S is a subset of the cartesian product $S \times S$.

We say that an element $x \in S$ is related to an element $y \in S$ by the relation R if and only if the ordered pair $(x, y) \in R$.

For example, on the set of integers $\mathbb{Z}$, the order relation $(>)$ is the set R of ordered pairs (x, y) for which x is greater than y. There is no distinction between the statements

$$x > y \quad \text{and} \quad (x, y) \in R$$

Another relation on $\mathbb{Z}$ is division. We write

$$a \mid b \qquad \text{if } a \text{ is related by division to } b$$

In this way the division relation is the subset of $\mathbb{Z} \times \mathbb{Z}$ consisting of the ordered pairs (a, b) where a divides b.

One of the most fundamental and pervasive notions in all of mathematics is a particular type of relation called an *equivalence* relation. Such a relation is used whenever a classification among the elements of some set as being "equal" is desired for some purpose.

For example, if we are interested in the magnitude of integers it is natural to regard a and $-a$ as "equal," certainly equal in absolute value. That is, a and $-a$ are equivalent if we wish to classify integers by their absolute value.

There are many uses for equivalence relations in algebra. The following example is a special case of one of the most important uses.

2.1.1 Example.   Congruence Modulo 10 on $\mathbb{Z}$

We shall classify integers by their remainder when divided by 10. For any integer m the division algorithm gives

$$m = 10 \cdot q + r \qquad \text{where } r = 0 \text{ or } 0 < r < 10$$

The remainder r depends only on m. If we classify integers by their remainder upon division by 10, we shall have 10 classes, one for each possible remainder.

All integers having the same remainder when divided by 10 will be put into one set and regarded, in a sense to be made precise, as "equal."

Note that if a positive integer m is written in base 10 notation the remainder r upon division by 10 is simply the last digit of the expansion. On the other hand, if m is negative, the positive remainder of m is

$$10 - (\text{the last digit of the expansion of } -m)$$

For example, integers that go into the same set as 7 include

$$7, 17, 117, -3, -23 \qquad \text{and } m \in \mathbb{Z} \text{ such that } 10 \mid (m - 7)$$

Here are the sets obtained from this classification:

$$[0] = \{b \in \mathbb{Z}: 10 \mid b\} \qquad\qquad [5] = \{b \in \mathbb{Z}: 10 \mid (b - 5)\}$$
$$[1] = \{b \in \mathbb{Z}: 10 \mid (b - 1)\} \qquad [6] = \{b \in \mathbb{Z}: 10 \mid (b - 6)\}$$
$$[2] = \{b \in \mathbb{Z}: 10 \mid (b - 2)\} \qquad [7] = \{b \in \mathbb{Z}: 10 \mid (b - 7)\}$$
$$[3] = \{b \in \mathbb{Z}: 10 \mid (b - 3)\} \qquad [8] = \{b \in \mathbb{Z}: 10 \mid (b - 8)\}$$
$$[4] = \{b \in \mathbb{Z}: 10 \mid (b - 4)\} \qquad [9] = \{b \in \mathbb{Z}: 10 \mid (b - 9)\}$$

Each integer belongs to one of these classes. Since no integer belongs to two of these classes we see that

$$[0], [1], \ldots, [9]$$

partition $\mathbb{Z}$.

The sets are called variously congruence classes, residue classes, and equivalence classes. The last term is the most general; see Definition 2.1.2.

Now we consider this same example using a relation that has many of the properties of equality.

Let a, b be integers. Define

$$a \equiv b \qquad \text{if and only if } 10 \mid (a - b)$$

This is a relation on $\mathbb{Z}$. The subset of $\mathbb{Z} \times \mathbb{Z}$ defining the relation is

$$\{(a, b): 10 \mid (a - b)\}$$

Here are the equality-like properties that ($\equiv$) enjoys. For all a, b, and $c \in \mathbb{Z}$,

   i. $a \equiv a$.
   ii. If $a \equiv b$ then $b \equiv a$.
   iii. If $a \equiv b$ and $b \equiv c$ then $a \equiv c$.

These properties are easy to prove. For example, here is a proof of iii. By hypothesis

$$10 \mid (a - b) \quad \text{and} \quad 10 \mid (b - c)$$

Hence $10 \mid (a - b) + (b - c)$ and since $(a - b) + (b - c) = a - c$ it follows that

$$10 \mid (a - c) \quad \text{and hence} \quad a \equiv c$$

The connection between the classification of integers with respect to their remainder upon division by 10 and the relation $\equiv$ is this: Integers a and b belong to the same congruence class if and only if $a \equiv b$.

To prove this, suppose first that $a \equiv b$ or that $10 \mid a - b$; thus

$$a - b = 10m \quad \text{or} \quad a = b + 10m$$

Now divide b by 10,

$$b = 10q + r \quad \text{where} \quad r = 0 \quad \text{or} \quad 0 < r < 10$$

and substitute

$$a = 10q + r + 10m = 10(q + m) + r$$

Thus the remainders arising when a and b are divided by 10 are equal. Conversely if

$$a = 10p + r \quad \text{and} \quad b = 10q + r$$

then

$$a - b = 10(p - q)$$

and so $10 \mid (a - b)$.

Now we extend the ideas in this example to a general set.

2.1.2 Definition. Equivalence Relation

An equivalence relation $\sim$ on a set S is a relation such that for all a, b, and c in S the following hold:

   i. $\sim$ is reflexive: $a \sim a$.
   ii. $\sim$ is symmetric: If $a \sim b$ then $b \sim a$.
   iii. $\sim$ is transitive: If $a \sim b$ and $b \sim c$ then $a \sim c$.

Many old friends can be recognized as equivalence relations on an appropriate set:

   1. Let S be the set of triangles in the plane. Define

$$a \sim b$$

   to mean that a and b are congruent. Properties i, ii, and iii hold.
   2. Let S be the set of triangles in the plane. Define

$$a \sim b$$

   to mean that a and b are similar triangles. Again i, ii, and iii hold.
   3. Let S be the set of fractions a/b where $b \neq 0$. Define $a/b = c/d$ if and only if $ad = bc$.
   Equality of fractions is an equivalence relation on $\mathbb{Z} \times (\mathbb{Z} - \{0\})$.

4. Let S be the set of differentiable functions on the open interval $(0, 1)$. Define two functions to be equivalent if they have the same derivative on $(0, 1)$.

2.1.3 Definition.   Equivalence Classes

If $\sim$ is an equivalence relation on a set S then the equivalence classes of $\sim$ are the sets

$$[a] = \{s \in S: s \sim a\}$$

2.1.4 Theorem.   Equivalence Relations and Partitions

If $\sim$ is an equivalence relation on a set S the equivalence classes partition S; that is, S is the set union of all the equivalence classes and for each $s \in S$ there is one and only one equivalence class to which s belongs.

Conversely, if S is partitioned by a set of subsets then the definition

$$x \sim y \qquad \text{if and only if } x \text{ and } y \text{ belong to the same subset}$$

is an equivalence relation on S whose equivalence classes are the subsets of the partition.

Proof. Suppose first that $\sim$ is an equivalence relation on S. If $s \in S$ then $s \sim s$ so that

$$s \in [s] = \{x \in S: x \sim s\}$$

This means that each member of S is in some equivalence class.

Next we prove that no two different equivalence classes have an element in common. Indeed, suppose that $c \in [a] \cap [b]$. We shall show that $[a] = [b]$. Suppose that $x \in [a]$. Then

$$x \sim a \quad \text{and} \quad c \sim a \qquad \text{since both } x \text{ and } c \in [a]$$

Hence by symmetry $a \sim c$ and then by transitivity $x \sim c$. Also

$$c \sim b \qquad \text{since } c \in [b]$$

Hence transitivity again gives $x \sim b$ and so $x \in [b]$. Thus

$$[a] \subseteq [b]$$

Interchanging the roles of a and b yields

$$[b] \subseteq [a]$$

and so $[a] = [b]$.

Now to prove the converse. Suppose that S has been partitioned by a family E of subsets of S. Define

$$x \sim y \qquad \text{if and only if } x \text{ and } y \text{ belong to the same subset in } E$$

We verify the three properties of an equivalence relation. The reflexive and symmetric laws are obvious.

To prove transitivity suppose that

$$x \sim y \quad \text{and} \quad y \sim z$$

Since E partitions S, there is exactly one subset $A \in E$ such that $y \in A$. Now $x \sim y$ means that $x \in A$ also. Similarly $z \in A$. But then, by definition, $x \sim z$.

Now we return to discuss congruences on the integers modulo any positive integer. These examples and theorem play a very important role in our work in group theory, ring theory, field theory, and their applications.

2.1.5 Definition.   Congruence Modulo *n* on $\mathbb{Z}$

Let $n \in \mathbb{Z}^+$. Integers a and b are *congruent modulo n* if and only if $n \mid (a - b)$. In symbols:

$$a \equiv b \pmod{n} \qquad \text{if and only if} \quad n \mid (a - b)$$

Comment.   It is good to remember that $n \mid (a - b)$ if and only if $b = a + mn$ for some $m \in \mathbb{Z}$.

2.1.6 Theorem.   Equality Properties of Congruence

Congruence modulo n is an equivalence relation. Moreover congruences may be added and multiplied:

$$a \equiv b \pmod{n} \quad \text{and} \quad c \equiv d \pmod{n}$$

imply

$$a + c \equiv b + d \pmod{n} \quad \text{and} \quad a \cdot c \equiv b \cdot d \pmod{n}$$

Proof.   The verification that congruence modulo n is an equivalence relation for any positive integer n is no more difficult than for the case $n = 10$, and we omit the details.

To prove the last part of the theorem we prove a special case first. Working modulo n,

$$\text{If } a \equiv b \quad \text{then } a + z \equiv b + z \quad \text{and} \quad a \cdot z \equiv b \cdot z \tag{1}$$

This is so because if $n \mid (a - b)$ then

$$n \mid \{(a + z) - (b + z)\} \quad \text{and} \quad n \mid (az - bz)$$

Now (1) together with transitivity of ($\equiv$) yields the second part of the theorem. Thus from (1)

If $a \equiv b \pmod{n}$ then $a + c \equiv b + c \pmod{n}$ and $ac \equiv bc \pmod{n}$ and if $c \equiv d \pmod{n}$ then $c + b \equiv d + b \pmod{n}$ and $cb \equiv db \pmod{n}$.

Now using commutativity of addition and multiplication and the transitivity of $\equiv$, the result claimed in the theorem is achieved;

$$a + c \equiv b + c \equiv d + b \quad \text{and} \quad ac \equiv bc \equiv db$$

The import of this theorem is that we can work with congruence modulo n very much as we work with equality. In particular, every equality among integers is also a congruence modulo n. Thus the basic axioms (associativity, commutativity, distributivity) of arithmetic hold for congruences.

Of course, not all properties of equality hold for congruences. Cancellation must be watched like a hawk! For example,

$$2 \cdot 6 \equiv 2 \cdot 2 \,(\mathrm{mod}\,8) \quad \text{but} \quad 6 \not\equiv 2 \,(\mathrm{mod}\,8)$$

Again

$$2 \not\equiv 0 \,(\mathrm{mod}\,6) \quad \text{and} \quad 3 \not\equiv 0 \,(\mathrm{mod}\,6) \quad \text{but} \quad 2 \cdot 3 = 6 \equiv 0 \,(\mathrm{mod}\,6)$$

The warning is that the product of integers not congruent to 0 mod n may be congruent to 0 mod n.

Another difference between congruence and integers is that equations having no solutions in integers may have solutions in certain congruences. Here is an example:

$5x = 1$ has no solution in $\mathbb{Z}$.

$5x \equiv 1 \,(\mathrm{mod}\,6)$ has a solution; $x = 5$.

The next theorem clarifies the situation. First we prove a useful lemma relating congruences for different moduli.

2.1.7 Lemma.   Changing Moduli of Congruences

    i. If $a \equiv b \,(\mathrm{mod}\,n)$ then $am \equiv bm \,(\mathrm{mod}\,mn)$ for all m.
    ii. If $a \equiv b \,(\mathrm{mod}\,n)$ and $d \mid n$ then $a \equiv b \,(\mathrm{mod}\,d)$.
    iii. If $a \equiv b \,(\mathrm{mod}\,n)$ and $d \mid n$, $d \mid a$, and $d \mid b$ then
      $a/d \equiv b/d \,(\mathrm{mod}\,n/d)$.

Proof.   All of these follow easily from the definition of congruence and the division properties. By the definition of congruence, $n \mid (b - a)$.

Thus, $mn \mid (ma - mb)$ also and so i holds.

If in addition, $d \mid n$ then $d \mid (b - a)$ and ii holds.

Moreover, if $d \mid a$ and $d \mid b$ as well then the equation $n = u(b - a)$ implies

$$\frac{n}{d} = u \left(\frac{a}{d} - \frac{b}{d} \right)$$

and hence iii holds.

2.1.8 Theorem.   Solution of Linear Congruences

Let $n \in \mathbb{Z}^+$. If a and n are relatively prime ($\gcd(a, n) = 1$) then

    i. If $ab \equiv 0 \,(\mathrm{mod}\,n)$ then $b \equiv 0 \,(\mathrm{mod}\,n)$.
    ii. If $ab \equiv ac \,(\mathrm{mod}\,n)$ then $b \equiv c \,(\mathrm{mod}\,n)$.
    iii. The congruence $ax \equiv b \,(\mathrm{mod}\,n)$ has a unique solution modulo n.

More generally, if $\gcd(a, b) = d$,

    iv. The congruence $ax \equiv b \pmod{n}$ has a solution if and only if $d \mid b$. If $d \mid b$ then the congruence has exactly d solutions that are distinct modulo n.

    v. (Chinese remainder theorem) If $\gcd(n, m) = 1$ then the pair of congruences

$$x \equiv a \pmod{n}$$
$$x \equiv b \pmod{m}$$

has a solution. The solution is unique modulo nm.

Proof. Result i is seen to be a special case of ii by taking $c = 0$. In fact ii is contained in iii since the congruence $ax \equiv ac$ has the unique solution $x \equiv c$. Thus if $ab \equiv ac$ it must be that $b \equiv c$.

    However, it is instructive to prove each part separately.

Proof of ii. By definition if $ab \equiv ac \pmod{n}$ then $n \mid (ab - ac)$ and thus $n \mid a(b - c)$. Now since $\gcd(a, n) = 1$ it follows from Property iii of the greatest common divisor theorem 1.3.7 that $n \mid (b - c)$ and thus that

$$b \equiv c \pmod{n}$$

Proof of iii. Begin with the euclidean algorithm: If $\gcd(a, n) = 1$ there exist integers u and v such that

$$1 = au + bv$$

and hence, multiplying through by b,

$$b = aub + nvb$$

Now we regard this equation as a congruence modulo n. Note that $nvb \equiv 0 \pmod{n}$ so that

$$aub \equiv b \pmod{n}$$

and we have established a solution.

    Now suppose that

$$ax \equiv b \pmod{n} \quad \text{and} \quad ay \equiv b \pmod{n}$$

Then $ax \equiv ay$ and by ii it follows that $x \equiv y \pmod{n}$. This establishes the uniqueness of the solution to within multiples of n; we say the solution is unique modulo n.

Proof of iv. First suppose that for some integer x

$$ax \equiv b \pmod{n}$$

Then $n \mid (ax - b)$ and since $d \mid n$ and $d \mid a$ it follows that $d \mid b$.

    Conversely suppose that $\gcd(a, n) = d$ and that $d \mid b$. Then

$$\gcd\left(\frac{a}{d}, \frac{n}{d}\right) = 1$$

and hence by iii the congruence

$$\frac{a}{d}x \equiv \frac{b}{d} \left(\mathrm{mod}\ \frac{n}{d}\right)$$

has a solution x. Then by Lemma 2.1.7 we conclude that $ax \equiv b \pmod{n}$. Now suppose that there are two solutions

$$ax \equiv b \pmod{n} \quad \text{and} \quad ay \equiv b \pmod{n}$$

so that $ax \equiv ay \pmod{n}$ and so

$$\frac{a}{d}x \equiv \frac{a}{d}y \left(\mathrm{mod}\ \frac{n}{d}\right)$$

and hence by ii

$$x \equiv y \left(\mathrm{mod}\ \frac{n}{d}\right)$$

since $\gcd(a/d, n/d) = 1$. Thus any two solutions are congruent modulo n/d.

Now to count the number of solutions distinct modulo n.

Let x_0 be one solution. We claim that for each $k = 0, 1, \ldots, d-1$

$$x_0 + k\left(\frac{n}{d}\right) \tag{2}$$

is a solution and no two are congruent modulo n. In any event

$$ax_0 + ak\frac{n}{d} = ax_0 + \left(\frac{a}{d}\right)kn \equiv ax_0 \equiv b \pmod{n}$$

So the integers (2) are d solutions to the congruence.

Any two are distinct modulo n. Indeed if

$$x_0 + k\left(\frac{n}{d}\right) \equiv x_0 + h\left(\frac{n}{d}\right) \pmod{n}$$

then

$$k\frac{n}{d} \equiv h\frac{n}{d} \pmod{n}$$

and hence

$$n \mid (h - k)\frac{n}{d}$$

so that

$$(h - k)\frac{n}{d} = un \quad \text{or} \quad (h - k)n = und$$

and so

$$h - k = ud \quad \text{or} \quad h \equiv k \pmod{d}$$

Thus we have shown that any two solutions are congruent modulo n/d and hence all solutions have the form (2). On the other hand, we have exhibited d solutions distinct modulo n.

Proof of v.   Consider the integers of the form

$$a + yn$$

Modulo n, all of these are congruent to a. We want to find an integer y such that

$$a + yn \equiv b \pmod{m}$$

This is the same as solving the congruence

$$yn \equiv (b - a) \pmod{m}$$

and since $\gcd(n, m) = 1$ we can conclude from iii that a solution exists.

The solution is unique modulo mn, for if y were another solution then we would have

$$x \equiv y \pmod{n} \quad \text{and} \quad x \equiv y \pmod{m}$$

and hence that both m and n divide $(x - y)$. But since $\gcd(n, m) = 1$ it follows that $mn \mid (x - y)$ and so

$$x \equiv y \pmod{mn}$$

Comment.   Property v is called the "Chinese remainder theorem" because it was supposed to have been known to the Chinese long before its "discovery" in western civilization.

Note that the solutions to these congruences may all be obtained by careful use of the euclidean algorithm.

Now we want to take another point of view toward congruences. We want to make a mathematical system out of the set of congruence classes modulo n.

2.1.9 Definition.   Operations on Congruence Classes

Let $n \in \mathbb{Z}^+$. A congruence class modulo n is a set

$$[r] = \{x \in \mathbb{Z} : x \equiv r \pmod{n}\}$$

Because of the division algorithm we may choose the representative r so that $0 \leq r < n$. If we do we may describe $[r]$ as the set of integers whose remainder, when divided by n, is r. Now we define addition and multiplication of congruence classes modulo n as follows:

$$[a] + [b] = [c] \quad \text{if} \quad a + b \equiv c \pmod{n}$$

and

$$[a] \cdot [b] = [c] \quad \text{if} \quad ab \equiv c \pmod{n}$$

There is an important logical gap in this definition that we must discuss and fill. When we define $[a] + [b]$ we are trying to define a binary operation on elements that are *sets*. But the definition is made in terms of certain elements, namely $a \in [a]$ and $b \in [b]$. The elements a and b have been selected, almost by accident, to represent the congruence classes. What if other representatives of the classes were chosen?

For example, consider congruence classes modulo 6. We find that
$$[3] = [9] \quad \text{since} \quad 3 \equiv 9 \,(\text{mod}\,6)$$
and
$$[10] = [-2] \quad \text{since} \quad 10 \equiv -2 \,(\text{mod}\,6)$$
Now the definition of addition of congruence classes gives on the one hand
$$[3] + [10] = [3 + 10] = [13]$$
and on the other
$$[9] + [-2] = [9 + (-2)] = [7]$$
So the crucial question is
$$[13] = [7]?$$
Well, the answer is "yes" since $13 \equiv 7 \,(\text{mod}\,6)$, but we need a guarantee that no ambiguities will ever occur if we are to permit the addition of congruence classes to be defined by the addition of selected representatives. We need to prove the following.

2.1.10 Lemma.   Operations Are Well Defined

If $[a]$, $[b]$, $[c]$, and $[d]$ are congruence classes modulo n and
$$[a] = [b] \quad \text{and} \quad [c] = [d]$$
then
$$[a + c] = [b + d] \quad \text{and} \quad [ac] = [bd]$$

Proof.   The proof is not difficult. Simply translate equality of congruence classes into congruences. We find then that the lemma is the content of the properties of equality for congruences; Theorem 2.1.6.

Comment.   We shall have frequent occasion to take some equivalence relation on a set S, construct the equivalence classes, and make the resulting equivalence classes into an algebraic system. Each time we do this by using representatives (members) of the classes we shall be obligated to prove that our definitions are well defined, that is, are independent of the representatives chosen.

Notation.   Let $n \in \mathbb{Z}^+$. The algebraic system of congruence classes under Definition 2.1.9 will be denoted $\mathbb{Z}_n$.
The important properties of the system $\mathbb{Z}_n$ are collected in the next theorem. Note especially that the cancellation law of multiplication *does* hold when the modulus n is a prime.

2.1.11 Theorem.   Arithmetic Properties of $\mathbb{Z}_n$

The system of congruence classes modulo n under the definitions of $(+)$ and $(\cdot)$ in Definition 2.1.9 satisfy the associative, commutative, and distributive axioms of the integers together with the existence of an additive and multiplicative identity. Moreover every congruence class has an additive inverse.

If n is a prime, say, $n = p$, then every residue class $[q] \neq [0]$ in $\mathbb{Z}_p$ has a multiplicative inverse.

More generally, $[a]$ has a multiplicative inverse in $\mathbb{Z}_n$ if and only if $\gcd(a, n) = 1$.

Proof. Since the operations on congruence classes are defined by representatives that satisfy the associativity, commutativity, and distributivity axioms of $\mathbb{Z}$ it follows that the axioms hold for the congruence classes of $\mathbb{Z}_n$.

Moreover

[0] is the additive identity.

[1] is the multiplicative identity.

Check to see that for all integers a

$$[a] + [-a] = [0]$$

so that $-[a]$ is the additive inverse of $[a]$.

Now let $\gcd(a, n) = 1$. We can solve the congruence

$$ax \equiv 1 \pmod{n}$$

and so find

$$[a][x] = [1]$$

Conversely if $[a][x] = [1]$ then $[ax] = [1]$ and so $ax \equiv 1 \pmod{n}$. Thus $ax = 1 + un$, or $1 = ax - nu$, and it follows that $\gcd(a, n) = 1$.

Now suppose that $n = p$ is a prime. If $[a] \neq [0]$ then $p \nmid a$ and so $\gcd(a, p) = 1$. Thus the existence of a multiplicative inverse is assured.

Comment. It is important to realize that the inverse congruence class for $[a]$, if it exists, may be computed with the euclidean algorithm. If $\gcd(a, n) = 1$ write $1 = ax + ny$ and thus have $1 \equiv ax \pmod{n}$.

We close this section by introducing the useful Euler phi function (φ-function). But first here is a useful lemma.

2.1.12 Lemma.   Congruence Classes and gcds

Let n be a positive integer. Consider the congruence classes modulo n.

   i. If $[a] = [b]$ then $\gcd(a, n) = \gcd(b, n)$.
   ii. If $\gcd(a, n) = \gcd(b, n) = 1$ then $\gcd(ab, n) = 1$, and conversely.

Proof of i.   If $[a] = [b]$ then $a \equiv b \pmod{n}$ and so

$$b = a + nu$$

for some integer u. But then by Lemma 1.3.4, $\gcd(a, n) = \gcd(b, n)$.

Proof of ii.   Write

$$1 = au + nv$$

and

$$1 = bw + nx$$

Multiply:

$$1 = (au + nv)(bw + nx) = abuw + n(vbw + xau + nvx)$$

and thus $\gcd(ab, n) = 1$ by Theorem 1.3.7.

Conversely, $(ab, n) = 1$ means

$$1 = abh + nk \qquad \text{for integers } h \text{ and } k$$

and so $(a, n) = 1$ and $(b, n) = 1$.

As a result of this lemma we may define a function φ, the so-called Euler phi function, as follows:

2.1.13 Definition. Euler φ-Function

For each positive integer n let $\varphi(n)$ denote the *number* of congruence classes $[a]$ modulo n such that $[a]$ has a multiplicative inverse in $\mathbb{Z}_n$.

Comment. Because we can choose a representative for $[a]$ in the range

$$1 \leq r \leq n$$

$\varphi(n)$ is equal to the number of positive integers r less than or equal to n and relative prime to n.

An important property of the Euler φ-function is proved in the next theorem.

Small values of $\varphi(n)$ are easy to tabulate. It is not hard to see that if p is a prime then $\varphi(p) = p - 1$ and

$$\varphi(p^n) = p^n - p^{n-1} = (p - 1)p^{n-1}$$

since the integers less than p^n which are *not* relative prime to p are precisely the multiples of kp of p and there are p^n/p choices for k. The other numbers are relative prime to p^n. There are thus $p^n - p^{n-1}$ positive integers less than or equal to p^n and relatively prime to p^n.

2.1.14 Theorem. A Multiplicative Property

The Euler phi function is multiplicative: If $\gcd(n, m) = 1$ then $\varphi(mn) = \varphi(n)\varphi(m)$.

Proof. We shall need to distinguish between congruence classes modulo n, modulo m, and modulo mn. In this proof we shall do this with a subscript:

$[k]_n$ denotes a congruence class modulo n.

$[k]_m$ denotes a congruence class modulo m.

$[k]_{nm}$ denotes a congruence class modulo nm.

We show that a function

$$F: \mathbb{Z}_{nm} \to \mathbb{Z}_n \times \mathbb{Z}_m$$

can be defined by

$$F: [k]_{nm} \to ([k]_n, [k]_m)$$

and that F is one-to-one and onto, a bijection.

Since the definition of F depends ostensibly on the representative k chosen from the congruence class modulo nm we must first verify that if

$$k \equiv h \pmod{nm}$$

then $k \equiv h \pmod{n}$ and $k \equiv h \pmod{m}$. But this is a result of Lemma 2.1.7. Thus F is a function from $\mathbb{Z}_{nm}$ into $\mathbb{Z}_n \times \mathbb{Z}_m$.

By the Chinese remainder theorem F is onto. Indeed, to find what class $[x]_{nm}$ maps to $([a]_n, [b]_m)$ under F we have only to solve the pair of congruences

$$x \equiv a \pmod{n} \quad \text{and} \quad x \equiv b \pmod{m}$$

Because this solution is unique modulo nm we have also shown that F is one-to-one.

Now to complete the proof we need only show that

$$\gcd(k, nm) = 1 \quad \text{if and only if} \quad \gcd(k, n) = 1 \text{ and } \gcd(k, m) = 1 \quad (3)$$

for then, under F, a congruence class modulo nm with an inverse corresponds to a pair of congruence classes with inverses modulo n and modulo m, respectively. But (3) was established in part ii of Lemma 2.1.12.

EXERCISE SET 2.1

2.1.1 Prove or disprove: Division is an equivalence relation on $\mathbb{Z}^+$.

2.1.2 Find all the equivalence relations on $S = \{1, 2, 3\}$. Compare with $P(S)$.

2.1.3 Find all the equivalence relations on $T = \{1, 2, 3, 4\}$. Compare with $P(T)$.

2.1.4 What is wrong with the following argument to show that the symmetric and transitive properties of equivalence imply the reflexive property: "If $a \sim b$ then $b \sim a$ by symmetry. Now by transitivity, $a \sim a$. Hence $\sim$ is reflexive."

2.1.5 Let S be the coordinate plane

$$S = \{(x, y): x \text{ and } y \in \mathbb{R}\}$$

Let p and q be points $p = (x, y)$ and $q = (u, v)$. Define a relation $\blacktriangle$ by

$$p \blacktriangle q \quad \text{if and only if} \quad |x| = |u| \quad \text{and} \quad |y| = |v|$$

Show that $\blacktriangle$ is an equivalence relation and give a geometric description of the equivalence classes.

2.1.6 Let S be the same set as in Problem 2.1.5. Define

$$p \bowtie q \quad \text{if and only if} \quad x - u = y - v$$

Show that ⋈ is an equivalence relation and give a geometric description of the equivalence classes.

2.1.7 Let S be the same set as in Problem 2.1.5. Define

$$p \odot q \qquad \text{if and only if} \quad x^2 + y^2 = u^2 + v^2$$

Show that $\odot$ is an equivalence relation and give a geometric description of the equivalence classes.

2.1.8 In $\mathbb{Z}_{10}$ compute

$$[7] \cdot [7], \qquad [8] + [5] - [9], \qquad [7] \cdot ([8] - [-2])$$

2.1.9 In $\mathbb{Z}_{18}$ find the least positive integer in $[-7]$. Find s such that

$$[-7][s] = [1]$$

in $\mathbb{Z}_{18}$.

2.1.10 Show that for each nonnegative integer k, $10^k \equiv 1 \pmod 3$. Show that $3 \mid n$ if and only if the sum of the digits in the decimal expansion of n is divisible by 3.

2.1.11 For each nonnegative integer k determine a, $-5 \le a \le 5$, such that

$$10^k \equiv a \pmod{11}$$

Express a as a simple function of k. Find a divisibility criterion for 11 analogous to the one for 3 in Problem 2.1.10.

2.1.12 Make a $+$ table for $\mathbb{Z}_6$ and a $\cdot$ table for $\mathbb{Z}_7$.

2.1.13 If 2 and -2 belong to the same congruence class modulo n, what is n?

2.1.14 For what integers n is there an integer k such that

$$k \equiv -k \pmod n \quad \text{and} \quad k \not\equiv 0 \pmod n$$

When there is, determine k in terms of n.

2.1.15 Find all solutions to $x^2 \equiv 1 \pmod 5$. Determine all solutions to $x^2 \equiv 1 \pmod p$ where p is prime.

2.1.16 Find all solutions to $x^2 \equiv 1 \pmod{10}$. Find all solutions to $x^2 \equiv 1 \pmod{2p}$ where p is an odd prime.

2.1.17 Find all solutions to $x^2 \equiv 1 \pmod{15}$. Find all solutions to $x^2 \equiv 1 \pmod{pq}$ where p and q are primes. *Hint*: Use the Chinese remainder theorem if $p \ne q$. What if $p = q$? How should this result be generalized?

2.1.18 Let n be a positive integer greater than 1. Suppose that for some integer a, $a \not\equiv 0 \pmod n$, there is a positive integer k such that

$$a^k \equiv 0 \pmod n$$

What can be said about the prime factorization of n? It may be helpful to see whether such a's exist for $n = 10$, $n = 12$, $n = 16$, $n = 36$. In any event determine in each of these cases whether such an a does exist.

2.1.19 Find a formula for the Euler φ-function $\varphi(n)$ based on the prime factorization on n.

2.1.20 Solve the pair of congruences

$$3x \equiv 1 \pmod 7 \quad \text{and} \quad 2x \equiv 5 \pmod 9$$

2.2 PARTIALLY ORDERED SETS AND LATTICES

In this section we introduce a generalization of the inclusion $\supseteq$ relation on subsets of a set. As we shall see, the subsystems of the algebraic structures of groups, rings, fields, vector spaces, and modules which we shall study fit this generalization. Most importantly for us, the language we use and the diagrams we can draw give us added insight and intuition concerning the interaction of the important subsystems of these algebraic systems.

In Section 1.1 we saw that inclusion $\supseteq$ was reflexive, transitive, and antisymmetric.

In Section 1.2 we reviewed the order relation on the integers (or on the rational numbers $\mathbb{Q}$, or the real numbers $\mathbb{R}$). There we singled out the properties of transitivity and trichotomy for special attention.

In Section 1.3 we recall the division relation on $\mathbb{Z}^+$. The division properties of reflexivity and transitivity hold. Although the full trichotomy law does not hold for division, we do know that on $\mathbb{Z}^+$, division is antisymmetric; that is,

$$\text{If } a \mid b \text{ and } b \mid a \text{ then } a = b$$

Now we formalize these properties in the following definition.

2.2.1 Definition.   Partially Ordered Set (POS)

A partially ordered set (POS) is a nonempty set together with a relation $\supseteq$ such that

    i. Reflexivity: For all $a \in S$ $a \supseteq a$.
    ii. Antisymmetry: If $a \supseteq b$ and $b \supseteq a$ then $a = b$.
    iii. Transitivity: If $a \supseteq b$ and $b \supseteq c$ then $a \supseteq c$.

If $a \supseteq b$ or $b \supseteq a$ we say that a and b are comparable. Otherwise we say they are incomparable.

Notation.   If P is a POS under a relation $\supseteq$ we often write

$$\langle P, \supseteq \rangle$$

to indicate both the set and the symbol used to denote the ordering. As examples of partially ordered sets we have already found:

    1. The integers are partially ordered by "greater than or equal." We write $\langle \mathbb{Z}, \geq \rangle$
    2. The positive integers $\langle \mathbb{Z}^+, \mid \rangle$, are partially ordered by division.
    3. The power set of a set S, $\langle P(S), \supseteq \rangle$ is partially ordered by inclusion.

Comment.   It is very important to note that if P is a partially ordered set by a relation $\supseteq$ then any nonempty subset of P is also a partially ordered set under the same relation.

Example.   Let *P* be the set of positive integers

$$\{1, 2, 3, \ldots, 12\}$$

partially ordered by division. It is easy, maybe a little tedious, to list all the relationships between elements. Thus we find

1 divides every member.

2 divides 2, 4, 6, 8, and 10.

3 divides 3, 6, 9, and 12.

and so on.

Diagrams.   It is often convenient to make a Hasse diagram for partially ordered sets. As with the $\supseteq$ relation for containment of subsets we represent an element by a dot and if $a \supseteq b$ in the partially ordered set we locate *a* above *b* on the page and draw a line from *a* to *b*, deleting any lines made redundant by transitivity. Figure 2.1 shows the integers $1, \ldots, 12$ ordered by division. It is often convenient to create examples of a POS by giving a diagram. The lines between the dots representing the elements give the ordering relation together with transitivity. If there is no line or sequence of line segments between elements then the elements are not comparable.

Figure 2.2 shows all the possible ways to partially order a set of three elements, up to a labeling of the elements. Subsets of a partially ordered set in which any two elements are comparable are important.

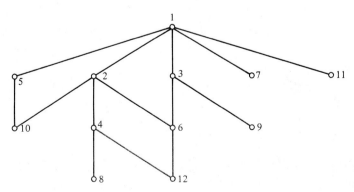

Figure 2.1 The Hasse diagram of $P = \{1, 2, \ldots, 12\}$ as a POS under division ($|$). An example of a *chain* is the subset $\{2, 4, 12\}$. The only *maximal* element of *P* is 1. The *minimal* elements are 8, 9, 10, 12, 7, and 11. The subset $\{6, 8\}$ has 2 and 1 as upper bounds but this subset has no lower bounds. The set $\{4, 6\}$ has 2 as its least upper bound and 12 as its greatest lower bound.

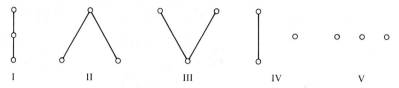

Figure 2.2 The possible diagrams for a POS consisting of three elements. (I) is the chain of three elements. (II) is the set with a top element and two noncomparable elements. (III) is the dual of (II); it is the set with a bottom element and two noncomparable elements. (IV) is the set with one chain of length two and one element not comparable with either of the elements in the chain. (V) is the POS of three mutually noncomparable elements. POS (I), (IV), and (V) are self dual.

2.2.2 Definition. Chain

A subset C of a POS is called a *chain* if any two elements of C are comparable. The *length* of the chain is $|C|$. Here are some examples of chains:

1. A single element is a chain.
2. Any subset of a chain is a chain.
3. The set of integers $\mathbb{Z}$ is a chain under $\geq$.
4. The chains of length 4 in the set of integers $\{1, 2, \ldots, 12\}$ under division are

$$\{1, 2, 4, 12\} \qquad \{1, 2, 4, 8\} \qquad \{1, 2, 6, 12\} \qquad \{1, 3, 6, 12\}$$

Terminology. If $\langle P, \supseteq \rangle$ is a partially ordered set and is also a chain we say that $\supseteq$ *totally orders* P or that P is a totally ordered set.

2.2.3 Definition. Cover

Let p and q be elements of the partially ordered set $\langle P, \supseteq \rangle$. The element p is said to *cover* q if $p \supseteq q$ and there is no element of P between p and q; that is,

$$p \supseteq x \supseteq q \text{ implies } p = x \quad \text{or} \quad x = q.$$

In this case we also say that q is covered by p.

For example, in Figure 2.1 we see that 2 covers 4, 6, and 10 while 6 is covered by 2 and 3.

2.2.4 Definition. Maximal and Minimal Elements

Let $\langle P, \supseteq \rangle$ be a POS. An element m is called a maximal element of P if there is no element $p \in P$ such that $p \neq m$ and $p \supseteq m$. An element n is called a minimal element of P if there is no element $q \in P$ such that $q \neq n$ and $n \supseteq q$. There is no necessity for a set to have any maximal or minimal elements.

Examples.

1. The integers $\mathbb{Z}$ under $\geq$ have no maximal or minimal elements.
2. The positive integers $\mathbb{Z}^+$ under division have 1 as a maximal element but no minimal element.
3. The set $P = \{1, 2, \ldots, 12\}$ under division has 1 as a maximal element and 7, 11, 8, 9, 10, and 12 as minimal elements.

2.2.5 Definition. Bounds

If S is a subset of a POS $\langle P, \supseteq \rangle$, an element $u \in P$ is called an upper bound for S if

$$u \supseteq s \qquad \text{for all } s \in S$$

Dually, an element v is called a lower bound for S if

$$s \supseteq v \qquad \text{for all } s \in S$$

Comment. Please note that the subset S need not have either an upper or a lower bound. Even if it does have an upper bound, the upper bound need not belong to the subset S. For example (look at Fig. 2.1 in the set $P = \{1, 2, \ldots, 12\}$ under division), the subset $\{4, 10\}$ has 1 and 2 as upper bounds but it has no lower bound. If S is a finite chain then it always has an upper and a lower bound.

We can now state an important logical principle, also referred to in the literature as *Zorn's lemma.*

2.2.6 Axiom. Maximal Principle

If $\langle P, \supseteq \rangle$ is a POS in which every chain has an upper bound then P has at least one maximal element.

Comment.   Some condition is necessary to guarantee the existence of maximal elements. As we have already noted, $\langle \mathbb{Z}, \geq \rangle$ has no maximal element.

The maximal principle turns out to be a very powerful axiom in formal set theory. The names of some of the greatest logicians—Zermelo, Frankel, Gödel, and Cohen—are associated with the problems arising from this principle. We shall have only a few occasions to use it.

Most of the examples of POS that we shall study consist of certain subsets of a set. Furthermore, our examples of POS will satisfy additional properties that make them even more useful. These properties are called *lattice* properties.

2.2.7 Definition.   Least Upper Bound (lub) and Greatest Lower Bound (glb)

Let $\langle P, \supseteq \rangle$ be a POS. Let S be a subset of P. An element u is said to be a *least upper bound* (lub) for S if u is an upper bound for S, and if u' is any other upper bound for S, then $u' \supseteq u$.

Dually, v is said to be a *greatest lower bound* (glb) for S if v is a lower bound, and if v' is any other lower bound for S then $v \supseteq v'$. Note that

$$\text{lub}(a, b) = a \qquad \text{if and only if } a \geq b$$

$$\text{glb}(a, b) = b \qquad \text{if and only if } a \geq b$$

2.2.8 Examples.   lubs and glbs in $\langle \mathbb{Z}^+, \mid \rangle$

An upper bound for integers a, b under division is a common divisor. Then $\text{lub}\{a, b\}$ is the greatest common divisor.

Dually a lower bound for $\{a, b\}$ is a common multiple; the $\text{glb}(a, b)$ is the least common multiple of a and b.

Confused? Well, just reorder $\mathbb{Z}^+$ by turning everything around: thus $a \supseteq b$ if $b \mid a$. This makes 1 the minimal element, and now the words "greatest lower bound" and "greatest common divisor" have the same meaning, as do "least upper bound" and "least common multiple."

In fact, any POS can be turned upside down in this way. Here is the definition.

2.2.9 Definition.   Dual Partial Ordering

Let $\langle P, \supseteq \rangle$ be a partially ordered set. The dual partially ordered set to P is $\langle P, \geq \rangle$ where the relation $\geq$ is defined:

$$p \geq q \qquad \text{if and only if } q \supseteq p$$

We omit the verification that this gives a partial ordering of P. Each is said to be the dual of the other. The Hasse diagram for $\langle P, \geq \rangle$ is obtained from that for $\langle P, \supseteq \rangle$ by simply turning the page upside down!

Look back at Figure 2.2. The first POS is a chain. Its dual gives the same diagram; we say it is self-dual. The second and third partially ordered sets in Figure 2.2 are duals of each other.

Now we are ready to define a lattice.

2.2.10 Definition.   Lattice

A *lattice* is a POS $\langle L, \supseteq \rangle$ in which every pair of elements has both a greatest lower bound and a least upper bound.

If L is a lattice denote the least upper bound and greatest lower bound of pairs of elements as follows:

$$a \cup b = \text{lub}\{a, b\} \quad \text{and} \quad a \cap b = \text{glb}\{a, b\}$$

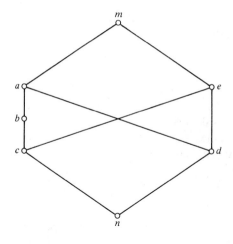

Figure 2.3 A POS showing a subset $\{c, d\}$ having upper bounds but no least upper bound. The upper bounds of $\{c, d\}$ are a, e, and m. This POS is not a lattice, although every set has both an upper and a lower bound.

Note that

$$a \cup b \supseteq a \quad \text{and} \quad a \supseteq a \cap b$$

for all a and b.

Of course, not every POS is a lattice. $P = \{1, 2, \ldots, 12\}$ is not a lattice under division since $\{5, 7\}$ have no lowest bound. Figure 2.3 shows another nonlattice.

2.2.11 Examples of Lattices

Here are some examples of lattices:

1. Any chain is a lattice.
2. The positive integers under division. We have

$$a \cup b = \gcd(a, b) \quad \text{and} \quad a \cap b = \operatorname{lcm}(a, b)$$

3. The power set of set S under set inclusion. We have

$$a \cup b \text{ is the set union of } a \text{ and } b.$$
$$a \cap b \text{ is the set intersection of } a \text{ and } b$$

4. See Figure 2.4.

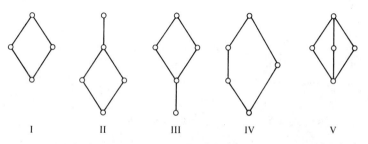

I   II   III   IV   V

Figure 2.4 These are the Hasse diagrams of the lattices having four or five elements that are not chains. Notice that II and III are dual and I, IV, and V are self-dual.

5. Let L consist of the *finite* subsets of the positive integers under set inclusion. L is a lattice because the set union and set intersection of finite sets are finite sets and so any two finite sets possess lubs and glbs that are in L.

Notice that L is a subset of the power set of $\mathbb{Z}^+$. Since the lattice operations in L are the same as the operations in $P(\mathbb{Z}^+)$ we say that L is a *sublattice* of $P(\mathbb{Z}^+)$. The term sublattice is defined in Problem 2.2.16.

Notice too that while every pair of elements of L has a least upper bound, not every subset of L has one. For example, each singleton set $\{a\}$ is a member of L but the set of all singleton sets has no upper bound in L since $\mathbb{Z}^+$ is not a member of L. In view of this example, the following definition identifies an important class of lattices.

2.2.12 Definition. Complete Lattice

A lattice is called *complete* if every nonempty subset has a least upper and greatest lower bound.

Any finite lattice is complete (see Problem 2.2.12).

The lattice of the power set of a set under set inclusion is also a complete lattice since it is easy to see that if E is a collection of subsets of a set S then

$$\text{lub}(E) = \bigcup_{S \in E} S \quad \text{and} \quad \text{glb}(E) = \bigcap_{S \in E} S$$

Terminology. An element u of a partially ordered set $\langle P, \supseteq \rangle$ is called the *top* element of P if $u \supseteq x$ for all $x \in P$. There can be at most one top element, for if u' were another, then $u \supseteq u'$ and $u' \supseteq u$, hence $u = u'$.

Dually we can define a *bottom* element, if it exists, as an element z such that $x \supseteq z$ for all x in P.

The next theorem is extremely useful to us to show that a given partially ordered set is not only a lattice but is a complete lattice.

2.2.13 Theorem. Sufficient Condition for a Complete Lattice

If $\langle P, \supseteq \rangle$ is a partially ordered set with a top element u, and if every nonempty subset of P has a greatest lower bound in P, then P is a complete lattice.

Proof. Under the hypothesis every nonempty set S of P has a greatest lower bound. We need only show that S has a least upper bound. We are given that S has at least one upper bound, namely, the top element. Let U be the set of its upper bounds. Thus

$$U = \{x \in P: x \supseteq s \text{ for all } s \in S\}$$

(The relationships of the sets are shown schematically in the sketch of a general lattice in Figure 2.5.)

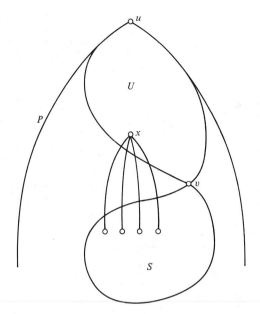

Figure 2.5 This diagram shows the relation of the elements in the proof of Theorem 2.2.13. Start with the set S. The set U shows the upper bounds of S and the element v is its greatest lower bound. The proof demonstrates that v is also the least upper bound of S.

Now U is nonempty because the top element $u \in U$. By hypothesis, U has a greatest lower bound, call it v.

$$v = \text{glb } U$$

We shall show that

$$v = \text{lub } S$$

First we need to show that v is an upper bound for S. In any event each $s \in S$ is a *lower bound* for U. Indeed, if $s \in S$ then

$$x \supseteq s \qquad \text{for all } x \in U$$

since x is an upper bound for S. Hence

$$v = \text{glb } U \supseteq s$$

for all $s \in S$ and so v is an upper bound for S. By the definition of v, $x \supseteq v$ for any upper bound x of S, so

$$v = \text{lub } S$$

Next we establish the algebraic properties of lub and glb as binary operations in a lattice. These are definitive in the sense that an algebraic system with two binary operations having these properties is a lattice.

2.2.14 Theorem. Lattice Properties

If $\langle L, \supseteq \rangle$ is a lattice then the glb ($\cap$) and the lub ($\cup$) operations satisfy the following properties. For all a, b, and c in L:

Idempotence: $a \cup a = a$ and $a \cap a = a$.

Associativity: $a \cup (b \cup c) = (a \cup b) \cup c$ and
$$a \cap (b \cap c) = (a \cap b) \cap c.$$

Commutativity: $a \cup b = b \cup a$ and $a \cap b = b \cap a$.

Absorptivity: $a \cup (a \cap b) = a$ and $a \cap (a \cup b) = a$.

On the other hand, if L is a nonempty set on which two binary operations $\cap$ and $\cup$ have been defined and satisfy the above properties then the following definition for $\geq$ makes $\langle L, \geq \rangle$ into a lattice,

$$a \geq b \qquad \text{if and only if } a = a \cup b$$

in which case

$$a \cup b = \text{lub}\{a, b\} \quad \text{and} \quad a \cap b = \text{glb}\{a, b\}$$

Comment. Notice that the properties listed are those satisfied by the operations of set union and intersection as detailed in Theorem 1.4.4 except for the distributive law. Not all lattices satisfy the distributive property; in particular the lattices 4 and 5 of Figure 2.4 are not distributive. For details see Figure 2.6 and Problem 2.2.8.

Note too that the properties are dual in the sense that interchanging $\cap$ and $\cup$ does not change the list of properties. A proof given for one property may be converted into the proof of the dual property by interchanging $\cap$ and $\cup$, $\supseteq$ and $\subseteq$, and glb and lub throughout. The Hasse diagram is just turned upside down.

Proof of the Lattice Properties. The proofs are not difficult; simply translate the symbols $\cap$ into $\cup$ into glb and lub. Thus idempotence and commutativity are transparent. Here are the details for one absorptive law.

To show that $a = a \cup (a \cap b)$ we are to show that $a = \text{lub}\{a, \text{glb}\{ab\}\}$. Because $a \supseteq \text{glb}\{a, b\}$ the general principle that the lub of comparable

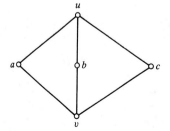

Figure 2.6 A Lattice in which the distributive law fails. Observe that $a \vee (b \wedge c) \neq (a \vee b) \wedge (a \vee c)$. because $(a \vee b) \wedge (a \vee c) = u \wedge u = u$ while $a \vee (b \wedge c) = a \vee v = a$.

elements is the one above implies that

$$a = \text{lub}\{a, \text{glb}\{a, b\}\}$$

Proof That the Properties Define a Lattice.   As in the statement of the theorem let a relation be defined by

$$a \geq b \qquad \text{if and only if } a = a \cup b$$

We shall show that $\langle L, \geq \rangle$ is a lattice.

First we shall establish a dual form of the relation by proving that

$$a = a \cup b \qquad \text{if and only if } b = b \cap a$$

from the absorptive and commutative laws. We have

$$a = a \cup b \text{ implies } b \cap a = b \cap (a \cup b) = b \cap (b \cup a) = b$$

Conversely

$$b = b \cap a \text{ implies } a \cup b = a \cup (b \cap a) = a \cup (a \cap b) = a$$

Now we show that $\geq$ is a partial ordering.

Reflexivity: Since $a = a \cup a$ we conclude $a \geq a$.

Antisymmetry: If $b \geq a$ and $a \geq b$ then $b = b \cup a$ and $a = a \cup b$ so that, with commutativity, $a = b$.

Transitivity: If $a \geq b$ and $b \geq c$ then $a = a \cup b$ and $b = b \cup c$ so that

$$a = a \cup b = a \cup (b \cup c) = (a \cup b) \cup c = a \cup c$$

so $a \geq c$.

Next we must show that $\text{lub}\{a, b\} = a \cup b$ and $\text{glb}\{a, b\} = a \cap b$. We have, using idempotence, associativity, and commutativity, that

$$a \cup b = (a \cup a) \cup b = a \cup (a \cup b) = (a \cup b) \cup a$$

so that $a \cup b \geq a$. Similarly $a \cup b \geq b$ so that $a \cup b$ is an upper bound for $\{a, b\}$.

Now suppose that u is any upper bound for $\{a, b\}$. Then

$$u = u \cup a \quad \text{and} \quad u = u \cup b$$

so that

$$u = u \cup b = (u \cup a) \cup b = u \cup (a \cup b)$$

and so $u \geq (a \cup b)$; so $a \cup b$ is the least upper bound of $\{a, b\}$.

By a dual argument $a \cap b$ is the greatest lower bound of $\{a, b\}$.

2.2.15 Example.   Lattice of Equivalence Relations

A very interesting class of examples of lattices is provided by the equivalence relations of a set. This class of lattices plays a very important role in the theory of lattices, but our interest here is simply as a source of examples.

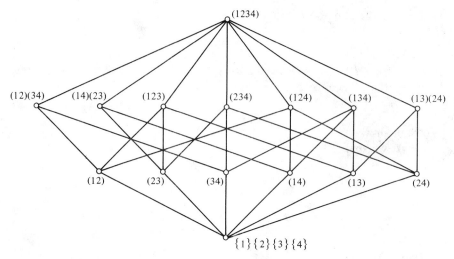

Figure 2.7 The lattice of equivalence relations on $\{1, 2, 3, 4\}$. An equivalence relation is labeled by the equivalence classes that contain more than one element. Thus (1 2) designates the equivalence relation consisting of the partition $\{1, 2\}, \{3\}, \{4\}$ of $\{1, 2, 3, 4\}$.

Let S be a nonempty set. Let π and σ denote equivalence relations of S. We define

$$\pi \geq \sigma \qquad \text{if } x\sigma y \text{ implies } x\pi y$$

that is, when two elements are equivalent under σ they must be equivalent under π.

This ordering of equivalence relations makes the set of all equivalence relations on S into a complete lattice; we ask for a proof in Problem 2.2.10.

The proof is not difficult if you recall that a relation is just a subset of the cartesian product $S \times S$ and the definition of $\geq$ is simply that of set inclusion on these subsets. There is a top element, namely, S itself, the relation in which any two elements are equivalent. Now all that has to be done is to show that the set intersection of a collection of equivalence relations is again an equivalence relation. Then apply Theorem 2.2.13.

Figure 2.7 shows the lattice of equivalence relations on $\{1, 2, 3, 4\}$.

EXERCISE SET 2.2

2.2.1 Draw the possible Hasse diagrams for POS having four elements. (There are 16 different ones.)

2.2.2 Draw the Hasse diagram for the set of positive divisors of 16 ordered by division. What sort of a lattice do they form?

2.2.3 Draw the Hasse diagram for the set of positive divisors of 30 ordered by division. Compare this with the diagram of the subsets of $\{1, 2, 3\}$. Explain the similarity.

2.2.4 Draw the diagram for the set of positive divisors of 24 ordered by division. Compare this lattice in Problem 2.2.3.

2.2.5 Let n be a positive integer. Prove that the set of positive divisors of n under division form a lattice. For what n will it be a chain? For what n will it be the same as the subsets of $\{1, 2, 3, \ldots, k\}$? And how is n related to k?

2.2.6 Prove that the lattice of Problem 2.2.5 is distributive.

2.2.7 Draw the diagram for the POS of all equivalence relations π on $\{1, 2, 3, 4, 5\}$ such that either $1\pi2$ or $3\pi4\pi5$; that is, either 1 and 2 are equivalent under π or $\{3, 4, 5\}$ are equivalent under π. Is this a lattice?

2.2.8 Show that lattice 4 of Figure 2.4 is not distributive.

2.2.9 Show that the set of all ideals in $\mathbb{Z}$ is a complete lattice under set inclusion.

2.2.10 Show that the partially ordered set of equivalence relations of a set S is a complete lattice.

2.2.11 Show that in any lattice $(\{[(a \cap b) \cup c] \cap a\} \cup c) \cap a = [(a \cap b) \cup c] \cap a$.

2.2.12 Prove that every finite lattice is complete.

2.2.13 Is the set of rational numbers under $\geq$ the usual "greater than or less than" a complete lattice?

2.2.14 Let $\langle P, \supseteq \rangle$ be a finite partially ordered set. Show that there exists a one-to-one function f from P into the integers $\mathbb{Z}$ such that $a \supseteq b$ implies $f(a) \geq f(b)$.

2.2.15 Let $\langle P, \supseteq \rangle$ be a partially ordered set whose elements are in one-to-one correspondence with the integers and can thus be denoted $a_1, a_2, \ldots, a_n, \ldots$. Show that there exists a one-to-one function from P into the rational numbers $\mathbb{Q}$, such that $a \supseteq b$ in P implies $f(a) \geq f(b)$ in $\mathbb{Q}$. Can $\mathbb{Q}$ be replaced by $\mathbb{Z}$?

2.2.16 A *sublattice* of a lattice $\langle L, \cup, \cap \rangle$ is a nonempty subset S of L such that for each $a, b \in S$ the elements $a \cup b$ and $a \cap b$ (these operations are in L) are also members of S. In the lattice of Figure 2.8 find all sublattices consisting of five elements.

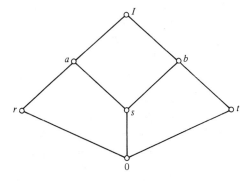

Figure 2.8

2.2.17 State and prove the dual of Theorem 2.2.13.

2.2.18 Draw the lattice diagrams for all lattices with six elements. (There are 15 of them!)

Chapter 3

Introduction to Groups

The first class of abstract algebras we study is the class of groups. Groups were first used effectively by Augustin Louis Cauchy (1789–1857) and by Evariste Galois (1811–1832) in their studies of the theory of equations.

Groups are used in the theory of equations to describe the effect of mapping the different roots of a polynomial equation onto themselves. It turns out that the number systems used to express these roots can be characterized in terms of the elements fixed under these mappings.

Groups are also used to describe the symmetries of geometric figures. In this way they help to classify two- and three-dimensional shapes. These include crystals found in nature, and so groups play an important role in studying the chemistry of molecules. Groups are also useful to physicists in their study of quantum mechanics and the nature and interactions of elementary particles.

Other applications abound in combinatorics, the study of problems that require sophisticated counting procedures. Groups and combinatorics are the basis of algebraic coding theory that provides error-free data transmission and secure encipherment.

The first axiomatic formulation for abstract groups was given in 1870 by Leopold Kronecker (1823–1891). An early comprehensive treatment of group theory was given by W. Burnside (1852–1927). His classic entitled *The Theory of Finite Groups* was first published in 1897. A later edition is still available from Dover Publications.

3.1 GROUP AXIOMS

To begin at the beginning, we first list the axioms of group theory.

3.1.1 Definition. Group

A group is a nonempty set of elements G together with a binary operation defined on G, here denoted $\diamondsuit$, which satisfies the following three axioms.

Associativity: For all a, b, and c in G, $(a \diamondsuit b) \diamondsuit c = a \diamondsuit (b \diamondsuit c)$.

Right identity: There exists an element $e \in G$ such that for all $a \in G$, $a \diamondsuit e = a$.

Right inverse: For some right identity element e there is, for every $a \in G$, at least one element a^* such that $a \diamondsuit a^* = e$.

Notation. We write $\langle G, \diamondsuit \rangle$ to denote a set G with a binary operation defined on it.

Remark. In any mathematical system $\langle G, \diamondsuit \rangle$, an element e such that
$$a \diamondsuit e = a$$
for all $a \in G$ is called a right identity for G under $\diamondsuit$. We shall prove shortly that the three group axioms ensure that there is but one right identity element in a group and that this right identity is also a left identity, that is, that $e \diamondsuit a = a$ for all $a \in G$.

In any mathematical system, an element a^* such that
$$a \diamondsuit a^* = e$$
is called a right inverse of a with respect to e under $\diamondsuit$. We shall shortly prove that in a group the right inverse a^* depends only on a; that is, a right inverse is unique and moreover it is a left inverse,
$$a^* \diamondsuit a = e$$
with respect to the identity e.

Notice that a set of elements alone is not a group. A group is the mathematical system obtained by defining a binary operation on a set of elements. To give an example of a group we must specify the set of elements and the operation. We shall speak loosely of the group elements and the group operation.

Many groups, especially the early examples we consider, satisfy another axiom.

3.1.2 Definition. Abelian Group

A group is called commutative or *abelian* if for every pair of elements
$$a \diamondsuit b = b \diamondsuit a$$

The name "abelian" is in honor of the Norwegian mathematician Niels Henrik Abel (1802–1829). In 1824, Abel showed the impossibility of expressing the roots of polynomial equations of degree five and higher in terms of radicals. His further investigations concerned the solution of equations in terms of radicals for those equations that could be associated with a commutative group of permutations of the roots of these equations. Thus the adjective "abelian" is an appropriate one for these groups. Abel is well known for his many important contributions to number theory and the theory of functions. However, it was the young genius Galois who first obtained necessary and sufficient conditions for a polynomial equation to have its roots expressible in terms of radicals. The criterion is a condition on the group he associated with the equation. More of that in Chapter 14.

The number of elements in the set G is an important characteristic of the group; it is called the *order* of the group.

3.1.3 Definition.   Order of a Group

The *order* of a group $\langle G, \Diamond \rangle$ is the cardinal number of elements in G. The order of G is denoted $|G|$.

3.1.4 Examples of Groups

1. The integers $\langle \mathbb{Z}, + \rangle$ under addition are an abelian group of infinite order. The identity is 0 and the inverse of a is $-a$.
2. For each positive integer n, the congruence classes modulo n, $\langle \mathbb{Z}_n, + \rangle$ under addition are an abelian group of order n.
3. The nonzero rational numbers $\langle \mathbb{Q}^\times, \cdot \rangle$ under multiplication are an abelian group of infinite order. The identity is 1 and the inverse of a is $1/a = a^{-1}$.
4. For each positive integer n, $\langle \mathbb{Z}_n^\times, \cdot \rangle$, the congruence classes $[k]$ modulo n, such that gcd $(k, n) = 1$ under multiplication, is an abelian group whose order is $\varphi(n)$. Recall Lemma 2.1.10. There it is proved that gcd $(a, n) = 1$ and gcd $(b, n) = 1$ imply gcd $(ab, n) = 1$ and thus multiplication is a binary operation on the set of classes $[k]$ where gcd $(k, n) = 1$. The identity is $[1]$ and the inverse of $[a]$ is the congruence class $[x]$ where x solves the congruence $ax \equiv 1 \pmod{n}$.

The group tables for $\langle \mathbb{Z}_6, + \rangle$ and $\langle \mathbb{Z}_9^\times, \cdot \rangle$ are given in Figure 3.1. A group table is just an array like a multiplication table in which the group operation may be recorded. The rows and columns of the array are labeled with the names of the group elements. In the upper left-hand corner the symbol for the group operation may be recorded. Thus suppose G is a group and the group operation is denoted $\Diamond$ and that $a \Diamond b = c$ in G. Then its

+	0	1	2	3	4	5
0	0	1	2	3	4	5
1	1	2	3	4	5	0
2	2	3	4	5	0	1
3	3	4	5	0	1	2
4	4	5	0	1	2	3
5	5	0	1	2	3	4

$\langle \mathbb{Z}_6, + \rangle$

·	1	2	4	5	7	8
1	1	2	4	5	7	8
2	2	4	8	1	5	7
4	4	8	7	2	1	5
5	5	1	2	7	8	4
7	7	5	1	8	4	2
8	8	7	5	4	2	1

$\langle \mathbb{Z}_9^\times, \cdot \rangle$

Figure 3.1 Group tables for two groups of order 6. On the left is $\langle Z_6, + \rangle$. On the right is $\langle \mathbb{Z}_9^\times, \cdot \rangle$. To save notation we have omitted the brackets to designate congruence classes. Thus, for example, to find $5 \cdot 7$ in $\langle \mathbb{Z}_9^\times, \cdot \rangle$ look in the row labeled 5 and the column labelled 7 to read 8. In full notation this says $[5][7] = [8]$. And this is so because $5 \cdot 7 = 35 \equiv 8 \pmod 9$. We shall learn that these two groups are isomorphic; that is, the elements of $\langle \mathbb{Z}_9^\times, \cdot \rangle$ can be renamed and the rows and columns of the table on the right can be reordered so that the two tables are identical. Can you guess how the element 1 on the right should be renamed? Then which element on the right will act like the element 1 on the left?

group table would have these entries:

◇		b
a		c

5. Let S be a nonempty set and let G be the set of all one-to-one functions from S onto S. If ◇ denotes composition of functions then $\langle G, ◇ \rangle$ is a group. If $|S| = n$ then $|G| = n!$. These groups are very important. The identity element is the identity function, which we denote by I. Thus

$$I: x \to x$$

We have already shown that composition of functions is associative. Each function f in G has an inverse because f is "onto" and one-to-one:

$$f^{-1}(x) = y \qquad \text{if and only if } f(y) = x$$

We now give more details for the case $n = 3$.

Let $S = \{1, 2, 3\}$. The elements of G are given below by giving a table for each function. The domain S is listed in the first row; the value of the function at the corresponding point is listed in the second row. These functions are called *permutations*.

The identity function is denoted by I, another function by a, and another by b. Then the remaining functions are given in terms of these under the composition.

$$I = \begin{pmatrix} 1 & 2 & 3 \\ 1 & 2 & 3 \end{pmatrix} \qquad a = \begin{pmatrix} 1 & 2 & 3 \\ 2 & 3 & 1 \end{pmatrix} \qquad b = \begin{pmatrix} 1 & 2 & 3 \\ 1 & 3 & 2 \end{pmatrix}$$

$$a \lozenge a = \begin{pmatrix} 1 & 2 & 3 \\ 3 & 1 & 2 \end{pmatrix} \qquad a \lozenge b = \begin{pmatrix} 1 & 2 & 3 \\ 3 & 2 & 1 \end{pmatrix} \qquad a \lozenge a \lozenge b = \begin{pmatrix} 1 & 2 & 3 \\ 2 & 1 & 3 \end{pmatrix}$$

Although permutations are functions, we shall follow the convention of computing compositions of permutations by proceeding from left to right. For example, the image of 1 under the composition $a \lozenge b$ is

$$1 \overset{a \lozenge b}{\rightarrow} 3 \quad \text{since} \quad 1 \overset{a}{\rightarrow} 2 \overset{b}{\rightarrow} 3$$

Notationally we compute $(1)a \lozenge b = (2)b = 3$. Similarly $(1)a \lozenge a \lozenge b = (2)a \lozenge b = (3)b = 2$. Now check the computations of the three permutations listed in the second row above.

To compress our notation we abbreviate using the exponential notation, thus

$$a \lozenge a = a^2 \quad \text{and} \quad a \lozenge a \lozenge b = a^2 \lozenge b$$

The group of all six permutations is called the full symmetric group on three symbols and is denoted $\mathbb{S}_3$. The group table is shown in Figure 3.2. Convention provides a further economy in listing a group table. By convention we place the identity in the first row and column and use the same order for the elements. We may then omit the labeling of the rows and columns and use instead the first column and row themselves as labels! We do this in the table of Figure 3.2. It is good exercise to verify all the entries. Note that the group is not abelian. The table shows

$$b \lozenge a \neq a \lozenge b$$

In fact $b \lozenge a = a^2 \lozenge b$, as you can verify by computing these two compositions.

In the next section we also see this group as the group of symmetries of an equilateral triangle.

I	a	a^2	b	$a \diamond b$	$a^2 \diamond b$
a	a^2	I	$a \diamond b$	$a^2 \diamond b$	b
a^2	I	a	$a^2 \diamond b$	b	$a \diamond b$
b	$a^2 \diamond b$	$a \diamond b$	I	a^2	a
$a \diamond b$	b	$a^2 \diamond b$	a	I	a^2
$a^2 \diamond b$	$a \diamond b$	b	a^2	a	I

Figure 3.2 The group table of the one-to-one functions from $S = \{1, 2, 3\}$ to itself. These functions are also called permutations, and this group is known as the full symmetric group of three symbols. We shall denote it as $\mathbb{S}_3$.

More Examples of Groups

6. Here are two groups each of order 1: Let 0 and 1 be the "zero" and "one" of the integers. Then $\langle \{0\}, + \rangle$ and $\langle \{1\}, \cdot \rangle$ are groups. The verification is more tedious than difficult!

7. Here is a group of order 2. Let $\{1, -1\}$ be the set consisting of the integer 1 and its negative. Then $\langle \{1, -1\}, \cdot \rangle$ is a group of order 2.

Remark.   Examples 6 and 7 are the only finite groups that can be constructed from finite subsets of $\mathbb{Z}$ using the usual operations of addition or multiplication.

8. Here is an example of a "nongroup." Let $S = \{0, 1, -1\}$. Let the operation on S be multiplication. Show that S is closed under multiplication. The element 1 is an identity element but there is no (multiplicative) inverse for 0.

9. Here is a group of matrices. Let G be the set of 2×2 matrices with integer entries and determinant ± 1.

Thus a typical element is

$$\begin{pmatrix} a & b \\ c & d \end{pmatrix} \quad \text{where } \det \begin{pmatrix} a & b \\ c & d \end{pmatrix} = ad - bc = \pm 1$$

Let the group operation be matrix multiplication which we denote by juxtaposition. Thus

$$\begin{pmatrix} a & b \\ c & d \end{pmatrix} \begin{pmatrix} u & v \\ w & x \end{pmatrix} = \begin{pmatrix} au + bw & av + bx \\ cu + dw & cv + dx \end{pmatrix}$$

We want to show that G is closed under matrix multiplication. It is routine to show that the determinant of the product is the product of the determinants:

The product of the determinants on the left-hand side is

$$(ad - bc)(ux - vw) = adux + bcvw - advw - bcux$$

while the determinant of the matrix on the right-hand side is

$$(au + bw)(cv + dx) - (av + bx)(cu + dw) = audx + bwcv - avdw - bxcu$$

This means that the determinant of the matrix product of two matrices whose determinants are ± 1 is again ± 1.

Thus matrix multiplication is a binary operation on G. A tedious check shows that this multiplication is associative. Don't do it; later we shall have a better way of looking at matrix multiplication and the associativity will in fact be the result of a composition of functions.

A less tedious check shows that

$$I = \begin{pmatrix} 1 & 0 \\ 0 & 1 \end{pmatrix}$$

is the identity and that if

$$A = \begin{pmatrix} a & b \\ c & d \end{pmatrix} \quad \text{then} \quad A^* = \begin{pmatrix} d/\det A & -b/\det A, \\ -c/\det A & a/\det A \end{pmatrix}$$

is the inverse of A. Simply compute

$$AA^* = I$$

No, we didn't check the determinant of A^*, but since $\det(A) = \pm 1$ and

$$\det(A)\det(A^*) = \det(I) = 1$$

it follows that $\det(A^*) = \pm 1$.

This group is infinite because for each integer n

$$\begin{pmatrix} 1 & n \\ 0 & 1 \end{pmatrix}$$

is in G.

Now that we have seen a number of examples of groups, both finite and infinite, both abelian and nonabelian, it is time to derive the first consequences of the axioms. An unstated hypothesis in the next few lemmas is that we deal with a group $\langle G, \diamondsuit \rangle$.

3.1.5 Definition. Idempotent Element

In any mathematical system $\langle G, \diamondsuit \rangle$ an element a such that $a \diamondsuit a = a$ is called an idempotent element under $\diamondsuit$.

The idempotent element for the integers under $+$ is 0. The idempotent element for the integers under multiplication is 1. In $\langle \mathbb{Z}_6, \cdot \rangle$ the integers modulo 6 under multiplication, the idempotents are [0], [1], [3], and [4]. (Verify that $3 \times 3 \equiv 3 \pmod{6}$ and $4 \times 4 \equiv 4 \pmod{6}$.)

3.1.6 Lemma. Idempotence Implies Identity

Let e be a right identity element such that every element has right inverse with respect to it. If x is an idempotent,

$$x \diamondsuit x = x$$

then $x = e$.

Proof. Let x^* be a right inverse of x with respect to e. Then using the idempotence of x and associativity of $\diamondsuit$ we make this computation:

$$e = x \diamondsuit x^* = (x \diamondsuit x) \diamondsuit x^* = x \diamondsuit (x \diamondsuit x^*) = x \diamondsuit e = x$$

We shall use this criterion to verify that the right identity element is unique.

3.1.7 Lemma. Uniqueness of the Right Identity Element

There is only one right identity element in a group.

Proof.   Let e be a right identity for which there is a right inverse for each element of the group. Now let f be any other right identity element of G. We prove that $f = e$.

Since f is a right identity element

$$x \Diamond f = x$$

for all elements x in the group. Set $x = f$ and we find $f \Diamond f = f$. By Lemma 3.1.6 this means $f = e$.

3.1.8 Lemma.   A Right Inverse Is a Left Inverse

If $a \Diamond a^* = e$ then $a^* \Diamond a = e$.

Proof.   We shall show that $a^* \Diamond a$ is an idempotent and then apply Lemma 3.1.6 to conclude that $a^* \Diamond a = e$.

We compute, using associativity freely:

$$(a^* \Diamond a) \Diamond (a^* \Diamond a) = ((a^* \Diamond a) \Diamond a^*) \Diamond a = (a^* \Diamond (a \Diamond a^*)) \Diamond a$$

$$= (a^* \Diamond e) \Diamond a = a^* \Diamond a$$

and we're done!

3.1.9 Lemma.   A Right Identity Is a Left Identity

If e is a right identity then also

$$e \Diamond x = x$$

for all elements in the group.

Proof.   Let x^* be a right inverse for x. Calculate

$$e \Diamond x = (x \Diamond x^*) \Diamond x = x \Diamond (x^* \Diamond x) = x \Diamond e = x$$

Note that we have used not only the associative law but the fact that (Lemma 3.1.8) a right inverse is a left inverse.

3.1.10 Lemma.   Uniqueness of Each Inverse

Let e be the identity element of a group. If a, b, and c are elements of a group such that

$$a \Diamond b = a \Diamond c = e$$

then $b = c$.

Proof.   From Lemma 3.1.7 we also have $b \Diamond a = e$ and $c \Diamond a = e$. Now compute

$$b = b \Diamond e = b \Diamond (a \Diamond c) = (b \Diamond a) \Diamond c = e \Diamond c = c$$

the last equality holding because a right identity is also a left one.

Comment. Without loss of generality we can replace the identity and inverse axioms for a group by these stronger ones:

i. There exists a unique element e such that $x \Diamond e = e \Diamond x = x$ for all $x \in G$.

ii. For each element x there is one and only one element x^* such that $x \Diamond x^* = x^* \Diamond x = e$.

So why didn't we just state these as axioms and avoid the last sequence of lemmas? There are two compelling reasons. One is a question of mathematical elegance. It is nice to know what you can prove from the least assumptions. And you have gained some experience with proofs in group theory!

Another good reason is that should you wish to verify that a particular operation on a set G would make G into a group then you need only verify associativity and existence of a right identity and a right inverse. Conditions i and ii require more verification.

An important consequence of these properties is that solutions of simple equations exist in a group.

3.1.11 Theorem. Solution of Group Equations

In a group the equations

$$a \Diamond x = b \quad \text{and} \quad y \Diamond a = b$$

have unique solutions.

Proof. Let a^* denote the inverse of a and let b^* denote the inverse of b. Then we may test the solutions

$$x = a^* \Diamond b \quad \text{and} \quad y = b \Diamond a^*$$

to see that they "work." Here is a sample calculation:

$$a \Diamond (a^* \Diamond b) = (a \Diamond a^*) \Diamond b = e \Diamond b = b$$

On the other hand, this is the only possible solution, for if

$$a \Diamond x = b$$

then

$$a^* \Diamond (a \Diamond x) = a^* \Diamond b$$

On the left hand side of this equation calculate

$$a^* \Diamond (a \Diamond x) = (a^* \Diamond a) \Diamond x = x$$

Thus $x = a^* \Diamond b$ is the only possible solution.

Comment.   Of course, if the group is abelian then $x = y$, but in general the solutions are not equal. Try it in the group of permutations on $\{1, 2, 3\}$ described in group example 5, page 70.

3.1.12 Definition.   Latin Square

An $n \times n$ square array of n distinct symbols is called a latin square of size n if in each row and column each symbol occurs once and only once.

3.1.13 Theorem.   Latin Square Property of Group Tables

A group table is a latin square of the symbols representing the elements of the group. The size of the table is the order of the group.

Proof.   Consider a row of the table indexed by an element a. We ask, does b occur in that row? This is equivalent to asking whether there is an element x such that $a \diamondsuit x = b$. If so, b occurs in row a and column x. In view of Theorem 3.1.11 this is so. A similar argument shows that each element occurs once and only once in each column.

Typical examples are provided by the tables of Figures 3.1 and 3.2.

Notice that the identity e occurs in row a and column a^* where a^* is the inverse of a. Since we follow the convention of using the same order of elements to denote the rows as columns, the identity appears on the main diagonal for those elements t such that

$$t \diamondsuit t = e$$

Of course, the identity itself is one such element, always in the upper left-hand corner. Find as many other elements in the tables of Figures 3.1 and 3.2 as you can that satisfy $t \diamondsuit t = e$.

Not every latin square is a group table. We can reorder the rows so that the same order is used for rows and columns and thus we can make the element in the upper left-hand corner the identity. Then the latin square property assures us that for each symbol there will be an inverse symbol. The time-consuming part is to decide whether the associative law holds. There are n^3 occurrences to verify. See Problems 3.1.10, 3.1.11, and 3.1.12.

Notation.   It is not convenient to have to use a special symbol for the group operation. Hereafter we denote a group operation by juxtaposition unless we need to differentiate between different group operations or want to make special emphasis. In those cases we shall be explicit about the symbols used. For now we write

$$ab \text{ for } a \diamondsuit b$$

We denote the identity by 1 or by I unless there is a special situation, in which case we shall be explicit.

We use the exponential notation

$$a^0 = 1$$

$$a^n = a \Diamond a \Diamond \cdots \Diamond a \qquad (n \text{ factors})$$

We denote the inverse of a by a^{-1} and

$$a^{-n} = \left(a^{-1}\right)^n = a^{-1} \Diamond \cdots \Diamond a^{-1} \qquad (n \text{ factors})$$

and we develop further properties of our notation in the next theorem.

3.1.14 Theorem. Elementary Group Computations

If G is a group then

$$\left(a_1 a_2 \ldots a_k\right)^{-1} = a_k^{-1} a_{k-1}^{-1} \ldots a_1^{-1} \qquad (1)$$

In particular $(a^n)^{-1} = (a^{-1})^n$. For integers n and m:

$$a^n a^m = a^{n+m} \qquad (2)$$

$$\left(a^n\right)^m = a^{nm} \qquad (3)$$

$$\text{If } ab = ba \quad \text{then} \quad ab^n = b^n a \qquad (4)$$

$$\text{If } ab = ba \quad \text{then} \quad (ab)^n = a^n b^n \qquad (5)$$

Proof.   The proofs are all by weak induction, and we shall omit most of the details.

For (1) the induction begins with $k = 2$. To prove

$$(ab)^{-1} = b^{-1} a^{-1}$$

we verify that $b^{-1} a^{-1}$ solves $(ab)x = 1$. This uses a lot of associativity:

$$(ab)(b^{-1}a^{-1}) = ((ab)b^{-1})a^{-1} = (a(bb^{-1}))a^{-1} = (a1)a^{-1} = aa^{-1} = 1$$

We omit the details of the inductive step. The important thing to note is that the inverse of a product is the product of the inverses *in reverse order*!

The proofs of (4) and (5) depend heavily on the fact that $ab = ba$, and without this added hypothesis the conclusions are false in general; see the "caution" below.

The proof of (4) is by induction on k for positive k. The case $k = 1$ is the hypothesis. The inductive step is established:

$$ba^k = baa^{k-1} = aba^{k-1} = aa^{k-1}b = a^k b$$

using the induction hypothesis.

Now we can prove (5) by induction on n. For $n = 1$ this is the hypothesis. The induction step goes as follows:

$$(ab)^n = (ab)(ab)^{n-1} = aba^{n-1}b^{n-1} = aa^{n-1}bb^{n-1} = a^n b^n$$

and uses (4) as well as the induction hypothesis. We have suppressed explicit mention of the associative law, and hereafter we shall not pay attention to its use.

We have shown that (4) and (5) hold for positive intergers n. The extension to negative integers is through the role of inverses. For example if n is a positive integer

$$\left(ab^{-n}\right)^{-1} = b^n a^{-1} = a^{-1} b^n = \left(b^{-n} a\right)^{-1}$$

and hence $ab^{-n} = b^{-n} a$.

Caution.   In general

$$\left(ab\right)^2 \neq a^2 b^2$$

For an example where the inequality holds, see group example 5 and take a for a and b for b. As shown in the group table, $(ab)^2 = I$ but

$$a^2 b^2 = a^2 I = a^2 \neq I$$

We have stressed that a group is a set of elements together with an operation. Change the set and the group is changed. However, there may be a corresponding change in the group operation so that the new set is, from an algebraic point of view, essentially the same.

Here is a simple example of this. Consider the group, call it G for now,

$$\langle \{1, -1\}, \cdot \rangle$$

consisting of the integers 1 and -1 under multiplication.

In contrast consider the group

$$\langle \mathbb{Z}_2, + \rangle = \langle \{[0], [1]\}, + \rangle$$

of congruence classes modulo 2 under addition.

Both groups have order 2. A group table for each group will have the form

I	x
x	I

where I is the identity element and x is the nonidentity element. Thus these two groups, while having different sets of elements and different operations, are the same from an algebraic point of view. We call such groups isomorphic. Here is how we make that notion precise.

3.1.15 Definition.   Group Isomorphism

Two groups $\langle G, \Diamond \rangle$ and $\langle H, * \rangle$ are said to be isomorphic if there is a one-to-one function μ from G onto H such that for every ordered pair of elements x, y in G

$$\mu(x \Diamond y) = \mu(x) * \mu(y) \tag{6}$$

It may be convenient to think of μ as creating a change in the labeling of the elements. Condition (6) makes sure that the label change is consistent with the group operation.

The function μ is called an *isomorphism*. We write $G \cong H$.

It will be important for us to decide whether two groups are or are not isomorphic. This is not always easy. Usually it will be possible to give some argument based on the properties of the two groups to decide whether or not they are isomorphic. We look for properties that are preserved under isomorphism. Here is the first result.

3.1.16 Theorem.   Preservation of Identity and Inverse under Isomorphism

If $\langle G, \diamondsuit \rangle$ and $\langle H, * \rangle$ are isomorphic groups, then

$$\mu(I) \text{ is the identity of } H$$

and

$$\mu(a^{-1}) = (\mu(a))^{-1}$$

Proof.   We show that the idempotence of $\mu(I)$ follows from the idempotence of I and the isomorphism property (5):

$$\mu(I) = \mu(I \diamondsuit I) = \mu(I) * \mu(I)$$

Thus $\mu(I)$ is the identity of H. Inverses work much that same way. Since

$$\mu(I) = \mu(a \diamondsuit a^{-1}) = \mu(a) * \mu(a^{-1})$$

we see that

$$(\mu(a))^{-1} = \mu(a^{-1})$$

As an example we find that the groups $\langle \mathbb{Z}_6, + \rangle$ and $\langle \mathbb{Z}_9^{\times}, \cdot \rangle$ whose tables appear in Figure 3.1 are isomorphic. The isomorphism will send 0 into 1 since they are the identities. What will be the image of $1 \in \mathbb{Z}_6$? Call it x. We know that

$$\mu(1 + \cdots + 1) = \mu(1) \cdot \cdots \cdot \mu(1)$$

so that

$$\mu(k) = x^k$$

Since $0, 1, \ldots, 5$ are distinct in $\mathbb{Z}_6$ it follows that

$$1, x, x^2, \ldots, x^5$$

will be distinct in $\mathbb{Z}_9^{\times}$. This limits the choices for x, and a little searching shows that $x = 2$ or $x = 5$ are possibilities. Choose $x = 2$. Thus we can list the function μ:

	0	1	2	3	4	5
μ:	⇓	⇓	⇓	⇓	⇓	⇓
	1	2	4	8	7	5

Now relabel the table on the left-hand side accordingly and reorder the rows and columns to discover the table on the right-hand side!

Remark. Isomorphism is an equivalence relation on any set of groups. We often speak of determining a certain class of groups *up to isomorphism*. By this we mean that from the class we select a set of representative groups so that any member of the class is isomorphic to one of them.

For example, suppose we want to study the groups of order n. If we label the elements $a_1, \ldots, a_n$ then there are only a finite number (albeit a huge number) of latin squares on these n symbols; hence, up to isomorphism, there are only a finite number of groups of order n.

EXERCISE SET 3.1

3.1.1 Define an operation $\odot$ on the integers $\mathbb{Z}$ by $a \odot b = a + b - 2$. Is $\langle \mathbb{Z}, \odot \rangle$ a group?

3.1.2 Let $\mathbb{Q}^*$ be the nonzero rational numbers. Is $\langle \mathbb{Q}^*, \div \rangle$ a group?

3.1.3 Let $S = \{(a, b): 0 \neq a,\ b \in \mathbb{Q}\}$ be the set of ordered pairs of rational numbers whose first member is not zero. Show that operation $\diamondsuit$ defined below makes S into a group:

$$(a, b) \diamondsuit (c, d) = (ac, bc + d)$$

Is the group abelian?

3.1.4 Let S be the set of real numbers excluding 0 and 1. Let f and g be the two real-valued functions defined on S by

$$f(x) = 1/x \quad \text{and} \quad g(x) = 1/(1 - x)$$

Using the operation of composition, find the smallest group of functions defined on S that contains f and g.

3.1.5 Let $\langle H, \diamondsuit \rangle$ be a group. Show that $(ab)^{-1} = a^{-1}b^{-1}$ if and only if $ab = ba$.

3.1.6 Show that if $x^2 = 1$ for all x in a group then the group is abelian.

3.1.7 Let K be a group. Show that for elements a, b in H and all positive integers k, h, and n
i. $(a^{-1}ba)^k = a^{-1}ba$ if and only if $b^k = b$.
ii. If $a^{-1}ba = b^h$ then $a^{-n}ba^n = b^{(h^n)}$.

3.1.8 Let S be the set of 2×2 matrices with rational number entries of the form

$$\begin{pmatrix} 0 & b \\ 0 & d \end{pmatrix}$$

Let $\diamondsuit$ be matrix multiplication; thus

$$\begin{pmatrix} 0 & b \\ 0 & d \end{pmatrix} \diamondsuit \begin{pmatrix} 0 & w \\ 0 & y \end{pmatrix} = \begin{pmatrix} 0 & by \\ 0 & dy \end{pmatrix}$$

Show that
i. There is right identity in S.
ii. There is a left inverse for each $A \in S$.
iii. S is not a group under $\diamondsuit$.

3.1.9 Let G be a group and let $a, b \in G$. Suppose that $a^{-1}ba = b^{-1}$ and $b^{-1}ab = a^{-1}$. Show that $a^2 = b^{-2}$ and $a^4 = b^4 = 1$.

3.1.10 Construct all possible group tables for a group of order 3 whose elements are $\{1, a, b\}$. Let the first row and column be indexed 1, a, and b in that order. Assume that 1 is the identity. Show that any two groups of order 3 are isomorphic.

3.1.11 Construct all possible group tables for a group of order 4 whose elements are $\{1, a, b, c\}$ where 1 is the identity. Index the first row and column by 1, a, b, c. Show that there are at most two nonisomorphic groups of order 4.

3.1.12 Is the latin square given below a group table?

e	a	b	c	d
a	e	c	d	b
b	d	e	a	c
c	b	d	e	a
d	c	a	b	e

3.1.13 Is $\langle \mathbb{Z}_7^\times, \cdot \rangle$ isomorphic to the groups whose tables appear in Figure 3.1?

3.1.14 Show that no two of the following groups are isomorphic:
  i. The integers under addition.
  ii. The rationals under addition.
  iii. The positive rationals under multiplication.

3.1.15 Show that the group of positive real numbers under multiplication is isomorphic to the group of all real numbers under addition. (It is legend that this was Galois' answer to Cauchy's question "What is a logarithm?")

3.1.16 Suppose that G is a group of even order. Show that the identity must appear at least twice on the main diagonal of its group table. More generally, show that the identity element appears an even number of times on the diagonal of a group table of a group of even order. What can be said for a group of odd order?

3.1.17 Let $\langle G, \Diamond \rangle$ be a group. Prove that the function β from G into G defined by $\beta(x) = x^{-1}$ for all $x \in G$ is always one-to-one and onto. Prove that it is an isomorphism of $\langle G, \Diamond \rangle$ onto itself if and only if G is abelian.

3.1.18 Let $\langle G, \cdot \rangle$ be a group. Let $a \in G$ be an element, not the identity. For what groups G will the following defined operations on the set of elements of G make G into a group? When a group is defined, is the original group $\langle G, \cdot \rangle$ isomorphic to the new one, $\langle G, \Diamond \rangle$?
Definition i: $x \Diamond y = axay$
Definition ii: $x \Diamond y = x^{-1}y^{-1}$
Definition iii: $x \Diamond y = xay$
Definition iv: $x \Diamond y = yx$
(Your answer many depend upon G or upon a special property of a as a group element of G.)

3.1.19 Let $\Diamond$ be an associative binary operation defined on a nonempty set S. Suppose that for each $a, b \in S$ there are elements x and y such that $a \Diamond x = b$ and $y \Diamond a = b$. Show that $\langle S, \Diamond \rangle$ is a group.

3.2 SUBGROUPS

3.2.1 Example.   Symmetries of the Square

In how many ways can a square slide be inserted in a projector? Well, a slide has two sides, call one *F* (front), the other *B* (back). One side must face toward the lens. If side *F* faces the lens, there are four possible choices for the corner in the upper left position. If side *B* faces the lens, there are four more. So there are eight ways. We can go from one position to another by a rotation of 90, 180, or 270 degrees and from *F* to *B* and back again by a flip. Many of us who have struggled with slide projectors have run through all eight possibilities before finding the correct one. If the slide had the letter J on it, Figure 3.3 shows how the eight possibilities would look. We begin with a correct placement instead of ending with it! We define a symmetry of a square as a one-to-one mapping of the vertices of the square so that distance is preserved.

Number the vertices $1, 2, 3, 4$ in clockwise fashion. Under a symmetry, wherever the vertex 1 is mapped, the vertex 2 must be mapped to an adjacent vertex since the image of vertex 2 must be just one unit away from the image of vertex 1. Once the images of 1 and 2 are known, the "distance preserving" feature determines the location of the other vertices.

This analysis permits us to count the symmetries. There are four choices for the image of vertex 1; there are then two choices for the vertex 2. The total number of symmetries is thus at most $4 \cdot 2 = 8$.

On the other hand, a rotation or a reflection will define a symmetry. There are four rotations about the center of the square. There are four reflections about the two diagonals and the two central axes, one horizontal, one vertical. No rotation is a reflection. (Can you give a proof?) Thus there are at least eight symmetries of the square.

These two observations tell us that there are exactly eight symmetries of the square. Here is a list of the symmetries. Each is represented as a

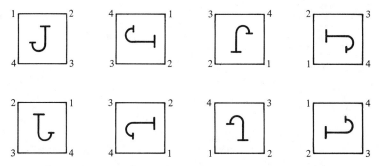

Figure 3.3 The eight possible configurations for a slide of the letter "J". Each possibility corresponds to a symmetry of the square.

permutation of the vertices

$$I = \begin{pmatrix} 1 & 2 & 3 & 4 \\ 1 & 2 & 3 & 4 \end{pmatrix} = \text{identity}$$

$$a = \begin{pmatrix} 1 & 2 & 3 & 4 \\ 2 & 3 & 4 & 1 \end{pmatrix} = \text{rotation of } 90° \text{ clockwise}$$

$$a^2 = \begin{pmatrix} 1 & 2 & 3 & 4 \\ 3 & 4 & 1 & 2 \end{pmatrix} = \text{rotation of } 180° \text{ clockwise}$$

$$a^3 = \begin{pmatrix} 1 & 2 & 3 & 4 \\ 4 & 1 & 2 & 3 \end{pmatrix} = \text{rotation of } 270° \text{ clockwise}$$

$$b = \begin{pmatrix} 1 & 2 & 3 & 4 \\ 1 & 4 & 3 & 2 \end{pmatrix} = \text{reflection about the } (1 - 3) \text{ diagonal}$$

$$ab = \begin{pmatrix} 1 & 2 & 3 & 4 \\ 4 & 3 & 2 & 1 \end{pmatrix} = \text{reflection about horizontal axis}$$

$$a^2b = \begin{pmatrix} 1 & 2 & 3 & 4 \\ 3 & 2 & 1 & 4 \end{pmatrix} = \text{reflection about the } (2 - 4) \text{ diagonal}$$

$$a^3b = \begin{pmatrix} 1 & 2 & 3 & 4 \\ 2 & 1 & 4 & 3 \end{pmatrix} = \text{reflection about the vertical axis}$$

These symmetries form a group under composition. We compute composition of permutations left to right; recall $\mathbb{S}_3$ in Examples 3.1.4. The notation in the table above suggests the compositions of a and b used to obtain each symmetry.

The group is called the dihedral group of order 8, denoted D_8. For each even integer $2n$ ($n \geq 2$) there is a dihedral group of order $2n$; we study these in Section 3.3. The group table for D_8 is shown in Figure 3.4. Notice that the

I	a	a^2	a^3	b	ab	a^2b	a^3b
a	a^2	a^3	I	ab	a^2b	a^3b	b
a^2	a^3	I	a	a^2b	a^3b	b	ab
a^3	I	a	a^2	a^3b	b	ab	a^2b
b	a^3b	a^2b	ab	I	a^3	a^2	a
ab	b	a^3b	a^2b	a	I	a^3	a^2
a^2b	ab	b	a^3b	a^2	a	I	a^3
a^3b	a^2b	ab	b	a^3	a^2	a	I

Figure 3.4 The group table for the symmetries of the square. A rotation of 90° is denoted by a. A reflection about the left-right diagonal is denoted by b. This group is called the dihedral group of order 8 and is denoted D_8. Only two rows are needed to derive the whole table. From the "a" row and "b" row all the other information can be derived. The important relations are $a^4 = b^2 = I$ and $ba = a^3b$.

subset

$$A = \{I, a, a^2, a^3\}$$

forms a group by itself. The table for this group appears as a subarray in the upper left-hand corner of the table in Figure 3.4. By inspection you can see that only these elements occur and that it has the latin square property and contains the identity.

There are other subsets that form groups by themselves. An easy example is

$$B = \{I, b\}$$

This is isomorphic to a group of order 2 since $b^2 = I$.

A less obvious example is the subset

$$C = \{I, b, a^2, a^2b\}$$

To view this group as a subarray of the table in Figure 3.4 you may circle the elements. In this example each element is its own inverse.

3.2.2 Definition.   Subgroup

Let $\langle G, \Diamond \rangle$ be a group. A subset $S \subseteq G$ is a *subgroup* of G if $\langle S, \Diamond \rangle$ is a group *under the same operation used in G.*

Notice that the subset consisting of the identity is a subgroup of every group, as is the group itself. It is possible that these are the only two subgroups in G (see Problem 3.2.16), but more often the subgroup structure is rich and provides added understanding of the whole group. Later we shall prove that under set inclusion the set of all subgroups of a group forms a complete lattice.

If a subset S is to be a subgroup of G then for the operation of G, $\Diamond$ to be an operation on S we must have that

$$a \Diamond b \in S \qquad \text{whenever } a \in S \text{ and } b \in S$$

This condition is expressed by saying that S is *closed* under the operation $\Diamond$ of G.

This condition alone is not sufficient to guarantee that S is a subgroup. For example, if $G = \langle \mathbb{Z}, + \rangle$ and S is the subset of positive integers, then S is closed under addition but S is not a subgroup of G; it lacks the identity and inverses. Correct criteria are given in the next theorem.

3.2.3 Theorem.   Criteria for Subgroups

Let $\langle G, \Diamond \rangle$ be a group and let S be a nonempty subset. The following conditions are equivalent:

    i. S is a subgroup of G.
    ii. Whenever a and b are in S then a^{-1} and ab are in S.
    iii. Whenever a and b are in S then $a^{-1}b$ is in S.

Proof. We continue to use 1 to denote the identity of G. We prove a circle of implications.

Proof that i *implies* ii.   First suppose that S is a subgroup and suppose that a and b are elements of S. Since S is a group under the operation of G we know that ab is in S. As a group, S will have an identity element. This will be idempotent, of course, and since G has only one idempotent, the identity elements of G and S are the same. Similarly, the inverse of a in S is the inverse of a in G since the equation $ax = 1$ has a unique solution in G.

Proof that ii *implies* iii.   If ii holds and a and b are in S then a^{-1} is in S by ii and then the product $a^{-1}b$ is in S by ii.

Proof that iii *implies* i.   First we show that S contains the identity 1. Simply select $a = b$ in S (S was assumed nonempty) and apply iii to have $a^{-1}a = 1$ is in S.

Second we show that if $a \in S$ then $a^{-1} \in S$. We apply iii with $a = a$ and $b = 1$; apply iii to have

$$a^{-1} \Diamond 1 = a^{-1}$$

in S.

Finally we show that S is closed under the group operation. Let a and b belong to S. We know that $a^{-1} \in S$. Apply iii with a^{-1} in the role of a and b in the role of b to find

$$\left(a^{-1}\right)^{-1}b = ab$$

in S.

Notice that we never have to check the associative law since that holds in all of G, hence for any subset closed under the group operation.

Next we introduce a most important kind of subgroup, the powers of a fixed group element.

3.2.4 Definition.   Cyclic Groups and the Order of an Element

If $\langle G, \Diamond \rangle$ is a group and $a \in G$ then

$$\langle a \rangle = \{ a^k : k \in \mathbb{Z} \}$$

is a subgroup of G called the *cyclic* subgroup generated by $\langle a \rangle$.

If there is an element $g \in G$ such that $G = \langle g \rangle$ then G is called a *cyclic* group.

The *order* of an element $a \in G$ is the order of the cyclic group $\langle a \rangle$ it generates. We write

$$|a| = |\langle a \rangle| = \text{the order of } a$$

We do need to prove that the set of powers of a do satisfy one of the criteria of Theorem 3.2.3; but it is immediate that criterion ii is satisfied by the set of powers.

As an example return to the group D_8 of Figure 3.4. Using the notation developed there

$\langle a \rangle$ is a subgroup of order 4. $|a| = 4$.

$\langle ab \rangle$ is a subgroup of order 2. $|ab| = 2$.

$\langle a^2 \rangle$ is a subgroup of $\langle a \rangle$ and has order 2; $|a^2| = 2$.

Note that subgroups that are cyclic may have several generators. In D_8 we find

$$\langle a \rangle = \langle a^3 \rangle$$

Other examples of cyclic groups are the groups $\langle \mathbb{Z}_6, + \rangle$ and $\langle \mathbb{Z}_9^{\times}, \cdot \rangle$. The congruence class [1] is a generator of $\langle \mathbb{Z}_6, + \rangle$. The congruence class [2] is a generator of $\langle \mathbb{Z}_9^{\times}, \cdot \rangle$. Both groups have order 6 and, as we find in the next theorem, that is enough to guarantee that cyclic groups are isomorphic.

3.2.5 Theorem. Subgroups of Cyclic Groups

  i. Every subgroup of a cyclic group is cyclic.
  ii. A cyclic group is isomorphic either to $\langle \mathbb{Z}, + \rangle$ or to $\langle \mathbb{Z}_n, + \rangle$ for some positive integer n.
  iii. If G is a finite cyclic group then there is exactly one subgroup for each divisor of the order of G. Moreover: If $G = \langle g \rangle$ has order n and $n = dm$ then the subgroup of order d is $\langle g^m \rangle$.

Proof.   Let $G = \langle g \rangle$ and let S be a subgroup. Each element in G has the form g^k. We begin by showing that the set of integers that appear as exponents of elements in the subgroup S form an ideal in $\mathbb{Z}$. We use this property to establish all the results of the theorem. Let

$$A = \{ k \in \mathbb{Z} : g^k \in S \}$$

We show that A is an ideal in $\mathbb{Z}$ by a straightforward verification of conditions i and ii of Definition 1.3.13.

Verification of Condition i   If h and k are in A we are to show that $h - k$ is in S. If g^h and g^k are in S then so is

$$g^h g^{-k} = g^{h-k}$$

and so $h - k$ is in A.

Verification of Condition ii   If $k \in A$ then we are to show that $kn \in S$ for all n. If $g^k \in S$ then so is

$$\left(g^k \right)^n = g^{nk}$$

and so $kn \in A$.

Now from Theorem 1.3.15 we know that A is the set of multiples of some integer m;

$$A = \langle m \rangle$$

But this means that

$$S = \langle g^m \rangle$$

since $m \in A$ means that $g^m \in S$. Conversely, if $g^k \in S$ then $k \in A$ and so $m \mid k$ or

$$k = md \quad \text{and} \quad g^k = (g^m)^d$$

This proves i.

We now apply these "ideal considerations" to the identity subgroup $\langle I \rangle$. Let the ideal

$$A = \{ k \in \mathbb{Z} : g^k = I \} = \langle n \rangle$$

Two cases arise; either $n = 0$ or $n > 0$.

Case 1. $n = 0$. This means that the only power of g that equals the identity is 0. In turn this means

$$g^h = g^k \qquad \text{if and only if } h = k$$

since otherwise $h - k$ is a nonzero element of A.

In this case the function α from $\langle \mathbb{Z}, + \rangle$ to G defined by

$$\alpha(k) = g^k$$

is a group isomorphism. There are two easy details. First, the one-to-one nature of the function α has just been determined. Second, check to see that α preserves the group operation.

Case 2. $n > 0$. In this case

$$g^h = g^k \qquad \text{if and only if } h \equiv k \ (\text{mod } n) \tag{1}$$

since

$$g^{h-k} = I \qquad \text{if and only if } h - k \in A,$$
$$\text{if and only if } n \mid (h - k)$$

This implies that the elements

$$\{ I = g^0, g, g^2, \dots, g^{n-1} \}$$

are all distinct and that any element in G is equal to one of them since, using the division algorithm $k = qn + r$ where $0 \le r < n$ and thus $g^k = g^r$.

Now we can prove that an isomorphism β from $\mathbb{Z}_n$ to G can be defined by

$$\beta([k]) = g^k$$

To prove this we must first ensure that a function has been defined since it is a function from a set $[k]$ to an element g^k defined in terms of a member of the set. Thus we must show that if

$$[h] = [k] \quad \text{then} \quad g^h = g^k$$

and to show the one-to-one nature of β we must also show the converse. However, this is easy since we already know that

$$[h] = [k] \quad \text{if and only if } n \text{ divides } h - k$$

and

$$g^h = g^k \quad \text{if and only if } n \text{ divides } h - k$$

Finally we have to check the preservation of the group operation, which is also easy since multiplication of group elements as powers of g is by addition of exponents. All this proves ii.

Now to prove iii.

Suppose that G is cyclic and has order n and that d is a positive divisor of n; say $n = dm$. We claim that g^m has order d and that any subgroup of order d is $\langle g^m \rangle$.

Observe that the integers

$$0, m, 2m, \ldots (d-1)m$$

are all less than $dm = n$ so that

$$\{ I = g^0, g^m, g^{2m}, \ldots, g^{(d-1)m} \}$$

are d distinct powers of g^m and hence belong to $\langle g^m \rangle$. On the other hand, for any u, g^{mu} must be equal to one of these since we can write

$$u = qd + r \quad \text{where } 0 \le r < d$$

and thus

$$g^{mu} = g^{mqd}g^{mr} = g^{qn}g^{mr} = 1g^{mr} = g^{mr}$$

This proves that the order of g^m is d.

To prove that any subgroup of order d in G is $\langle g^m \rangle$ we argue as follows: We know from part i that if S is a subgroup then

$$S = \langle g^t \rangle$$

where t has been defined by the ideal

$$A = \{ k \in \mathbb{Z} : g^k \in S \} = \langle t \rangle$$

We have only to show that $t = m$. In any event, $g^n = 1 \in S$ so that $t \mid n$ and by the argument above $\langle g^t \rangle$ has order n/t. Thus

$$|S| = d = n/t \quad \text{and} \quad |S| = |\langle g^t \rangle| = n/m$$

and so $t = m$.

We collect some of the most useful results of this theorem in the next corollary. There we apply this theorem to an element g of finite order n by considering the subgroup $\langle g \rangle$ of G generated by g.

3.2.6 Corollary.   Properties of the Order of an Element

If $g \in G$ has finite order n then

    i. The least positive integer m such that $g^m = 1$ is n, $g^m = 1$ if and only if $n \mid m$; more generally $g^h = g^k$ if and only if $h \equiv k \pmod{n}$.

    ii. $\langle g \rangle = \{ I = g^0, g, \ldots, g^{n-1} \}$.

    iii. The order of g^s is $n/\gcd(s, n)$.

    iv. The number of generators for $\langle g \rangle$ is $\varphi(n)$; where φ is the Euler φ-function. The elements g^s where $\gcd(s, n) = 1$ generate $\langle g \rangle$.

Proof.   Part i is statement (1) in case 2 of the proof of Theorem 3.2.5. Part ii then follows immediately in the proof.

To prove iii, let $m = \gcd(s, n)$. We shall prove that

$$\langle g^s \rangle = \langle g^m \rangle$$

and thus from part iii of Theorem 3.2.5 we have that

$$|\langle g^s \rangle| = n/m$$

Now since s is a multiple of m it follows that

$$g^s \in \langle g^m \rangle \quad \text{and} \quad \text{so} \langle g^s \rangle \subseteq \langle g^m \rangle$$

On the other hand, the congruence

$$sx \equiv m \pmod{n}$$

has a solution since $m = \gcd(s, n)$ and $m \mid m$ (recall Theorem 2.1.8, iv). But this means by i that

$$g^{sx} = g^m \quad \text{or} \quad g^m \in \langle g^s \rangle$$

Hence $\langle g^m \rangle \subseteq \langle g^s \rangle$. To prove iv we need only observe that g^s generates $\langle g \rangle$ if and only if $\langle g^s \rangle = \langle g \rangle$ which happens if and only if they have the same order. But the order of g^s is $n/\gcd(s, n)$, which is equal to n if and only if $\gcd(s, n) = 1$. The number of such s is the Euler φ-function.

Exercise 3.2.17 asks for a proof of the following useful number theoretic result:

$$n = \sum_{d \mid n} \varphi(d)$$

where φ is the Euler φ-function.

We turn now to the lattice of subgroups of a group.

3.2.7 Theorem.   Lattice of Subgroups

Let G be a group. The set $L(G)$ of all subgroups of G forms a complete lattice under set inclusion.

Proof.   We employ the criterion of Theorem 2.2.13. Certainly $L(G)$ is a partially ordered set under set inclusion. The subgroup G itself is the top element of $L(G)$ under set inclusion.

We must now show that every set of subgroups of G has a glb in $L(G)$. We prove that if C is any set of subgroups then the set intersection

$$\cap C = \bigcap_{S \in C} S$$

of all the subgroups S in the set C is a subgroup and is the greatest lower bound of C under the set inclusion relation.

First we prove that $\cap C$ is a subgroup. We use criterion iii of Theorem 3.2.3. Suppose that $a, b \in \cap C$. Then a, b belong to all subgroups S in C and hence

$$a^{-1}b \in S \qquad \text{for all } S \in C$$

Therefore, $a^{-1}b$ belongs to $\cap C$, the set intersection of the subgroups S in C. Thus the criterion iii has been verified for $\cap C$. Now to verify that $\cap C$ is the greatest lower bound of all the sets S in C. Certainly it is a lower bound under set inclusion, but because $\cap C$ is the greatest lower bound of C in the partially ordered set of all subsets of G, it follows that no subset, let alone subgroup, of G could be a greater lower bound for C.

Notation. We use the following notation for the lattice operations on $L(G)$. For subgroups A and B of G

$$\text{glb}(A, B) = A \cap B = \text{set intersection of } A \text{ and } B$$

$$\text{lub}(A, B) = A \vee B$$

Remark. The subgroup $A \vee B$ is almost never the set union of A and B, and that is why a symbol other than $\cup$ which often denotes set union is used. Remember that the proof of Theorem 2.2.13 says that $A \vee B$ is the glb of the set of all subgroups of G containing both A and B. In particular $A \vee B$ contains, but seldom equals, the set union $A \cup B$. The next theorem gives a more constructive description of $A \vee B$.

3.2.8 Theorem. A Characterization of Subgroup Join

Let A and B be subgroups of a group G. The lattice join $A \vee B$ is the set of elements of G that can be expressed as the product of a finite number of elements in either A or B. In symbols

$$A \vee B = \{x \in G : x = u_1 u_2 \ldots u_r, \text{ where } u_i \in A \text{ or } u_i \in B\}$$

Proof. Let U denote the set of elements $x \in G$ that can be expressed in the form

$$x = u_1 u_2 \ldots u_r$$

for some positive integer r and where for each i, $u_i \in A$ or $u_i \in B$. Note that in any event $A \subseteq U$ and $B \subseteq U$ since we can let $r = 1$ and choose x to be an element in A or B as we please.

Note also that any subgroup containing A and B must contain all the elements x having this form by the closure condition for subgroups.

To show that U is a subgroup we verify criterion ii of Theorem 3.2.3. First we show that if $x \in U$ then so is x^{-1}. That is easy, for we compute

$$x^{-1} = (u_1 u_2 \ldots u_r)^{-1} = u_r^{-1} \ldots u_2^{-1} u_1^{-1}$$

and since each u_i^{-1} is in A or B according as u_i is A or B, it follows that x^{-1} can be expressed in the appropriate form.

Second we show that U is closed under the group operation of G. Suppose

$$x = u_1 u_2 \ldots u_r \quad \text{and} \quad y = v_1 v_2 \ldots v_s$$

where the u's and v's belong either to A or to B. Then, just by concatenation,

$$xy = u_1 u_2 \ldots u_r v_1 v_2 \ldots v_s$$

and thus xy can be expressed in the appropriate form.

So U is a subgroup and it is an upper bound for the two groups $\{A, B\}$. Is it the *least* upper bound? It is indeed; we've already observed that any subgroup of G that contains both A and B would contain all the elements in U. This means that U is contained in every upper bound for $\{A, B\}$. So the proof is complete.

This description permits us to determine the lattice of subgroups of many groups. Notice that any subgroup contains the cyclic subgroups generated by its elements. Then it follows that any subgroup is the lattice join of the cyclic subgroups it contains. As an example we determine all the subgroups of D_8, the symmetries of the square, and draw its lattice diagram.

3.2.9 The Subgroups of D_8

Begin by determining the cyclic subgroups. Here is the list:

$$I = \langle I \rangle$$

$$A = \langle a \rangle = \{I, a, a^2, a^3\} = \langle a^3 \rangle$$

$$A^2 = \langle a^2 \rangle = \{I, a^2\}$$

$$B = \langle b \rangle = \{I, b\}$$

$$D = \langle ab \rangle = \{I, ab\}$$

$$E = \langle a^2 b \rangle = \{I, a^2 b\}$$

$$F = \langle a^3 b \rangle = \{I, a^3 b\}$$

Notice that $A^2 \subset A$ but that the other cyclic subgroups are noncomparable under set inclusion.

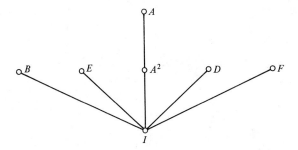

Figure 3.5 Partial construction of the lattice of subgroups of the group D_8, the symmetries of the square. This is the partially ordered set of the cyclic subgroups of D_8.

Now draw the partially ordered set consisting of these cyclic subgroups. This is done in Figure 3.5. Now begin to construct the subgroup joins $S \vee T$ for the cyclic subgroups. In most cases this join will be all of D_8 because we shall see that $\{a, b\} \in S \vee T$. Thus:

$$A \vee B = D_8 \quad \text{since } A \vee B \supset \{a, b\}$$
$$A \vee D = D_8 \quad \text{since } A \vee D \supset \{a, ab\}$$

and hence $(a^{-1})(ab) = b \in A \vee D$. So $A \vee D \supset \{a, b\}$. Similarly

$$A \vee E = A \vee F = D_8$$

Also

$$B \vee D = D_8$$

since $B \vee D \supset \{b, ab\}$ and hence $(ab)b = a \in B \vee D$. Similarly

$$B \vee F = E \vee D = E \vee F = D_8$$

Thus we've checked all the joins involving A, B, D, E, and F. The joins with A^2 are interesting: $A^2 \vee B \supset \{I, a^2, b\}$ and hence also a^2b. Now we've already seen that the four elements

$$\{I, a^2, b, a^2b\} = C$$

form a subgroup of D_8. Hence

$$A^2 \vee B = C = \{I, a^2, b, a^2b\}$$

Note that it follows by interchanging the roles of b and a^2b that also

$$C = A^2 \vee E$$

It is also true that

$$C = B \vee E$$

You can verify this by noting that $(a^2b)(b) = a^2b^2 = a^2 \in B \vee E$ or by

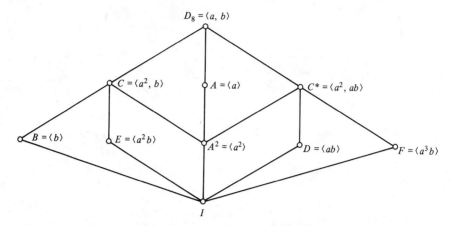

Figure 3.6 The lattice of subgroups of the group D_8, the symmetries of the square. D_8 is generated by two elements, a and b. The element a has order 4, the element b has order 2, and the two elements are related by $ba = a^3b$.

noting that B and E are contained in C and in no smaller subgroup. Similarly $A^2 \vee D = C^* =$

$$\{ I, a^2, ab, a^3b \} = A^2 \vee F$$

You should verify that C^* is also a subgroup to tidy up the details. Note too that C and C^* are isomorphic.

Putting all this together gives the diagram of Figure 3.6, which is the lattice of subgroups of D_8.

Finally check to see that forming $C \vee S$ and $C^* \vee S$ for all previously constructed subgroups S are shown in the lattice diagram. You don't need to check a subgroup S that is comparable with C or with C^*. For the others observe that

$$C \vee S = C^* \vee S = D_8$$

Now we can claim that we have determined all subgroups. Indeed, any subgroup is the lattice join of the cyclic subgroups it contains. And, since the subgroups form a lattice, this means that the intersections $S \cap T$ must be correct as shown.

3.2.10 Subgroup Generated by a Set

If $\{ a_1, \ldots, a_r \}$ is a subset of a group G we let

$$\langle a_1, \ldots, a_n \rangle = \langle a_1 \rangle \vee \cdots \vee \langle a_r \rangle$$

denote the smallest subgroup of G containing the elements $\{ a_1, \ldots, a_r \}$. It coincides with the subgroup join of the cyclic subgroups $\langle a_1 \rangle, \ldots, \langle a_r \rangle$. We say that the set $\{ a_1, \ldots, a_r \}$ *generates* the subgroup $\langle a_1, \ldots, a_r \rangle$. For example, we say that $\{ a, b \}$ generates D_8. Notice that a cyclic group is one that can be generated by a single element.

EXERCISE SET 3.2

3.2.1 List the elements of the group D_6 of symmetries of an equilateral triangle as permutations of the vertices $1, 2, 3$. Write its group table and compare it with the group in Figure 3.2. Are the groups isomorphic?

3.2.2 Determine all the cyclic subgroups of D_6. Draw the lattice of all subgroups of D_6.

3.2.3 List the elements of the group of symmetries of a rectangle that is not a square. Write its group table. Draw its lattice of subgroups.

3.2.4 Prove the following about the orders of elements of a finite order in a group G.

$$|a| = |a^{-1}|.$$

$$|a^{-1}ba| = |b|.$$

If $\gcd(|a|, |b|) = 1$ and $ab = ba$ then $|ab| = |a||b|$.

Show by example that $|ab| \neq |a||b|$ in general.

3.2.5 Draw the lattice of subgroups of $\langle \mathbb{Z}_6, + \rangle$ and $\langle \mathbb{Z}_9^\times, \diamondsuit \rangle$, the groups of Figure 3.1.

3.2.6 Let S be a subset of a finite group G. Suppose that S is closed under the operation of G. Show that S is a subgroup of G. Show by constructing an example that this is not a sufficient condition for S to be a subgroup in general.

3.2.7 Show that if G is an abelian group and A and B are subgroups then

$$A \vee B = \{ ab \colon a \in A \text{ and } b \in B \}$$

3.2.8 In the group $\langle \mathbb{Q}, + \rangle$ of rational numbers under addition find these subgroup joins:

$$\langle \mathbb{Z}, + \rangle \vee \langle a, b \rangle \quad (\text{assume } \gcd(a, b) = 1)$$

$$\langle 1/a \rangle \vee \langle 1/b \rangle$$

and

$$\langle 1/a \rangle \cap \langle 1/b \rangle$$

3.2.9 Show that two cyclic groups are isomorphic if and only if they have the same order. Show that a group of even order has an element of order 2.

3.2.10 Draw the subgroup lattice of $\langle \mathbb{Z}_{16}^\times, \cdot \rangle$. This is a group of order 8. *Hint:* List the elements and their orders. Find all the cyclic subgroups and draw their partially ordered set. Complete the lattice.

3.2.11 List the elements of $\langle \mathbb{Z}_{27}^\times, \cdot \rangle$. Draw its lattice of subgroups.

3.2.12 For each positive integer n and each prime number p describe the lattice of subgroups of a cyclic group of order p^n.

3.2.13 Draw the subgroup lattices of cyclic groups of order 6, 10, 15, 21, and 35. State a theorem generalizing these cases.

3.2.14 Find a set of generators for the symmetries of the square consisting of two elements of order 2.

3.2.15 Show that if there were a finite group whose lattice of subgroups is as shown then the group would be cyclic. Show that there is no finite group having this as its lattice of subgroups.

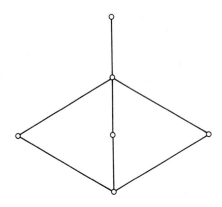

3.2.16 Suppose that the only subgroups of a group G are G and I. Show that the order of G is finite and is a prime.

3.2.17 Show that for each positive integer n,

$$n = \sum_{d \mid n} \varphi(d)$$

3.2.18 When, if ever, is the set union of two subgroups a subgroup?

3.2.19 Let G be an abelian group. Let A, B, and C be subgroups. Prove the modular law, if $A \supseteq B$ then

$$A \cap (B \vee C) = B \vee (A \cap C)$$

3.2.20 Show that if G is a cyclic group the lattice of subgroups of G is distributive, that is

$$A \cap (B \vee C) = (A \cap B) \vee (A \cap C)$$

for all subgroups A, B, and C.

❊3.2.21 Show that if the lattice of subgroups of a finite group is distributive then G is a cyclic group.

3.2.22 Suppose that G is an abelian group. For each positive integer m show that

$$K_m = \{ x \in G : x^m = 1 \}$$

is a subgroup of G. Show that this may be false if G is not abelian. Show that if $\gcd(n, m) = 1$ then $K_n \cap K_m = $ identity.

3.2.23 Suppose that G is an abelian group. For each positive integer m show that

$$H_m = \{ x^m : x \in G \}$$

is a subgroup of G.

3.2.24 Continuing the notation of Exercises 3.2.22 and 3.2.23 find K_3, K_2, and K_4 in $\mathbb{Z}_{45}^{\times}$. Show that $\mathbb{Z}_{45}^{\times} = K_3 \vee K_4$. Find H_3, H_2, and H_4 for $\mathbb{Z}_{45}^{\times}$.

3.3 DIHEDRAL GROUPS

We have studied cyclic groups and their subgroups. Another important class of groups are the finite *dihedral* groups. These groups arise, and may be defined, as the symmetries of a regular plane polygon of n sides, an "n-gon" for $n \geq 3$. We have already mentioned two specific examples from this class: the symmetries of the equilateral triangle and the square. The group of symmetries of the equilateral triangle turned out to be isomorphic to the group of all one-to-one functions from a set of three elements onto itself, a group of order 6. The group of symmetries of the square is a group of order 8 and is only a subgroup of the group of all one-to-one functions from a set of four elements onto itself.

3.3.1 Theorem.   The Group of Symmetries of a Regular n-gon

If $n \geq 3$, the symmetries of a regular n-gon are a group of order $2n$.

Proof.   Figure 3.7 shows a schematic picture of a regular n-gon whose vertices are numbered $0, 1, \ldots, (n-1)$. We use this notation in the proof.

The symmetries may be counted as follows. A symmetry preserves distance so the locations of 0 and 1 determine the entire symmetry. There are n choices for the vertex 0; having made that choice there are two choices for the vertex 1 since it must go into a vertex adjacent to the one chosen for 0. Hence there are at most $2n$ possible symmetries. All these possibilities do arise, as the next table of symmetries shows.

As in Figure 3.7, let a denote a clockwise rotation of $2\pi/n$ radians. Let b denote the reflection in the diameter through the vertex zero, shown as a

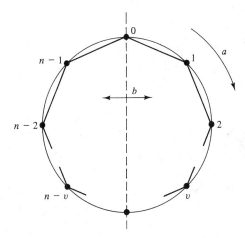

Figure 3.7 A schematic diagram of a regular n-gon, a polygon of n sides. The rotation labeled a is a rotation of $2\pi/n$ radians clockwise. The rotation a takes vertex k to vertex $k+1$ (modulo n). Thus $0 \to 1$, $1 \to 2$, $2 \to 3, \ldots,$ and $(n-1) \to 0$. The reflection labeled b is across the diameter through 0. The reflection b fixes 0 and interchanges vertex v and vertex $(n-v)$. Thus $1 \leftrightarrow (n-1)$, $2 \leftrightarrow (n-2), \ldots$. If n is odd only 0 is fixed. If n is even the vertex $n/2$ is also fixed.

dotted line in Figure 3.7. We record these symmetries

$$a = \begin{pmatrix} 0 & 1 & 2 & \cdots & (n-1) \\ 1 & 2 & 3 & \cdots & 0 \end{pmatrix}$$

$$b = \begin{pmatrix} 0 & 1 & 2 & \cdots & (n-1) \\ 0 & (n-1) & (n-2) & \cdots & 1 \end{pmatrix}$$

and then we can compute, for $0 \le k \le (n-1)$, the compositions

$$a^k = \begin{pmatrix} 0 & 1 & \cdots & (n-1) \\ k & (k+1) & \cdots & (k-1) \end{pmatrix}$$

$$a^k b = \begin{pmatrix} 0 & 1 & \cdots & (n-1) \\ (n-k) & (n-k-1) & \cdots & (n-k+1) \end{pmatrix}$$

The indices must be interpreted modulo n. The symmetry a^k is a rotation clockwise of $2k\pi/n$ radians and fixes no vertex (unless $k = 0$ when all vertices are fixed). The symmetry $a^k b$ fixes one vertex if n is odd and two or none if n is even.

From a group standpoint, the key rules the symmetries obey are these:

$$a^n = I \qquad b^2 = I \quad \text{and} \quad ba = a^{-1}b$$

From these relations and the group properties alone we can determine the group table. In fact

$$a^h b a^k b = a^{h-k}$$

which gives $(a^k b)^2 = I$.

3.3.2 Definition. Dihedral Group of Order 2n ($n \ge 2$)

A group generated by two elements a, b such that

    i. a has order n: $a^n = I$.
    ii. b has order 2: $b^2 = I$.
    iii. The relation $ba = a^{-1}b$ holds.

is called a dihedral group of order $2n$.

Notation. Let D_{2n} denote a dihedral group of order $2n$.

Remark. We have just shown that for $n \ge 3$ there is such a group, the group of symmetries of a regular n-gon. Since there is no 2-gon, the geometric picture fails, but it is easy to see that in this case the definition describes a group of order 4 whose elements may be denoted $\{I, a, b, ab\}$ and whose table is

I	a	b	c
a	I	c	b
b	c	I	a
c	b	a	I

This group is often referred to simply as the "four-group" or sometimes as the Klein group in honor of Felix Klein (1849–1925). Klein was a German geometer who studied geometry from the point of view of a group of transformations.

3.3.3 Theorem. Order of a Dihedral Group

The order of the dihedral group D_{2n} is $2n$.

Proof.   Let $G = \langle a, b \rangle$ where a has order n ($n \geq 2$), b has order 2, $ba = a^{-1}b$, and $a \neq b$.

First we show that $\langle a \rangle \cap \langle b \rangle = \{1\}$. Since b has order 2, the intersection is at most $\{1, b\}$ and so if the intersection is not just $\{1\}$ it must contain b. Thus for some integer k, $1 \leq k \leq n$,

$$a^k = b$$

Hence a and b commute. This means that

$$a^{-1}b = ba = ab$$

Now cancel b and infer that $a^{-1} = a$ or that $a^2 = 1$. Hence $\langle a \rangle = \{1, a\}$ and so the only possibility for k is $k = 1$ and thus $a = b$ contrary to hypothesis.

Now we show that $|D_{2n}| \geq 2n$ by showing that the $2n$ elements

$$1, a, \ldots, a^{n-1}$$
$$b, ab, \ldots, a^{n-1}b$$

are distinct. We ask for the details in Exercise 3.3.1 at the end of this section.

Finally we show that $|D_{2n}| \leq 2n$ by showing that the set of $2n$ elements listed above are closed under the operation of G. The key step is showing that for positive h,

$$ba^h = a^{-h}b$$

This follows from h applications of the relation $ba = a^{-1}b$:

$$ba^h = ba\, a^{h-1} = a^{-1}ba^{h-1} = \cdots = a^{-h}b$$

(Alternatively you make an easy induction on h.) Again the details are asked as a part of Exercise 3.3.1.

EXERCISE SET 3.3

3.3.1 Complete the details of the proof of Theorem 3.3.3.

3.3.2 Determine the orders of all the elements in D_{2n}.

3.3.3 Here is a diagram for the subgroup lattice of D_{12}. Let a have order 6, b have order 2, and $ba = a^{-1}b$. (This group is also the symmetries of a regular hexagon.) Determine all the subgroups and correctly complete the labeling of the subgroups by giving generators for each. As a hint, three key subgroups are identified.

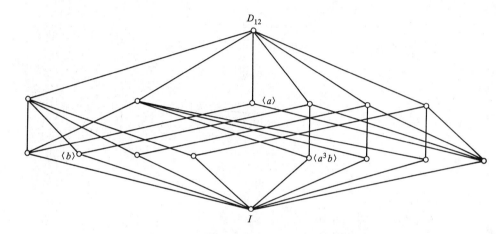

3.3.4 Let G be a group with elements x, y such that $x^2 = y^2 = 1$. Suppose that xy has finite order m. Show that $\langle x, y \rangle$ is a dihedral group D_{2m}.

3.3.5 Let D_{2n} be a dihedral group of order $2n$. Show that a subgroup of D_{2n} is either of order 2, a subgroup of the cyclic group of order n, or is isomorphic to D_{2m} where m divides n. Determine the number of the subgroups isomorphic to D_{2m} for each divisor m of n.

3.4 COUNTING THEOREMS

Now we know the axioms for a group and we know about a most important aspect of groups, their subgroups. But, aside from the examples in the previous section we have little hold on the subject. One yardstick with which to measure our progress is the determination of the groups of a given order. By this we mean that given a positive integer n, we should like to have a list of nonisomorphic groups of order n such that any other group of order n is isomorphic to one in the list. When we have done so, we say that we have determined the groups of order n *up to isomorphism*.

In theory this is a possible task, for there are but a finite number of group tables for a group of order n. Such a brute force attack, although successful in Exercise Set 3.1 for groups of order 1, 2, 3, and 4, is out of bounds for large n. The task we have just set is formidable and we shall not conquer it. But we shall make some headway on it. For example, one general yet easy result comes immediately from the theorem of Lagrange: All groups of prime order are cyclic and hence isomorphic.

Certainly we want to know all the groups of small order.

To be specific, for this chapter we shall obtain a list of groups of order 6 so that given a group of order 6 we shall be able to say that it is isomorphic to one of the groups on this list. In the case of order 6 this is a short list; only two groups are on it. We have already seen these groups from the examples in Section 3.1. The two groups are a cyclic group of order 6 and the symmetries of the equilateral triangle S_3. But we are not yet able to show that this list is complete.

Some of the important tools in this endeavor are theorems that relate the order of a subgroup or of an element in the group to the order of the group.

In this section we establish the first of several arithmetic theorems related to groups. These theorems help us determine the possible orders of subgroups of a group whose order is known. The principal result is due to the French mathematician Joseph Louis Lagrange (1736–1813).

3.4.1 Theorem of Lagrange

For finite groups, the order of each subgroup divides the order of the group.

Discussion. Let G be a group and S a subgroup. The idea of the proof is to "list" the elements of G, beginning with those in S, in disjoint subsets each of size $|S|$, the order of S. Actually we obtain a partition of G into subsets (which are called *cosets*) each having $|S|$ elements. Thus G is the disjoint set union of these cosets, and so $|G|$ is the product of the number of these cosets in the partition and the number of elements $|S|$ in each coset.

Here is an idea of how the list, or partition, is constructed. Let $|S| = n$ and let the elements of S be listed:

$$S = \{I = s_1, s_2, \ldots, s_n\}$$

Now if $|S| = |G|$ our list is complete and $S = G$. All is trivial.

If $S \neq G$, let x be an element not in the list; $x \in G - S$. Now list these elements:

$$xS = \{x = xs_1, xs_2, \ldots, xs_n\}$$

First these are n distinct elements. Indeed, if $xs_i = xs_j$ then cancellation gives $s_i = s_j$. Second, no element in the subset xS is in S. Indeed, if

$$xs_i = s_j, \qquad \text{then } x = s_j s_i^{-1} \in S$$

contrary to the choice of x.

Now if $S \cup xS = G$ then $|G| = 2|S|$ and the theorem is proved. If not, let $y \in G - (S \cup xS)$. List the elements

$$yS = \{ y = ys_1, ys_2, \ldots, ys_n \}$$

As before, these are n distinct elements, disjoint from S. They are also disjoint from xS. Indeed, if

$$ys_i = xs_j \qquad \text{then } y = xs_j s_i^{-1} \in xS$$

since $s_j s_i^{-1} \in S$. So the sets S, xS, yS are disjoint. If

$$G = S \cup xS \cup yS$$

then $|G| = 3|S|$ and the proof is complete. If not, continue . . . !

Here are the formal details that we present in a manner applicable to infinite groups. First a definition, some examples, and a lemma.

3.4.2 Definition. Coset

Let S be a subgroup of a group G. A left coset of S in G is a subset of G whose elements may be expressed

$$xS = \{ xs: s \in S \}$$

The element x is called a representative of the coset. Similarly, a right coset of S in G is a subset that may be expressed

$$Sy = \{ sy: s \in S \}$$

Remarks

1. Notice that the subgroup S itself is always both a left and a right coset; $S = IS = SI$.
2. If G is abelian each left coset aS equals the right coset Sa.
3. It turns out, as we prove in Lemma 3.4.4, that any element of a coset may be used as its representative. In fact

   $$xS = yS \qquad \text{if and only if } x \in yS$$

4. If the group operation is addition $(+)$ then a left coset is denoted

   $$x + S = \{ x + s: s \in S \}$$

3.4.3 Examples of Cosets

In examples 1, 2, and 3 to follow the group G will be the dihedral group D_6 of order 6. (We know this group to be isomorphic to the group whose table appears in Fig. 3.1.) Here is a list of its elements:

$$D_6 = \{ I, a, a^2, b, ab, a^2b \}$$

where

$$a^3 = b^2 = I \quad \text{and} \quad ba = a^{-1}b = a^2b$$

1. Let the subgroup be $S = \langle a \rangle = \{I, a, a^2\}$. The left cosets of S in G are

$$S = IS \quad \text{and} \quad bS = \{b, ba, ba^2\} = \{b, a^2b, ab\}$$

Note that $G = S \cup bS$. The right cosets of S in G are

$$S = SI \quad \text{and} \quad Sb = \{b, ab, a^2b\}$$

In this case note that $bS = Sb$.

2. Let the subgroup be $T = \langle b \rangle = \{I, b\}$. The left and right cosets of T in D_6 in addition to T itself are:

Left Cosets	**Right Cosets**
$aT = \{a, ab\}$	$Ta = \{a, ba\} = \{a, a^2b\}$
$a^2T = \{a^2, a^2b\}$	$Ta^2 = \{a^2, ba^2\} = \{a^2, ab\}$

Note that

$$D_6 = T \cup aT \cup a^2T = T \cup Ta \cup Ta^2$$

Note that in this case $aT \neq Ta$ and $a^2T \neq Ta^2$. Also $aT \neq Ta^2$.

3. Let the subgroup be $V = \langle ab \rangle = \{I, ab\}$. The left and right cosets of V in D_6 aside from V are:

Left Cosets	**Right Cosets**
$aV = \{a, a^2b\}$	$Va = \{a, aba\} = \{a, b\}$
$a^2V = \{a^2, b\}$	$Va^2 = \{a^2, aba^2\} = \{a^2, a^2b\}$

Again

$$D_6 = V \cup aV \cup a^2V = V \cup Va \cup Va^2$$

And again in this case:

$$aV \neq Va \quad \text{and} \quad aV \neq Va^2$$

4. Let the group be $\langle \mathbb{Z}, + \rangle$, the integers under addition. Consider the subgroup N of $\mathbb{Z}$ consisting of the multiples of the integer n:

$$N = \langle n \rangle = \{kn : k \in Z\}$$

Then the cosets of N in $\mathbb{Z}$ are the congruence classes

$$[0], [1], \ldots, [n-1]$$

We have

$$[r] = r + N = \{r + kn : k \in Z\}$$

Thus $[r]$ is precisely the set of integers that give the remainder r when divided by n. (Recall Example 2.1.1.) Since $(+)$ is commutative left and right cosets with the same representative are equal.

3.4.4 Lemma.   Coset Partitions

The left cosets of S in G partition G into subsets each having $|S|$ elements. If R is a set of left coset representatives then

$$G = \bigcup_{x \in R} xS$$

Proof.   First, notice that every element $x \in G$ belongs to some left coset; in fact, $x \in xS$ since $x = xI$ and $I \in S$.

Second, two left cosets xS and yS are either equal as sets or are disjoint. Indeed, suppose that $z \in xS \cap yS$ so that

$$z = xr = yt$$

where r and t both belong to S. This means that $x = ytr^{-1}$ and so, if $s \in S$,

$$xs = ytr^{-1}s \in yS$$

since $tr^{-1}s$ in S. Thus $xS \subseteq yS$. Similarly, $ys = xrt^{-1}s \in xS$ and so $yS \subseteq xS$.

Finally we can show that the coset xS has the same number of elements as does S by showing that the mapping $s \to xs$ is a one-to-one function from S onto xS. Thus $|S| = |xS|$.

Remark.   We may replace the word "left" with the word "right" throughout Lemma 3.4.4 and its proof to show that the right cosets partition G and that the number of elements in a right coset is also equal to the number of elements of S.

If $|G|$ is finite, then since G is the disjoint union of subsets each with $|S|$ elements, the number of left or right cosets of S in G is the quotient $|G|/|S|$.

If G is finite or infinite, the following lemma gives an interesting way to show that the number of left and right cosets of S in G are equal.

3.4.5 Lemma.   Index of a Subgroup

If S is a subgroup of a group G then the number of left cosets equals the number of right cosets. This number is called the index of S in G and is denoted $[G : S]$.

Proof.   First we show that the set of inverses of the elements in a left coset xS is the right coset Sx^{-1}. The following chain of equalities does it!

$$(xS)^{-1} = \{(xs)^{-1}: s \in S\} = \{s^{-1}x^{-1}: s \in S\} = \{sx^{-1}: s \in S\} = Sx^{-1}$$

The first equality is really a definition, the second is an equality of elements, the third recognizes that the set of inverses of a subgroup is the subgroup itself, and the fourth equality is the definition of a right coset.

Second verify that the function from left cosets to right cosets given by

$$xS \to Sx^{-1}$$

is a one-to-one function. We omit the details.

Now we can state the stronger version of the theorem of Lagrange.

3.4.6 Theorem.   Second Version of the Theorem of Lagrange

If S is a subgroup of G the order of G is the product of the order of S and the index of S in G.

$$|G| = |S|[G:S]$$

Proof.   Our work is really done. Lemma 3.4.5 makes unambiguous the meaning of "index" as the number of left or right cosets. Lemma 3.4.4 states that G is the disjoint union of cosets each having the same number of elements. The definition of multiplication in either the finite or infinite case then yields the result.

3.4.7 Theorem.   Applications and Extensions of the Theorem of Lagrange

1. In a finite group the order of each element divides the order of the group.
2. Any group of prime order is cyclic.
3. If A and B are finite subgroups, $|A \cap B|$ divides $|A|$ and $|B|$.
4. If A and B are finite subgroups whose orders are relatively prime, $A \cap B = \{I\}$.
5. If $B \subseteq A \subseteq G$ then $[G:B] = [G:A][A:B]$.

Proof of 1.   Put $S = \langle g \rangle$, the cyclic group generated by g. Then $|g| = |\langle g \rangle| = |S|$ and the result follows from the theorem of Lagrange.

Proof of 2.   Let p be the order of a group G and let $g \neq I$ be an element of G. The order of g divides p and hence must be p. Thus $|\langle g \rangle| = p$ and so $\langle g \rangle$ is the whole group.

Proof of 3.   Apply the theorem first with $A \cap B$ as the subgroup and A as the group. Then replace A by B and repeat!

Proof of 4.   From (3), $|A \cap B|$ divides both $|A|$ and $|B|$. Since these numbers are relatively prime, it follows that $|A \cap B| = 1$ and so $A \cap B = \{I\}$.

Proof of 5.   For finite groups this comes out of the arithmetic of the theorem of Lagrange. For infinite groups we can argue as follows: Let R be a set of left coset representatives for A in G. Thus

$$G = \bigcup_{x \in R} xA$$

Let S be a set of left coset representatives for B in A. Thus

$$A = \bigcup_{y \in S} yB$$

and so

$$xA = \bigcup_{y \in S} xyB$$

Thus each coset xA of A in G is partitioned into cosets xyB of B in A. Hence

$$G = \bigcup_{\substack{x \in R \\ y \in S}} xyB$$

where the union is over all pairs (x, y) where $x \in R$ and $y \in S$. That number is the product

$$|R||S| = [G : A][A : B]$$

so all that is left to do is to show the cosets xyB are disjoint.

Suppose that for x_1, x_2 in R and y_1, y_2 in S

$$x_1 y_1 B = x_2 y_2 B$$

Then $x_1 y_1 = x_2 y_2 b$ for some $b \in B$. Since $B \subseteq A$ and $y_2 \in A$, it follows that $y_2 b \in A$, so that x_1 and x_2 belong to the same coset of A. Since R is a set of representatives, one from each coset of A in G, it follows that $x_1 = x_2$. But then $y_1 = y_2 b$, so that y_1 and y_2 belong to the same coset of B in A. Since S is a set of coset representatives, one from each coset of B in A, it follows that $y_1 = y_2$.

3.4.8 Definition.   Group Complex of Two Subgroups

If A and B are subgroups of a group G, the set of elements

$$AB = \{ ab: a \in A \text{ and } b \in B \}$$

is called the complex of A and B.

3.4.9 Theorem.   The Number of Elements in a Complex

If A and B are finite subgroups of a group G, the number of elements in the complex AB is

$$|AB| = \frac{|A| \cdot |B|}{|A \cap B|}$$

Proof.   The number of products ab where $a \in A$ and $b \in B$ that can be formed is $|A| \cdot |B|$ since there are $|A|$ choices for a and b choices for b. However, some of these products may be equal as elements of G.

How many distinct elements arise in this form? The fact is that if an element $g \in G$ can be written as a product of the form

$$g = ab \qquad \text{where } a \in A \quad \text{and} \quad b \in B$$

then it can be written in that form in $|A \cap B|$ different ways. Indeed, if

$x \in A \cap B$ then

$$g = (ax)(x^{-1}b) = a_1 b_1$$

since $ax \in A$ and $x^{-1}b \in B$. For each $x \in A \cap B$ we get a different form for g. Thus each element in AB appears at least $|A \cap B|$ times.

On the other hand, if

$$g = ab = a_1 b_1$$

then $a_1^{-1}a = b_1 b^{-1} = x$. Now $x = a_1^{-1}a$ says that $x \in A$, while $x = b_1 b^{-1}$ says that $x \in B$; hence $x \in A \cap B$. Thus a repetition can only come from an element in $A \cap B$.

Remember that in Problem 3.2.7 we asked you to show that if G were abelian then AB is a subgroup. This is often true even when G is not abelian, and we shall give several different criteria for this to be so. Watch for them; it is very useful to know when the complex of subgroups is a subgroup.

3.4.10 Application.   Groups of Order 6

Given a group of order 6, we want to know that it is isomorphic to one of a few (actually only two) possibilities. It is the determination of these possibilities we seek.

Often it is helpful to classify the elements of a group by their orders. In this case the possible orders must divide 6 and so must be 1, 2, 3, or 6. An element of order 1 is the identity. If the group has an element of order 6, then that element generates the group; the group is cyclic. Now we consider the other possibilities.

Suppose first that all the elements except the identity had order 2. Then, as told by Problem 3.1.6, the group G is abelian. Moreover, then, if a, b are distinct nonidentity elements, the set

$$V = \{I, a, b, ab\}$$

constitutes a subgroup, a four-group, and 4 does not divide 6. Thus there must be elements of order 3 in G.

We use our counting principle to show that there can be but one subgroup of order 3. Suppose there were two, say A and B. Since A and B are distinct subgroups of prime order, $A \cap B = \{I\}$ so that

$$|AB| = \frac{|A||B|}{|A \cap B|} = |A||B| = 9$$

But 9 is greater than 6 and so G cannot contain two distinct subgroups of order 3.

Thus a group of order 6 must contain exactly one subgroup of order 3. In a noncyclic group of order 6, the remaining 3 elements must be of order 2.

Let a be an element of order 3 and let b be an element of order 2. From (4) in Applications 3.4.7, we conclude that $\langle a \rangle \cap \langle b \rangle = \{I\}$. Now the counting principle says

$$|\langle a \rangle \langle b \rangle| = |\langle a \rangle||\langle b \rangle| = 6$$

So $G = \langle a \rangle \langle b \rangle$ and the elements of G are

$$\{I, a, a^2, b, ab, a^2b\}$$

The group is not yet determined because we don't yet know which element in the list is equal to ba. And so we cannot construct a group table. The element ba is one of the following:

$$ba \overset{?}{=} I, a, a^2, b, ab, \quad \text{or} \quad a^2b$$

Of the six possibilities the first three are easily excluded. For example, if $ba = a^2$ then cancellation says $b = a$, a contradiction. Two cases remain:

$$ba = ab \quad \text{and} \quad ba = a^2b$$

In the first case, a and b commute, their orders are 3 and 2, thus relatively prime, and so ab generates the cyclic group of order 6. In the second case we obtain the dihedral group of order 6 (Theorem 3.3.3). Hence, up to isomorphism, there are just two groups of order 6:

The cyclic group of order 6, C_6.

The dihedral group of order 6, D_6.

The next theorem extends to any prime the result of Problem 3.1.16 which assured that if 2 divides the order of a group then there is an element of order 2 in the group. The result is extremely useful, and the proof we give here is a natural generalization of the idea behind a solution to Problem 3.1.16. We shall give another proof of this theorem when we have more group theoretic machinery!

3.4.11 A Theorem of Cauchy

If p is a prime dividing the order of a group then there is an element of order p in the group.

Proof (McKay).   The idea of this proof is to construct a set of p-tuples whose size is easily computed and known to be divisible by p. This set is then partitioned into equivalence classes of either size 1 or size p. There will be at least one of size 1, and hence by a division argument, there will be at least $(p - 1)$ more equivalence classes of size 1. The classes of size 1 have been defined so that they arise from solutions of $x^p = I$ in the group. Thus there must be a nonidentity element x such that $x^p = I$ and so x has order p. Here are the details.

Let G be the group. By hypothesis p divides $|G|$. Consider the set W of ordered p-tuples of elements in G whose product is I. Thus

$$W = \{(a_0, a_1, \ldots, a_{p-1}): a_0a_1 \ldots a_{p-1} = I\}$$

Notice that the p-tuple $(I, \ldots, I)$ of all I's is in W, so W is not empty. Notice too that

$$a_0 = (a_1 a_2 \ldots a_{p-1})^{-1}$$

In fact, to construct a sequence in W, the $(p-1)$ elements,

$$a_1, \ldots, a_{p-1}$$

can be chosen freely for then choosing $a_0 = (a_1 a_2 \ldots a_{p-1})^{-1}$ puts

$$(a_0, a_1, a_2, \ldots, a_{p-1})$$

in W. This means that we can count the number of p-tuples in W. There are $|G|$ choices for each of $a_1, \ldots, a_{p-1}$ and then a_0 is determined. So the number of p-tuples in W is just $|G|^{p-1}$, a number that is divisible by p since $|G|$ is divisible by p. Thus

$$|W| = |G|^{p-1} \quad \text{and} \quad p \text{ divides } |W|$$

Note too that if there were an element a, such that $a^p = I$ then the constant p-tuple $(a, \ldots, a)$ is in W.

The next thing to observe is that if

$$(a_0, a_1, \ldots, a_{k-1}, a_k, \ldots, a_{p-1})$$

is in W then so is a cyclic shift of this p-tuple by k places. That is, so is

$$(a_k, \ldots, a_{p-1}, a_0, \ldots, a_{k-1})$$

in W. This requires the verification that the product

$$a_k \ldots a_{p-1} a_0 \ldots a_{k-1} = I$$

and this follows by observing that $(a_k \ldots a_{p-1})$ and $(a_0 a_1 \ldots a_{k-1})$ are inverses of each other since by the insertion of parentheses

$$(a_0 \ldots a_{k-1})(a_k \ldots a_{p-1}) = I$$

Now we are ready to partition W. Put into the same equivalence class with $(a_0, \ldots, a_{p-1})$ all its cyclic shifts $(a_k, \ldots, a_{p-1}, a_0, \ldots, a_{k-1})$ for $k = 0, 1, 2, \ldots, (p-1)$ and only these. The next lemma shows that no two of these p shifts are identical or they are *all* identical.

3.4.12 Lemma.   Cyclic Shifts

If p is a prime, two cyclic shifts of $(a_0, \ldots a_{p-1})$ are identical if and only if $a_0 = a_1 = \cdots = a_{p-1}$.

Proof.   Suppose for $0 \le h < k \le p - 1$, the two p-tuples,

$$(a_h, a_{h+1}, \ldots, a_{h-1})$$

and

$$(a_k, a_{k+1}, \ldots, a_{k-1})$$

are equal. Then

$$a_h = a_k, \ a_{h+1} = a_{k+1}, \ldots, \ a_{h-1} = a_{k-1}$$

or

$$a_{h+u} = a_{k+u} \qquad \text{for } u = 0, 1, \ldots, p-1$$

where the indices must be interpreted modulo p. Thus when $u = p - h$ we get

$$a_0 = a_{k-h}$$

and when $u = k - 2h$ we get

$$a_{k-h} = a_{2(k-h)}$$

and when $u = vk - (v+1)h$ we get

$$a_{v(k-h)} = a_{(v+1)(k-h)}$$

so that, substituting $r = k - h$, we get this chain of equalities:

$$a_0 = a_r = a_{2r} = \cdots = a_{(p-1)r}$$

We claim that each a_j, $j = 0, 1, \ldots, p - 1$ occurs in this chain of equalities. This is so because p is prime and hence the congruence

$$rx \equiv j \pmod{p}$$

can be solved. Thus the lemma is proved.

To complete the proof of the theorem, we see that we have partitioned W into equivalence classes having either one p-tuple (which occurs when all the a_j's are equal, so $(a_j)^p = 1$) or having size p (which occurs when not all the a_j's are equal). Thus

$$|W| = |G|^{p-1} = (\text{number of classes of size 1}) + p \cdot (\text{number of classes of size } p)$$

and so the number of classes of size 1 must be divisible by p. That number could not be zero, since, as we observed, the constant p-tuple $(I, \ldots, I)$ is one such class. Thus there is a class consisting of $(a, \ldots, a)$ with $a \neq I$ and so G contains an element of order p.

3.4.13 Application.   Groups of Order 15

We use these counting theorems to determine the structure of a group of order 15. First, from Cauchy's theorem there must be an element a of order 3 and one b of order 5. Second, there cannot be two subgroups B and C of order 5 since then

$$15 = |G| \geq |BC| = \frac{|B| \, |C|}{|B \cap C|} = 25$$

a contradiction.

Now it follows from Exercise 3.2.4 that the order of aba^{-1} is also 5. Hence

$$aba^{-1} \in \langle b \rangle$$

so that

$$aba^{-1} = b^k$$

for some k and hence (Exercise 3.1.7) that

$$a^3ba^{-3} = b^{(k^3)}$$

Since $a^3 = I$, we conclude $b = b^{(k^3)}$ and so by Corollary 3.2.6, i it follows that

$$k^3 \equiv 1 \ (\text{mod } 5)$$

This means that the order of $[k]$ in $\langle \mathbb{Z}_5^\times, \cdot \rangle$ divides 3 and $4 = |\mathbb{Z}_5^\times|$. Thus $k \equiv 1 \ (\text{mod } 5)$. Alternatively check all possibilities $k = 0, 1, 2, 3,$ and 4. Thus $aba^{-1} = b$, or $ab = ba$. Thus a and b commute and since their orders are relatively prime, it follows Problem 3.2.4 that ab has order 15. But this means that the group is cyclic. Thus any group of order 15 is the cyclic group C_{15}.

EXERCISE SET 3.4

3.4.1 Let S be a subgroup of G. Show that the relation $\sim$ defined on G by

$$a \sim b \qquad \text{if and only if} \qquad a^{-1}b \in S$$

is an equivalence relation. What are the equivalence classes?

3.4.2 Show that if a subgroup S has index 2 in G then each left coset is also a right coset.

3.4.3 Let A and B be subgroups of G. Show that the relation $\oplus$ on the cartesian product $A \times B$ defined by

$$(a_1, b_1) \oplus (a_2, b_2) \qquad \text{if and only if} \qquad a_1 b_1 = a_2 b_2$$

is an equivalence relation. Determine the number of elements in each equivalence class.

3.4.4 A subgroup of a group G is called *proper* if it is neither the identity subgroup nor the whole group G. Show that every infinite group has a proper subgroup. Determine those groups having no proper subgroups.

3.4.5 Let D_8 be the group of all symmetries of the square. Following the notation of Example 3.2.1, show that if S is a subgroup of $\langle a \rangle$, then every left coset is a right coset of S in D_8. Show that if $T = \langle b \rangle$ then only T and a^2T are both left and right cosets of T in D_8.

3.4.6 Let $D_{2n} = \langle a, b \rangle$ be a dihedral group of order $2n$ with

$$a^n = b^2 = I \quad \text{and} \quad ba = a^{-1}b$$

If $S = \langle a^k \rangle$ show that every left coset of S is also a right coset of S. Determine the k's such that the left coset $a^k b \langle b \rangle$ is also a right coset of $\langle b \rangle$. (Your answer will differ when n is even and when n is odd.)

3.4.7 Let G be a group and A and B subgroups. Show that if AB is a subgroup then $AB = A \vee B$. Find examples in D_8 of subgroups A and B such that the complex AB is and is not a subgroup. Prove that if $A = \langle a \rangle$ in D_8 then $AB = D_8$ unless B is a subgroup of A.

3.4.8 Let the cyclic group $\langle a \rangle$ have odd order. Show that a appears on the diagonal of the group table of $\langle a \rangle$.

3.4.9 Let G be an abelian group of even order. Show that if $a \in G$ appears on the diagonal of a group table for G then it does so an even number of times.

3.4.10 Show that if A and B are subgroups of G then $[A \vee B : A] \geq [B : A \cap B]$. Show that equality holds if AB is a subgroup.

3.4.11 If A and B are subgroups of finite index in G, show that $A \cap B$ is also of finite index in G.

3.4.12 Let G be an abelian group of order $2s$ where s is odd. Show that G has exactly one element of order 2 and that the product of all the elements in G is equal to that unique element of order 2.

In Problems 3.4.13 to 3.4.16 you are asked to find all groups of a certain order. For each order n, this means that you are to find a set of groups, no two isomorphic, but such that any group of order n is isomorphic to one of them. For example, for $n = 6$ such a set consists of C_6 and D_6.

3.4.13 Find all groups of order 10.

3.4.14 Find all groups of order 14.

3.4.15 Find all groups of order 85.

3.4.16 Find all groups of order 21. (A straightforward application of the arguments used in Section 3.4 should show that there are at most three possibilities. One will be the cyclic group; the other two are isomorphic. It is a more difficult task, from what you know now, to construct a noncyclic group of order 21.)

Special Exercises on Number Theory. Here are some important and interesting results from elementary number theory that can be proved using our present knowledge of group theory. Most of the exercises depend on selecting an appropriate group and applying appropriate theorems. Useful groups to try are $\langle \mathbb{Z}_n, + \rangle$ and $\langle \mathbb{Z}_n^{\times}, \cdot \rangle$ for suitable n.

3.4.17 (Fermat) If p is a prime, show that $a^p \equiv a \pmod{p}$ for all $a \in \mathbb{Z}$.

3.4.18 (Euler) For all positive integers n and all a such that $\gcd(a, n) = 1$, $a^{\varphi(n)} \equiv 1 \pmod{n}$ where φ is the Euler φ-function.

3.4.19 (Wilson) If p is a prime, $(p - 1)! \equiv -1 \pmod{p}$.

3.4.20 Let p be a prime. Show that for all positive integers n, $n \mid \varphi(p^n - 1)$ where φ is the Euler φ-function.

3.4.21 Show that if p is a prime then $2^p - 1$ is divisible only by primes larger than p. Conclude that there are an infinite number of primes! (Numbers of the form $2^n - 1$ are called Mersenne numbers. Show that a Mersenne number is prime only if n is a prime. The converse is false. Can you find a counterexample? It is not known whether there are an infinite number of Mersenne primes. A large one is $2^{19937} - 1$.)

Chapter 4

Group Homomorphisms

4.1 REGULAR TETRAHEDRONS AND THE ALTERNATING GROUP $\mathcal{A}_4$

A tetrahedron is a three-dimensional solid with four vertices, four faces, each a triangle, and six edges (see Figure 4.1 on the next page). It is *regular* if all the faces are equilateral triangles. We consider the group of those of its symmetries which are the rotations in 3-space, sometimes called the group of rigid motions.

This group has order 12 and has many interesting properties that we cite as examples in later sections. In truth its properties turn out to play the key role in solutions of fourth-degree polynomial equations in terms of radicals.

4.1.1 Physical Model

A physical model may help you visualize the tetrahedron and the rotations that comprise the group. A model is easily constructed from a standard letter envelope. Score the diagonals; cut out and discard the triangular piece containing the flap. Now flex the envelope along the scored diagonals until it is easy to fold one of the top corners (shown as corner A in Figure 4.2) inside the corner diagonally opposite (corner B). Done! Label the vertices 1, 2, 3, 4. To follow the notation used in this text label as follows: Label any vertex 4. Place the tetrahedron on a flat surface with vertex 4 pointing up. Now looking down on vertex 4, number the vertices of the base as 1, 2, 3 in clockwise fashion; see Figure 4.1.

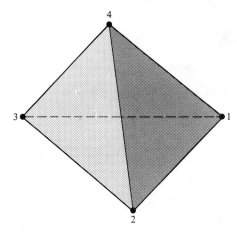

Figure 4.1 A regular tetrahedron. Its group of rotations is a group of order 12 called, for other reasons, the alternating group $\mathscr{A}_4$.

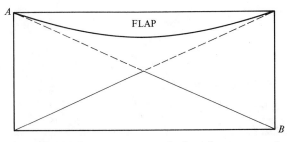

Figure 4.2 Make your own tetrahedron from an envelope. Cut along dotted diagonal lines. Score along solid diagonal lines. Fold corner "*A*" inside corner "*B*."

4.1.2 The Group of Rotations of a Regular Tetrahedron

The exercise of labeling the tetrahedron gives us a bound on the number of symmetries in the group. There are four choices for vertex 4. For each such choice there are three choices for vertex 1. The "clockwise" prescription then determines the location of the other vertices. Thus there are at most $12 = 4 \times 3$ symmetries in all.

Can all 12 symmetries be achieved by a rotation in 3-space?

Begin with the rotations fixing a vertex: Consider, for example, an axis through vertex 4 perpendicular to the opposite face. Let a be the rotation of 120° clockwise about this axis. Here is the effect of a and of a^2, a rotation of 240° clockwise:

$$a = \begin{pmatrix} 1 & 2 & 3 & 4 \\ 2 & 3 & 1 & 4 \end{pmatrix} \qquad a^2 = \begin{pmatrix} 1 & 2 & 3 & 4 \\ 3 & 1 & 2 & 4 \end{pmatrix}$$

Similarly, fixing vertex 3 and rotating 120° clockwise:

$$b = \begin{pmatrix} 1 & 2 & 3 & 4 \\ 4 & 1 & 3 & 2 \end{pmatrix} \qquad b^2 = \begin{pmatrix} 1 & 2 & 3 & 4 \\ 2 & 4 & 3 & 1 \end{pmatrix}$$

and fixing vertex 2:

$$c = \begin{pmatrix} 1 & 2 & 3 & 4 \\ 3 & 2 & 4 & 1 \end{pmatrix} \qquad c^2 = \begin{pmatrix} 1 & 2 & 3 & 4 \\ 4 & 2 & 1 & 3 \end{pmatrix}$$

and fixing vertex 1:

$$d = \begin{pmatrix} 1 & 2 & 3 & 4 \\ 1 & 4 & 2 & 3 \end{pmatrix} \qquad d^2 = \begin{pmatrix} 1 & 2 & 3 & 4 \\ 1 & 3 & 4 & 2 \end{pmatrix}$$

Together with the identity this gives nine rotations. Another is provided by the product (composition of rotations):

$$ab^2 = \begin{pmatrix} 1 & 2 & 3 & 4 \\ 2 & 3 & 1 & 4 \end{pmatrix}\begin{pmatrix} 1 & 2 & 3 & 4 \\ 2 & 4 & 3 & 1 \end{pmatrix} = \begin{pmatrix} 1 & 2 & 3 & 4 \\ 4 & 3 & 2 & 1 \end{pmatrix}$$

which fixes no vertex. This is a rotation of 180° about the axis through the midpoint of the edge connecting vertex 1 and vertex 4 and the midpoint of the opposite edge connecting vertex 2 with vertex 3. For future reference call this axis α. (To carry out this rotation on your model simply put your middle finger on one midpoint, your thumb on the other and spin!)

Similarly

$$ac^2 = \begin{pmatrix} 1 & 2 & 3 & 4 \\ 2 & 1 & 4 & 3 \end{pmatrix}$$

is a rotation about the axis, denoted β, through the edges connecting vertices 1 and 2 and vertices 3 and 4. Finally

$$ad^2 = \begin{pmatrix} 1 & 2 & 3 & 4 \\ 3 & 4 & 1 & 2 \end{pmatrix}$$

is a rotation about the axis denoted γ through the opposite sides connecting vertices 1 and 3 and 2 and 4.

As you may verify, the axes α, β, γ are mutually orthogonal, and as any symmetry is performed, these axes are permuted among themselves.

Note that a, b, c, and d have order 3 while ab^2, ac^2, and ad^2 have order 2.

The 12 rotations we have just enumerated do form a group as we now argue. First, any rotation or symmetry can be recorded, as we have done, as a one-to-one function of $\{1, 2, 3, 4\}$ onto itself. There are $24 = 4!$ such possible functions. The group of rotations is a subgroup of order at least 12; hence by the theorem of Lagrange, its order is either 12 or 24. Second, it can't be 24 because not all the one-to-one functions are rotations. For example, no rotation will fix two vertices and interchange the other two.

A most distinctive feature of this group is the subgroup

$$V = \{ I, ab^2, ac^2, ad^2 \}$$

consisting of all the elements of order 2 and the identity.

The group of the rotations of a regular tetrahedron has another life; it is more commonly called the *alternating* group on four letters. Later we study alternating groups on n letters. At that time we explain the word "alternating"; meanwhile we use the notation $\mathscr{A}_4$ to denote this important group.

4.1.3 Definition.   The Alternating Group $\mathcal{A}_4$

This group is isomorphic to the rotations of a regular tetrahedron. The 12 elements of $\mathcal{A}_4$ can be represented as the permutations on $\{1, 2, 3, 4\}$ listed below

$$a = \begin{pmatrix} 1 & 2 & 3 & 4 \\ 2 & 3 & 1 & 4 \end{pmatrix} \qquad b = \begin{pmatrix} 1 & 2 & 3 & 4 \\ 4 & 1 & 3 & 2 \end{pmatrix}$$

$$c = \begin{pmatrix} 1 & 2 & 3 & 4 \\ 3 & 2 & 4 & 1 \end{pmatrix} \qquad d = \begin{pmatrix} 1 & 2 & 3 & 4 \\ 1 & 4 & 2 & 3 \end{pmatrix}$$

together with the identity, the squares of these elements, and the three elements of order 2 belonging to the subgroup,

$$ab^2 = \begin{pmatrix} 1 & 2 & 3 & 4 \\ 4 & 3 & 2 & 1 \end{pmatrix} \qquad ac^2 = \begin{pmatrix} 1 & 2 & 3 & 4 \\ 2 & 1 & 4 & 3 \end{pmatrix} \qquad ad^2 = \begin{pmatrix} 1 & 2 & 3 & 4 \\ 3 & 4 & 1 & 2 \end{pmatrix}$$

4.1.4 Theorem.   Subgroup Lattice of $\mathcal{A}_4$

The subgroup lattice of $\mathcal{A}_4$ is shown in Figure 4.3.

Proof.   Verify that the subgroups shown exist. The most surprising one is the subgroup V of order 4. It is a Klein four-group.

The subgroup join $A \vee B$ of two distinct subgroups of order 3 must be $\mathcal{A}_4$ since in this case the equation

$$|AB| = \frac{|A| \, |B|}{|A \cap B|} = 9$$

and so $|A \vee B|$ is at least 9 and divides 12; hence it must be 12.

We now see that the subgroup join of a subgroup of order 3 and a subgroup of order 2 is also $\mathcal{A}_4$. Without loss of generality (i.e., by an

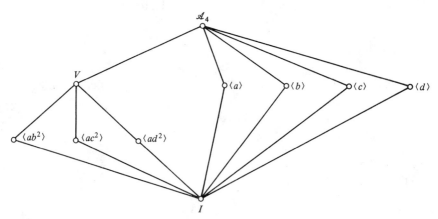

Figure 4.3 The subgroup lattice of $\mathcal{A}_4$, the group of rotations of a regular tetrahedron.

appropriate numbering) we suppose that the subgroup of order 3 is $\langle a \rangle$ and the subgroup of order 2 is $\langle ab^2 \rangle$. Then

$$\langle a \rangle \vee \langle ab^2 \rangle$$

contains a and ab^2 and thus contains

$$a^{-1}ab^2 = b^2$$

and so all of $\langle b \rangle$. Hence

$$\langle a \rangle \vee \langle ab^2 \rangle \supseteq \langle a \rangle \vee \langle b \rangle = \mathscr{A}_4$$

Thus all lattice joins are correct and since every subgroup of a group is the lattice join of the cyclic subgroups it contains, the subgroup lattice as shown in Figure 4.3 is correct.

Remark. The group $\mathscr{A}_4$ has no subgroup of order 6. This shows that the converse of Lagrange's theorem is false.

EXERCISE SET 4.1

4.1.1 Show that the alternating group $\mathscr{A}_4$ can be generated by the elements a and b by expressing the elements c and d in terms of a and b.

4.1.2 In the alternating group $\mathscr{A}_4$ find elements x, y, and z such that

$$b = xax^{-1} \qquad c = yay^{-1} \qquad d = zaz^{-1}$$

4.1.3 In the alternating group $\mathscr{A}_4$ determine the number of elements of the form

$$xax^{-1}$$

as x ranges over $\mathscr{A}_4$. *Hint:* Remember Problem 3.2.4.

4.1.4 Let V be the four 3×3 diagonal matrices of the form

$$\begin{pmatrix} x & 0 & 0 \\ 0 & y & 0 \\ 0 & 0 & z \end{pmatrix}$$

where either $x = y = z = 1$ or exactly two of $\{x, y, z\}$ are equal to -1 and the third is 1. Show that, under matrix multiplication, V is isomorphic to the Klein four-group.

Let

$$a = \begin{pmatrix} 0 & 1 & 0 \\ 0 & 0 & 1 \\ 1 & 0 & 0 \end{pmatrix}$$

Show that under matrix multiplication a has order 3.

Show that $\langle a \rangle \vee V$ is isomorphic to the alternating group $\mathscr{A}_4$.

4.2 GROUP HOMOMORPHISMS

This section introduces a useful class of functions from one group to another. The image of one of these functions (its range) produces a model of the original group. Although the model may not have all the properties of its parent, it often inherits particular information that is helpful in studying the parent. In algebra these functions are called homomorphisms, and we shall study them for each of the algebraic systems we consider.

We begin with an example.

4.2.1 Example.   A Group Model of $\mathscr{A}_4$

In studying the rotation group of the tetrahedron each element was represented as a one-to-one function of $\{1, 2, 3, 4\}$ onto itself. We found the axes $\{\alpha, \beta, \gamma\}$ that were mapped onto themselves by these rotations. If we replace each rotation by its effect on $\{\alpha, \beta, \gamma\}$ we get a model of the group $\mathscr{A}_4$. Denote by $\mathscr{A}^*$ the one-to-one functions of $\{\alpha, \beta, \gamma\}$ induced by the group of rotations $\mathscr{A}_4$. Using this notation, the effect on the axes of the rotation a is denoted by a^*.

Recall that an axis, say α, was defined as the line connecting the midpoints of the edge $(1, 4)$ and the edge $(2, 3)$. In a brief but suggestive notation

$$\alpha = (1, 4); (2, 3)$$

$$\beta = (1, 2); (3, 4)$$

$$\gamma = (1, 3); (2, 4)$$

Here is the effect of the rotations on $\{\alpha, \beta, \gamma\}$. You can verify this from your physical model or by calculating the image of the vertices under each rotation. For example:

$$a = \begin{pmatrix} 1 & 2 & 3 & 4 \\ 2 & 3 & 1 & 4 \end{pmatrix} \rightarrow \begin{pmatrix} \alpha & \beta & \gamma \\ \gamma & \alpha & \beta \end{pmatrix} = a^*$$

$$a^2 = \begin{pmatrix} 1 & 2 & 3 & 4 \\ 3 & 1 & 2 & 4 \end{pmatrix} \rightarrow \begin{pmatrix} \alpha & \beta & \gamma \\ \beta & \gamma & \alpha \end{pmatrix} = (a^2)^*$$

$$b = \begin{pmatrix} 1 & 2 & 3 & 4 \\ 4 & 1 & 3 & 2 \end{pmatrix} \rightarrow \begin{pmatrix} \alpha & \beta & \gamma \\ \gamma & \alpha & \beta \end{pmatrix} = b^*$$

$$b^2 = \begin{pmatrix} 1 & 2 & 3 & 4 \\ 2 & 4 & 3 & 1 \end{pmatrix} \rightarrow \begin{pmatrix} \alpha & \beta & \gamma \\ \beta & \gamma & \alpha \end{pmatrix} = (b^2)^*$$

$$ab = \begin{pmatrix} 1 & 2 & 3 & 4 \\ 1 & 3 & 4 & 2 \end{pmatrix} \rightarrow \begin{pmatrix} \alpha & \beta & \gamma \\ \beta & \gamma & \alpha \end{pmatrix} = (ab)^*$$

$$ab^2 = \begin{pmatrix} 1 & 2 & 3 & 4 \\ 4 & 3 & 2 & 1 \end{pmatrix} \rightarrow \begin{pmatrix} \alpha & \beta & \gamma \\ \alpha & \beta & \gamma \end{pmatrix} = (ab^2)^*$$

Now here is an important fact: Since the axes are a physical part of the tetrahedron, a composition of two rotations of the tetrahedron must induce the *same composition* on the axes α, β, γ of the tetrahedron. In terms of the group operation this means, for example,

$$(a^2)^* = (a^*)^2 \qquad (b^2)^* = (b^*)^2 \qquad (ab)^* = a^*b^* \quad \text{and} \quad (ab^2)^* = a^*(b^*)^2$$

Since each element of $\mathscr{A}_4$ can be expressed in terms of a and b and since the above equations hold, this group model $\mathscr{A}^*$ is generated by a^* and b^*. Since $b^* = a^*$ we find that $\mathscr{A}^*$ is a cyclic group of order 3.

4.2.2 Definition.   Homomorphism

A homomorphism is a function π from a group G into a group G^* such that the *homomorphism condition*

$$\pi(xy) = \pi(x)\pi(y)$$

holds for all elements x, y in G.

We say that π preserves the group operation although the group operations of G and G^* may be quite different; see Example 4.2.4.

Two subsets that will be shown to be subgroups in the next lemma are important, the *kernel* and the *image* of π. The kernel of π is the set of elements of G that are mapped into the identity of G^* by π,

$$\text{Kernel} = \{ x \in G : \pi(x) = I^*, \text{the identity of } G^* \}$$

The image of π is the range, denoted $\pi(G)$, of the function π,

$$\text{Image} = \pi(G) = \{ x^* \in G^* : \text{There exists } x \in G \text{ with } x^* = \pi(x) \}$$

If $\pi(G) = G^*$ then π is a homomorphism of G onto G^* and G^* is called a *homomorphic image* of G.

Remark.   A group isomorphism (recall Definition 3.1.15) is a special example of a homomorphism.

The *trivial* homomorphism maps every element of G to the identity I^* of G^*. Other special cases of homomorphisms are important and carry special names.

4.2.3 Definition.   Isomorphism, Endomorphism, and Automorphism

An *isomorphism* is a homomorphism that is one-to-one.

An *endomorphism* is a homomorphism of a group into itself.

An *automorphism* is an isomorphism of a group *onto* itself.

Remark.   We have used juxtaposition to denote the group operation of G and G^*. In practice the operations in the two groups may be quite different. Here is a simple example.

4.2.4 A Homomorphism of the Four-Group

Let G be the Klein four-group of Definition 3.3.2. Let G^* be the cyclic group of order 4, $\langle \mathbb{Z}_4, + \rangle$. We list the elements of each group:

$$G = \{I, a, b, c\} \qquad G^* = \{[0], [1], [2], [3]\}$$

Define π as follows:

$$I \to [0]$$
$$a \to [2]$$
$$b \to [2]$$
$$c \to [0]$$

Now verify that the homomorphism condition is satisfied. Here is one case:

$$\pi(ab) \overset{?}{=} \pi(a) + \pi(b)$$

Since $ab = c$ in G, this is equivalent to asking $\pi(c) \overset{?}{=} \pi(a) + \pi(b)$. Plugging in from the table, this is equivalent to asking

$$[0] \overset{?}{=} [2] + [2] = [0]$$

And the answer is "Yes." In this case the homomorphism condition is met.

The other possibilities are equally easy to verify, although it is a tedious process at this point. We need to build some theorems to show how to construct, use, and verify that functions from one group to another are homomorphisms.

Pause to note that the kernel of π is $\{I, c\}$ and the range of π is $\{[0], [2]\}$.

4.2.5 Example. A Homomorphism of $\mathscr{A}_4$

Here we cast Example 4.2.1 in the terminology of the preceding definition. Let $G = \mathscr{A}_4$ and let G^* be a cyclic group of order 3, say the one generated by the permutation of $\{\alpha, \beta, \gamma\}$ that we called a^*,

$$a^* = \begin{pmatrix} \alpha & \beta & \gamma \\ \gamma & \alpha & \beta \end{pmatrix}$$

In Example 4.2.1 we verified the homomorphism condition. The kernel of this homomorphism is, as you should verify, the subgroup $V = \{I, ab^2, ac^2, ad^2\}$. The image is the cyclic group of order 3, $\langle a^* \rangle$.

4.2.6 Lemma. Elementary Properties of Homomorphisms

Let π be a homomorphism from G to G^* with kernel K.

1. The image of the identity of G is the identity of G^*. In symbols,

$$\pi(I) = I^*$$

2. The image of an inverse is the inverse of the image. In symbols,

$$\pi(a^{-1}) = (\pi(a))^{-1}$$

3. The image of π is a subgroup of G^*.
4. The kernel K is a subgroup of G.
5. For all a, b in G the following are equivalent:

$$\pi(a) = \pi(b)$$

$$ab^{-1} \in K$$

$$a^{-1}b \in K$$

6. For each $a \in G$, the left and right cosets of K are equal and are equal to the set of preimages of $\pi(a)$. That is,

$$aK = Ka = \{x: \pi(x) = \pi(a)\}$$

7. A homomorphism is an isomorphism if and only if the kernel is just the identity.

Proof of 1.   Verify that $\pi(I)$ is an idempotent in G^* and hence, by Lemma 3.1.6, is the identity of G^*. Here we go, using the homomorphism condition:

$$\pi(I) = \pi(II) = \pi(I)\pi(I)$$

Hereafter denote the identity of G^* by I^*.

Proof of 2.   Calculate

$$I^* = \pi(I) = \pi(aa^{-1}) = \pi(a)\pi(a^{-1})$$

so that

$$(\pi(a))^{-1} = \pi(a^{-1})$$

Proof of 3.   Verify the subgroup criterion iii of Theorem 3.2.3. Let a^* and b^* be in the image of π. We want to show that $(a^*)^{-1}b^*$ is in the image of π. Let a and b be elements of G such that $a^* = \pi(a)$ and $b^* = \pi(b)$. Now calculate

$$(a^*)^{-1}b^* = (\pi(a))^{-1}\pi(b) = \pi(a^{-1})\pi(b) = \pi(a^{-1}b)$$

and so $(a^*)^{-1}b^*$ is in the image of π.

Proof of 4.   Again verify the same subgroup criterion. This time, suppose that $I^* = \pi(a) = \pi(b)$ now calculate

$$\pi(ab^{-1}) = \pi(a^{-1})\pi(b) = (\pi(a))^{-1}\pi(b) = I^{*-1}I^* = I^*$$

Proof of 5.   Since $\pi(a) = \pi(b)$ if and only if $I^* = \pi(a)(\pi(b))^{-1} = \pi(ab^{-1})$, it follows that $\pi(a) = \pi(b)$ if and only if $(ab^{-1}) \in K$. A similar argument shows that $a^{-1}b \in K$ if and only if $\pi(a) = \pi(b)$.

Proof of 6.   Recall that if x belongs to the coset Ka then $x = ka$ for some $k \in K$. This is equivalent to $xa^{-1} \in K$. Using this and 5, we conclude that $x \in Ka$ if and only if $\pi(a) = \pi(x)$. That is,

$$Ka = \{x: \pi(x) = \pi(a)\}$$

Thus Ka is the set of preimages of $\pi(a)$. A similar argument shows that aK is also the set of preimages of $\pi(a)$.

Proof of 7.   If π is an isomorphism, the only possible preimage of the identity is the identity; and so $K = \{I\}$.

Conversely if kernel K is just the identity then $aK = \{a\}$ and so there is just one element in the preimage of a; hence π is one-to-one.

One of the most important parts of the lemma is part 6, which says that for each a, the cosets aK and Ka of the kernel K are equal. In Example 3.4.3 we saw that this is not always true of every subgroup. When it occurs we say the subgroup is a normal subgroup. There are three important and equivalent ways of defining a normal subgroup. We incorporate them in the next theorem.

4.2.7 Theorem.   Three Equivalent Properties

Let N be a subgroup of a group G. The following three conditions on N are equivalent. A subgroup satisfying these conditions is called *normal*.

1. For all $a \in G$, $aN = Na$.
2. For all $a \in G$ and all $n \in N$, $ana^{-1} \in N$.
3. For all $a \in G$, $aNa^{-1} = \{ana^{-1}: n \in N\} = N$.

Proof that 1 Implies 2.   If the cosets aN and Na are equal, then for each $n \in N$ there is an $m \in N$ such that

$$an = ma$$

or

$$ana^{-1} = m \in N$$

Proof that 2 Implies 3.   Condition 2 implies, for all $a \in G$, that the set $aNa^{-1} \subseteq N$. Conversely, we show that if $m \in N$ then there exists $n \in N$ such that $ana^{-1} = m$. So, solve for n:

$$n = a^{-1}ma$$

Is this n an element of N? To answer this question, use condition 2 with a^{-1} playing the role of a and m playing the role of n. Thus condition 2 says

$$a^{-1}ma \in N$$

and thus $n \in N$; and indeed

$$a(a^{-1}ma)a^{-1} = m$$

Thus $N \subseteq aNa^{-1}$; and so finally,

$$aNa^{-1} = N$$

Proof that 3 Implies 1.   Let $an \in aN$. From condition 3, $ana^{-1} \in N$ and so

$$an \in Na$$

Thus
$$aN \subseteq Na$$

Conversely, again using condition 3 with a^{-1} in the role of a, we find
$$a^{-1}na \in N$$
so
$$na \in aN$$
and so
$$Na \subseteq aN$$
and thus
$$aN = Na$$

4.2.8 Definition.   Normal Subgroup

A subgroup N of a group G is called normal if any one (and so all three) of the properties of Theorem 4.2.7 hold.

The term "normal" comes from the special role these subgroups play in Galois theory. We shall have to wait until Chapter 13 for a proper explanation. Meanwhile, remember that "normal" has the special properties of Theorem 4.2.7.

4.2.9 Examples of Normal Subgroups

1. Any group has the identity subgroup and itself as normal subgroups.
2. A subgroup of index 2 is always normal in G. This is due to the fact that in a coset decomposition, if N is of index 2 in G, there are only two cosets, N and another, say, aN. Thus
$$G = N \cup aN = N \cup Na$$
   Since
$$N \cap aN = \varnothing = N \cap Na$$
   we see that the set complement of N in G is, on the one hand, aN, and on the other is Na. Thus $aN = Na$ and since $1N = N1$ also, N is normal in G.
3. In an abelian group every subgroup is normal; commutativity of the group operation makes condition 1 easy.
4. If no other subgroup of G has the same order as N, then N is normal in G. The argument here rests on the fact that
$$aNa^{-1} = \{ ana^{-1} : n \in N \}$$
   is a subgroup of G and its order is the same as the order of N because the correspondence
$$n \rightarrow ana^{-1}$$
   is one-to-one. These facts are easily proved; see Problem 4.2.1. Thus the subgroup aNa^{-1} must be N.
5. The subgroup V of $\mathscr{A}_4$ is normal because V is the only subgroup of order 4 in $\mathscr{A}_4$.

Next we learn how to construct homomorphic images of groups. We also find that any normal subgroup will serve as the kernel of some homomorphism. The key to the construction is in part 6 of Lemma 4.2.6. This part states that the coset aK acts like the element $a^* = \pi(a)$ in $\pi(G)$. Moreover, the homomorphism condition

$$\pi(a)\pi(b) = \pi(ab)$$

suggests a definition for a group operation on the left cosets in a corresponding manner:

$$aKbK = abK$$

Thus the idea will be to define a group operation on, say, the left cosets of K.

4.2.10 Theorem.   A Normal Subgroup Is a Kernel

Let N be a normal subgroup of a group G. The set of left (or right) cosets of N in G may be made into a group, called the factor group of G by N, denoted G/N, by defining

$$(aN)(bN) = abN \tag{1}$$

The map defined by $a \to aN$ is a homomorphism of G onto G/N with kernel N. This homomorphism is called the *natural* homomorphism of G onto G/N.

If $\pi: G \to G^*$ is a homomorphism of G into G^* with kernel N, then the image of G under π is isomorphic to the factor group G/N.

Proof.   First, we must discuss the most subtle point in the proof. Has a group operation been defined by equation (1)? The situation is similar to the definition of operations on congruence classes of integers treated in Definition 2.1.9. Here equation (1) defines an operation on cosets in terms of specific representatives. We must know that the operation has been well-defined, that is, we must first verify that the definition is independent of the coset representative chosen. We must verify that if

$$x \in aN \quad \text{and} \quad y \in bN \quad \text{then} \quad xy \in abN$$

or that, in terms of cosets, if

$$xN = aN \quad \text{and} \quad yN = bN \quad \text{then} \quad xyN = abN$$

It is in proving this that we use the normality of N. Here is the argument:

If $x \in aN$ then $x = an$ for some $n \in N$.

If $y \in bN$ then $y = bm$ for some $m \in N$.

Thus

$$xy = anbm = ab(b^{-1})nbm = ab(b^{-1}nb)m$$

Now $b^{-1}nb \in N$ since N is normal (use condition 2) and so $b^{-1}nbm \in N$. Thus $xy \in abN$.

Second we now prove that the set of left cosets

$$G/N = \{ aN: a \in G \}$$

forms a group under the operation defined in equation (1). The definition makes the verification of the associative axiom easy. The identity is the coset $IN = N$. The inverse of aN is the coset $a^{-1}N$ since

$$aNa^{-1}N = aa^{-1}N = IN = N$$

Third, we show that the function ν defined by

$$\nu: a \rightarrow aN$$

yields a homomorphism of G into G/N with kernel N. This too, is easy to verify because the operation defined in equation (1) is the homomorphism condition. The function ν is onto because each coset does arise! The kernel of ν is N since $aN = N$ if and only if $a \in N$.

Finally, suppose that $\pi: G \rightarrow G^*$ is a homomorphism of G into G^* with kernel N. We want to define a function λ from G/N onto the image of π by

$$\lambda: aN \rightarrow \pi(a)$$

Again the definition uses a specific coset representative so we must again verify that the definition of λ does not depend on the particular coset representative chosen. That is, we must verify that

$$\text{If } x \in aN \quad \text{then} \quad \pi(x) = \pi(a)$$

but of course it does, because of Lemma 4.2.6, part 6.

Having done this, it is easy to verify that λ is one-to-one, because, again,

$$\pi(a) = \pi(b) \qquad \text{if and only if } aN = bN$$

is the result of part 6.

The homomorphism condition for λ follows from the definition given in equation (1), and the homomorphism condition for π, as these equalities show:

$$\lambda(aNbN) = \lambda(abN) = \pi(ab) = \pi(a)\pi(b) = \lambda(aN)\lambda(bN)$$

This completes the proof.

4.2.11 Definition.   Factor Group

If N is a normal subgroup of group G the factor group of G by N, denoted G/N, is the set of cosets of N in G under the operation

$$(aN)(bN) = abN$$

as discussed in Theorem 4.2.10.

4.2.12 Example.   The Factor Group $\mathscr{A}_4 / V$

We have already verified that V is a normal subgroup of $\mathscr{A}_4$. Now we list the elements in the notation of Example 4.1.2.

$$V = \{ I, ab^2, ac^2, ad^2 \}$$
$$aV = \{ a, a^2b^2, a^2c^2, a^2 \}$$
$$a^2V = \{ a^2, b^2, c^2, d^2 \}$$

so

$$\mathscr{A}_4/V = \{ V, aV, a^2V \}$$

Recall that V was the kernel of the homomorphism of $\mathscr{A}_4$ onto the cyclic group of rotations of the axes generated by

$$a^* = \begin{pmatrix} \alpha & \beta & \gamma \\ \gamma & \alpha & \beta \end{pmatrix}$$

and indeed, as in the proof of Theorem 4.2.11,

$$\lambda \colon aV \to \pi(a) = a^*$$

Remark.   Different normal subgroups may lead to isomorphic factor groups. That is, it is possible that G have distinct normal subgroups N and M with G/N isomorphic to G/M.

Here is a simple example: Let g be a four-group, say,

$$G = \{ I, a, b, ab \}$$

Since G is abelian each subgroup of order 2 is normal and each factor group has order 2 and so all factor groups are isomorphic.

Figure 4.4 illustrates the relation between the lattice of subgroups of a homomorphic image with the lattice of subgroups of the group that contain the kernel.

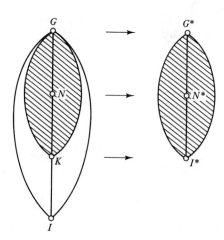

Figure 4.4 The subgroup lattice of G is indicated on the left. On the right is the subgroup lattice of G^*, a homomorphic image of G. The cross-hatching indicates the import of Theorem 4.2.13. The lattice of subgroups of G^* is the same as the sublattice of subgroups of G that contain the kernel K of the homomorphism.

4.2.13 Theorem. Subgroup Properties of Homomorphism

Let π be a homomorphism of G into G^*.

1. There is a one-to-one correspondence between subgroups of G containing the kernel K and subgroups of $\pi(G)$, the image of G under π. Specifically, if $K \subseteq N \subseteq G$ then the function defined by

$$N \to \pi(N)$$

   is a one-to-one function from these subgroups N onto the subgroups of $\pi(G)$.
2. If $K \subseteq N \subseteq G$ then
   2.1. $N \subseteq M$ if and only if $\pi(N) \subseteq \pi(M)$.
   2.2. $\pi(N \cap M) = \pi(N) \cap \pi(M)$.
   2.3. $\pi(N \vee M) = \pi(N) \vee \pi(M)$.
3. If $K \subseteq N \subseteq G$ then N is normal in G if and only if $\pi(N)$ is normal in $\pi(G)$.
4. If $K \subseteq N \subseteq G$ and N is normal in G then G/N is isomorphic to $\pi(G)/\pi(N)$.

Proof of 1. First, we prove that the function defined by $N \to \pi(N)$ is one-to-one. Suppose that $K \subseteq N$, $M \subseteq G$ and that $\pi(N) = \pi(M)$. We shall show that $N \subseteq M$. Then we shall see that the reverse inclusion follows simply by interchanging the roles of N and M. To this end, suppose that $n \in N$. Since $\pi(N) \subseteq \pi(M)$ it follows that $\pi(n) \in \pi(M)$ and so that

$$\pi(n) = \pi(m) \qquad \text{for some } m \in M$$

But then, Lemma 4.2.6, part 5 yields that $nm^{-1} \in K \subseteq M$ so that $nm^{-1} \in M$ and so $n \in M$. Thus $N \subseteq M$. Interchanging the roles of N and M yields $M \subseteq N$, and so $N = M$.

Second, we prove that every subgroup A^* of $\pi(G)$ arises as the image of some subgroup N, with $K \subseteq N \subseteq G$. Simply consider

$$N = \{ g \in G : \pi(g) \in A^* \}$$

We claim that N is a subgroup of G and that $K \subseteq N \subseteq G$. Well, certainly $K \subseteq N$ since all the elements of the kernel map into the identity of $\pi(G)$, a well-known element of every subgroup of $\pi(G)$. Now verify the subgroup condition iii of Theorem 3.2.3: If g and h are members of N then $\pi(g)$ and $\pi(h)$ are members of A^* and thus

$$(\pi(h))^{-1} \pi(g) \in A^*$$

and hence

$$(\pi(h))^{-1} \pi(g) = \pi(h^{-1}) \pi(g) = \pi(h^{-1}g)$$

Thus $h^{-1}g \in N$. Finally, $\pi(N) = A^*$ by definition of N.

Proof of 2.1. Suppose $K \subseteq N \subseteq M \subseteq G$. Then by definition $\pi(N) \subseteq \pi(M)$. In view of the one-to-one correspondence of 1, N and M are distinct subgroups if and only if $\pi(N)$ and $\pi(M)$ are distinct. Conversely, suppose that $\pi(N) \subseteq \pi(M)$. We must show that $N \subseteq M$. Let $n \in N$. Thus $\pi(n) \in \pi(M)$; so for some $m \in M$

$$\pi(n) = \pi(m)$$

Hence $nm^{-1} \in K \subseteq M$. So it follows that $n \in M$ and hence $N \subseteq M$.

Proof of 2.2. First, $K \subseteq N \cap M \subseteq N$ and $N \cap M \subseteq M$ implies

$$\pi(N \cap M) \subseteq \pi(N) \quad \text{and} \quad \pi(N \cap M) \subseteq \pi(M)$$

Hence $\pi(N \cap M) \subset \pi(N) \cap \pi(M)$.

To prove $\pi(N) \cap \pi(M) \subseteq \pi(N \cap M)$ we may argue as follows: Since $\pi(N) \cap \pi(M)$ is a subgroup of $\pi(G)$ it follows from 1 that there is some subgroup P such that $K \subseteq P \subseteq G$ and

$$\pi(P) = \pi(N) \cap \pi(M)$$

Thus

$$\pi(P) \subseteq \pi(N)$$

and so

$$P \subseteq N$$

and

$$\pi(P) \subseteq \pi(M)$$

and so

$$P \subseteq M$$

hence

$$P \subseteq N \cap M$$

hence

$$\pi(P) \subseteq \pi(N \cap M)$$

or

$$\pi(N) \cap \pi(M) \subseteq \pi(N \cap M)$$

Proof of 2.3. The proof can be made by copying the proof of 2.2 by replacing $\cap$ by $\vee$ and $\subseteq$ by $\supseteq$. (This technique works in lattices because the lattice properties listed in Theorem 2.2.14 are unchanged by switching $\cup$ and $\cap$. This important principle is called duality in lattice theory.)

Proof of 3. Suppose that N is a normal subgroup in G. To show that $\pi(N)$ is a normal subgroup of $\pi(G)$ we'll verify condition 2 of Theorem 4.2.7. Let n^* be any element of $\pi(N)$ and let h^* be any element of $\pi(G)$. Then there exist elements $n \in N$, $h \in G$ such that

$$n^* = \pi(n) \quad \text{and} \quad h^* = \pi(h)$$

and now calculate

$$h^* n^* h^{*-1} = \pi(h)\pi(n)(\pi(h))^{-1} = \pi(h)\pi(n)\pi(h^{-1}) = \pi(hnh^{-1})$$

Since N is normal, $hnh^{-1} \in N$ so that

$$h^*n^*h^{*-1} \in \pi(N)$$

The proof of the converse essentially reverses the logic and the equalities. This time let n be any element of N and let h be any element of G. Then

$$\pi(hnh^{-1}) = \pi(h)\pi(n)(\pi(h))^{-1} \in \pi(N)$$

since $\pi(N)$ is assumed normal in $(G)\pi$.

Proof of 4. Suppose $K \subseteq N$ and that N is normal in G. Then $\pi(N)$ is normal in $\pi(G)$. Consider the composition of the homomorphisms $\pi: G \to \pi(G)$ and the natural homomorphism σ of $\pi(G)$ onto $\pi(G)/\pi(N)$ defined by

$$\sigma: g^* \to g^*\pi(N)$$

where g^* is any element of $\pi(G)$. We will prove that the composition $\sigma\pi$ is a homomorphism of G onto $\pi(G)/\pi(N)$ with kernel N. Here is the verification of the homomorphism condition

$$\sigma\pi(ab) = \sigma(\pi(ab)) = \sigma(\pi(a)\pi(b)) = \sigma(\pi(a))\sigma(\pi(b)) = \sigma\pi(a)\sigma\pi(b)$$

Finally, $\sigma\pi(a)$ is the identity of $\pi(G)/\pi(N)$ if and only if $\pi(a)$ is in the kernel of σ which is $\pi(N)$ and that happens if and only if $a \in N$; hence the kernel of the composite homomorphism is N. Thus, from Theorem 4.2.10, G/N is isomorphic with the image of the composite homomorphism which is $\pi(G)/\pi(N)$.

That concludes the proof of Theorem 4.2.13.

EXERCISE SET 4.2

4.2.1 Let a be an element of a group G. Show that function $\theta_a: G \to G$, defined by

$$n \to ana^{-1}$$

is an automorphism of G.

Show that θ_{ab} is the composition $\theta_a\theta_b: x \to \theta_a\theta_b(x)$.

4.2.2 Suppose that G^* is a homomorphic image of G and that A^* is a subgroup of G^*. Express the order of A, the preimage of A^* in the homomorphism, in terms of the order of A^* and the order of the kernel.

4.2.3 If G^* is a homomorphic image of G, show that the order of $a^* \in G^*$ divides the order of all its preimages in G. What can you say about the order of G^* relative to the order of G? Show that if G^* has an element of order n then so does G.

4.2.4 Show that the homomorphic image of a cyclic group is cyclic.

4.2.5 Show that the homomorphic image of a dihedral group is either a dihedral group or is cyclic.

4.2.6 The center of a group G is defined as

$$Z = \{z \in G: zg = gz \text{ for all } g \in G\}$$

Show that Z is a normal subgroup of G.

4.2.7 Let G^* be the homomorphic image of a group G. Show that G^* is abelian if and only if $xyx^{-1}y^{-1} \in$ kernel for all x, y in G. (The subgroup of G generated by elements of this form is called the *commutator* subgroup of G. For more on commutators see Exercise 4.3.6.)

4.2.8 Give an induction proof of "If G is an abelian group and if p is prime dividing $|G|$ then g has an element of order p" along these lines: Let $a \in G$. If p divides the order of a, then ————— (complete the sentence). If p doesn't divide the order of a, form the factor group $G^* = G/\langle a \rangle$. Apply induction to G^* and relate the element of order p in G^* to an element of order p in G. Do not use Cauchy's theorem (3.4.11).

4.2.9 Let G be a group of order pm where p is a prime and $p > m$. Show that G has a normal subgroup of order p. (Do cite Cauchy's theorem!)

4.2.10 Let G be the multiplicative group of 2×2 matrices

$$\begin{pmatrix} a & b \\ c & d \end{pmatrix}$$

where each entry is in $\mathbf{Z}_3$, the integers modulo 3 and for which

$$\det\begin{pmatrix} a & b \\ c & d \end{pmatrix} = ad - bc = 1$$

Show that G has 24 elements. Show that $-I = \begin{pmatrix} -1 & 0 \\ 0 & -1 \end{pmatrix}$ is in the center of G. Find the elements A^* in $G/\langle -I \rangle$ of order 2. (*Hint*: If A is a preimage of A^* in G, $A^2 = I$ or $-I$.) Determine the center of G. Save your solution for Problem 4.3.4.

4.3 NORMAL SUBGROUPS

In this section we investigate the lattice properties of normal subgroups. We begin with the determination of all the normal subgroups of two groups, one familiar, one new.

4.3.1 Example.   The Normal Subgroups of $\mathscr{A}_4$

Any group has itself and the identity as normal subgroups. The alternating group $\mathscr{A}_4$ has only one other normal subgroup, V. In the notation of Example 4.1.2,

$$V = \{ I, ab^2, ac^2, ad^2 \}$$

We have already (see Example 4.2.9) observed that V is normal in $\mathscr{A}_4$. To see that it is the only one, we essentially test every other subgroup.

We begin by showing that no subgroup of order 2 is normal. Here are the subgroups of order 2:

$$X = \{ I, ab^2 \} \qquad Y = \{ I, ac^2 \} \qquad Z = \{ I, ad^2 \}$$

We find

$$a^{-1}ab^2a = b^2a = ad^2 \quad \text{so} \quad a^{-1}Xa = Z$$
$$a^{-1}ad^2a = d^2a = ac^2 \quad \text{so} \quad a^{-1}Za = Y$$
$$a^{-1}ac^2a = c^2a = ab^2 \quad \text{so} \quad a^{-1}Ya = X$$

Next we show that no subgroup of order 3 is normal. The subgroups of order 3 are

$$\langle a \rangle, \quad \langle b \rangle, \quad \langle c \rangle, \quad \text{and} \quad \langle d \rangle$$

We compute

$$aba^{-1} = c \quad \text{so} \quad a\langle b \rangle a^{-1} = \langle c \rangle$$
$$aca^{-1} = d \quad \text{so} \quad a\langle c \rangle a^{-1} = \langle d \rangle$$
$$ada^{-1} = b \quad \text{so} \quad a\langle d \rangle a^{-1} = \langle b \rangle$$

Thus the normality condition is violated for each of these subgroups.

We have already observed that in an abelian group every subgroup is normal. The next example shows that it is possible for a nonabelian group to have all its subgroups normal. The group in this example is called the quaternion group. This group also serves as the basis of an important number system that extends the complex numbers. This number system is also called the quaternions. This system was invented by William R. Hamilton (1805–1865), who used them in his studies of mathematical physics.

4.3.2 Example. The Quaternion Group

The quaternion group which we call Q is a group generated by two distinct elements a, b, each of order 4, which satisfy these further relations:

$$a^2 = b^2 \quad \text{and} \quad ba = a^{-1}b \tag{1}$$

First we investigate the consequences of these relations. Then we give a construction for this group.

To begin, condition (1) implies that $b \notin \langle a \rangle$. Indeed, if $b = a$ then replacing b by a in (1) implies that $a^2 = I$. If $b = a^3$, replacing b by a^3 in (1) implies that $a^4 = a^2$. Either is a contradiction of the requirement that a have order 4. Next since a has order 4, $a^{-1} = a^3$ and since $a^2 = b^2$ it follows that

$$a^{-1}b = a^3b = ab^3 = ab^{-1}$$

Next we show that $(ab)^2 = a^2 = b^2$ and so the order of ab is also 4. Here are the details of that computation:

$$(ab)(ab) = a(ba)b = aa^{-1}bb = b^2 = a^2$$

and so

$$(ab)^4 = \left((ab)^2\right)^2 = a^4 = I$$

This shows that the order of (ab) divides 4, but it can't be 2 so it is 4. A similar calculation shows that $(ab)^3 = a^3b$. Thus there are six elements of order 4:

$$a, a^3, b, b^3, (ab), \quad \text{and} \quad (ab)^3$$

There is one element of order 2, namely, a^2 ($a^2 = b^2 = (ab)^2$). An eighth element is the identity I. Not surprisingly it turns out (see Exercise 4.3.5) that the order of the quaternions is 8. The subgroup lattice of the quaternions can now be drawn. It is shown in Figure 4.5.

Here is a simple construction for the quaternion group using 2×2 matrices with complex entries. Let

$$a = \begin{pmatrix} i & 0 \\ 0 & -i \end{pmatrix} \quad b = \begin{pmatrix} 0 & -1 \\ 1 & 0 \end{pmatrix} \quad c = \begin{pmatrix} 0 & -i \\ -i & 0 \end{pmatrix}$$

Verify that $ab = c$ and that

$$a^2 = b^2 = c^2 = \begin{pmatrix} -1 & 0 \\ 0 & -1 \end{pmatrix} = -I$$

We claim that, under matrix multiplication, the eight matrices

$$Q = \{I, -I, a, -a, b, -b, c, -c\}$$

form a group. To verify this we do not have to reverify the associative law since it is true in general for matrix multiplication. We do have to verify closure and the existence of inverses. The latter comes for free since we have verified that for all $x \in Q$, $x^4 = I$. Verification of closure is easy once we have noted that the matrix $-I$ commutes with all matrices and have verified

$$ab = c \qquad bc = a \quad \text{and} \quad ca = b$$
$$ba = -Ic = -c \qquad cb = -Ia = -a \quad \text{and} \quad ca = -Ib = -b$$

Assuming that we have done this, we can see that this group Q satisfies the properties laid down for the quaternions. It is generated by the two elements a, b, each of order 4 which satisfy the relations (1). It would be a simple

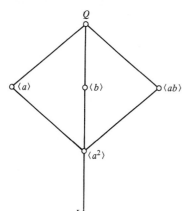

Figure 4.5 The lattice of subgroups of the quaternion group Q. All of its subgroups are normal. The factor group $Q/\langle a^2 \rangle$ is the Klein four-group.

matter to construct the table showing the multiplications in Q. You are hereby urged to do so!

Now, to show that the subgroups are all normal. First the ones of order 4 (which are all cyclic) are normal because they are of index 2 in Q. The one of order 2 is normal because it is the only subgroup of that order in Q. There is also another reason for this which comes from Theorem 4.3.4.

We now turn to the lattice structure of normal subgroups.

4.3.3 Lemma.   Sufficient Condition for a Complex *AB* to be a Subgroup

If A is a normal subgroup of G then, for any subgroup $B \subseteq G$, the complex AB,

$$AB = \{ ab: a \in A, b \in B \}$$

is a subgroup and $AB = A \vee B$.

Proof.   First we verify closure: Let $a, a_1 \in A$ and let $b, b_1 \in B$. We want to show that $(ab)(a_1 b_1) \in AB$. We calculate

$$aba_1 b_1 = a(ba_1 b^{-1})bb_1$$

and since A is normal, $ba_1 b^{-1} \in A$. Thus $a(ba_1 b^{-1}) \in A$, and since $bb_1 \in B$ it follows that the product

$$a(ba_1 b^{-1})bb_1 = (ab)(a_1 b_1) \in AB$$

Second we show that $(ab)^{-1} \in AB$. Calculate

$$(ab)^{-1} = b^{-1} a^{-1} = (b^{-1} a^{-1} b) b^{-1}$$

The normality of A implies that $b^{-1}ab \in A$. Thus $(ab)^{-1} \in AB$ and we have shown that AB is a subgroup.

Finally it is clear that $A \vee B \supseteq AB$. To show the converse, note that since AB is a subgroup containing both A and B, it must be that $AB \supseteq A \vee B$. Hence $AB = A \vee B$.

4.3.4 Theorem.   Lattice Properties of Normal Subgroups

If A and B are normal subgroups of a group G then their meet $A \cap B$ and join $A \vee B$ are normal subgroups of G and so the set of normal subgroups forms a sublattice of the lattice of all subgroups of G.

Proof.   To verify that $A \cap B$ is normal we use criterion 2 of Theorem 4.2.7. Let $n \in A \cap B$ and let $x \in G$. Then $xnx^{-1} \in A$ since $n \in A$ and A is normal. Similarly, $xnx^{-1} \in B$ and so $xnx^{-1} \in A \cap B$. Thus $A \cap B$ is normal in G.

To verify $A \vee B$ is normal we use the same criterion and that $A \vee B = AB$. Thus if $m \in A \vee B = AB$ we must have $m = ab$ and so for any $x \in G$

$$xabx^{-1} = xax^{-1}xbx^{-1}$$

Since $xax^{-1} \in A$ and $xbx^{-1} \in B$ it follows that $xabx^{-1} \in AB = A \vee B$. Thus $A \vee B$ is a normal subgroup of G.

4.3.5 Theorem.   The Diamond Isomorphism

Let N be a normal subgroup of G and let A be a subgroup of G. Then $A \cap N$ is a normal subgroup of A and $A/(A \cap N)$ is isomorphic to $(A \vee N)/N$.

See Figure 4.6 for the "diamond."

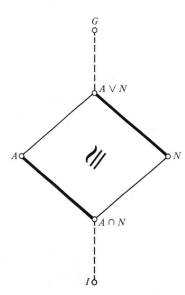

Figure 4.6 The Diamond Isomorphism. The diamond is drawn in solid lines. The subgroup N is normal in G. The opposite sides of the diamond drawn in heavy lines represent the isomorphic factor groups $A/(A \cap N) \cong (A \vee N)/N$.

Proof.   To verify that $A \cap N$ is a normal subgroup of A, let $a \in A \cap N$ and let $x \in A$. We are to show that $xax^{-1} \in A$.

Since a and x are in A, it follows that xax^{-1} is in A.

Since $a \in N$ and N is normal in G it follows that $xax^{-1} \in N$. Thus $xax^{-1} \in A \cap N$. Thus we can form the factor group $A/(A \cap N)$.

Since N is normal in G, N is normal in the subgroup $A \vee N$ and by Lemma 4.3.3, $A \vee N = AN$.

Thus if $x \in AN$ then

$$x = an \qquad \text{for some } a \in A \quad \text{and} \quad n \in N$$

Now we shall try to define a homomorphism π from $A \vee N$ to $A/A \cap N$ via

$$\pi(x) = \pi(an) = a(A \cap N)$$

We have many things to prove. The first is that π is a function; the definition above does seem to depend upon the particular way x is expressed as an element in AN.

Suppose then that

$$x = an = a_1 n_1$$

with $a, a_1 \in A$ and $n, n_1 \in N$. Then $a_1^{-1}a = n_1 n^{-1} \in A \cap N$. Since $a = a_1 n_1 n^{-1}$ it follows that

$$a_1(A \cap N) = a(A \cap N)$$

Thus π is a well-defined function on $A \vee N = AN$. Clearly its range is all of the factor group $A/(A \cap N)$ since every coset $a(A \cap N)$ of $A \cap N$ in A arises as the image of $a = aI \in AN$.

The homomorphism condition holds too. To see this, suppose that

$$x = a_1 n_1 \quad \text{so that} \quad \pi(x) = a_1(A \cap N)$$

and

$$y = an \quad \text{so that} \quad \pi(y) = a(A \cap N)$$

Then

$$xy = a_1 n_1 an = a_1 aa^{-1} n_1 an = a_1 an_2$$

where $n_2 = a^{-1} n_1 an \in N$. Thus

$$\pi(xy) = a_1 a(A \cap N) = \pi(x)\pi(y)$$

Next we determine the kernel of π. Suppose that $x = an$ belongs to the kernel. Then

$$\pi(x) = a(A \cap N) = A \cap N$$

Since $a(A \cap N) = A \cap N$ if and only if $a \in A \cap N$ and since $a \in A$ anyway, it follows that $a \in A \cap N$ if and only if $a \in N$. Thus the kernel of π is N.

Finally, Theorem 4.2.10 tells us that $(A \vee N)/N$ is isomorphic to $A/(A \cap N)$.

We conclude this section with a useful criterion to show that a subgroup generated by a finite set of elements is normal in G. Remember that a subgroup H is generated by a set $\{a_1, \ldots, a_k\}$ if every element of H can be expressed in the form

$$x_1 x_2 \ldots x_n$$

where each x_i is in one of the cyclic subgroups $\langle a_j \rangle$.

4.3.6 Lemma. A Normality Criterion

Suppose that $N = \langle a_1, \ldots, a_k \rangle$ and suppose that for each $g \in G$

$$ga_i g^{-1} \in N$$

Then N is a normal subgroup of G.

Proof.   In the notation of 3.2.10

$$\langle a_1, \ldots, a_k \rangle = \langle a_1 \rangle \vee \cdots \vee \langle a_k \rangle$$

and by an extension of Theorem 3.2.8 to more than two subgroups, as indicated above,

$$N = \left\{ x_1 \ldots x_n \colon x_i \in \langle a_j \rangle \text{ for some } j \right\}$$

Thus the elements of N are the elements of G that can be expressed as the product of a finite number of factors, each in one of the subgroups $\langle a_1 \rangle, \ldots, \langle a_k \rangle$.

Now let $y \in G$ and notice that

$$z = yx_1 \ldots x_n y^{-1} = \left(yx_1 y^{-1} \right)\left(yx_2 y^{-1} \right) \ldots \left(yx_n y^{-1} \right)$$

We shall show for each i that $yx_i y^{-1} \in N$, and hence $z \in N$. Now for each i, $x_i \in \langle a_j \rangle$ for some j; hence $x_i = a_j^m$ for some integer m and so

$$yx_i y^{-1} = \left(ya_j y^{-1} \right)^m$$

But $ya_j y^{-1} \in N$ by hypothesis! The proof is complete.

EXERCISE SET 4.3

4.3.1   Show that the sublattice of normal subgroups of a group is a complete lattice.

4.3.2   Find all the normal subgroups of the dihedral group D_{2n}. Your answer will depend upon n.

4.3.3   Show that the quaternion group Q has order 8. Write out the group table for Q.

4.3.4   This problem continues Problem 4.2.10. Let G be the multiplicative group of 2×2 matrices

$$\begin{pmatrix} a & b \\ c & d \end{pmatrix}$$

where each entry is in $\mathbb{Z}_3$, the integers modulo 3 and for which

$$\det\begin{pmatrix} a & b \\ c & d \end{pmatrix} = ad - bc = 1$$

Let

$$-I = \begin{pmatrix} -1 & 0 \\ 0 & -1 \end{pmatrix}$$

Show that $G/\langle -I \rangle$ is isomorphic to the alternating group $\mathscr{A}_4$. Let Q be the subgroup of G generated by

$$\begin{pmatrix} 1 & 1 \\ 1 & -1 \end{pmatrix} \quad \text{and} \quad \begin{pmatrix} 0 & -1 \\ 1 & 0 \end{pmatrix}$$

Show that Q is isomorphic to the quaternion group and is a normal subgroup of G. Find a subgroup S of order 3 so that $G = Q \vee S$.

4.3.5 A subgroup $H \subseteq G$ is called maximal if $H \neq G$ and if for no other subgroup K is $H \subset K \subset G$. Find the maximal subgroups of $\langle \mathbf{Z}, + \rangle$. If G is abelian determine a necessary and sufficient condition on the index $[G : H]$ for H to be a maximal subgroup of G. What if G is not abelian?

4.3.6 The *commutator* subgroup of a group G (usually denoted G') is the subgroup of G generated by the set of elements $\{ xyx^{-1}y^{-1} \}$. Show that G' is a normal subgroup of G. Show that G/G' is abelian. Show that if G/N is abelian then $N \supseteq G'$. (It may be useful to recall Problem 4.2.7.)

4.3.7 Find the commutator subgroup of $\mathscr{A}_4$, the alternating group on four letters. Find the commutator subgroup of the quaternions. Find the commutator subgroup of the dihedral group D_{2n}.

4.3.8 Show that any subgroup of the center of G is also a normal subgroup of G. Suppose a group G contains a normal subgroup C and C contains a subgroup B having the property that no other subgroup of C has order equal to $|B|$. Show that B is normal in G. (Example 4.2.9 only shows that B is normal in C.)

4.3.9 Construct an example of a group G having three subgroups $A \subset B \subset C$ such that A is a normal subgroup of B and B is a normal subgroup of C, but A is not a normal subgroup of C.

4.3.10 Suppose that A is a normal subgroup of G. Then for all subgroups B and C of G this restricted modular law holds:

$$\text{If } A \subseteq B \quad \text{then} \quad A \vee (B \cap C) = B \cap (A \vee C)$$

(Show that the left-hand side is always contained in the right-hand side, regardless of the normality of A. The reverse containment holds if A is normal in G.)

4.3.11 Let A, B, C, and D be subgroups of a group G. Prove

If A is a normal subgroup of B and C is a normal subgroup of D, then $A \cap C$ is a normal subgroup of $B \cap D$.

If $\overset{\cdot}{A}$ is a normal subgroup of B and C, then A is a normal subgroup of $B \vee C$.

If N is a normal subgroup of G and A is a normal subgroup of B then $A \vee N$ is a normal subgroup of $B \vee N$.

4.3.12 An *inner automorphism* of a group G is one of the form (recall Problem 4.2.1)

$$\pi_a(x) = axa^{-1}$$

Show that a subgroup N is normal in G if and only if $\pi(N) = N$ for all inner automorphisms of G.

4.3.13 Show that the set of all automorphisms of a group G form a group under composition. Show that the inner automorphisms of G form a normal subgroup of the group of all automorphisms.

4.3.14 Find all the automorphisms of the Klein four-group. Find all the automorphisms of a cyclic group of order n.

4.3.15 Show that the mapping $a \rightarrow \pi_a$ where π_a is the inner automorphism defined in Problem 4.3.12 is a homomorphism of a group into its group of inner automorphisms.

Describe the kernel of this homomorphism.

4.3.16 (Macdonald) Let Z be the center of a group G. Show that the mapping

$$(Za, Zb) \to a^{-1}b^{-1}ab$$

is well defined from the ordered pairs of cosets of the center in G to the set of commutators. Conclude that if $|G/Z|^2 < |G'|$, then the set of commutators do not themselves form a subgroup of G.

For further information on this problem see

I. G. Macdonald, Commutators and Their Products,
American Mathematics Monthly, vol. 93, pp. 440–444, 1986.

P. J. Cassidy, Products of Commutators Are Not Always Commutators,
American Mathematics Monthly, vol. 86, p. 772, 1979.

(These papers will be easier to read after you have read more about groups and more about polynomials! However, they serve to point out that the product of commutators is not always a commutator.)

Chapter 5

Direct Products

We have seen what defines a group, lots of examples of groups, their sub-groups, and their homomorphic images. In this chapter we see how we may build new groups from old ones, very much as new integers may be built from products of primes. This process, which is the simplest of several different ways to build new groups from old, is called a direct product. Conversely, we see how we may recognize a group as the result of our construction process. The analogy here is with the unique factorization theorem for integers.

For this to be a powerful process we must have lots of groups at our fingertips. So we begin this chapter with a summary of what we know about groups of small order and what we can prove with the theorems we now have.

5.1 GROUPS OF LOW ORDER

We use our theorems to determine the possible nonisomorphic groups of low order. In doing so we meet some examples of direct products.

First we establish the nature of groups of prime order and order twice a prime.

Notation. Let C_n denote a cyclic group of order n.

5.1.1 Theorem.   Groups of Order p and $2p$

Let p be a prime.

1. A group of order p is cyclic.
2. A group of order $2p$ is either cyclic or D_{2p}, the dihedral group of order $2p$.

Proof.   If a group has prime order, any element not the identity must generate the whole group since its order is a divisor of p and so must in fact be p, since p is a prime.

Now consider a group of order $2p$. If $p = 2$ then we already know the groups of order 4 and have included the four-group as a dihedral group. Hereafter we assume that p is an odd prime. From Cauchy's Theorem 3.4.11 such a group has an element of order p. Let a be an element of order p. The cyclic subgroup $\langle a \rangle$ has index $2p/p = 2$ and hence (Example 4.2.9) is normal. Again the group must have an element of order 2, call it b, that is not in $\langle a \rangle$ since p is odd. Therefore,

$$bab^{-1} = bab \in \langle a \rangle$$

so

$$bab^{-1} = a^k \qquad \text{where } 0 \le k < p$$

Nice things happen when this equation is multiplied on the left by b and on the right by b^{-1}:

$$b(bab^{-1})b^{-1} = ba^k b^{-1}$$

Since $b^2 = I$, the left-hand side simplifies. Thus

$$a = ba^k b^{-1} = (bab^{-1})^k = (a^k)^k = a^{(k^2)}$$

Recall Corollary 3.2.6, i: The equation $a = a^x$ means that

$$x \equiv 1 \ (\text{mod } p)$$

in this case

$$k^2 \equiv 1 \ (\text{mod } p)$$

Thus p divides $(k^2 - 1) = (k - 1)(k + 1)$ and since p is a prime this means

$$p \mid (k - 1) \quad \text{or} \quad p \mid (k + 1)$$

hence

$$k \equiv 1 \quad \text{or} \quad k \equiv -1 \text{ modulo } p$$

Hence

$$a^k = a \quad \text{or} \quad a^k = a^{-1}$$

So two cases arise. Either

$$bab^{-1} = a \quad \text{or} \quad bab^{-1} = a^{-1}$$

From the first alternative it follows that $ba = ab$, so a and b commute. This means that ab has order $2p$ and so $\langle a, b \rangle = \langle ab \rangle$ is the cyclic group C_{2p}. The second alternative gives the definition of the dihedral group D_{2p}.

This theorem determines, for example, the nature of the groups of order 2, 3, 4, 5, 6, 7, 10, 11, 13, 14, 17, 19, 22, and 23.

The next theorem studies the groups of order 8. We anticipate the notation of direct products given in the next section to name these groups. But first we give two examples.

5.1.2 Example.   An Abelian Group of Order 8

Consider the group $\mathbb{Z}_{20}^{\times}$ of residue classes modulo 20 which are relatively prime to 20. The elements are

$$\{[1], [3], [7], [9], [11], [13], [17], [19]\}$$

A computation shows that [3] has order 4 and so $[-3] = [17]$ has order 4 also. We have

$$\langle [3] \rangle = \{[1], [3], [9], [7]\}$$

and

$$\langle [-3] \rangle = \{[1], [17], [9], [13]\}$$

The other elements $[11] = [-9]$ and $[19] = [-1]$ have order 2.

Let $a = [3]$ and $b = [19] = [-1]$. We have that a has order 4, b has order 2, and $b \notin \langle a \rangle$.

Notice that the factor group $G/\langle b \rangle$ is cyclic of order 4 because a must map into an element of order 4 since a^2 is not in $\langle b \rangle$. Just this information tells a lot about the lattice of subgroups. We have the cyclic subgroup $\langle a \rangle$ of order 4, the cyclic subgroup $\langle b \rangle$ of order 2 and its factor group. Figure 5.1 shows what these arguments tell us about the lattice of subgroups of G. The information about the subgroups of G that contain $\langle b \rangle$ comes from the factor group $G/\langle b \rangle$ which is isomorphic to $\langle a \rangle$ as presented in Theorem 4.2.13.

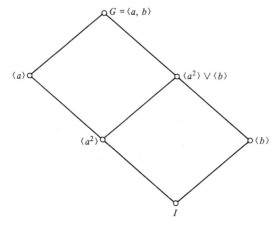

Figure 5.1 Partial information about the lattice of subgroups of the abelian group $\mathbb{Z}_{20}^{\times}$. This group is generated by two elements: an element a of order 4 and an element b of order 2. Shown here is the information given by the subgroups $\langle a \rangle$ and $\langle b \rangle$ and their factor groups.

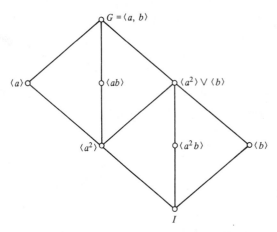

Figure 5.2 The lattice of subgroups of the abelian group of order 8 generated by elements a, b such that a has order 4, b has order 2, and $ab = ba$. Note that this group has a subgroup and a factor group isomorphic to the four-group. In the next section this group is identified as the direct product $C_4 \times C_2$.

The remainder of the lattice can now be filled in. There is an element of order 2 missing; it is a^2b and is in the four-group $\langle a^2, b \rangle$. Next the subgroups containing $\langle a^2 \rangle$ can be determined by observing that the factor group $G/\langle a^2 \rangle$ is another four-group since the square of any element in G is either a^2 or I. Thus we obtain the lattice in Figure 5.2.

5.1.3 Example.   The Elementary Abelian Group of Order 8

The group is made from the cartesian product

$$\mathbb{Z}_2 \times \mathbb{Z}_2 \times \mathbb{Z}_2 = \{(u, v, w): u, v, w \in \mathbb{Z}_2\}$$

The group operation is addition, modulo 2, of each component:

$$(u, v, w) + (x, y, z) = (u + x, v + y, w + z)$$

and we omit the details of verifying the abelian group axioms. Do notice that every element except the identity $(0, 0, 0)$ has order 2.

The subgroup lattice is the most complicated one we have drawn up to this time because there are so many subgroups. There are seven elements of order 2 and so there are seven subgroups of order 2. It turns out that there are also seven subgroups of order 4, each a four-group.

To build the lattice we can use some of the same techniques we've used before. First some notation to make the computations a little easier. Let

$$a = (1, 0, 0) \qquad b = (0, 1, 0) \qquad c = (0, 0, 1)$$

Begin with the subgroup of order 4, $V = \langle a, b \rangle$. The subgroup V is the Klein four-group whose subgroup lattice we know. Next the subgroup $\langle c \rangle$ is of course normal and so $G/\langle c \rangle$ is a group of order 4. We claim that $G/\langle c \rangle$ is a

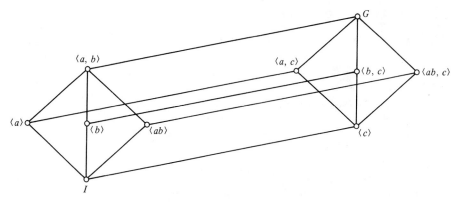

Figure 5.3 Partial information about the lattice of subgroups of the elementary abelian group of order 8 in which every element has order 2. It is generated by three elements, a, b, and c. In this portion of the lattice we show the subgroup $\langle a, b \rangle$ and the subgroups of G containing $\langle c \rangle$, which is isomorphic to $G/\langle c \rangle$ and is a four-group. The lattice is completed in Figure 5.4.

Klein four-group. Indeed we may take as distinct coset representatives the elements of $V = \langle a, b \rangle$. Thus we may build as much of the lattice as in Figure 5.3.

To complete it, note that any single element a, b, or ab in V combined with the element c builds another Klein four-group. There are three of these,

$$\langle a, c \rangle \qquad \langle b, c \rangle \quad \text{and} \quad \langle ab, c \rangle$$

and so we must find three more elements of order 2, ac, bc, and abc and the corresponding subgroups of order 2. Finally add the subgroups of order 4 to complete the necessary lattice joins. The complete result is in Figure 5.4.

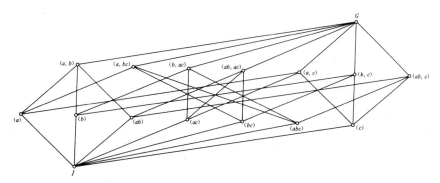

Figure 5.4 The subgroup lattice of the elementary abelian group of order 8. It is the direct product of three cyclic groups of order 2. The diagram omits the lines from I to every covering subgroup and the lines from G to every subgroup it covers. Note that each subgroup covered by G is a four-group and every subgroup covering I is a cyclic group of order 2.

5.1.4 Theorem. The Groups of Order 8

Any group of order 8 is isomorphic to one of the following five groups:

The cyclic group of order 8, $\langle \mathbb{Z}_8, + \rangle$.

The group $\mathbb{Z}_{20}^{\times}$.

The elementary abelian group of order 8.

The dihedral group D_8.

The quaternion group Q.

Proof. The proof is in two parts. In the first part we determine all the abelian groups; in the second we determine the nonabelian groups of order 8.

Part 1. Suppose G is an abelian group of order 8.

Case 1. G has an element of order 8. Then G is isomorphic to $\langle \mathbb{Z}_8, + \rangle$, a cyclic group of order 8. Hereafter in the proof we suppose that G has no element of order 8.

Case 2. G is abelian and has an element a of order 4 and an element b of order 2 which is not in $\langle a \rangle$. In this case $\langle a \rangle \cap \langle b \rangle = \{ I \}$. Moreover, since G is abelian, it follows from Theorem 4.3.3 that the subgroup generated by a and b is the complex:

$$\langle a, b \rangle = \langle a \rangle \langle b \rangle = \langle a \rangle \vee \langle b \rangle$$

We count the number of elements in this subgroup

$$|\langle a, b \rangle| = \frac{|\langle a \rangle| \cdot |\langle b \rangle|}{|\langle a \rangle \cap \langle b \rangle|} = 8$$

so we now know $G = \langle a \rangle \langle b \rangle$, and that every element can be written in one and only one way in the form $a^i b^j$. Now it is a straightforward verification to show that the mapping

$$a \rightarrow [3]$$
$$b \rightarrow [-1]$$

extends to an isomorphism of G onto $\mathbb{Z}_{20}^{\times}$ mapping $a^i b^j$ onto $[3]^i [-1]^j$.

Case 3. G is abelian and has an element of order 4. Suppose, however, there is no element of order 2 outside $\langle a \rangle$. Then the only element of order 2 is a^2. Thus if $b \notin \langle a \rangle$, the order of b is 4 and $b^2 = a^2$. But since G is abelian

$$(ab)^2 = a^2 b^2 = a^2 a^2 = a^4 = I$$

and so $(ab) \in \langle a \rangle$. So either $ab = 1$ or $ab = a^2$. The first alternative implies $b = a^{-1}$; the second implies that $b = a$. Both are contradictions. Hence no such group of order 8 exists; case 3 does not arise.

Case 4. Every element not the identity has order 2. (In this case G must be abelian.) Let a and b be distinct elements of G. It is easy to see these elements generate a subgroup of order 4 isomorphic to the Klein four-group:

$$\langle a, b \rangle = \langle a \rangle \langle b \rangle = \{ I, a, b, ab \}$$

Let c be any other element of G not in $\langle a, b \rangle$. As in case 2 the complex $\langle a, b \rangle \langle c \rangle$ is a subgroup

$$\langle a, b, c \rangle = \langle a, b \rangle \langle c \rangle = \langle a \rangle \langle b \rangle \langle c \rangle$$

And its order is

$$|\langle a, b \rangle \langle c \rangle| = \frac{|\langle a, b \rangle| \cdot |\langle c \rangle|}{|\langle a, b \rangle \langle c \rangle|} = \frac{4 \times 2}{1} = 8$$

so that $G = \langle a, b, c \rangle$.

It is easy to verify that $G = \{a^i b^j c^k : i, j, k \in \langle \mathbb{Z}_2, + \rangle\}$ and that the mapping $a^i b^j c^k \to (i, j, k)$ is in fact an isomorphism from G onto the group $\mathbb{Z}_2 \times \mathbb{Z}_2 \times \mathbb{Z}_2$ of Example 5.1.3.

Part 2. G is a nonabelian group of order 8. In this case G is not isomorphic to a cyclic group of order 8 and so there can be no element of order 8 in G. Neither can all the elements not equal to the identity of G have order 2. Thus we can assume that G has an element of order 4. Call it a and note that the subgroup $\langle a \rangle$ has index 2 in G and so is normal in G. We complete our discussion in two more cases, depending on whether there is an element b of order 2 not in $\langle a \rangle$ or whether every element of order 2 belongs to $\langle a \rangle$ and hence is equal to a^2.

Case 5. G is a nonabelian group of order 8. There is an element a of order 4 and an element $b \notin \langle a \rangle$ whose order is 2. Again the normality of $\langle a \rangle$ guarantees that the complex $\langle a \rangle \langle b \rangle$ is a subgroup of G, and a count of its elements shows

$$|\langle a, b \rangle| = |\langle a \rangle \langle b \rangle| = \frac{|\langle a \rangle| \cdot |\langle b \rangle|}{|\langle a \rangle \cap \langle b \rangle|} = 8$$

so that $G = \langle a, b \rangle$.

Now since $\langle a \rangle$ is normal in G, $bab^{-1} \in \langle a \rangle$, say, $bab^{-1} = a^k$. Pre- and postmultiply this equation by b and b^{-1}, respectively, to obtain

$$b^2 a b^{-2} = b a^k b^{-1} = (bab^{-1})(bab^{-1}) \cdots (bab^{-1}) = (a^k)^k = a^{(k^2)}$$

Since $b^2 = I$ we may conclude from the extreme ends of this chain of equalities that

$$a = a^{(k^2)}$$

and thus that $k^2 \equiv 1 \pmod 4$. A check of the possible values for k shows that

$$k \equiv 1 \pmod 4 \quad \text{or that} \quad k \equiv -1 \pmod 4$$

If $k \equiv 1$, that means that $bab^{-1} = a$ and then $ba = ab$. But since G is generated by these two commuting elements it means that G is abelian. Hence, for case 5 it must be that $k \equiv -1 \pmod 4$ and so

$$bab^{-1} = a^{-1} \quad \text{and so} \quad ba = a^{-1}b$$

Now we can see that these are the conditions given in Definition 3.3.2 and so the group is the dihedral group D_8.

Case 6. G is a nonabelian group of order 8 containing an element a of order 4 and such that every element of order 2 is in $\langle a \rangle$ and hence equals a^2. Now let $b \notin \langle a \rangle$. Thus b has order 4 and b^2, having order 2, is equal to a^2. Many of the details are like the arguments of case 5. As in case 5, $\langle a \rangle$ is a normal subgroup of G and hence the complex $\langle a \rangle \langle b \rangle$ is a subgroup and $|\langle a \rangle \langle b \rangle| = 8$. The normality of $\langle a \rangle$ implies $bab^{-1} \in \langle a \rangle$, say, $bab^{-1} = a^k$. This in turn implies

$$b^2 a b^{-2} = a^{(k^2)}$$

While $b^2 \neq I$, it is true that $b^2 = a^2$ and so commutes with a; hence we can conclude that

$$a = a^{(k^2)}$$

and so that $k^2 \equiv 1 \pmod 4$. We conclude $k \equiv -1 \pmod 4$. But now we have verified all the conditions of Example 4.3.2 for G to be isomorphic to Q, the quaternion group.

This completes the discussion of all possible cases, and by collecting the results we obtain Theorem 5.1.4.

5.1.5 Example. The Groups of Order 9

There are just two nonisomorphic groups of order 9. One is the cyclic group C_9. The other is made from two groups of order 3 using a simple process that is discussed in detail in the next section. But for now consider the cartesian product

$$\mathbb{Z}_3 \times \mathbb{Z}_3 = \{(a, b) : a \in \langle \mathbb{Z}_3, + \rangle \text{ and } b \in \langle \mathbb{Z}_3, + \rangle\}$$

Since $|\mathbb{Z}_3| = 3$ there are nine elements in the cartesian product.

The cartesian product $\mathbb{Z}_3 \times \mathbb{Z}_3$ is made into a group by defining an operation $(+)$ as follows:

$$(a, b) + (c, d) = (a + c, b + d)$$

where the additions on the right-hand side of the equality take place in $\langle \mathbb{Z}_3, + \rangle$. Thus, for example,

$$([1], [2]) + ([2], [2]) = ([0], [1])$$

Upon a little reflection you can see that the identity is the element $([0], [0])$ and that the inverse of $([a], [b])$ is $([-a], [-b])$. Associativity and commutativity of the operation $(+)$ on $\mathbb{Z}_3 \times \mathbb{Z}_3$ are derived from those properties of addition in $\langle \mathbb{Z}_3, + \rangle$. Later we shall prove that any group of order 9 is isomorphic either to C_9 or to $\langle \mathbb{Z}_3 \times \mathbb{Z}_3, + \rangle$. In particular note that a group of order $9 = 3^2$ is abelian. These results extend: If p is a prime, a group of order p^2 is isomorphic to either the cyclic group C_{p^2} or to $\langle \mathbb{Z}_p \times \mathbb{Z}_p, + \rangle$ where $(+)$ is defined componentwise as above.

EXERCISE SET 5.1

5.1.1 Determine, up to isomorphism, all groups of order 26.

5.1.2 Determine all the abelian groups of order 12. This will be much easier after Section 5.3, but it is instructive to do it with the tools at hand. Determine all abelian groups of order 21.

5.1.3 Determine all abelian groups of order 24. (You know there is an element of order 3; hence a subgroup of order 3. Consider the factor group of order 8.)

5.1.4 (This problem generalizes Wilson's theorem, Problem 3.4.19. Let G be an abelian group whose elements are

$$I = a_1, \ldots, a_n$$

Let $u = a_1 a_2 \cdots a_n$ be the product of all the elements in G. If n is odd, prove that $u = I$. If $n = 2m$ where m is odd, determine u.
* If $n = 2^k m$ where m is odd, determine u.

5.1.5 List the elements of the group $\mathbb{Z}_{16}^\times$. Find an element of order 4. Call this element a. Find an element of order 2 not in $\langle a \rangle$. Call it b. Determine the group $\mathbb{Z}_{16}^\times / \langle b \rangle$. What is the group $\mathbb{Z}_{16}^\times / \langle a^2 \rangle$? To which group of order 8 is $\mathbb{Z}_{16}^\times$ isomorphic?

5.1.6 The Pauli spin matrices are defined to be

$$P_1 = \begin{pmatrix} 0 & 1 \\ 1 & 0 \end{pmatrix} \qquad P_2 = \begin{pmatrix} 0 & -i \\ i & 0 \end{pmatrix} \qquad P_3 = \begin{pmatrix} 1 & 0 \\ 0 & -1 \end{pmatrix}$$

where the entries are complex numbers. They generate a group of order 16 under matrix multiplication called the Pauli spin group $\mathbb{P}$. Here is an argument: Let $A = P_3 P_1$, $B = P_2 P_3$. Determine the group $\langle A, B \rangle$. Show that

$$P_1 B = iI = \begin{pmatrix} i & 0 \\ 0 & i \end{pmatrix}$$

(We write $cI = \begin{pmatrix} c & 0 \\ 0 & c \end{pmatrix}$ if $c \in \mathbb{C}$ and cM for $(cI)M$ if M is a matrix.) Show that the 16 matrices

$$\{ \pm I, \pm A, \pm B, \pm AB, \pm iI, \pm iA, \pm iB, \pm iAB \}$$

constitute the group $\mathbb{P}$. Find the center of $\mathbb{P}$. Find the group $\mathbb{P} / \langle -I \rangle$. Show that each of the four types of noncyclic groups of order 8 appear either as a subgroup or as a factor group of $\mathbb{P}$.

5.1.7 Draw the subgroup lattice of Pauli spin group $\mathbb{P}$. (There are seven elements of order 2, seven subgroups of order 4, and seven subgroups of order 8.)

5.1.8 The object of this problem is to determine the automorphism group of the dihedral group D_8. To fix notation let the generators of D_8 be a and b where

$$a^4 = 1 \qquad b^2 = 1 \quad \text{and} \quad ba = a^{-1}b$$

If α is an automorphism what is the order of $\alpha(a)$? What are the possibilities for $\alpha(a)$? What are the possibilities of $\alpha(b)$? Show that if r has order 4 and s is not in $\langle a \rangle$ then the map defined by $a^k b^h \to r^k s^h$ is an automorphism of D_8. Determine all automorphisms of D_8. Determine the inner automorphisms of D_8. Identify the group of all automorphisms of D_8 as one of the groups of order 8.

5.2 DIRECT PRODUCTS OF GROUPS

In this section we give an easy construction for building new groups from known ones. This technique is used later to determine the structure of abelian groups. In particular we shall see that every finite abelian group can be constructed from cyclic groups in this way. The method is called the direct product. And it is not complicated!

5.2.1 Definition. Direct Product of Two Groups

The direct product of two groups A and B is the cartesian product $A \times B$,

$$A \times B = \{(a, b): a \in A \text{ and } b \in B\}$$

where the group operation is performed component by component:

$$(a, b)(a_1, b_1) = (aa_1, bb_1)$$

It must of course be proved that a bona fide group operation has been defined. Note that $A = B$ is permissible!

5.2.2 Theorem.   Properties of the Direct Product

Let A and B be groups.

  1. The direct product $A \times B$ is a group under the componentwise operation given in Definition 5.2.1.
  2. The identity is (I, I).
  3. The inverse of (a, b) is (a^{-1}, b^{-1}).
  4. The order of $A \times B$ is $|A||B|$.
  5. The subsets $\hat{A} = \{(a, I): a \in A\}$ and $\hat{B} = \{(I, b): b \in B\}$ are normal subgroups of $A \times B$; $\hat{A}$ is isomorphic to A and $\hat{B}$ is isomorphic to B.

$$\hat{A} \vee \hat{B} = A \times B \quad \text{and} \quad \hat{A} \cap \hat{B} = \text{the identity}$$

  6. The factor group $(A \times B)/\hat{A}$ is isomorphic to B, and $(A \times B)/\hat{B}$ is isomorphic to A.

Proof.   The proofs are easy because the operation is done on each component, where of course the group axioms hold. It is routine to verify that (I, I) is the identity and that (a^{-1}, b^{-1}) is the inverse of (a, b).

The proof that $\hat{A}$ is a subgroup is also easy, given the observation that the second component is never changed from $I \in B$. That $\hat{A}$ is normal in $A \times B$ follows from another simple verification. Let a typical element in $A \times B$ be (a, b) and a typical element in $\hat{A}$ be $(a', 1)$. Then

$$(a, b)(a', I)(a, b)^{-1} = (a, b)(a', I)(a^{-1}, b^{-1}) = (aa'a^{-1}, bIb^{-1})$$

$$= (aa'a^{-1}, I) \in \hat{A}$$

The groups $\hat{A}$ and A are isomorphic through the natural mapping

$$(1, a) \leftrightarrow a$$

The join $\hat{A} \vee \hat{B} = A \times B$ since

$$(a, b) = (a, I)(I, b)$$

Similar proofs hold for $\hat{B}$.

Finally the easiest way to see that $(A \times B)/\hat{A}$ is isomorphic to B is to apply the diamond isomorphism, Theorem 4.3.5, which gives us this chain of equalities and isomorphisms:

$$(A \times B)/\hat{A} = (\hat{A} \vee \hat{B})/\hat{A} \cong \hat{B}/(\hat{A} \cap \hat{B}) = \hat{B} \cong B$$

Alternatively it is easy to argue from first principles. It is easy to show that the map from $A \times B$ onto B defined by

$$(a, b) \rightarrow b$$

is a group homomorphism of $A \times B$ onto B with kernel $\hat{A}$. The map is certainly onto B and the only elements of $A \times B$ mapped onto I are those pairs (a, I), so indeed the kernel will be $\hat{A}$. Thus the only verification we need is that the homomorphism condition is satisfied. We omit the details.

5.2.3 Examples of Direct Products

1. The four-group is the direct product of two cyclic groups each of order 2. Here we let $A = B = \{I, a\}$ and then

$$A \times B = \{(I, I), (I, a), (a, I), (a, a)\}$$

The multiplication gives, for example, that $(a, I)(I, a) = (a, a)$.

2. If $\gcd(n, m) = 1$ then the cyclic group of order nm is isomorphic to the direct product of a cyclic group of order n and a cyclic group of order m.

   To prove this let $C_n = \langle a \rangle$ be a cyclic group of order n and let $C_m = \langle b \rangle$ be a cyclic group of order m. All we need to show is that $C_n \times C_m$ is cyclic of order nm. To do this we need to find a generator. One comes to mind immediately; put the generators together in an ordered pair! We shall prove that

   $$\text{if } x = (a, b) \text{ then } x \text{ has order } nm.$$

   The calculational rules give

   $$(a, b)^k = (a^k, b^k) = (I, I)$$

   if and only if $a^k = I$ in C_n and $b^k = I$ in C_m. But this happens if and only if n divides k and m divides k. Since $\gcd(n, m) = 1$, this happens if and only if nm divides k. All this shows that the order of (a, b) is $nm = |C_n \times C_m|$.

3. The elementary abelian group of order 8. Form the direct product of a four-group and C_2, a cyclic group of order 2. Since a four group is

$C_2 \times C_2$ we are considering

$$(C_2 \times C_2) \times C_2 = C_2 \times C_2 \times C_2$$
$$= \{(x, y, z): x \in C_2, y \in C_2, z \in C_2\}$$

where of course the operations are done componentwise. This gives a group of order 8, and the multiplication shows that it is abelian and that every element except the identity (I, I, I) has order 2. The details are in Example 5.1.3.

We now show how we can sometimes factor a group into a direct product.

5.2.4 Theorem.   A Criterion for a Group to Be a Direct Product

If G has two subgroups A and B such that

  1. A and B are normal.
  2. $A \cap B = I.$
  3. $A \vee B = G.$

then G is isomorphic to the direct product $A \times B$.

Proof.   First note that the normality of one of the subgroups, say A, guarantees (recall Lemma 4.3.3) that

$$A \vee B = AB$$

Second, the normality of both A and B, and the condition $A \cap B = I$ guarantees that

$$ab = ba \qquad \text{for } a \in A \text{ and } b \in B$$

To see this we show that

$$aba^{-1}b^{-1} = I$$

This is so because, using the normality of A,

$$a(ba^{-1}b^{-1}) = aa^* \in A$$

and using the normality of B,

$$(aba^{-1})b^{-1} = b^*b^{-1} \in B$$

Thus

$$aba^{-1}b^{-1} \in A \cap B = \{I\}$$

and so

$$aba^{-1}b^{-1} = I$$

Now with these facts it is straightforward to verify that the mapping from $A \times B$ into G defined by

$$(a, b) \rightarrow ab$$

is an isomorphism of $A \times B$ onto G. We omit all the details. Briefly, the ontoness is a result of $AB = G$, the homomorphism condition is a result of the rule $ab = ba$ for $a \in A$ and $b \in B$, and the one-to-one property is the result of $A \cap B = I$.

Here is a useful consequence of this theorem.

5.2.5 The Groups of Order 9

First there is the cyclic group C_9. We now show that any noncyclic group of order 9 is isomorphic to the direct product $C_3 \times C_3$.

If G is a group of order 9 and is not cyclic, then every element except I has order 3. Let $A = \langle a \rangle$ and $B = \langle b \rangle$ where b is not in A, and hence where $A \cap B = I$. Thus the intersection of two distinct subgroups of order 3 is the identity.

The usual counting argument shows that the complex AB has nine elements and so

$$G = AB = \left\{ a^i b^j : 0 \le i, j < 3 \right\}$$

We claim that A and B are normal. Once we show this, Theorem 5.2.4 shows that G is isomorphic to the direct product as alleged.

First we show that A is normal. To do this it suffices, by Lemma 4.3.6, to show that $\langle a \rangle$ contains all the elements

$$\left(a^i b^j \right)^{-1} a \left(a^i b^j \right) = b^{-j} a b^j = \left(b^{-1} a b \right)^j$$

Thus it suffices to show that

$$b^{-1} a b \in \langle a \rangle$$

We know that $b^{-1} a b = a^i b^j$ for some i and j.

The alternative $i = 0$ is impossible, for then $b^{-1} a b = b^j$ and this implies that $a \in \langle b \rangle$.

The alternative $j = 0$ is what we desire to show.

The alternative $b^{-1} a b = ab$, or $a^{-1} b$ lead, after a cancellation, to $b^{-1} \in A$, a contradiction.

The alternative $b^{-1} a b = ab^{-1}$ implies that

$$ab^{-1} ab^{-1} = ab^{-1} ab^2 = a \left(b^{-1} ab \right) (b) = a \left(ab^{-1} \right) b = a^2$$

and so $\langle (ab^{-1}) \rangle \cap \langle a \rangle \ne I$, hence $\langle ab^{-1} \rangle = \langle a \rangle$, a contradiction. Finally the alternative $b^{-1} a b = a^{-1} b^{-1}$ implies that (ab^{-1}) has order 2, a contradiction, since

$$ab^{-1} ab^{-1} = ab^{-1} ab^2 = ab^{-1} abb = aa^{-1} b^{-1} b = I$$

Hence we conclude that $\langle a \rangle$ is normal in G. By the symmetry between a and b it follows that $\langle b \rangle$ is normal in G. Finally apply Theorem 5.2.4 to complete the proof.

5.2.6 Application.   A Direct Product Decomposition of $\mathbb{Z}_{nm}^{\times}$

Let m and n be relatively prime integers. The group $\mathbb{Z}_{nm}^{\times}$ is isomorphic to the direct product

$$\mathbb{Z}_n^{\times} \times \mathbb{Z}_m^{\times}$$

The Euler φ-function is multiplicative:

$$\text{If } \gcd(n, m) = 1 \quad \text{then } \varphi(nm) = \varphi(n)\varphi(m)$$

Proof.   Let A be the subgroup of $\mathbb{Z}_{nm}^{\times}$ defined by

$$A = \{[x] \in \mathbb{Z}_{nm}^{\times}: x \equiv 1 \,(\text{mod } n)\}$$

and let

$$B = \{[x] \in \mathbb{Z}_{nm}^{\times}: x \equiv 1 \,(\text{mod } m)\}$$

We omit the details that these definitions do not depend upon the coset representative x chosen.

The fact that A and B are subgroups of $\mathbb{Z}_{nm}^{\times}$ is routine to verify. They are normal subgroups because $\mathbb{Z}_{nm}^{\times}$ is abelian. We have $A \cap B = I$ because

$$x \equiv 1 \,(\text{mod } n) \quad \text{and} \quad x \equiv 1 \,(\text{mod } m)$$

means that mn divides $(x - 1)$, hence that $x \equiv 1 \,(\text{mod } mn)$.

It is a little trickier to show that

$$\mathbb{Z}_{nm}^{\times} = A \vee B = AB$$

Because $\gcd(n, m) = 1$ we can write

$$x - 1 = nu + mv \quad \text{so that } x = 1 + nu + mv$$

Let $a = 1 + nu$ and let $b = 1 + mv$. Then a computation shows that

$$ab \equiv 1 + nu + mv = x \,(\text{mod } nm)$$

Thus every element in $\mathbb{Z}_{nm}^{\times}$ appears in the form ab with $a \in A$ and $b \in B$.

From Theorem 5.2.4 we conclude that $\mathbb{Z}_{nm}^{\times}$ is isomorphic to $A \times B$. What remains is to show, for example, that A is isomorphic to $\mathbb{Z}_m^{\times}$. Note the reversal of the roles of n and m in the definition for A and in the selection of the group to which it is isomorphic.

The isomorphism is, as we shall verify, given by

$$[x]_{nm} \rightarrow [\tilde{x}]_m \quad \text{where } x \equiv \tilde{x} \,(\text{mod } m)$$

The homomorphism condition is routine to verify. The mapping is onto because, by the Chinese remainder theorem (Theorem 2.1.8, v), we can solve the pair of congruences

$$x \equiv 1 \,(\text{mod } n) \quad \text{and} \quad x \equiv \tilde{x} \,(\text{mod } m)$$

for any $\tilde{x}$. The kernel of the mapping is just [1] since the pair of congruences $x \equiv 1 \,(\text{mod } n)$ and $x \equiv 1 \,(\text{mod } m)$ have, modulo nm, the unique solution $x = 1$.

Now the assertion about the multiplicative nature of the Euler φ-function follows since the order of $\mathbb{Z}_{nm}^{\times}$ is just the product of the orders of $\mathbb{Z}_n^{\times}$ and $\mathbb{Z}_m^{\times}$.

There is an extension of the definition of the direct product of two groups to the direct product of finitely many groups.

5.2.7 Definition.   Direct Product of Several Groups

If $G_1, \ldots, G_n$ are groups their direct product is the cartesian product

$$G_1 \times G_2 \times \cdots \times G_n = \{(g_1, g_2, \ldots, g_n) : g_i \in G_i, 1 \leq i \leq n\}$$

where the group operation is done component by component. Theorem 5.2.4 then extends as follows:

5.2.8 Theorem.   A Criterion for a Group to Be the Direct Product of Several Subgroups

If a group G has subgroups $N_1, \ldots, N_k$ such that

1. Each subgroup N_i is a normal subgroup of G.

2. For each i, $N_i \cap \left(\bigvee_{h \neq 1} N_h \right) = I$.

3. $\bigvee_{1 \leq h \leq k} N_h = G$.

then G is isomorphic to the direct product

$$N_1 \times \cdots \times N_k$$

Sketch of the Proof.   The conditions ensure that for each element $g \in G$, there is a unique way of writing g as a product of elements in $N_1, \ldots, N_n$,

$$g = n_1 \cdots n_k \qquad \text{where } n_h \in N_h$$

We ask for the details of the proof in the exercises.

Note that condition 2 implies that

$$N_i \cap N_j = I$$

if $i \neq j$ and so

$$n_i n_j = n_j n_i$$

if $n_i \in N_i$ and $n_j \in N_j$, as in the proof of Theorem 5.2.4. Then it is easy to prove that

$$n_1 \cdots n_k \rightleftarrows (n_1, \ldots, n_k) \tag{1}$$

is the desired isomorphism.

Remark.   When $n > 2$, condition 2 is much stronger than the condition $N_i \cap N_j = I$. For example, in the four-group the three subgroups of order 2 satisfy the pairwise disjointness condition, but not $N_1 \cap (N_2 \vee N_3) = I$.

The full strength of condition 2 is used in proving that the mapping (1) is one-to-one.

In view of Theorems 5.2.6 and 5.2.8 we can completely determine the structure of $\mathbb{Z}_n^\times$ provided we know $\mathbb{Z}_{p^k}^\times$ for the various prime power factors of n. Thus, by repeated applications of 5.2.6, we obtain

$$\mathbb{Z}_n^\times \cong \mathbb{Z}_{p_1^{k_1}}^\times \times \cdots \times \mathbb{Z}_{p_t^{k_t}}^\times$$

EXERCISE SET 5.2

5.2.1 Show that the direct product of abelian groups is abelian. When is the direct product of two cyclic groups cyclic?

5.2.2 Show that the multiplicative group of nonzero real numbers is isomorphic to the direct product of the multiplicative group of positive real numbers and a cyclic group of order 2. Show that the multiplicative nonzero complex numbers are the direct product of the positive real numbers and an appropriately chosen subgroup. Find the subgroup.

5.2.3 In this exercise let V denote a four-group and let C_k be a cyclic group of order k. Which of the following are isomorphic?

$$C_4 \times C_3 \qquad C_6 \times C_2 \qquad V \times C_3 \qquad D_6 \times C_2$$

5.2.4 Find all the automorphisms of $C_4 \times C_2$.

5.2.5 For what integers n is the dihedral group D_{2n} a direct product of two proper subgroups?

5.2.6 Suppose that a group is isomorphic to the direct product of two of its subgroups A and B. Thus

$$G = A \vee B \text{ is isomorphic to } A \times B$$

Show that if N is a normal subgroup of A then

$$G/N \text{ is isomorphic to } (A/N) \times B$$

5.2.7 Let G have normal subgroups A, B such that $A \cap B = I$. Show that G is isomorphic to a subgroup of $(G/A) \times (G/B)$.

5.2.8 Let G have normal subgroups A, B such that $A \vee B = G$ and such that $ab = ba$ for all $a \in A$ and $b \in B$. Show that G is a homomorphic image of $A \times B$.

5.2.9 Find a nonabelian subgroup of order 12 in $S_3 \times C_4$ which is isomorphic neither to D_{12} nor to $\mathscr{A}_4$. *Hint:* Try to find one with a cyclic subgroup of order 4.

5.2.10 Complete the details of the proof of Theorem 5.2.8.

5.2.11 Let G be a group with normal subgroups $N_1, \ldots, N_k$ such that

$$\bigvee_{1 \le h \le k} N_h = G$$

and for each i, $2 \le i \le h$

$$N_i \cap \left(\bigvee_{1 \le h \le i} N_h \right) = I$$

Show that G is isomorphic to the direct product

$$N_1 \times \cdots \times N_k$$

5.3 FINITE ABELIAN GROUPS

In many intriguing ways the structure of finite abelian groups extends the arithmetic structure of the positive integers. The key theorem that can be extended is the unique factorization theorem. Just as every positive integer can be expressed as a product of prime powers in essentially one and only one way, every finite abelian group can be expressed as the direct product of groups of prime power order in essentially only one way. This result is a direct consequence of the powerful Sylow theorem, which is proved in Chapter 11. But there is a more direct argument that is sketched in the exercises at the end of this section.

The basic building blocks for abelian groups are cyclic groups. The so-called fundamental theorem of finite abelian groups states that every finite abelian group is the direct product of cyclic groups. In this book we give two proofs of this important result. The first proof is done with a minimal amount of machinery, and the prerequisites for it have already been presented. However, the proof is lengthy; there is one tedious technical lemma. For this reason it is postponed to Chapter 11. The proof tells nothing of the possible uniqueness of the decomposition. The inherent difficulty is demonstrated by the cyclic group of order 6. It is itself cyclic, but it is also the direct product of a group of order 2 and a group of order 3. So much for uniqueness of an elementary sort. There is a very strong uniqueness result, and it applies not only to finite abelian groups but also to "finitely generated" abelian groups. (A group is said to be finitely generated if there exists a finite number of elements $a_1, \ldots, a_n$ such that the group is the join $\langle a_1 \rangle \vee \cdots \vee \langle a_n \rangle$. Thus, for example, the infinite cyclic group $\langle \mathbb{Z}, + \rangle$ is finitely generated by 1. However, $\langle \mathbb{Q}, + \rangle$ is not finitely generated.) The proof of this theorem comes as an application of the theory of modules developed in Chapter 16. But for completeness, and a preview of things to come, here is a statement of the most definitive result on finite abelian groups.

5.3.1 Theorem.   Structure of Finite Abelian Groups

A finite abelian group G is the direct product of cyclic groups $T_1, \ldots, T_u$ such that $1 < |T_i|$ and $|T_i|$ divides $|T_{i+1}|$ for $i = 1, \ldots, u - 1$.

This representation is unique in the following sense: If G has a second representation as a direct product of nonidentity cyclic groups $S_1, \ldots, S_v$ and if $|S_i|$ divides $|S_{i+1}|$ for $i = 1, \ldots, v - 1$ then $u = v$, and for each i, $T_i \cong S_i$.

Now a cyclic group is determined by its order alone. Thus to determine all the finite groups of a given order, say n, the theorem tells us to determine all the finite sequences of positive integers $(t_1, \ldots, t_u)$ such that $2 \leq t_1, t_i | t_{i+1}$, and $n = t_1, \ldots, t_u$.

As an example let us use this result to determine the abelian groups of order $20 = 4 \times 5$. We must determine all the ways it is possible to have a sequence of cyclic groups whose orders satisfy the divisibility conditions of Theorem 5.3.1. The product of these orders will be the order of the group, in this case 20. A little experimentation shows that there are just two sequences that meet the divisibility conditions of Theorem 5.3.1:

$$(20) \quad \text{and} \quad (2, 10)$$

Because of the divisibility condition, the prime factor 5 must appear in the last term of the sequence. Then in the last term of the sequence either 4 appears with 5 or 2 appears with 5. In the first case there is just one term to the sequence: it is 20. In the second case the first term of the sequence is 2. The first case gives rise to C_{20}, the cyclic group of order 20, and the second to the abelian group $C_2 \times C_{10}$.

Now we are going to prove some useful but more elementary results having to do with commutativity.

5.3.2 Lemma.   Expressing an Element as the Product of Commuting Elements

Let g be an element of finite order n in a group G.

1. If $n = ab$ where $\gcd(a, b) = 1$ then there exist elements u and v in G such that
$$g = uv = vu$$
   and u has order a and v has order b.
2. If $n = p_1^{e_1} \cdots p_k^{e_k}$ is the prime factorization of n, then g is the product of k permuting elements h_i of order $p_i^{e_i}$ where each h_i is a power of g.

Proof.   The safest place to look for commuting elements is among the powers of g. We prove 1 first. Since $\gcd(a, b) = 1$ there are integers s and t such that
$$1 = sa + tb$$
and hence
$$g = g^1 = g^{sa+tb} = g^{sa}g^{tb}$$
Let $u = g^{sa}$ and $v = g^{tb}$. The elements u and v commute because they are powers of g. The order of g^{sa} is b since $\gcd(s, b) = 1$ and similarly the order of g^{tb} is a.

Now we can prove 2 by k applications of 1. Simply split off each prime power with each application: By 1, $g = h_1 g_1$, where h_1 is a power of g and has order $p_1^{e_1}$ and g_1 is a power of g and has order $p_2^{e_2} \cdots p_k^{e_k}$. Continue, replacing g by g_1, or use an obvious induction.

The next lemma is used to show that in a finite abelian group there is an element whose order is the least common multiple of all the orders of all the elements in the group.

5.3.3 Lemma.   An Element of Order the lcm of Two Orders

If a group G contains commuting elements x and y each of finite order, then G has an element whose order is the least common multiple of the orders of x and y.

Proof.   First suppose the orders of x and y are relatively prime. Exercise 3.2.4 shows that the order of xy is the product of the orders of x and y.

We can reduce the lemma to this case using Lemma 5.3.2. Using this lemma we may construct powers of x that have order p^e for every prime power dividing the order of x. Similarly construct powers of y that have prime power order for every prime power dividing the order of y. Since x and y commute, so do their powers. Now the least common multiple of the order of x and the order of y is the product of the prime powers p^v where p divides either the order of x or the order of y, and if p divides both of these numbers then v is the maximum of the exponents. So now for each prime dividing the order of x or y we have determined an element whose order is the maximal prime power dividing either x or y. These elements commute, and because their orders are relatively prime, their product is an element whose order is the product of the orders. But this product is, by construction, just the least common multiple of the orders of x and y.

Now we can extend this lemma once more in the case of a finite abelian group.

5.3.4 Theorem.   Exponent of an Abelian Group

If G is a finite abelian group then there exists an element $x \in G$ whose order is the least common multiple m of the orders of the elements in G. In addition $g^m = 1$ for all elements in G.

Proof.   Simply apply the previous lemma to a sequence listing the elements of G. Remember

$$\mathrm{lcm}(a, b, c) = \mathrm{lcm}(\mathrm{lcm}(a, b), c)$$

More generally the following definition is useful.

5.3.5 Definition.   Exponent of a Group

The exponent of a group is the minimal positive integer m, if it exists, such that $x^m = 1$ for all elements x in the group. If no such m exists, the group is said to have infinite exponent.

In a finite group an exponent always exists because of the theorem of Lagrange. At worst the exponent is the order of the group. (The worst often occurs in nonabelian groups!) For finite abelian groups we have just proved that the exponent is the least common multiple of the orders of the elements in the group. Notice in the definition there is no requirement that there be an element whose order is the exponent of the group. For example, it is easy to see that the exponent of the dihedral group of order 6 is 6, but there is no element of order 6. On the other hand, we have just proved that in a finite abelian group there is an element whose order is the exponent of the group.

A famous question raised by Burnside asks whether, if G has a finite exponent and can be generated by a finite number of elements, it must follow that G be finite. It is known that if the exponent is 2, 3, 4, or 6 then the group must be finite. Russian mathematicians Novikov and Adian showed in 1968 that if the exponent is an odd number sufficiently large (≥ 4381) then such a group may be infinite. These interesting questions are described in an accessible article by Narain Gupta (On Groups in Which Every Element Has Finite Order, *American Mathematical Monthly*, vol. 96, pp. 297–308, 1989).

We end this section with a very useful criterion to guarantee that a group be cyclic.

5.3.6 Theorem. Criterion for a Group to Be Cyclic

If G is a finite group and if for each positive integer n there are at most n elements $x \in G$ such that $x^n = 1$, then G is cyclic.

Proof. The theorem of Lagrange implies that if n is the order of an element $a \in G$ then each of the n elements of the cyclic subgroup $\langle a \rangle$ are solutions of $x^n = 1$ in G. Hence, by the hypothesis of this theorem, $\langle a \rangle$ contains all the solutions of $x^n = 1$. In particular, there can be no other cyclic subgroup of order n in G. As we show next, this in turn implies that every cyclic subgroup of G is normal in G. [Moreover, although this is irrelevant to the proof, since any subgroup is the join of its cyclic subgroups, it then follows from the lattice properties of normal subgroups (Theorem 4.3.4) that every subgroup of G is normal in G.]

To prove that the cyclic subgroups of G are normal in G, we use the normality criterion of Lemma 4.3.6. Let $A = \langle a \rangle$ by a cyclic subgroup of G of order n. Let $g \in G$. The element gag^{-1} has order n; hence $gag^{-1} \in A$. Hence A is normal in G.

Now let $b \in G$ be chosen so that its order is maximal among all elements of G. We now show that $G = \langle b \rangle$. Let $c \in G$ have order r. If r divides m then $\langle b \rangle$ contains an element of order r and hence all the possible solutions of $x^r = 1$ in G. Thus $c \in \langle b \rangle$. We now show that r must divide m. Suppose, to the contrary, that r does not divide m. This means that for some prime power p^k in the prime power factorization of r, p^k does not divide m. Let $m = p^h s$ and $r = p^k t$ where p divides neither s nor t and where $0 \le h < k$. Now $\langle b \rangle$ contains an element u of order s and $\langle c \rangle$ contains an element v of order p^k. Because $p \nmid s$ it follows that $\langle u \rangle \cap \langle v \rangle = I$, and because these cyclic subgroups are normal in G it follows (Theorem 5.2.4) that their join is isomorphic to their direct product: $\langle u \rangle \vee \langle v \rangle = \langle u \rangle \times \langle v \rangle$. For this reason u and v commute and, since their orders are relatively prime, their product uv has order sp^k which is strictly greater than the order of b, contrary to the choice of b. Hence it must follow that $c \in \langle b \rangle$ as was to be shown.

EXERCISE SET 5.3

5.3.1 Without recourse to Theorem 5.3.1 determine all abelian groups of order p^2.

5.3.2 Let A be a finite abelian group. For each prime p dividing the order of A define

$$S(p) = \{ x \in A : \text{the order of } x \text{ is a power of } p \}$$

Show that $S(p)$ is a subgroup of A. Show that A is the direct product of the subgroups $S(p)$ for all the primes p dividing the order of A.

5.3.3 Use Theorem 5.3.1 to find all abelian groups whose order is at most 31.

5.3.4 Draw the lattice of subgroups of $C_4 \times C_4$.

5.3.5 Show that if p is a prime there are as many subgroups of index p in $C_{p^2} \times C_{p^2}$ as there are subgroups of order p.

5.3.6 Give a simple proof for Theorem 5.3.6 in the case that the order of G is a power of a prime p.

5.3.7 Show that any finitely generated group of exponent 2 is finite. (Exercise 3.1.6 shows that a group of exponent 2 is abelian.)

5.3.8 Show that if a group, not necessarily abelian, can be generated by a finite number of elements and has exponent 4 then G is finite.

5.3.9 Show that $\langle \mathbf{Q}, + \rangle$ is not the direct product of two proper subgroups.

5.3.10 Show that $\langle \mathbf{Q}, + \rangle / \langle \mathbf{Z}, + \rangle$ is isomorphic to the group of complex numbers that are nth roots of unity for $n = 1, 2, 3, \dots$. Find a direct factor of this group.

Chapter 6

Permutation Groups

Permutations provide a concrete model for any group. Historically, groups were first studied in this representation almost exclusively. In this book we have given a more abstract presentation because groups arise in contexts other than permutations. The abstract point of view prepares us to recognize groups when they appear in new and unfamiliar circumstances. But concrete representations are useful, and indeed many deep results for groups are best achieved through these representations. We introduce these representations in Section 6.3.

6.1 TRANSPOSITIONS AND THE ALTERNATING GROUP

We begin our study of permutations with some basic definitions. Then we study a special kind of permutation, a transposition. We prove that any permutation can be represented as a product of these. Those which are a product of an even number of transpositions constitute an important group, the alternating group.

6.1.1 Definition. Permutation

Let S be a set. A permutation π of S is a one-to-one mapping from S onto S:

$$\pi: x \rightarrow (x)\pi$$

An element $x \in S$ is said to be *fixed* by π if $(x)\pi = x$; otherwise x is said to be *moved* by π. The identity permutation is the permutation that fixes all elements of S.

Remark.   Notice that no permutation moves exactly one element and if π moves x, say $(x)\pi = y$ then $(y)\pi \neq y$ and so π moves y as well.

Notice too that the notation here places the function name on the right of the element, not on the left as is customary for functional notation. The image of x under π is $(x)\pi$. This switch is traditional (but not universal, so beware if you are consulting other sources) in discussing permutations. This notation may have originated from the way permutations were written and in conjunction with the traditional western way of proceeding from left to right.

Traditionally permutations were recorded by writing the elements of S as a row of symbols. Then a second row gave the images under the permutation of the first row. For example, if $S = \{1, 2, 3,\}$ and π is the permutation mapping 1 into 3, 2 into 1, and 3 into 2, π would be recorded:

$$\pi = \begin{pmatrix} 1 & 2 & 3 \\ 3 & 1 & 2 \end{pmatrix}$$

Now suppose ρ is another permutation on this set, say

$$\rho = \begin{pmatrix} 1 & 2 & 3 \\ 1 & 3 & 2 \end{pmatrix}$$

and we want to express the product (composition of mappings) $\pi\rho$. Well, notation suggests

$$\pi\rho = \begin{pmatrix} 1 & 2 & 3 \\ 3 & 1 & 2 \end{pmatrix}\begin{pmatrix} 1 & 2 & 3 \\ 1 & 3 & 2 \end{pmatrix}$$

To compute the image of, say, 1, begin on the left with π. Find that π maps 1 into 3. Now find 3 on the top row of ρ and find that ρ maps 3 into 2. Hence $\pi\rho$ maps 1 into 2. Indeed $(1)\pi\rho = ((1)\pi)\rho = (3)\rho = 2$. The arrows below show schematically how the computation takes place.

$$\pi\rho = \left(\begin{bmatrix} 1 \\ 3 \end{bmatrix}\begin{matrix} 2 \\ 1 \end{matrix}\begin{matrix} 3 \\ 2 \end{matrix}\right)\left(\begin{matrix} 1 \\ 1 \end{matrix}\begin{matrix} 2 \\ 3 \end{matrix}\begin{bmatrix} 3 \\ 2 \end{bmatrix}\right)$$

6.1.2 Theorem.   Full Symmetric Group $\mathbb{S}_n$

The set of all permutations of a set S forms a group under composition of mappings. If $|S| = n$ the group is called the full symmetric group on n elements and is denoted by $\mathbb{S}_n$. The order of $\mathbb{S}_n$ is $n!$ (n factorial).

Proof.   We omit the details; but recall example 5 of Examples 3.1.4.

Examples of Full Symmetric Groups   We have already met $\mathbb{S}_3$. It is isomorphic to the dihedral group D_6. The mappings are explicitly listed in example 5 of Examples 3.1.4.

To construct $\mathbb{S}_4$ we may begin with the alternating group $\mathscr{A}_4$. The elements of $\mathscr{A}_4$ are listed as mappings of $\{1, 2, 3, 4\}$ in Definition 4.1.3. It turns out that the alternating group $\mathscr{A}_4$ is a subgroup of index 2 in $\mathbb{S}_4$. To list all of $\mathbb{S}_4$ just add the coset

$$\begin{pmatrix} 1 & 2 & 3 & 4 \\ 2 & 1 & 3 & 4 \end{pmatrix} \mathcal{A}_4$$

to obtain 12 more permutations. We have

$$\mathbb{S}_4 = \mathcal{A}_4 \cup \begin{pmatrix} 1 & 2 & 3 & 4 \\ 2 & 1 & 3 & 4 \end{pmatrix} \mathcal{A}_4$$

Remark. By a *group of permutations* we mean a set of permutations of a set S that form a group under composition. Thus a group of permutations of a set S is a subgroup of the full symmetric group on S.

6.1.3 Lemma. Permutations Fixing a Subset

Let G be a group of permutations of a set S. Let T be a subset of S. The set of permutations in G fixing each element of T forms a subgroup of G.

Proof. Observe that if $(x)\pi = x$ then $(x)\pi^{-1} = x$ and if

$$(x)\pi = x \quad \text{and} \quad (x)\rho = x \quad \text{then} \quad (x)\pi \circ \rho = x$$

now apply the subgroup criterion 3.2.3, ii.

6.1.4 Definition. Transposition

A transposition is a permutation τ that moves exactly two elements.
   Suppose that τ moves a, b, in which case

$$(a)\tau = b \quad \text{and} \quad (b)\tau = a$$

Thus τ has the form

$$\tau = \begin{pmatrix} \cdots & x & \cdots & a & \cdots & y & \cdots & b & \cdots & z & \cdots \\ \cdots & x & \cdots & b & \cdots & y & \cdots & a & \cdots & z & \cdots \end{pmatrix}$$

   For brevity we write

$$\tau = (a, b) \quad \text{or} \quad \tau = (b, a)$$

rather than the fuller function table we have used heretofore. Note that a transposition τ is an element of order 2:

$$\tau^2 = I$$

6.1.5 Example. The Transpositions in $\mathbb{S}_4$

There are six transpositions in $\mathbb{S}_4$. If $S = \{1, 2, 3, 4\}$ the transpositions in $\mathbb{S}_4$ are

$$\begin{array}{ccc} (1, 2) & (2, 3) & (3, 4) \\ (1, 3) & (2, 4) & \\ (1, 4) & & \end{array}$$

In general for $n > 1$, there are $\binom{n}{2}$ transpositions in $\mathbb{S}_n$.

6.1.6 Theorem.   A Permutation Is a Product of Transpositions

Any permutation moving at most a finite number of elements can be written as a product of transpositions, in fact as a product of fewer transpositions than the number of elements moved.

Proof.   We induct on the number of symbols moved, using the strong induction principle. If the permutation π moves exactly two elements, then, by definition π is a transposition. The induction hypothesis may be taken to be the statement: If a permutation moves at most $k - 1$ $(k > 2)$ elements then that permutation may be written as the product of fewer than $(k - 1)$ transpositions.

Now suppose that π moves k elements. Let the elements moved by π be

$$\{ a = x_1, b = (a)\pi, x_3, \ldots, x_k \}$$

Let τ be the transposition (a, b) that interchanges a and b. We claim that the product $\pi\tau$ fixes fewer than k elements. To see this, argue first that $\pi\tau$ fixes all the elements fixed by π since the transposition τ fixes these elements too. In addition $\pi\tau = \pi(a, b)$ also fixes a since

$$a \xrightarrow{\ \pi\ } (a)\pi = b \xrightarrow{\ \tau\ } a$$

or

$$(a)\pi\tau = (b)\tau = (b)(a, b) = a$$

Hence the composition $\pi\tau$ can move at most $\{b, x_3, \ldots, x_k\}$; $(k - 1)$ elements. Thus, by induction

$$\pi(a, b) = \tau_1 \ldots \tau_m \qquad \text{where } m \le (k - 2)$$

and hence

$$\pi = \tau_1 \ldots \tau_m \tau$$

a product of fewer than k transpositions.

6.1.7 Definition.   Even and Odd Permutations

A permutation of S is called *even* if it can be written as the product of an even number of transpositions. A permutation is called *odd* if it can be written as the product of an odd number of transpositions.

It is conceivable that a permutation could be both even and odd. That is not so, as we show in Theorem 6.1.9. The heart of the argument is that the identity permutation that fixes all elements and is thus the product of zero transpositions is not odd. We prove this lemma first.

6.1.8 Lemma.   The Identity Is Not an Odd Permutation

If π is the product of an odd number of transpositions then

$$\pi \neq I$$

Proof.   We proceed by weak induction on the (odd) number n of transpositions used to express π. We show that π moves at least one element.

$$\text{If } n = 1 \qquad \text{then } \pi = (a, b) \neq I$$

Now the induction hypothesis:

$$\text{Suppose the lemma holds for } n = 2k - 1, \quad (n \geq 1).$$

Now the induction step: Suppose

$$\pi = \tau_1 \ldots \tau_n \tau_{n+1} \tau_{n+2} \tag{1}$$

Of course, $n + 2$ is odd since n is odd. We are to show

$$\pi \neq I$$

If two adjacent (consecutive) τ's are equal in (1) then they will cancel and π will be expressed as the product of n transpositions. Then, by the induction hypothesis, $\pi \neq I$.

Suppose then that this is not the case. Let $\tau_{n+2} = (a, b)$. If a is moved by no other transposition except τ_{n+2}, then π moves a and so $\pi \neq I$. So we can assume that a is moved by one or more transpositions before the last.

The strategy at this point is to show that either a is moved by π or that we can express π as the product of n transpositions and thus apply the induction hypothesis. We first show that we can express π as the product of $n + 2$ transpositions in which all those moving a occur at the right end of the product. Second we show that either π moves a or we can effect a cancellation in the transpositions moving a.

To get a form for π with all the transpositions involving a as the rightmost transpositions in the product for π we shift all the transpositions in (1) that move a to the right without changing the total number of transpositions used. We can do this by making repeated use of one of the following two calculations:

    i. $(a, x)(c, d) = (c, d)(a, x)$ if a, x, c, d are all distinct elements.
    ii. $(a, b)(b, c) = (b, c)(a, c)$ if $a, b,$ and c are distinct elements.

Use i and ii as follows: Start at the right with τ_{n+2} and proceed left to the first transposition that does not involve a. If there is none, we are done. If there is, go the next transposition to the left that does involve a. If there is none, we are done. If there is, call it τ_m. Then the combination $\tau_m \tau_{m+1}$ is of type i or type ii. Use it to shift the transposition τ_m involving a to the right. And repeat, always starting from the right. After all permutations involving a

have been moved to the right we obtain

$$\pi = \sigma_1 \ldots \sigma_h \rho_1 \ldots \rho_m \tag{2}$$

where $h + m = n + 2$ and where a is not moved in any σ and each ρ has the form

$$\rho = (a, x) \qquad \text{for some } x, \text{ say } \rho_1 = (a, d)$$

From (2) we see that π will move a unless d is moved by some ρ_i. So either π moves a or $\rho_i = (a, d)$ for some $i > 1$.

Let i be the first such index. Now we see that the transposition $\rho_1 = (a, d)$ can be shifted $(i - 1)$ places to the right and so cancel with ρ_i. Indeed, we need only observe the calculation

iii. $(a, d)(a, x) = (d, x)(a, d)$ if a, d, x are distinct elements.

Thus after $(i - 1)$ such operations

$$\pi = \sigma_1 \ldots \sigma_h \rho_2 \ldots (a, d)(a, d) \ldots \rho_m$$

and the adjacent transpositions (a, d) cancel. Thus π is the product of n transpositions and by induction $\pi \neq I$.

6.1.9 Theorem.   No Permutation Is Both Even and Odd

A permutation cannot be written as the product of an even number of transpositions in one way and as the product of an odd number of transpositions in another way.

Proof.   Suppose that

$$\pi = \tau_1 \ldots \tau_{2n} = \sigma_1 \ldots \sigma_{2m+1}$$

then

$$I = \tau_1 \ldots \tau_{2n} \sigma_{2m+1} \ldots \sigma_1$$

(remember that $\sigma^{-1} = \sigma$ if σ is a transposition) is the product of an odd number of transpositions. Tilt!

6.1.10 Definition.   The Alternating Group $\mathscr{A}_n$

The set of even permutations in $\mathbb{S}_n$ forms a subgroup called the alternating group denoted $\mathscr{A}_n$.

6.1.11 Theorem.   Two Properties of $\mathscr{A}_n$

The alternating group $\mathscr{A}_n$ has index 2 in $\mathbb{S}_n$. The order of $\mathscr{A}_n$ is $n!/2$.

Proof.   First we prove that $\mathscr{A}_n$ is a subgroup. Lemma 6.1.8 shows that the identity is an even permutation since it is the product of zero transpositions. If π is the product of an even number of transpositions, then so is π^{-1}. The product of two even permutations is again even since the composition of an even number of transpositions with an even number of transpositions is the product of an even number of transpositions.

The index of the alternating group is two because, if ρ is an odd permutation then $\rho(a, b)$ is an even permutation and so ρ belongs to the coset $\mathscr{A}_n(a, b)$. Thus there are at most two cosets $\mathscr{A}_n$ and $\mathscr{A}_n(a, b)$ in $\mathbb{S}_n$. By Theorem 6.1.9 each transposition, say, (a, b), is odd, so $\mathscr{A}_n$ is not equal to $\mathscr{A}_n(a, b)$.

Thus there are two distinct cosets and so the order of $\mathscr{A}_n$ is $n!/2$.

EXERCISE SET 6.1

6.1.1 Write each permutation in $\mathbb{S}_4$ as a product of transpositions and classify them as odd or even permutations.

6.1.2 List the $\binom{5}{2} = 10$ transpositions in $\mathbb{S}_5$. Find all permutations in $\mathscr{A}_5$ which are the product of two transpositions.

6.1.3 How many elements in $\mathscr{A}_5$ have order 5? Show that such a permutation must move all 5 elements.

6.1.4 Let $x_1, x_2, \ldots, x_{n-2}$ be an arbitrary sequence of $(n - 2)$ distinct elements from $\{1, 2, \ldots, n\}$. Show that there is a $\pi \in \mathscr{A}_n$ such that $(i)\pi = x_i$ for $i = 1, 2, \ldots, (n - 2)$.

6.1.5 Show that $\mathbb{S}_{n-1}$ is isomorphic to a subgroup $\mathbb{S}$ of $\mathbb{S}_n$. Find coset representatives $\pi_1, \ldots, \pi_n$ where π_i is a transposition and $\mathbb{S}_n = \mathbb{S}\pi_1 \cup \mathbb{S}\pi_2 \cup \cdots \cup \mathbb{S}\pi_n$.

6.1.6 Write a programmable algorithm to list all the permutations on $\{1, 2, \ldots, n\}$ in a sequence beginning with identity so that thereafter each permutation in the sequence is the composition of its predecessor and a transposition.

6.2 *n*-CYCLES AND THE SIMPLICITY OF $\mathscr{A}_n$ ($n \geq 5$)

Now we extend the idea of a transposition to a special class of permutations of order n called n-cycles.

A transposition is an example of an n-cycle; in this case $n = 2$. We use this notation to give a compact notation for permutations. This notation simplifies calculations and permits us to prove that the alternating group $\mathscr{A}_n$ for $n \geq 5$ is simple; that is, the only normal subgroups are the group itself and the identity.

By the way, as we suggested when we began our study of groups, a group gives a great deal of information about the roots of an equation. The fact that $\mathbb{S}_n$, for $n > 4$, has no proper normal subgroup is the key reason why polynomial equations of degree greater than 4 with rational coefficients cannot be solved in terms of radicals. This classic theorem, however, still requires a great deal of background information which will be developed as we proceed.

6.2.1 Definition. *n*-Cycle

Let n be a positive integer greater than 1. Suppose that π moves exactly n elements, $x_1, \ldots, x_n$, and that these elements can be arranged in a sequence so that

$$x_2 = (x_1)\pi$$

$$x_3 = (x_2)\pi = (x_1)\pi^2$$

$$\vdots$$

$$(x_k) = (x_{k-1})\pi = (x_1)\pi^{(k-1)}$$

$$\vdots$$

$$x_n = (x_{n-1})\pi = (x_1)\pi^{(n-1)}$$

$$x_1 = (x_n)\pi = (x_1)\pi^n$$

Then π is called an n-cycle (or a cycle of length n).

Notation.   If π is an n-cycle we write

$$\pi = (x_1, x_2, \ldots, x_n)$$

Two cycles $(x_1, \ldots, x_n)$ and $(y_1, \ldots, y_m)$ are called disjoint if no x_i equals a y_j.

Remark.   While it would be correct to define cycles of length 1, such cycles would consist of just one element and the element would be fixed. Later we shall express a permutation as a product of cycles. To avoid listing the fixed elements, cycles of length 1 have been omitted from the definition.

6.2.2 Examples. Cycles in $\mathbb{S}_4$

In $\mathbb{S}_4$

$$\pi = \begin{pmatrix} 1 & 2 & 3 & 4 \\ 2 & 3 & 4 & 1 \end{pmatrix} = (1, 2, 3, 4)$$

is a cycle of length 4.

$$\mu = \begin{pmatrix} 1 & 2 & 3 & 4 \\ 2 & 1 & 4 & 3 \end{pmatrix} = (1, 2)(3, 4)$$

is not a cycle but it is the product of two transpositions.

$$\lambda = \begin{pmatrix} 1 & 2 & 3 & 4 \\ 1 & 3 & 4 & 2 \end{pmatrix} = (2, 3, 4)$$

is a cycle of length 3.

6.2.3 Theorem.   Properties of n-Cycles

1. The following are equivalent ways of writing an n-cycle:

$$(x_1, x_2, \ldots, x_n) = (x_i, x_{i+1}, \ldots, x_{i+(n-1)})$$

where the indices are to be interpreted modulo n.

2. Each n-cycle has order n and for each i and k,

$$x_{i+k} = (x_i)\pi^k$$

where the indices are to be interpreted modulo n. In particular, if $\pi^k \neq I$ then π^k moves n elements.

3. Let $\pi = (x_1, \ldots, x_n)$. If $\gcd(k, n) = 1$ then

$$\pi^k = (x_1, x_{1+k}, x_{1+2k}, \ldots, x_{1+(n-1)k})$$

is an n-cycle. More generally, if $\gcd(k, n) = d > 1$ then π^k is the product of d disjoint cycles each of length n/d. For $j = 1, \ldots, d$, these cycles are

$$(x_j, x_{j+k}, \ldots, x_{j+(n/d-1)k})$$

4. Disjoint cycles commute.

5. If π is a n-cycle:

$$\pi = (x_1, x_2, \ldots, x_n)$$

then π^{-1} is an n-cycle:

$$\pi^{-1} = (x_1, x_n, x_{n-1}, \ldots, x_2)$$

6. A cycle of length n can be written as the product of $(n - 1)$ transpositions; thus an n-cycle is a member of the alternating group if and only if n is odd:

$$(x_1, x_2, \ldots, x_n) = (x_1, x_2)(x_1, x_2)\ldots(x_1, x_n)$$

Proof.   Most of these calculations are routine and the details can be left as exercises. The proof of 3 may need some elaboration.

Suppose that $\gcd(n, k) = 1$. Then the indices

$$1 + mk \qquad \text{for } 1 \leq m \leq n$$

are all distinct modulo n and thus π^k is an n-cycle.

If $\gcd(k, n) = d$ then $(n/d)k = (k/d)n \equiv 0 \pmod{n}$ and we obtain the n/d-cycles listed.

6.2.4 Theorem.   A Permutation Is the Product of Disjoint Cycles

In $\mathbb{S}_n$ each permutation $\pi \neq I$ can be written as the product of disjoint cycles. This form is unique except for the order in which the cycles appear. The order of π is the least common multiple of the lengths of the cycles.

Proof.   The proof is by induction on the number m of symbols moved. If $m = 2$ then π is a transposition and there is nothing to prove. Now let a be moved by π. Consider the sequence

$$x_1 = a, \; x_2 = (a)\pi, \ldots, x_k = (a)\pi^{k-1}, \ldots$$

Since π has finite order it must be that $x_i = x_j$ for some i and j and, by an argument now familiar, for some least positive integer n,

$$x_1 = a = (a)\pi^n$$

Thus $(x_1, \ldots, x_n)$ is an n-cycle. Next it is routine to show that

$$\pi_1 = (x_1, \ldots, x_n)^{-1}\pi$$

fixes all the elements fixed by π and, in addition, each x_i.

Thus, by induction π_1 can be expressed as the product of disjoint cycles. None of these involve the x's since these elements are fixed by π_1. Hence by using the decomposition for π_1 we get a decomposition of π as a product of disjoint cycles. These cycles commute and so any permutation of the cycles gives again a decomposition of π as a product of disjoint cycles.

Now to prove uniqueness. Again this is by induction, this time on the minimal number of disjoint cycles needed to express a permutation.

First, suppose that this minimal number is 1, that is, that π is an n-cycle. Then π cannot be written as the product of more than one disjoint cycle, say,

$$\pi = \sigma_1 \ldots \sigma_t$$

where the σ's are disjoint cycles. Since the total number of elements moved is n, some σ must have length $s < n$. Then π^s will fix all the elements moved by σ; yet π^s moves all elements, a contradiction. Now suppose that

$$\pi = (x_1, \ldots, x_n) = (y_1, \ldots, y_m)$$

are two representations as single cycles. Then $n = m$ since that is the number of symbols moved by π. Furthermore, then $y_1 = x_i$ for some i. But by Theorem 6.2.3, $y_{1+k} = (y_1)\pi^k = (x_i)\pi^k = x_{i+k}$ and so these cycles are equal.

Suppose then that the uniqueness holds for permutations that can be written as the product of fewer than h disjoint cycles. Now let

$$\pi = \alpha_1 \ldots \alpha_h = \beta_1 \ldots \beta_k$$

where each α and β is a cycle of length greater than one. Let a be an element moved by α_1. Since a is moved by π, it must appear in one of the cycles on the right-hand side of the equation above. By reordering the β's if necessary, we suppose a is in the cycle β_1. Then, using Theorem 6.2.3, we see that the cycles α_1 and β_1 are equal permutations since the sequence of the elements in the cycles is determined by the powers of π. By canceling $\alpha_1 = \beta_1$ and applying induction to the remaining permutation, uniqueness follows.

The order of the permutation π is the least common multiple of the lengths of the disjoint cycles in its decomposition because these cycles commute and the order of a cycle is equal to its length.

6.2.5 Example.   Cyclic Notation for Permutations

In $\mathbb{S}_6$ we find

$$\begin{pmatrix} 1 & 2 & 3 & 4 & 5 & 6 \\ 2 & 1 & 4 & 5 & 3 & 6 \end{pmatrix} = (1,2)(3,4,5)$$

$$\begin{pmatrix} 1 & 2 & 3 & 4 & 5 & 6 \\ 5 & 3 & 6 & 1 & 4 & 2 \end{pmatrix} = (1,5,4)(2,3,6)$$

The subgroup $V \subset \mathscr{A}_4$ on $\{1,2,3,4\}$ (recall Definition 4.1.3) consists of the elements

$$\{ I, (1,2)(3,4), (1,3)(2,4), (1,4)(2,3) \}$$

You are advised to write all 24 elements of $\mathbb{S}_4$ as the product of disjoint cycles.

Terminology. The *cycle structure* of a permutation is an accounting, for each positive integer n, of the number of n-cycles that appear in the cycle decomposition of the permutation. Thus the cycle structure of $(1,2)(3,4)$ is two 2-cycles. The cycle structure of $(1,2,3)(4,5)$ is a 3-cycle and a 2-cycle.

Next we prove two lemmas that are very helpful to us and give us some facility with computations with cycles. Be sure you follow the calculations; a pencil and paper may be handy!

6.2.6 Lemma.   Generators for the Alternating Group

If n is at least 3, the alternating group $\mathscr{A}_n$ is generated by its 3-cycles.

Proof. Since every element in $\mathscr{A}_n$ is the product of an even number of transpositions we need only show that each possible product of two transpositions

$$(x, y)(w, z)$$

can be achieved as a product of 3-cycles. There are three types, depending on the nature of the intersection of the set $\{x, y\}$ and $\{w, z\}$.

Type 1. If $\{x, y\} = \{w, z\}$ then $(x, y) = (w, z)$ and $(w, y)(w, z) = I$.

Type 2. If the two sets have one common element, say y, then

$$(x, y)(y, z) = (x, z, y)$$

is a 3-cycle.

Type 3. If the sets are disjoint then $n \geq 4$ and

$$(x, y)(w, z) = (x, y)(y, w)(y, w)(w, z) = (x, w, y)(y, z, w)$$

is the product of two 3-cycles.

6.2.7 Lemma.   Conjugates of Cycles

Let the n-cycle $\pi = (x_1, \ldots, x_n)$. For any permutation α, the conjugate of π defined as $\alpha^{-1}\pi\alpha$ is the following n-cycle:

$$\alpha^{-1}\pi\alpha = (y_1, \ldots, y_n)$$

where

$$y_i = (x_i)\alpha$$

Remark.   In other words, if π is an n-cycle, to calculate $\alpha^{-1}\pi\alpha$, simply replace each element in the cycle representation for π by its image under α. In particular, you don't have to know the cycle decomposition for α.

Proof.   Define, as suggested in the lemma, the elements y_i by

$$y_i = (x_i)\alpha$$

Therefore, we also have

$$x_i = (y_i)\alpha^{-1}$$

Thus we simply calculate

$$(y_i)\alpha^{-1}\pi\alpha = \left[(y_i)\alpha^{-1}\right]\pi\alpha = (x_i)\pi\alpha = (x_{i+1})\alpha = y_{i+1}$$

Moreover, if z is an element fixed by π, then $(z)\alpha$ is fixed by $\alpha^{-1}\pi\alpha$ since

$$(z)\alpha(\alpha^{-1}\pi\alpha) = (z)\pi\alpha = (z)\alpha$$

6.2.8 Corollary.   Invariance of Cycle Structure under Conjugation

If a permutation π is written as the product of disjoint cycles then the permutation

$$\alpha^{-1}\pi\alpha$$

is the product of the same number of disjoint cycles, and the lengths of the cycles in $\alpha^{-1}\pi\alpha$ are the same as those in π.

Proof.   Simply insert $\alpha^{-1}\alpha$ between the cycles of π and use Lemma 6.2.7. There is a partial converse of this invariance. The converse is sure to hold in the full symmetric group $\mathbb{S}_n$, but it may not hold if G is just a subgroup of $\mathbb{S}_n$.

6.2.9 Lemma.   Conjugacy of k-Cycles in $\mathbb{S}_n$

In the full symmetric group $\mathbb{S}_n$ for all $k > 2$, any two k-cycles are conjugate; that is,

$$\pi = (x_1, x_2, \ldots, x_k) \quad \text{and} \quad \rho = (y_1, y_2, \ldots, y_k)$$

then there is a permutation α such that $\rho = \alpha^{-1}\pi\alpha$.

Proof.   Just let α be any permutation in $\mathbb{S}_n$ such that

$$y_i = (x_i)\alpha \qquad \text{if } 1 \le i \le k$$

There may be several possibilities for α.

We can extend this result and its proof to a product of *disjoint* cycles.

Corollary.   Two permutations in $\mathbb{S}_n$ are conjugate if and only if they have the same cycle structure.

We now want to show that $\mathscr{A}_n$ is simple if $n \ge 5$; that is, $\mathscr{A}_n$ has no normal subgroups except itself and the identity. The proof is by induction and begins by recalling from Example 4.3.1 that $\mathscr{A}_4$ has exactly one proper normal subgroup,

$$V = \{I, (1,2)(3,4), (1,3)(2,4), (1,4)(2,3)\}$$

Two technical lemmas ease the details of the proof.

6.2.10 Lemma.   A Normal Subgroup of $\mathscr{A}_n$ Contains a Permutation Fixing an Element

If $n \ge 5$ and $N \neq I$ is a normal subgroup of $\mathscr{A}_n$ then there exists $\tau \in N$, $\tau \neq I$, such that τ fixes at least one element.

Proof.   Take any permutation $\theta \in N$, $\theta \neq I$. Consider the decomposition of θ into cycles. Say

$$\theta = \alpha_1 \ldots \alpha_h$$

Suppose that the lengths of these cycles are not all equal. Suppose, therefore, that u, the length of α_1, is less than or equal to the lengths of all the cycles and is strictly less than the length of some one cycle. Then

$$\rho = \theta^u = \alpha_1^u \alpha_2^u \ldots \alpha_h^u = \alpha_2^u \ldots \alpha_h^u$$

is not the identity but ρ does belong to N and ρ fixes at least the elements of the cycle α_1.

Thus either the lemma holds or N consists only of permutations τ with a cycle decomposition consisting of cycles of equal length. We now assume this second alternative.

Suppose first that $\tau \in N$ has at least two disjoint cycles of lengths at least 3. For this, $n \ge 6$. Say

$$\tau = (a, b, c, \ldots)(d, e, f, \ldots) \ldots$$

Then let $\alpha = (a, b)(c, e)$. We have $\alpha \in \mathscr{A}_n$ and we compute

$$\beta = \alpha^{-1}\tau\alpha = (b, a, e, \ldots)(d, c, f, \ldots) \ldots$$

Since N is normal in $\mathscr{A}_n$, $\beta \in N$ and so $\tau\beta \in N$ also. However, note that $\tau\beta$ fixes the element a. Moreover $\tau\beta \neq I$ since $(b)\tau\beta = f$.

Second, suppose that $\pi \in N$ is the product of disjoint transpositions. Since these cycles have length 2 and since we are assuming that no element is fixed by π, it follows that n must be even. Since $n \geq 5$, it must be that n is at least 6. Thus

$$\pi = (a, b)(c, d)(e, f)\ldots$$

Now let $\alpha = (a, b, c) \in \mathscr{A}_n$ so that

$$\beta = \alpha^{-1}\pi\alpha = (b, c)(a, d)(e, f)\ldots$$

and thus $\beta\pi \in N$ fixes e and f. Again $\beta\pi \neq I$ since $(a)\beta\pi = c$.

Finally, suppose that $\pi \in N$ is a single n-cycle. Since $n \geq 5$ we may suppose that

$$\pi = (a, b, c, d, e, \ldots)$$

Now let $\alpha = (a, b)(c, d)$ so that

$$\beta = \alpha^{-1}\pi\alpha = (b, a, d, c, e, \ldots)$$

and now find that $\pi\beta$ fixes the element a.

6.2.11 Lemma.   Sufficient Condition for Normality in $\mathscr{A}_n$ to Imply Equality to $\mathscr{A}_n$

If a normal subgroup N of the alternating group $\mathscr{A}_n$ contains a 3-cycle then $N = \mathscr{A}_n$.

Proof. Since we already know the alternating groups for $n = 2, 3$, and 4, we know that the lemma holds there and so we assume that n is at least 5.

Since the 3-cycles generate $\mathscr{A}_n$ it suffices to show that if, say,

$$(a, b, c) \in N$$

then all 3-cycles belong to N. We prove this.

Let $\beta = (c, d, e)$ where the five elements a, b, c, d, e are all different. Then

$$\beta^{-1}(a, b, c)\beta = (a, b, d)$$

is in N. Thus N contains all 3-cycles of the form (a, b, x).

Similarly, starting with $(a, c, b) = (a, b, c)^2$, it follows that N contains all 3-cycles of the form (a, c, x).

Again this calculation, starting with $(a, x, b) = (a, b, x)^2$, which we now know to be in N, tells us that all cycles of the form (a, x, y) belong to N.

Finally, starting with $(x, y, a) = (a, x, y)$ we obtain all cycles of the form (x, y, z) as members of N.

6.2.12 Theorem.   Alternating Group $\mathscr{A}_n$ Simple if $n \geq 5$

If n is greater than or equal to 5 the alternating group $\mathscr{A}_n$ is simple.

Proof.   We suppose that all the permutations are on the set $\{1, 2, 3, 4, 5, \dots\}$. The proof is by induction on n. The induction begins with $n = 5$. However, the arguments for the inductive step are so similar to those for the case $n = 5$ that we give the proof for the induction step, modifying it when $n = 5$.

Suppose then that $N \neq I$ is a normal subgroup of $\mathscr{A}_{n+1}$. By Lemma 6.2.10, N contains an element $\pi \neq I$, which fixes an element, say, $(n + 1)$. Now the subgroup of $\mathscr{A}_{n+1}$ consisting of those even permutations fixing $(n + 1)$ is the group $\mathscr{A}_n$ on $\{1, 2, \dots, n\}$ and so $\pi \in \mathscr{A}_n$. Thus

$$M = N \cap \mathscr{A}_n \neq I$$

and M is a normal subgroup (recall Theorem 4.3.5) of $\mathscr{A}_n$.

If $n = 4$, then $M = \mathscr{A}_4$ or $M = V$. If $M = V$, then $(a, b)(c, d)$ is in N. Let e be a fifth distinct element and let

$$\alpha = (a, b, e) \quad \text{and} \quad \beta = \alpha^{-1}(a, b)(c, d)\alpha = (b, e)(c, d)$$

Thus β is in N, so that

$$(a, b)(c, d)(b, e)(c, d) = (a, e, b)$$

is a 3-cycle in N.

Otherwise, $M = \mathscr{A}_n$, $(n \geq 4)$ and so $M \subset N$ contains a 3-cycle. Thus N contains a 3-cycle in all cases and so by Lemma 6.2.11, $N = \mathscr{A}_{n+1}$.

EXERCISE SET 6.2

6.2.1 Write all the permutations of $\mathbb{S}_4$ as products of disjoint cycles.

6.2.2 In $\mathbb{S}_5$, how many permutations are there whose cycle decomposition consists of a 3-cycle and a disjoint transposition?

6.2.3 What are the possible cycle structures of elements in $\mathscr{A}_5$? How many of each? (The total must be 60, the order of $\mathscr{A}_5$.)

6.2.4 Determine the dihedral subgroups of $\mathscr{A}_4$ and $\mathscr{A}_5$.

6.2.5 Determine the order and number of the cyclic subgroups in $\mathscr{A}_5$ and $\mathbb{S}_5$.

6.2.6 Let P be a group of permutations of a set S. Define a relation $\sim$ on S by $x \sim y$ if and only if there exists $\pi \in P$ such that $(x)\pi = y$.
   1. Show that $\sim$ is an equivalence relation on S. (These equivalence classes are called the orbits of P.)
   2. Relate the orbits of the cyclic group $\langle \pi \rangle$ to the cycle decomposition of π.

6.2.7 In $\mathscr{A}_5$ find all the elements α that commute with $(1, 2, 3, 4, 5)$. The set of such permutations forms a subgroup C of $\mathscr{A}_5$. Verify that the number of distinct conjugates $\beta^{-1}\pi\beta$ is the index of C in $\mathscr{A}_5$.

6.2.8 If π is an n-cycle, express π^d as a product of disjoint cycles. What will their lengths be?

6.2.9 Compute $\alpha^{-1}(1,2,3,4)(5,6,7)\alpha$ where, in turn, $\alpha = (1,2)$, $\alpha = (1,5,6)$, $\alpha = (2,3)(5,7)$, and $\alpha = (1,3)(2,4)(5,6,7)$.

6.2.10 Find, when possible, an $\alpha \in \mathbb{S}_7$ such that

$$\alpha^{-1}(1,2,3,4)(5,6,7)\alpha = (1,2,3)(4,5,6,7)$$

$$\alpha^{-1}(1,2,3,4)(5,6,7)\alpha = (1,2,3,4,5,6,7).$$

When it is impossible, give arguments.

6.2.11 Find, when possible, an $\alpha \in \mathscr{A}_5$ such that

$$\alpha^{-1}(1,2,3)\alpha = (3,4,5)$$

$$\alpha^{-1}(1,2,3)\alpha = (3,5,4)$$

$$\alpha^{-1}(1,2,3,4,5)\alpha = (1,2,4,3,5)$$

When it is impossible, give arguments.

6.3 REPRESENTATIONS OF GROUPS BY PERMUTATIONS

When groups were first studied extensively, the groups often arose naturally as groups of permutations. See, for example, the splendid treatise by Burnside, *The Theory of Groups*, first published in England in 1897. The following theorem shows that there is no loss in generality in doing so. There is sometimes, however, a loss in understanding the essence of an argument if it is couched solely in terms of permutations. Both points of view are useful, and in this section we give several ways of constructing a permutation model of a group. The first theorem is by the English mathematician Arthur Cayley (1821–1895). The idea behind the theorem is the group table. Recall that a group table has the latin square property (Theorem 3.1.13) and in particular each row of the table is a permutation of the elements of the group. Cayley's idea was simply to use the permutation in row a for the element a itself.

6.3.1 Theorem (Cayley).   The Right Regular Representation

Any group is isomorphic to a group of permutations.

Proof.   Let G be a group. First, for each $a \in G$ the function π_a defined by

$$(g)\pi_a = ga$$

is a permutation of the elements of G. The details of this argument are familiar (see Theorem 3.1.11) and are omitted.

Furthermore, if π_a and π_b are two such permutations, then the composition $\pi_a\pi_b$ gives

$$((g)\pi_a)\pi_b = (ga)\pi_b = (ga)b = g(ab) = (g)\pi_{ab}$$

so, as permutations,

$$\pi_a \pi_b = \pi_{ab}$$

This means that if $\mathbb{P}$ is the group of all permutations on the set G, then the function Φ from G into $\mathbb{P}$ defined by

$$\Phi: a \rightarrow \pi_a$$

is a homomorphism of G *into* $\mathbb{P}$. In fact Φ is an isomorphism, as we now prove by showing that the kernel of Φ is just the identity of G.

Indeed, if π_a is the identity permutation then for all $g \in G$

$$(g)\pi_a = g$$

but by definition of π_a

$$(g)\pi_a = ga$$

so $ga = a$ and thus $a = I$ in G. Thus π_a is the identity permutation if and only if $a = I$.

Thus the image of Φ is a group, a subgroup of $\mathbb{P}$, which is isomorphic to G.

Terminology.   This is called the *right* regular representation because to get the permutation representing a, each element g is multiplied on its right side by a. There is also a left regular representation; see Problem 6.3.5.

Remark.   The permutations used in this theorem are on as many elements as the order of the group. For even a moderately large group this can be unwieldy. Next we present some alternatives.

6.3.2 Theorem.   Representation by Permutation of Cosets

Let S be a subgroup of a group G. For each $a \in G$ define the permutation ρ_a of the right cosets of S in G,

$$(Sg)\rho_a = Sga \qquad (1)$$

then

1. The mapping Φ defined by $\Phi(a) = \rho_a$ is a homomorphism of G into the group of all permutations of the right cosets of S in G.
2. The kernel of Φ is the largest subgroup of S that is normal in G; hence the homomorphism Φ is an isomorphism if no subgroup of S except I is normal in all of G.

Terminology.   This representation is called the *right coset permutation representation of G by S.*

Proof.   There is much to prove. The definition (1) given above for the permutation ρ_a uses coset representatives; so before we can speak of a permutation having been defined we must check to be sure that a function has been defined, that is, that the definition (1) is independent of the coset representative chosen. To do this, suppose that $Sg = Sh$. Then we must check (it is trivial, but we must check!) to see that $Sga = Sha$.

Now that this is done we must check to see that ρ_a is a *permutation* of the right cosets.

To show that ρ_a is one-to-one, suppose that $Sga = Sha$. Then, for some $s \in S$

$$ga = sha$$

hence $g = sh$, and hence $Sg = Sh$.

To show that ρ_a is onto, let Sg be a right coset. Then

$$\left(Sga^{-1} \right)\rho_a = Sga^{-1}a = Sg$$

Thus ρ_a is a permutation of the right cosets of S.

Now we must check the homomorphism condition. We calculate, for $a, b \in G$ and Sg any right coset of S:

$$(Sg)\rho_a\rho_b = (Sga)\rho_b = (Sga)b = Sgab = Sg(ab) = (Sg)\rho_{ab}$$

Thus the function Φ is a homomorphism of G.

The kernel of the homomorphism is of course a normal subgroup of G. If a belongs to the kernel, the permutation ρ_a is the identity and so, in particular

$$(S)\rho_a = Sa = S$$

so that a is in S.

On the other hand, suppose that K is a normal subgroup of G and $K \subseteq S$. Then for all $a \in K$

$$(Sg)\rho_a = Sga = Sgag^{-1}g$$

and since $gag^{-1} \in K$ and $K \subseteq S$, it follows that $Sgag^{-1} = S$ and $Sgag^{-1}g = Sg$. Thus ρ_a is the identity permutation and so a belongs to the kernal of Φ.

Remark. This theorem can be used to guarantee the existence of a normal subgroup of G. Suppose that $[G : S]!$ is less than the order of G. Then the representation cannot be an isomorphism since the order of the image, being a subgroup of the full symmetric group on $[G : S]$ elements, is at most $[G : S]!$. This means that if G has a subgroup whose order is sufficiently large, then G has a normal subgroup. As Theorem 6.3.3 shows, the inequality may be replaced by a divisibility requirement that is stronger. Later we have some theorems (the Sylow theorems) to determine the orders of some subgroups.

6.3.3 Theorem.   Criterion for a Proper Normal Subgroup

If S is a subgroup of G such that the order of G does not divide $[G : S]!$ then G has a proper normal subgroup contained in S.

Proof. Let G^* be the image of G under the homomorphism Φ of Theorem 6.3.2. Now G^* is a subgroup of the full symmetric group on $[G : S]$ elements whose order is $[G:S]!$ and so, by the theorem of Lagrange, must be divisible by the order of G^*. But if the order of G does not divide $[G:S]!$, then it cannot be equal to the order of G^* and so the homomorphism must have a proper kernel, which must be a proper normal subgroup of S.

6.3.4 Corollary. Groups of Order *pm*, *p* > *m*

If G is a group of order pm where p is a prime greater than m then G has a normal subgroup of order p.

Proof. Since p is a prime, Cauchy's theorem tells us that G has a subgroup S of order p. Now $m = [G:S]$ and since p is a prime larger than m it follows that p does not divide $m!$. Thus G has a proper normal subgroup contained in S. Since $|S|$ is a prime, the only nonidentity subgroup of S is S. Thus S is normal in G.

For an extension of this corollary and an application to determine all the groups of order 30, see Problems 6.3.8 and 6.3.9.

Here is another example of this technique.

6.3.5 Example. A Group of Order 24 Has a Proper Normal Subgroup

A group of order 24 must have a normal subgroup. We will show later, using the Sylow theorems, that a group of order 24 must have a group of order 8. Call the group S. We have $[G:S] = 3$ and since 24 does not divide $3! = 6$, it follows that G has a normal subgroup. Moreover the order of that subgroup must be a power of 2.

There is an even more dramatic example of the use of the technique of this theorem. The exercises to follow show that any group of order p^2, p a prime, has a proper normal subgroup. From this it follows:

6.3.6 Theorem. Groups of Order p^2

A group of order p^2, p a prime, is either cyclic or is isomorphic to the direct product $C_p \times C_p$ of two cyclic groups of order p.

Proof. See Problems 6.3.6 and 6.3.7.

EXERCISE SET 6.3

6.3.1 Find the right regular representation for a cyclic group of order n.

6.3.2 Find the right regular representation for $\mathbb{S}_3$. (Use the notation for the elements as in D_6.)

6.3.3 Find the right coset permutation representation for D_6 by $S = \langle b \rangle$, where b is an element of order 2 not in the cyclic subgroup of order 3. Is this an isomorphism?

6.3.4 Find the right coset permutation representation of D_8 by $S = \langle a^2 \rangle$, where a generates the cyclic subgroup of order 4. Is this an isomorphism?

6.3.5 Let $a \in G$ and define a permutation $_a\pi$ on the elements of G by $(g)_a\pi = a^{-1}g$. Show that the mapping $a \rightarrow {_a\pi}$ is a representation of G by permutations. It is called the left regular representation of G.

6.3.6 Let G be a noncyclic group of order p^2, p a prime. Let $A = \langle a \rangle$ be a subgroup of order p. Show that the image ρ_a of a in the right coset permutation representation of G by A has an order that divides $(p - 1)!$. On the other hand, argue that under any homomorphism the order of the image of a must divide p. Conclude that the order of ρ_a is 1 and that this means that A is normal in G.

6.3.7 Use Problem 6.3.6 to show that a noncyclic group of order p^2, p a prime, is the direct product of two cyclic groups of order p.

6.3.8 Let the order of a group be the product of distinct primes:

$$|G| = p_1 p_2 \cdots p_t$$

where

$$p_1 < p_2 < \cdots < p_t$$

Show that for each i there is a normal subgroup S_i in G such that

$$|S_i| = p_i p_{i+1} \cdots p_t$$

6.3.9 Determine all the groups of order 30.

Chapter 7

Introduction to Rings

7.1 RING AXIOMS AND ELEMENTARY PROPERTIES

A ring is an algebraic system with two binary operations, one called addition and denoted by $+$, the other called multiplication and denoted by $\cdot$ or by juxtaposition. The integers $\mathbb{Z}$ come immediately to mind as an example of such a system. Another system is the integers modulo n, $\mathbb{Z}_n$. The algebraic laws here resemble those for the integers but some, like cancellation, must be modified. Thus when we speak about rings we assume less than when we speak of integers.

This means that a ring is a more general construct than the integers. For example, the rational numbers also constitute a ring and have more properties —like the existence of multiplicative inverses—than do the integers. Other examples given in more detail below include the set of 2×2 matrices with entries from the real numbers.

Another way to think of a ring is to regard it as an abelian group whose operation is addition to which another operation (multiplication) has been added. The axioms demand that the multiplication be associative and also define just how multiplication relates to the addition of the underlying abelian group.

Here are the defining axioms.

7.1.1 Definition. Ring

A ring $\langle R, +, \cdot \rangle$ is a nonempty set R together with two binary operations $+$ and $\cdot$ such that

1. The system $\langle R, + \rangle$ is an abelian group.
   The identity under $+$ is usually denoted 0.
   The inverse of a under $+$ is usually denoted by $-a$.
   The operation $+$ is called *addition*.

2. The operation $\cdot$ is associative: For all a, b, c in R

$$(a \cdot b) \cdot c = a \cdot (b \cdot c)$$

The operation $\cdot$ is called *multiplication* and is usually denoted by juxtaposition.

3. The two distributive laws

$$a \cdot (b + c) = (a \cdot b) + (a \cdot c) \quad \text{and} \quad (b + c) \cdot a = (b \cdot a) + (c \cdot a)$$

hold for all a, b, c in R.

Notation. Just a reminder that, as with additive notation in any group, if n is a positive integer and r is an element of the ring, then

$$nr = r + r + \cdots + r \quad (n \text{ summands})$$

and

$$(-n)r = -nr$$

In particular nr does not denote the ring product of n (which may not even be in the ring) and r.

Remark. Note that the commutative law for multiplication is *not* assumed. Nor is the existence of multiplicative inverses assumed.

In some advanced work, the associative law for multiplication is not assumed; these are the so-called nonassociative rings. However, in this book, ring multiplication will always be associative.

There are many special classes of rings which will be described shortly. First some examples. The operations $+$ and $\cdot$ are the usual ones unless otherwise noted.

7.1.2 Examples. Familiar Rings

1. The integers $\mathbb{Z}$.
2. The even integers $2\mathbb{Z} = \{2z : z \in \mathbb{Z}\}$.
3. The integers modulo n, $\mathbb{Z}_n$.
4. The rational numbers $\mathbb{Q}$.
5. The real numbers $\mathbb{R}$.
6. The complex numbers $\mathbb{C}$.
7. The set of 2×2 matrices with entries from $\mathbb{Z}$. This example can be extended in many ways. Here are some:

7.1 Replace the size 2×2 by $n \times n$.
7.2 Replace $\mathbb{Z}$ by $\mathbb{Z}_n$ or by $\mathbb{Q}$, or by $\mathbb{R}$, or by $\mathbb{C}$.
7.3 Take only the set of upper (or lower) triangular matrices with the entries chosen from any of the alternatives already mentioned. Here is a typical and important special instance:

The Ring of Upper Triangular 2 × 2 Matrices over $\mathbb{Q}$

$$T = \left\{ \begin{pmatrix} a & b \\ 0 & c \end{pmatrix} : a, b, c \in \mathbb{Q} \right\}$$

The rules of addition and multiplication are those of matrix addition and multiplication:

$$\begin{pmatrix} a & b \\ 0 & c \end{pmatrix} + \begin{pmatrix} u & v \\ 0 & w \end{pmatrix} = \begin{pmatrix} a+u & b+v \\ 0 & c+w \end{pmatrix}$$

and

$$\begin{pmatrix} a & b \\ 0 & c \end{pmatrix}\begin{pmatrix} u & v \\ 0 & w \end{pmatrix} = \begin{pmatrix} au & av+bw \\ 0 & cw \end{pmatrix}$$

It is an easy routine to verify that the distributive laws hold here as well as the associative law of multiplication. It is also immediate that I,

$$I = \begin{pmatrix} 1 & 0 \\ 0 & 1 \end{pmatrix}$$

is a multiplicative identity. Moreover, if neither diagonal element is zero then an upper triangular matrix has a multiplicative inverse:

$$\begin{pmatrix} a & b \\ 0 & c \end{pmatrix}\begin{pmatrix} 1/a & -b/ac \\ 0 & 1/c \end{pmatrix} = \begin{pmatrix} 1 & 0 \\ 0 & 1 \end{pmatrix}$$

Notice our use of the word "over" as a preposition to indicate the elements that may appear in the matrices. This suggests that in some way this ring is larger than the ring of rational numbers. This is true in the sense that T contains an isomorphic copy of the integers. The correspondence is as follows:

$$x \leftrightarrow \begin{pmatrix} x & 0 \\ 0 & x \end{pmatrix}$$

Notation. If R is a ring we let $M_n(R)$ denote the ring of all $n \times n$ matrices over R using the standard addition and multiplication of matrices.

8. Here is a pathological example of a finite ring constructed from $M_2(\mathbb{Z}_2)$, the ring of 2×2 matrices over the integers modulo 2. Let R be the following set of four matrices:

$$\text{Let } R = \left\{ \begin{pmatrix} a & 0 \\ b & 0 \end{pmatrix} : a, b \in \mathbb{Z}_2 \right\}$$

$$= \left\{ 0 = \begin{pmatrix} 0 & 0 \\ 0 & 0 \end{pmatrix}, A = \begin{pmatrix} 1 & 0 \\ 0 & 0 \end{pmatrix}, \right.$$

$$\left. B = \begin{pmatrix} 0 & 0 \\ 1 & 0 \end{pmatrix}, C = \begin{pmatrix} 1 & 0 \\ 1 & 0 \end{pmatrix} \right\}$$

We find that the additive group $\langle R, + \rangle$ is the four-group. The multiplication table for R is

·	0	A	B	C
0	0	0	0	0
A	0	A	0	A
B	0	B	0	B
C	0	C	0	C

Note that $0X = X0 = 0$ for all X. The zero row and column will be omitted from tables in the future. Note too that A serves as a right identity since $XA = X$ for all X. However, A is not a left identity since, for example, $AB = 0$, instead of B as it should if A were a left identity.

9. The zero ring, $R = \{0\}$, consisting of one element, zero.

10. A null ring. Take any abelian group $\langle R, + \rangle$ and define multiplication by $xy = yx = 0$ for all x, y in R. Since all products are zero, all the ring axioms involving multiplication are satisfied.

11. The subset of the real numbers,

$$\mathbb{Z}[\sqrt{2}] = \{a + b\sqrt{2} : a, b \in \mathbb{Z}\}$$

Here we must check to see that this set is closed under addition and multiplication. It is easy to see for addition and it's not hard for multiplication; here is the calculation:

$$(a + b\sqrt{2})(c + d\sqrt{2}) = ac + 2bd + (ad + bc)\sqrt{2}$$

12. The Quaternion Ring: We consider 2×2 matrices with complex entries. Let

$$\text{QUAT} = \left\{ \begin{pmatrix} \alpha & -\beta \\ \bar{\beta} & \bar{\alpha} \end{pmatrix} : \alpha, \beta \in \mathbb{C} \right\}$$

Recall that a bar over a complex number denotes its complex conjugate. Thus if $\alpha = a + bi$ then $\bar{\alpha} = a - bi$. Complex conjugation satisfies these two elementary properties:

$$\overline{(\alpha + \beta)} = \bar{\alpha} + \bar{\beta} \quad \text{and} \quad \overline{(\alpha\beta)} = \bar{\alpha} \cdot \bar{\beta}$$

QUAT is a ring under matrix addition and matrix multiplication. The additive structure of $\langle \text{QUAT}, + \rangle$ is guaranteed if we can show it is an additive subgroup of the additive group of all 2×2 matrices with complex entries. Thus we must show that QUAT is closed under addition and that each matrix in QUAT has an additive inverse in QUAT. Similarly the associative property of multiplication and the distributive laws hold since they hold in general for $M_2(\mathbb{C})$. Thus we check on the closure property for addition. We do this now. The sum:

$$\begin{pmatrix} \alpha & -\beta \\ \bar{\beta} & \bar{\alpha} \end{pmatrix} + \begin{pmatrix} \gamma & -\delta \\ \bar{\delta} & \bar{\gamma} \end{pmatrix} = \begin{pmatrix} \alpha + \gamma & -(\beta + \delta) \\ \bar{\beta} + \bar{\delta} & \bar{\alpha} + \bar{\gamma} \end{pmatrix}$$

The additive inverse

$$-\begin{pmatrix} \alpha & -\beta \\ \bar{\beta} & \bar{\alpha} \end{pmatrix} = \begin{pmatrix} -\alpha & \beta \\ -\bar{\beta} & -\bar{\alpha} \end{pmatrix}$$

The product

$$\begin{pmatrix} \alpha & -\beta \\ \bar{\beta} & \bar{\alpha} \end{pmatrix} \begin{pmatrix} \gamma & -\delta \\ \bar{\delta} & \bar{\gamma} \end{pmatrix} = \begin{pmatrix} \alpha\gamma - \beta\bar{\delta} & -\alpha\delta - \beta\bar{\gamma} \\ \bar{\beta}\gamma + \bar{\alpha}\bar{\delta} & -\bar{\beta}\delta + \bar{\alpha}\bar{\gamma} \end{pmatrix}$$

The sum, the inverse, and the product have the correct form because of the elementary properties of the complex conjugate.

We have in fact showed that this ring is a subring (the term "subring" anticipates Definition 7.2.1) of the ring $M_2(\mathbb{C})$.

The ring QUAT satisfies another important property; nonzero elements have a multiplicative inverse and hence the nonzero elements in QUAT form a group under multiplication, in fact a *nonabelian* group.

We have

$$\det\begin{pmatrix} \alpha & -\beta \\ \bar{\beta} & \bar{\alpha} \end{pmatrix} = \alpha\bar{\alpha} + \beta\bar{\beta}$$

and this is zero if and only if $\alpha = \beta = 0$. This means that for a nonzero quaternion

$$\begin{pmatrix} \alpha & -\beta \\ \bar{\beta} & \bar{\alpha} \end{pmatrix}^{-1} = \frac{1}{\alpha\bar{\alpha} + \beta\bar{\beta}}\begin{pmatrix} \bar{\alpha} & \beta \\ -\bar{\beta} & \alpha \end{pmatrix}$$

The important thing to note is that the nonzero elements of QUAT form a multiplicative, nonabelian group. Just in passing, note that addition of matrices makes QUAT into a subgroup of the direct product of four copies of the additive group of the complex numbers $\langle \mathbb{C}, + \rangle$.

7.1.3 Special Classes of Rings

We distinguish several special classes of rings. We study mainly integral domains, indeed even a special subclass defined in the next chapter, and fields.

1. A *commutative ring* is one in which the multiplication satisfies the addition property

$$ab = ba \qquad \text{for all } a, b \in R$$

2. A *ring with identity* is a ring with a multiplicative identity, usually denoted by 1, that satisfies the property

$$a1 = 1a = a \qquad \text{for all } a \in R$$

In such a ring it is permissible to "confuse" the sum

$$nr = r + r + \cdots + r \qquad (n \text{ summands})$$

and the product

$$(n1) \cdot r = (1 + 1 + \cdots + 1) \cdot r = nr$$

3. An *integral domain* is a commutative ring with identity provided that

$$ab = 0 \qquad \text{only if} \quad a = 0 \text{ or } b = 0$$

or, equivalently, only if the cancellation law

$$\text{Whenever } ab = ac \quad \text{and} \quad a \neq 0 \quad \text{then} \quad b = c$$

holds.

4. A *division ring* is a ring D in which the nonzero elements

$$D^{\times}= D - \{0\}$$

form a group under multiplication. In particular, a division ring has a multiplicative identity and every nonzero element has a multiplicative inverse.

5. A *field* is a commutative ring F in which the nonzero elements

$$F^{\times}= F - \{0\}$$

form an abelian group.

Remark.   Note that the axioms for a division ring and a field require that sets $D^{\times}$ and $F^{\times}$ be nonempty. Thus division rings and fields must have at least two elements, the additive identity 0 and the multiplicative identity 1.

Figure 7.1 shows the relations of these classes. It is drawn as a partially ordered set ordered by class inclusion. Beside each class is an example of that class that does not belong to a class below it.

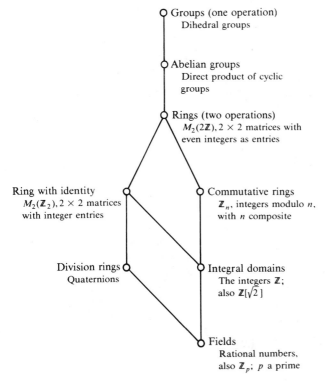

Figure 7.1 Classes of rings. The example given of each class is chosen so that it is not contained in the next class below.

7.1.4 Lemma.   Properties of Rings

Let $\langle R, +, \cdot \rangle$ be a ring.

1. For all $a \in R$, $a \cdot 0 = 0 \cdot a = 0$.
2. For all $a, b \in R$, $(-a) \cdot b = a \cdot (-b) = -(a \cdot b)$.
3. For all integers n and m and all $r, s \in R$.
    i. $(n + m)r) = nr + mr$.
    ii. $nr \cdot ms = nm(r \cdot s)$.
4. For all $a, b \in R$, $(-a) \cdot (-b) = a \cdot b$.
5. If R has an identity 1 then it is unique, and if R is not the zero ring then $1 \neq 0$.

Proof.   For 1, apply the distributive law to show that $a \cdot 0$ is an idempotent in the additive group $\langle R, + \rangle$.

For 2, apply the distributive laws to show that $(-a) \cdot b$ and $a \cdot (-b)$ act as additive inverses of $a \cdot b$.

For 3i, use the associative law of addition in the group $\langle R, + \rangle$.

For 3ii, treat the case that $n > 0$ and $m > 0$ by using the distributive law and then use 1 and 2 to treat the other cases.

For 4, use 2 with b replaced by $(-b)$.

For 5, suppose that 1 and e are two identities. Then, on the one hand,

$$1 \cdot e = e \qquad \text{since 1 is an identity}$$

and on the other hand,

$$1 \cdot e = 1 \qquad \text{since } e \text{ is an identity}$$

Thus $1 = e$. Now, if $1 = 0$ then $a = a \cdot 1 = a \cdot 0 = 0$ for all $a \in R$.

Remark.   It may not be obvious just from the definitions that a field is an integral domain. However, suppose that F is a field, a is a nonzero element, and

$$ab = 0$$

Multiplication on the left by a^{-1} gives $a^{-1}(ab) = a^{-1} \cdot 0 = 0$ and:

$$a^{-1}(ab) = (a^{-1}a)b = 1 \cdot b = b$$

hence $b = 0$.

Although Figure 7.1 gives examples that show the classes of rings are distinct, it would be nice to have examples that consist of only a finite number of elements. It is surprising that this is sometimes impossible. We state two results here. The first is easy to prove; the second is more difficult and a proof will have to be postponed to Section 12.2.

7.1.5 Theorem.   Finite Integral Domains

1. Any finite integral domain is a field.
2. (Wedderburn) Any finite division ring is a field.

Proof of 1.   The only thing missing from the verification that the nonzero elements in an integral domain form a multiplicative group is the existence of multiplicative inverses for nonzero elements. If the integral domain is *finite* this comes about as follows:

Let $a \neq 0$. Because the sequence of powers of a,

$$a, a^2, a^3, \ldots$$

can contain at most a finite number of distinct terms it must be that for some $n < m$,

$$a^{m-n}a^n = a^m = a^n = 1 \cdot a^n$$

By cancellation we find $a^{m-n} = 1$, and thus a has an inverse a^{m-n-1}.

Terminology.   Zero Divisors. In a ring, whenever there are nonzero elements a and b such that $ab = 0$ we say that a and b are zero divisors.

It is useful, in noncommutative rings, to say in this case that a is a left zero divisor and b a right zero divisor. One need not imply the other. See 8 in Examples 7.1.2. There A and C are left zero divisors since $AB = CB = 0$ but are not right zero divisors. However, the following is true.

7.1.6 Lemma.   Cancellation

If a is not a left zero divisor then $ax = ay$ implies $x = y$. If b is not a right zero divisor, then $xb = yb$ implies $x = y$.

Proof.   From $ax = ay$ it follows that $a(x - y) = 0$. Thus either $x - y = 0$ or a is a zero divisor.

7.1.7 Definition.   Units and Associates

Let $\langle R, +, \cdot \rangle$ be a ring with identity 1. An element u is called a unit if there is a v such that $uv = vu = 1$.

An element $a \in R$ is called an associate of $b \in R$ if there exists a unit u such that $b = au$.

Remark.   It is routine to show that the relation $\sim$ defined by $a \sim b$ if and only if a is an associate of b in R is an equivalence relation on R.

7.1.8 Lemma.   The Group of Units

In a ring, the set of units form a multiplicative group.

Proof.   We need only show that the product of two units is again a unit. Let u and s be units. Thus there exist elements v and t such that

$$uv = vu = st = ts = 1$$

Thus

$$(us)(tv) = u(st)v = uv = 1$$

and

$$(tv)(us) = t(vu)s = ts = 1$$

Thus the product of units us is a unit.

7.1.9 Examples.   Units in Familiar Rings

1. In the ring of upper triangular 2×2 matrices over $\mathbb{Q}$ in Examples 7.1.2 the units are the matrices in which neither diagonal element is zero.

   More generally in $M_2(\mathbb{Q})$, the ring of 2×2 matrices over the rational numbers, the units are the nonsingular matrices (the ones with determinant not equal to zero) for these matrices have multiplicative inverses. For more details see Exercise 7.1.11.

2. In a field, like the rational numbers, every nonzero element is a unit.

3. In the ring $\mathbb{Z}[\sqrt{2}]$ (recall 11 of Examples 7.1.2) the units include 1 and -1 and

$$(1 + \sqrt{2}) \quad \text{and} \quad (1 - \sqrt{2})$$

since their product is -1 and so the inverse of one is the negative of the other. From Lemma 7.1.8 it follows that powers of units are units. On the other hand, $\sqrt{2}$ is not a unit since its inverse cannot be written in the form of an element in $\mathbb{Z}[\sqrt{2}]$.

It is usually a very hard problem to determine all the units of a ring. In the case of integral domains like $\mathbb{Z}[\sqrt{2}]$ it becomes a challenging problem in number theory; see, for example, Uspensky and Heaslet, *Elementary Number Theory*, and look under "Fermat's equation." It turns out that all the units of this ring have the form

$$\pm(1 + \sqrt{2})^m$$

where $m \in \mathbb{Z}$.

EXERCISE SET 7.1

7.1.1 Let $\langle R, +, \cdot \rangle$ be a ring with three elements. Identify the group $\langle R, + \rangle$. Show that $x^2 = y^2$ if x and y are nonzero elements of R. Show that $\langle R, +, \cdot \rangle$ is either the null ring or an integral domain.

7.1.2 Show that if $r^2 = r$ for all elements r in a ring, then the ring is commutative.

7.1.3 Let S be a nonempty set. Show the set of all subsets of S may be made into a ring, called a boolean ring, by defining addition to be the symmetric difference

$$A + B = (A^c \cap B) \cup (A \cap B^c)$$

and multiplication

$$AB = A \cap B$$

Here A^c denotes the complement of the set A.

7.1.4 It is true in general that a ring R with identity in which $r^2 = r$ for all r can be represented as a boolean ring. If $|R|$ is finite prove this.

*7.1.5 (Herstein) Prove that if $r^3 = r$ for all $r \in R$ then R is commutative.

7.1.6 Show that if R is a ring with identity 1, then the commutative law for addition follows from the other axioms by considering

$$(1 + 1)(a + b)$$

7.1.7 Let D be an integral domain with operations of addition $(+)$ and multiplication $(\cdot)$. Let two new operations be defined on D:

$$a \oplus b = a + b - 1$$

and

$$a \odot b = ab - (a + b) + 2$$

Prove or disprove: $\langle D, \oplus, \odot \rangle$ is an integral domain.

7.1.8 Show that if R is a ring with identity the relation $\sim$ defined on R by $a \sim b$ if and only if a is an associate of b in R is an equivalence relation on R.

*7.1.9 (Kaplansky) Let R be a ring with identity. Show that if $u \in R$ has two distinct right inverses then it has infinitely many.

7.1.10 Let R be any ring and consider the cartesian product

$$R \times \mathbb{Z}$$

Define addition and multiplication of these ordered pairs as follows:

$$(r, n) + (s, m) = (r + s, n + m)$$
$$(r, n)(s, m) = (rs + mr + ns, nm)$$

Show that $R \times \mathbb{Z}$ is a ring with identity even if R has none.
Show that if R is not the zero ring, then $R \times \mathbb{Z}$ contains zero divisors.

7.1.11 Let R be a commutative ring with identity. Show that $M_2(R)$ defined to be the set of 2×2 matrices $\begin{bmatrix} a & b \\ c & d \end{bmatrix}$ with a, b, c, and d chosen as elements of R form a ring if addition and multiplication defined by

$$\begin{bmatrix} a & b \\ c & d \end{bmatrix} + \begin{bmatrix} w & x \\ y & z \end{bmatrix} = \begin{bmatrix} a + w & b + x \\ c + y & d + z \end{bmatrix}$$

$$\begin{bmatrix} a & b \\ c & d \end{bmatrix}\begin{bmatrix} w & x \\ y & z \end{bmatrix} = \begin{bmatrix} aw + by & ax + bz \\ cw + dy & cx + dz \end{bmatrix}$$

Show that $A = \begin{bmatrix} a & b \\ c & d \end{bmatrix}$ has a multiplicative inverse if det $A = ad - bc$ is a unit of R. Show that $\left\{ \begin{bmatrix} a & 0 \\ 0 & a \end{bmatrix} : a \in R \right\}$ is a subring of $M_2(R)$ isomorphic to R.

7.2 SUBRINGS, HOMOMORPHISMS, AND FACTOR RINGS

The study of the structure of rings patterns itself very much after that for groups. We shall study subrings, homomorphisms, factor rings, and direct products of rings. This is due in large part to the fact that a ring is, first of all, a group. A deeper reason is that these concepts are the general features of any abstract algebra.

7.2.1 Definition.   Subring

A subset S of a ring $\langle R, +, \cdot \rangle$ is a subring if it is a ring under the operations $+$ and $\cdot$ of R. In particular, $\langle S, + \rangle$ must be a subgroup of $\langle R, + \rangle$ and S must be closed under $\cdot$; that is,

$$ab \in S \qquad \text{whenever } a \text{ and } b \text{ are in } S$$

Many of the examples we saw in Section 7.1 were subrings of other rings. For instance, the even integers $2\mathbb{Z}$ is a subring of $\mathbb{Z}$.

Properties that can be defined by identities holding in the parent ring are inherited by subrings. Thus if a ring R is commutative then so is any subring. If R has a cancellation law:

$$\text{Whenever } ab = ac \quad \text{and} \quad a \neq 0 \quad \text{then} \quad b = c$$

then the same cancellation must hold in all its subrings.

However, some properties are not inherited. For example, the ring $\mathbb{Z}$ has an identity but its subring $2\mathbb{Z}$ of even integers has no identity. This example shows that a subring may not have an identity, even though the larger ring has.

Another property that is not inherited is that of being a zero divisor. Recall 8 of Examples 7.1.2. One verifies easily that $\{0, A\}$ is a subring. The element A is not a zero divisor in S since $A \cdot A \neq 0$ but A is a zero divisor in R since $AB = 0$.

7.2.2 Theorem.   Lattice of Subrings

The set of subrings of a ring form a complete lattice under set inclusion.

Proof.   The proof is almost word for word the same as the one given for groups. We apply the criterion of Theorem 2.2.13. Clearly the whole ring is a subring, so the partially ordered set of subrings has a top element. The proof that the set intersection of subrings is again a subring is routine. We already know it for the subgroup of the additive group; only the closure under multiplication need be checked. Thus, for two subrings A and B

$$A \vee B = \cap \{T: T \text{ is a subring and } T \supseteq A \text{ and } T \supseteq B\}$$

We need an alternate characterization of the elements in the join $A \vee B$. Recall Theorem 3.2.8 as you read the next lemma.

7.2.3 Lemma.   A Form for Elements in the Join of Subrings

The lattice join of two subrings A and B is the set of all elements in the ring that can be expressed as a finite sum of elements, each of which is a *finite product* of elements either in A or in B. Thus

$$x \in A \vee B \qquad \text{if and only if } x = \Sigma p_i$$

where

$$p_i = c_{i1}c_{i2}\dots c_{in(i)} \qquad c_{ij} \in A \text{ or } B$$

Proof. Clearly each element of A and B is included in this form, and, from the closure laws, any subring containing all the elements of A and B must contain all elements of this form. Thus all that needs to be done is to show that the totality of elements of this form do constitute a subring.

First, elements of this form are an additive subgroup: If x has this form so does $-x$. If x and y have this form, so does $x + y$, for the sum is just extended. Second, after a moment's thought, we see that the product xy also has this form. To see this, let

$$x = \Sigma p_i \quad \text{and} \quad \text{let } y = \Sigma q_j$$

then, making free use of the distributive law,

$$xy = \Sigma p_i q_j$$

where now, the product $p_i q_j$ is again a product of a finite number of elements from A or B.

Now we turn to perhaps the most useful algebraic tool in the theory of rings, homomorphisms. The theory is an analog of the theory for homomorphisms for groups with a little extra thrown in because of the multiplication.

7.2.4 Definition.   Ring Homomorphisms

A function ϕ from a ring R into a ring R' is a homomorphism provided

1. ϕ is a (additive) group homomorphism of $\langle R, + \rangle$ into $\langle R', + \rangle$.
2. The multiplicative homomorphism condition holds:

$$\phi(ab) = \phi(a)\phi(b)$$

for all a, b belong to R. The *kernel* of ϕ is the set of preimages of 0,

$$K = \{x \in R : \phi(x) = 0 \text{ in } R'\}$$

Thus the kernel of a ring homomorphism is the kernel of the group part of the homomorphism.

A homomorphism ϕ is called an

Isomorphism if ϕ is one-to-one and $\phi(R) = R'$. We write $R \cong R'$.
Endomorphism if $R' = R$.
Automorphism if $R' = R$ and ϕ is an isomorphism.

As with groups, the kernel plays a special role. It is a subring and because, if $k \in K$ then for all r in R,

$$\phi(rk) = \phi(r)\phi(k) = 0 \tag{1}$$

and

$$\phi(kr) = \phi(k)\phi(r) = 0 \tag{2}$$

we see that if $k \in K$ then ak and ka also belong to the kernel K. We isolate these properties of a kernel in the definition of an ideal (see Definition 7.2.5).

In addition, just because of the additive group part of the homomorphism, the set of preimages of an element $r' \in R'$ is an additive coset $r + K$ of K in R. Note that the cosets here are written additively and they are referred to variously as *congruence classes or residue classes*. An important example of a ring homomorphism is the mapping of $\mathbb{Z}$ into $\mathbb{Z}_n$ given by

$$u \to [u]_n$$

Although couched in slightly different language, Lemma 2.1.10 proves that the homomorphism conditions are satisfied. Look ahead, too, to Example 7.2.8.

7.2.5 Definition.   Ideal

A subring J of a ring R is called an *ideal* of R provided the following multiplicative condition is met:

If $a \in J$ then   for all $r \in R$ $ar \in J$ and $ra \in J$

Remark.   If R is a ring then the sets $\{0\}$ and R are always ideals of R.

Remark.   Suppose that J is an ideal in a ring with identity. If J contains a unit, then $J = R$.

Indeed, if u is a unit, say, $uv = 1$, and if $u \in J$, then $uv \in J$, so $1 \in J$ and so $r = r \cdot 1$ is also in J. Thus $J = R$.

In particular, this shows that in a field or a division ring, the only ideals are $\{0\}$ and the whole field, or division ring.

Remark.   Note that the multiplicative condition for an ideal is stronger than the multiplicative condition for J to be a subring. Thus to verify that a subset of a ring is an ideal, one may verify that it is an additive subgroup and has the stronger multiplicative condition. Thus an ideal "swallows up" products with other ring elements outside it. The condition in the definition is colloquially referred to as the "swallow up" condition.

As we have just observed, if K is the kernel of a ring homomorphism, then the group part of the homomorphism condition guarantees that K is an additive subgroup; equations (1) and (2) show that K is a subring and an ideal of R.

Just as with groups and normal subgroups, if J is an ideal then we can define a multiplication on the residue classes (also known as cosets) of J in R so that they become a ring, the factor ring $\overline{R} = R/J$, and there is a natural homomorphism of R onto $\overline{R}$. We collect these facts in Theorem 7.2.7, but first we must prove a simple lemma, the one that will show that the definition for the multiplication of two residue classes is independent of the representative. Please compare Lemma 7.2.6 and Theorem 7.2.7 with the one for groups, Theorem 4.2.11, that described how to multiply cosets of a normal subgroup.

7.2.6 Lemma.   Independence of Residue Class Representatives

If J is an ideal in a ring R and if $r_1 \in r + J$ and $s_1 \in s + J$ then $r_1 s_1 \in rs + J$.

Proof. Suppose $r_1 = r + h$ and $s_1 = s + k$ for some h, k belonging to J. Then

$$r_1 s_1 = rs + rk + hs + hk$$

Now, since J is an ideal, and h and k are in J, the products rk, hs, and hk all belong to J. Thus

$$r_1 s_1 = rs + t \quad \text{where} \quad t = rk + hs + hk \in rs + J$$

In particular, note that

$$r_1 s_1 + J = rs + J$$

since, as cosets of the additive subgroup J, they have common elements rs and $r_1 s_1$.

7.2.7 Theorem.   Factor Ring

Let R be a ring and let J be an ideal of R. The factor ring R/J is the ring whose elements are the residue classes of J:

$$R/J = \{r + J : r \in R\}$$

where the addition and multiplication are defined by

$$(r + J) + (s + J) = (r + s) + J$$

and

$$(r + J) \cdot (s + J) = rs + J$$

The function from R to R/J defined by

$$r \to r + J$$

is a homomorphism from R onto R/J whose kernel is J. It is called the natural homomorphism.

If J is the kernel of a homomorphism Φ from R into a ring R' then R/J is isomorphic to the range $\Phi(R)$; in symbols, $R/J \cong \Phi(R)$.

Proof. Because $\langle R, + \rangle$ is an abelian group the results of this theorem with respect to addition hold from the corresponding theorem for groups, Theorem 4.2.10. Lemma 7.2.6 guarantees that multiplication behaves correctly as well. Details are asked for in the exercises.

7.2.8 Example. $\mathbb{Z}_n$ as a Factor Ring

Recall Section 2.1. Now we can relate what happened there in more sophisticated terms. We know that $\langle \mathbb{Z}, +, \cdot \rangle$ is a ring and that the additive subgroup $\langle n \rangle = \{zn : z \in \mathbb{Z}\}$ is an ideal (Theorem 1.3.15). Thus the factor ring $\mathbb{Z}/\langle n \rangle$ is $\mathbb{Z}_n$.

Most of the rings we study will have a multiplicative identity. For these it is important to know the smallest subring containing the identity. Its characterization is given in the next theorem. The technique of proof is one that we reuse when we study polynomials. For this proof we need to distinguish between the multiplicative identity of the ring and the integer 1. For this proof we denote the multiplicative identity by I.

7.2.9 Theorem. The Subring $\langle I \rangle$

Let R be a ring with identity I. The smallest subring of R containing I is

$$\langle I \rangle = \{ nI : n \in \mathbb{Z} \}$$

Moreover, either

$$\langle I \rangle \text{ is isomorphic to } \mathbb{Z} \quad \text{or} \quad \langle I \rangle \text{ is isomorphic to } \mathbb{Z}_m$$

for some integer m.

Proof. We define a homomorphism ψ from $\mathbb{Z}$ onto $\langle I \rangle$ by

$$\psi : n \to nI$$

It is clear that a function has been defined; we have only to show the homomorphism properties. We use the two basic properties 3 of Lemma 7.1.4. First:

$$\psi(n + m) = (n + m)I = nI + mI = \psi(n) + \psi(m)$$

second

$$\psi(nm) = nmI = (nI)(mI) = \psi(n)\psi(m)$$

The kernel of ψ is thus some ideal $\langle m \rangle$ of $\mathbb{Z}$. The ring homomorphism Definition 7.2.4 and Example 7.2.8 give

$$\langle I \rangle \cong \mathbb{Z}/\langle m \rangle \cong \mathbb{Z}_m$$

We need only add that if $m = 0$ then ψ is an isomorphism and $\langle I \rangle$ is an isomorphic copy of the integers.

Remark. Notice that an alternative characterization is that the characteristic of R is the *additive* order of the multiplicative identity I (except that "infinite" order now becomes "0" characteristic!).

The notion of a congruence relation on the integers is also extended to general rings. We give the definitions and the theorems but omit the proofs since they are direct analogs of the integer case. Do, however, be careful not to use commutativity of multiplication in the arguments.

7.2.10 Definition. Characteristic of a Ring

The characteristic of a ring R with identity I is $m > 0$ if $\langle I \rangle$ is isomorphic to $\mathbb{Z}_m$. Otherwise $\langle I \rangle$ is isomorphic to $\mathbb{Z}$ and its characteristic is 0.

Notation.   We write char(R) for the characteristic of R.

Remark.   In view of Theorem 7.2.9 and the definition of the characteristic of a ring, if char(R) $= m > 0$ then m is the additive order of m and hence the least positive integer such that $mI = 0$. It follows then that

$$mr = (mI)r = 0$$

for all $r \in R$. Thus the additive order of each element in R divides m. On the other hand, if char(R) $= 0$ then $mI = 0$ in R implies that $m = 0$ in $\mathbb{Z}$.

7.2.11 Definition.   Congruence on a Ring

Let R be a ring. An equivalence relation $\equiv$ on R is called congruence relation if and only if whenever

$$a \equiv b \quad \text{and} \quad a_1 \equiv b_1$$

then

$$a + a_1 \equiv b + b_1 \quad \text{and} \quad a \cdot a_1 \equiv b \cdot b_1$$

7.2.12 Theorem.   Ideals and Congruence Relations

Let $\equiv$ be a congruence relation on a ring R. Then

$$K = \{r \in R : r \equiv 0\}$$

forms an ideal in R.
   The mapping

$$r \to \{x \in R : x \equiv r\} = \bar{r}$$

is the natural homomorphism of R onto R/K.
   Conversely, given an ideal K in R then the relation $\equiv_K$ defined by

$$a \equiv_K b \qquad \text{if and only if } (a - b) \in K$$

is a congruence relation on R and

$$K = \{r \in R : r \equiv_K 0\}$$

Proof.   We omit the details.

7.2.13 Theorem.   Lattice Properties under Homomorphisms

Let ϕ be a homomorphism from a ring R onto a ring $\bar{R}$ with kernel K.

1. The mapping

$$S \to \phi(S) = \bar{S}$$

is a one-to-one mapping of the subrings S of R such that $K \subseteq S \subseteq R$ onto the subrings of $\bar{R}$.

2. If S is an ideal of R containing K then $\overline{S}$ is an ideal of $\overline{R}$, and conversely, if $\overline{S}$ is an ideal of $\overline{R}$ then

$$S = \{ s \in R \colon \phi(s) \in \overline{S} \}$$

is an ideal containing K.

3. The lattice operations $\cap$ and $\vee$ are preserved:

$$\phi(A \cap B) = \phi(A) \cap \phi(B)$$
$$\phi(A \vee B) = \phi(A) \vee \phi(B)$$

4. If S is an ideal of R then the factor ring R/S is isomorphic to the factor ring $\overline{R}/\overline{S}$.

Proof. The theorem and its proof are analogous to the corresponding theorem for groups, Theorem 4.2.14. We leave the details as an exercise.

Direct products of rings work much the same way as do direct products of groups. We need only sort out how the multiplication works and replace "normal subgroup" with "ideal." There is a change in terminology as well. We speak of a direct sum of rings rather than a direct product. This reflects the emphasis on the additive nature of a ring. Here are the details.

7.2.14 Definition. Direct Sum of Rings

Let R and S be rings. The direct sum is the cartesian product

$$R \oplus S = \{ (r, s) \colon r \in R \text{ and } s \in S \}$$

on which an addition and multiplication are defined component by component:

$$(r, s) + (r', s') = (r + r', s + s')$$
$$(r, s) \cdot (r', s') = (rr', ss')$$

More generally, the direct sum of several rings R_i, $i = 1, \ldots, n$ is given

$$R_1 \oplus R_2 \oplus \cdots \oplus R_n = \{ (r_1, r_2, \cdots, r_n) \colon r_i \in R_i \}$$

where addition and multiplication are performed component by component.

It is routine to show that these operations defined on the cartesian product satisfy the ring axioms. The zero element is $(0, 0)$. The direct sum has a multiplicative identity if and only if each summand has a multiplicative identity. In this case the multiplicative identity is $(1_R, 1_S)$ where 1_R and 1_S are the multiplicative identities of R and S, respectively.

Notice the multiplication action forced by the special role of 0:

$$(r, 0)(0, s) = (0, 0)$$

for all $r \in R$ and $s \in S$. So there are lots of zero divisors in a nontrivial direct sum of rings.

Just as with groups, there are isomorphic copies of R and S in the direct sum $R \oplus S$:

$$\hat{R} = \{(r, 0): r \in R\} \quad \text{and} \quad \hat{S} = \{(0, s): s \in S\}$$

are subrings isomorphic to R and S, respectively.

Almost all of this has been proved under the aegis of groups. What is new is to check the multiplication action, but that is straightforward and we omit the details.

The criteria for a ring R to be the direct sum of two of its subrings is analogous to those for groups.

7.2.15 Theorem.   Criteria for a Ring to Be a Direct Sum

A ring R is isomorphic to the direct sum of subrings A and B provided these three conditions hold:

1. A and B are ideals of R.
2. $A \cap B = 0$.
3. $R = A \vee B$.

Proof. First observe that condition 2 means that products of the form ab or ba where $a \in A$ and $b \in B$ are always equal to 0. This is because A and B are ideals and so both ab and ba belong to $A \cap B$, which, from condition 2, is 0.

Now we can cite Theorem 5.2.4 and know that the mapping from the cartesian product $A \times B$ into R given by

$$(a, b) \rightarrow a + b$$

is an additive group isomorphism. We need only show that this mapping preserves multiplication. Let

$$(a', b') \rightarrow a' + b'$$

We are to show that

$$(a, b)(a'b') \rightarrow (a + b)(a' + b')$$

By definition

$$(a, b)(a', b') = (aa', bb')$$

and thus by definition

$$(a, b)(a', b') \rightarrow aa' + bb'$$

Thus we must show that

$$aa' + bb' = (a + b)(a' + b')$$

Expand

$$(a + b)(a' + b') = aa' + ab' + ba' + bb'$$

Since $ab' = b'a = 0$ by our first observation we have proved that multiplication has been preserved.

An extension to more than two ideals is readily obtainable. Simply quote Theorem 5.2.8 and replace "group" by "ring," "subgroup" by "subring," and "normal" by "ideal."

EXERCISE SET 7.2

7.2.1 Show that any subgroup of $\langle \mathbf{Z}, + \rangle$ is a subring of $\mathbf{Z}$. Find all the subrings of $\mathbf{Z}$. Find a subgroup of $\langle \mathbf{Q}, + \rangle$ that is not a subring of $\mathbf{Q}$.

7.2.2 Determine the ideals of $\mathbf{Z}_n$. Show that if n divides m then $\mathbf{Z}_n$ is a homomorphic image of $\mathbf{Z}_m$. Show that the units of $\mathbf{Z}_m$ map onto the units of $\mathbf{Z}_n$. (Use the theorems of this section as much as possible.)

7.2.3 Let $\Phi: R \rightarrow S$ be a ring homomorphism from the ring R onto the ring S. Prove or give a counterexample for the following:
1. If R is commutative then S is commutative.
2. If S is commutative then R is commutative.
3. If R has an identity then S has an identity.
4. If S has an identity then R has an identity.
5. If u is a unit of R then $\Phi(u)$ is a unit of S.
6. If $\Phi(u)$ is a unit of S then u is a unit of R.
7. If z is a zero divisor in R then $\Phi(z)$ is a zero divisor in S.
8. If $\Phi(z)$ is a zero divisor in S then z is a zero divisor in R.
9. If R is an integral domain then S is an integral domain.
10. If S is an integral domain then R is an integral domain.
11. What is the relation of the characteristic of R and the characteristic of S?

7.2.4 Prove Theorem 7.2.7 and Theorem 7.2.13.

Problems 7.2.5 to 7.2.7 concern nilpotent elements and nil ideals in a ring.

An element r in a ring R is called *nilpotent* if there is a positive integer n such that $r^n = 0$. (Different r's may have different n's; that is, the exponent n may depend on r.)

An ideal in R is called *nil* if every element in it is nilpotent.

7.2.5 Find all the nilpotent elements in $\mathbf{Z}_{18}$. Find all the nil ideals in $\mathbf{Z}_{18}$. Are there any ideals that are not nil in $\mathbf{Z}_{18}$?

7.2.6 Let r and s be two elements that are nilpotent and that commute. Show that $r + s$ is also nilpotent.

7.2.7 Let R be a commutative ring. Show that the set of all nilpotent elements in R forms an ideal K of R. Show that R/K has no nilpotent elements.

7.2.8 Let R be the ring of 2×2 diagonal matrices with integer entries:

$$R = \left\{ \begin{pmatrix} a & 0 \\ 0 & b \end{pmatrix} : a, b \in \mathbf{Z} \right\}$$

Find all the ideals in R. Prove or disprove: R isomorphic to $\mathbf{Z} \oplus \mathbf{Z}$. Find all the zero divisors in R.

7.2.9 Let T be the ring of 2×2 upper triangular matrices over $\mathbf{Q}$ defined in Example 7.1.2. Find all the ideals in T. Prove or disprove: T is the direct sum of two subrings.

*7.2.10 Show that the only ideals of $M_2(\mathbf{Q})$ are $\{0\}$ and the ring itself. Such rings are called "simple" rings. Show that in general $M_n(D)$ is always simple if D is a division ring.

7.2.11 Show that if A is an ideal of a ring R then for any subring B of R, $A \vee B = \{a + b : a \in A \text{ and } b \in B\} = A + B$.

7.2.12 Show that the set of all ideals of a ring form a complete lattice. Show that in this lattice if $A \subseteq B$ then, for all ideals C, $(A \vee C) \cap B = A \vee (C \cap B)$. This law is called the "modular law."

*7.2.13 State and prove a "diamond" lemma for rings.
(Try it; if you get stuck skip to Lemma 8.2.1.)

7.2.14 Show that the characteristic of an integral domain is 0 or a prime integer.

7.2.15 Show that an ideal A in $\mathbf{Z}[\sqrt{2}\,]$ has the form $a\alpha + b\beta$ where a and b are integers and α and β are in the ideal A.

Chapter 8

Integral Domains

8.1 ARITHMETIC IN AN INTEGRAL DOMAIN

In this section we extend to arbitrary integral domains the fundamental definitions of division, prime element, irreducible element, gcd, and lcm we have studied for the integers. Because much of what is true for integral domains is also true for any commutative ring, we make the definitions and some of the lemmas more inclusive. Please give some attention to the conditions assumed on the ring in what follows.

8.1.1 Definition. Division in Commutative Rings

Let R be a commutative ring and let a be a nonzero element in R. We say that a divides b, in symbols,

$$a \mid b$$

if there exists $c \in R$ such that

$$b = ac$$

We also say that b is a *multiple* of a if a divides b.

If a does not divide b we write

$$a \nmid b$$

Remark. To make the point of this definition clear, let us compare the integers $\mathbb{Z}$ and the even integers $2\mathbb{Z}$. Now 6 belongs to both these rings and of course $6 = 2 \cdot 3$ in $\mathbb{Z}$. Thus 2 divides 6 in $\mathbb{Z}$. However, 2 does not divide 6 in $2\mathbb{Z}$; there is no even integer a such that $6 = 2 \cdot a$.

Thus we see that divisibility is not inherited by subrings. On the other hand, divisibility is passed on to homomorphic images: If $\Phi\colon R \to S$ is a ring homomorphism and if $a \mid b$ in R then $\Phi(a) \mid \Phi(b)$ in S.

8.1.2 Lemma.   Properties of Division

Let R be a commutative ring.

> **1.** If $a \mid b$ then $a \mid bc$ and $a \mid -b$.
> **2.** If $a \mid b$ and $b \mid c$ then $a \mid c$.
> **3.** If $a \mid b$ and $a \mid c$ then $a \mid (b + c)$.

If in addition R has an identity:

> **4.** If u is a unit in R then $u \mid a$ for all $a \in R$.

If in addition R is an integral domain:

> **5.** If $a \mid b$ and $b \mid a$ then $b = au$ where u is a unit.
> **6.** If $a \mid b$ then there is only one c such that $b = ac$ and we write $c = b/a$.

Proof.   We omit the details.

8.1.3 Definition.   Irreducible and Prime Elements

Let R be an integral domain. An element $q \in R$ that is not zero and not a unit is called an *irreducible* if its only divisors are units and associates of q. Alternatively, q is irreducible provided that

$$\text{Whenever } q = ab \quad \text{then} \quad a \text{ or } b \text{ is a unit}$$

An element $p \in R$ that is not zero or a unit is called a *prime* provided that

$$\text{Whenever } p \mid ab \quad \text{then} \quad p \mid a \text{ or } p \mid b$$

In some integral domains the set of primes is different from the set of irreducibles. The next lemma shows that every prime is an irreducible. Example 8.1.5 shows that the converse is not so.

8.1.4 Lemma.   Primes Are Irreducibles

If p is a prime element of an integral domain then p is irreducible.

Proof.   Let p be a prime and suppose that $p = ab$. Thus $a \mid p$ and $b \mid p$. On the other hand, $p \mid ab$ and so $p \mid a$ or $p \mid b$. Thus for either a or b, say, a, it is true that

$$p \mid a \quad \text{and} \quad a \mid p$$

and by 5 of Lemma 8.1.2 it follows that p and a are associates so that $p = au$

where u is a unit. Since $p = ab$ it follows from 6 of Lemma 8.1.2 that $b = u$ and so b is a unit.

8.1.5 Example.   An Irreducible That Is Not Prime

In the integral domain

$$\mathbb{Z}[\sqrt{-5}] = \{a + b\sqrt{-5} : a, b \in \mathbb{Z}\}$$

the element 3 is irreducible but is not a prime.

Proof.   It is routine to verify that $\mathbb{Z}[\sqrt{-5}]$ is a subring of the complex numbers. Moreover, if $\alpha \in \mathbb{Z}[\sqrt{-5}]$ then its complex conjugate $\bar{\alpha} \in \mathbb{Z}[\sqrt{-5}]$, since if

$$\alpha = a + b\sqrt{-5} \quad \text{then} \quad \bar{\alpha} = a - b\sqrt{-5} \in \mathbb{Z}[\sqrt{-5}]$$

There are five steps to the argument; we leave the details to you. First determine the units of this integral domain.

1. Show that if α is unit then so is $\bar{\alpha}$.
2. Show that therefore $\alpha\bar{\alpha} = a^2 + 5b^2$ is a unit in $\mathbb{Z}$ and hence equal to 1. Conclude that the units of $\mathbb{Z}[\sqrt{-5}]$ are 1 and -1.
3. Next, show that 3 is irreducible. Note that if

$$3 = \alpha\beta$$

then, taking conjugates,

$$3 = \bar{\alpha}\bar{\beta}$$

and so

$$9 = (\alpha\bar{\alpha})(\beta\bar{\beta})$$

and thus

$$\alpha\bar{\alpha} = a^2 + 5b^2 = 1, 3, \text{ or } 9$$

Show that $\alpha\bar{\alpha} = 3$ is impossible and so conclude that α or β is a unit. Thus 3 is irreducible.
4. Now show that $(1 + 2\sqrt{-5})(1 - 2\sqrt{-5}) = 21 = 3 \cdot 7$.
5. Show that 3 divides neither $(1 + 2\sqrt{-5})$ nor its conjugate; hence 3 is not a prime.

Step 4 shows in effect that in $\mathbb{Z}[\sqrt{-5}]$ the number 21 has two distinct factorizations into irreducibles. True we haven't established that all the factors are irreducibles, but arguments like those above show this. Thus there can be no unique factorization theorem in $\mathbb{Z}[\sqrt{-5}]$ as there was in $\mathbb{Z}$. It turns out that the unique factorization theorem is highly desirable, and we shall go to some lengths to determine some integral domains with this property. For now we continue to extend the elementary arithmetical concepts to integral domains.

8.1.6 Definition.   Greatest Common Divisor (gcd) and Least Common Multiple (lcm)

Let R be an integral domain. Let $a, b \in R$ and suppose that one, say, a, is not zero.

A *greatest common divisor* of a and b, if it exists, is a divisor d of both a and b such that any other common divisor v of a and b divides d. Write

$$d = \gcd(a, b)$$

If gcd $(a, b) = 1$, we say that a and b are coprime or relatively prime.

A *least common multiple* of a and b, if one exists, is a multiple m of both a and b such that any other multiple of a and b is a multiple of m. Write

$$m = \text{lcm}(a, b)$$

In the next lemma we will see that in an integral domain a greatest common divisor and least common multiple are unique up to unit factors.

8.1.7 Lemma.   Uniqueness of gcd and lcm to Within Units

If R is an integral domain and d and d' are greatest common divisors of a and b, then d and d' are associates.

If m and m' are least common multiples of a and b then m and m' are associates.

Proof.   Suppose that d and d' are greatest common divisors of a and b. By definition, each divides the other and so $d = ed'$ where e is a unit.

A similar argument shows that two least common multiples can differ only by a unit factor.

Remark.   As dual as the two definitions of gcd and lcm sound, Example 8.1.8 shows that for certain integral domains there exist a and b such that the $\gcd(a, b)$ exists while the $\text{lcm}(a, b)$ does not exist.

8.1.8 Example.   Two Elements Having a gcd but Not an lcm

Let $R = \mathbb{Z}[\sqrt{-5}]$ and let $\alpha = 1 + 2\sqrt{-5}$. We have seen in Example 8.1.5 that 3 is irreducible but does not divide α. So $\gcd(3, \alpha) = 1$. However, an $\text{lcm}(3, \alpha)$ does not exist.

Proof.   To prove this, show first that if $\gamma = \text{lcm}(3, \alpha)$ then, to within a unit factor, $\gamma = 3\alpha$. However, this candidate for $\text{lcm}(3, \alpha)$ fails since 21 is a multiple of 3 and of α, but not of 3α.

To show that 3α is the only candidate for $\text{lcm}(3, \alpha)$ argue as follows: Suppose that $\gamma = \text{lcm}(3, \alpha)$. Then

$$3 \mid \gamma \quad \text{say} \quad \gamma = 3\lambda$$

and, since 3α is a common multiple,

$$\gamma \mid 3\alpha \quad \text{say} \quad 3\alpha = \gamma\delta$$

Thus taking complex conjugates and multiplying

$$9\alpha\bar{\alpha} = \gamma\bar{\gamma}\delta\bar{\delta} = 9\lambda\bar{\lambda}\delta\bar{\delta}$$

and since $9\alpha\bar{\alpha} = 9 \cdot 21$ we conclude

$$21 = \lambda\bar{\lambda}\delta\bar{\delta}$$

Now $\lambda\bar{\lambda}$ and $\delta\bar{\delta}$ are integers so each must equal 1, 3, 7, or 21. If either is 1, then λ or δ is a unit.

If λ were a unit then 3 would be the lcm$(3, \alpha)$, a contradiction.

If δ is a unit then 3α is the lcm$(3, \alpha)$, and that's what we want to show.

Finally the alternatives 3 and 7 are impossible, for no element of $\mathbb{Z}[\sqrt{-5}]$ has complex norm 3 or 7.

If either is 21 then the other is 1 and we are back to the first part of our argument. So we are done.

On the positive side there is this result.

8.1.9 Theorem.   A Relation between gcd and lcm

1. If nonzero elements a and b in an integral domain have a gcd and an lcm then, up to a unit factor,

$$ab = \gcd(a, b)\operatorname{lcm}(a, b)$$

2. If a and b have an lcm, then the gcd exists and

$$\gcd(a, b) = ab/\operatorname{lcm}(a, b)$$

Proof.   It suffices to prove part 2 of the theorem. Let lcm$(a, b) = m$. Since ab is a common multiple of a and b it follows that

$$ab = md$$

We shall prove that $d = \gcd(a, b)$.

First we may write

$$m = bb' \quad \text{and} \quad m = aa'$$

since m is a multiple of a and b.

Next we show that d divides both a and b. We have $ab = md = bb'd$ and, canceling, $a = b'd$ thus d divides a. Similarly, d divides b.

Finally, let u be any common divisor of a and b, say,

$$a = ux \quad \text{and} \quad b = uy$$

Then $bx = uyx$ is a common multiple of a and b. Hence $m \mid uyx$, say, $uyx = mz$. But then

$$md = ab = (ux)(uy) = u(uyx) = umz$$

canceling we find

$$d = uz$$

thus u divides d.

EXERCISE SET 8.1

8.1.1 Show that 2 is irreducible in $\mathbb{Z}[\sqrt{-5}]$ but is not a prime there. *Hint*: Let $\alpha = 1 + \sqrt{-5}$. Consider $\alpha\bar{\alpha}$.

8.1.2 Show that if a and b are elements in an integral domain and $\gcd(a, b) = d$ then $\gcd(a/d, b/d) = 1$.

8.1.3 Let p be a prime integer. Define the subset of the rational numbers $D_p = \{r = a/b: \gcd(a, b) = 1, b = p^k \text{ for some integer } k \geq 0\}$. Show that D_p is an integral domain. Find the units and the primes in D_p. Show that every irreducible is a prime.

8.1.4 Let S be a subset of the positive prime integers. Define the subset of the rational numbers $D_S = \{a/b: \gcd(a, b) = 1 \text{ and } b = 1 \text{ or is a product of primes in } S\}$. Show that D_S is a subring of the rational numbers. Find all the units, primes, and irreducible elements in D_S.

8.1.5 Continue the notation from Problem 8.1.4. Show that $D_{S \cap T} = D_S \cap D_T$ and $D_{S \cup T} = D_S \vee D_T$.

8.1.6 Continue the notation from Problems 8.1.4 and 8.1.5. Show that any subring of the rational numbers containing 1 is a D_S for suitable S. Conclude from Problem 8.1.5 that the lattice of subrings of the rational numbers containing 1 is isomorphic to the lattice (boolean algebra) of all the subsets of $\mathbb{Z}^+$.

8.1.7 Show that there is no integral domain with six elements. Can you generalize this result?

8.2 IDEALS IN COMMUTATIVE RINGS

We begin with a simple description of the lattice join of two ideals. And we prove the ring theoretic analog of the diamond isomorphism for groups.

8.2.1 Lemma. The Join of Ideals and the Diamond Lemma for Rings

If S is a subring of a ring R and A is an ideal of R then

$$S \vee A = \{s + a: s \in S \text{ and } a \in A\}$$

Moreover, A is an ideal of $S \vee A$, then $S \cap A$ is an ideal of S, and $(S \vee A)/A \cong S/(S \cap A)$.

Proof. From Lemma 7.2.3 we know that, in a subring join, a typical element in $S \vee A$ has the form

$$\sum p_i$$

where p_i is a product of elements in S or in A. If one of the products has a factor in A, then the swallow-up property of ideals ensures that the whole product is in A. On the other hand, if all the factors in a product are in S then the product is in S. Thus each p_i belongs either to S or to A. Now collect those terms in S; let their sum be called s for the moment. If there are none of these terms, let $s = 0$. Collect all the terms in A; let their sum be called a for the moment. If there are none of these terms, let $a = 0$. We have thus just shown that

$$\sum p_i = s + a$$

Now since A is an ideal of R, it is automatically an ideal in any subring containing it. The subring $S \cap A$ is an ideal of S because if $s \in S$ and $a \in S \cap A$ then $sa \in S$ since $a \in S$ and $sa \in A$ by the swallow-up property of A. Similarly $as \in S \cap A$. Thus $S \cap A$ is an ideal of S.

Next we prove the isomorphism of $(S \vee A)/A$ and $S/(S \cap A)$. We make things easy by recognizing that the additive groups of these two rings are isomorphic by the group theoretic diamond lemma. Hence we already know that the Φ from $S \vee A$ to $S \cap A$ defined by

$$\Phi: x \rightarrow [s] = s + (S \cap A)$$

if $x = s + a \in S \vee A$ is an additive homomorphism from $S \vee A$ onto $S/(S \cap A)$ whose kernel is A.

What is left is to show that Φ preserves multiplication. Let $\Phi(x) = [s]$ and $\Phi(y) = [t]$. We know that as residue classes in $S/(S \cap A)$, $[s][t] = [st]$. We will show that $\Phi(xy) = [st]$. We may suppose that $x = s + a$ for some $a \in A$ and $y = t + b$ for some $b \in A$. Then $xy = st + sb + at + ab$. Now, because A is an ideal, the terms sb, at, and ab are all in A. Hence $xy = st + a'$ where $a' \in A$ and so $\Phi(xy) = [st]$. Thus Φ is a ring homomorphism for $S \vee A$ onto $S/(S \cap A)$ whose kernel is A. Hence $(S \vee A)/A$ is isomorphic to $S/(S \cap A)$.

We want now to look at some of the elementary constructs and properties of ideals in a commutative ring. First, in analogy with a cyclic subgroup generated by a single element, there are ideals generated by a single element. The next lemma gives the construction; the ideal so generated is called a *principal* ideal. The principal ideal generated by an element a is the smallest ideal containing a.

8.2.2 Lemma.   Construction of an Ideal

Let R be a commutative ring, not necessarily having an identity. Let $a \in R$. The set

$$\langle a \rangle = \{ ra + na : r \in R \text{ and } n \in \mathbb{Z} \}$$

is an ideal. If R has an identity then

$$\langle a \rangle = \{ ra : r \in R \}$$

and we write $\langle a \rangle = Ra$.

Proof. The verification is routine: If $ra + na$ and $sa + ma \in \langle a \rangle$ then

$$(ra + na) - (sa + ma) = (r + s)a + (n - m)a \in \langle a \rangle$$

If $ra + na \in \langle a \rangle$ and $s \in R$ then

$$s(ra + na) = (sr + ns)a \in \langle a \rangle$$

If R has no identity we must allow for elements like

$$a + a + a = 3a$$

If R has an identity 1, we can write $3a = 3(1) \cdot a$, which has the form of a product of two elements in R, $3(1)$ and a. Failing this, elements like $3a$ have been included explicitly in the ideal $\langle a \rangle$.

Indeed, if there is an identity 1 in R then

$$\langle a \rangle = \{ ra \colon r \in R \}$$

since, for any integer n, we may write

$$na = (n1) \cdot a$$

and thus express the sum of n terms, each equal to a, as a product of n elements $n(1)$ and a in R. Thus, if R is a commutative ring with identity then

$$\langle a \rangle = \{ x \in R \colon a \mid x \}$$

8.2.3 Definition. Principal Ideal

Let R be a commutative ring. A principal ideal J is one for which there is an element a such that

$$J = \langle a \rangle$$

In most of the rings that we shall study all the ideals are principal. That results in great simplification. The major tool we have for establishing that a given ring has this property is the euclidean algorithm. This we take up in Section 9.3., Euclidean Domains and Principal Ideals.

8.2.4 Definition. Prime Ideal

Let R be a commutative ring. An ideal J in R is called *prime* if whenever a product of elements in R is in J then one of the factors is in J; in symbols

$$\text{Whenever } ab \in J \quad \text{then} \quad a \in J \text{ or } b \in J$$

The importance of prime ideals is that they help to determine certain integral domains. This is the context of the next theorem. But first we pause to give a criterion for a principal ideal to be prime in a commutative ring.

8.2.5 Lemma. Criterion for a Principal Ideal to Be Prime

Let R be a commutative ring with identity. If $p \neq 0$ the principal ideal $\langle p \rangle$ is prime if and only if p is prime in R.

Proof. First, suppose that $\langle p \rangle$ is a prime ideal. We shall show that p is a prime element. Suppose $p \mid ab$. Then $ab \in \langle p \rangle$ and so one of a or b is in $\langle p \rangle$, say, a. But that means that $p \mid a$.

Second, suppose that p is a prime element. Suppose that $ab \in \langle p \rangle$, hence $p \mid ab$ and hence p divides either a or b; and whichever p divides is in $\langle p \rangle$.

8.2.6 Theorem.   Criterion for a Factor Ring to Be an Integral Domain

Let R be a commutative ring with identity and let J be an ideal of R. The factor ring R/J is an integral domain if and only if J is a prime ideal.

Proof.   To begin with, note that any factor ring of R will be commutative and have an identity. Thus to prove that R/J is an integral domain we need only show that it has no zero divisors.

Suppose first that J is a prime ideal of R. Now suppose that in R/J the product of two residue classes is J, the zero element of R/J. That is, suppose

$$(a + J)(b + J) = J$$

By definition,

$$(a + J)(b + J) = ab + J$$

It follows that $ab + J = J$, and so $ab \in J$. If J is prime then either a or b is in J, say, a. Then the residue class

$$a + J = J$$

and so $a + J = 0$ in R/J. Thus R/J has no zero divisors and is an integral domain.

Conversely suppose that R/J is an integral domain. Now suppose that a product $ab \in J$. Then

$$(a + J)(b + J) = ab + J = J$$

Since R/J is an integral domain, it follows that one of the factors, say, $a + J$ is the 0-residue class. That means that $a + J = J$ or that $a \in J$. Hence J must be a prime ideal in R.

8.2.7 Example. $\mathbb{Z}_p$ Is a Field

Let the ring R be the integers $\mathbb{Z}$. We know from Theorem 1.3.15 that all the ideals of $\mathbb{Z}$ are principal. Thus the homomorphic images of $\mathbb{Z}$ that are integral domains are

$$\mathbb{Z}/\langle p \rangle = \mathbb{Z}_p$$

and being finite, Theorem 7.1.5 tells us that each one is a field. One reason that we obtain a field is that a prime ideal $\langle p \rangle$ in $\mathbb{Z}$ is *maximal*, in the sense that no ideal strictly contains $\langle p \rangle$ and is strictly contained in $\mathbb{Z}$. This result holds in general, as we prove next.

8.2.8 Definition.   Maximal Ideal

Let R be a ring. An ideal M of R is called maximal if whenever there is an ideal K such that

$$M \subseteq K \subseteq R \quad \text{then} \quad K = M \text{ or } K = R$$

8.2.9 Theorem.   Criterion for a Factor Ring to Be a Field

Let R be a commutative ring with identity. A factor ring R/J is a field if and only if J is a maximal ideal in R.

Proof.   The easy part is to show that if R/J is a field then J is maximal. From the lattice properties of ring homomorphisms (Theorem 7.2.13) we know that the ideals of R containing J are in one-to-one correspondence with those of R/J. Now if R/J is a field, then R/J has only the ideals zero and itself, R/J. But that means that there are no ideals of R strictly containing J and contained in R.

To prove the converse, and the useful part, we need only to prove that the nonzero elements of R/J have multiplicative inverses. That is, we need to show that if a is not in J then there is an element b in R such that

$$(a + J)(b + J) = ab + J = 1 + J$$

Remember, the residue class $1 + J$ is the multiplicative identity in R/J. An equivalent way of writing this is that if a is not in J there is an element b in R and an element j in J such that

$$1 = ab + j$$

This suggests constructing the ideal

$$\langle a \rangle \vee J = \langle a \rangle + J = \{ ra + t : r \in R \text{ and } t \in J \}$$

By Lemma 8.2.1 this is an ideal and it is larger than J since $a \in \langle a \rangle \vee J$ and a is not in J. Therefore, it must be R and so

$$1 \in \langle a \rangle \vee J$$

But this means that for some $j \in J$ and some $b \in R$,

$$1 = ab + j$$

EXERCISE SET 8.2

8.2.1   Let S be a subset of the positive prime integers. Define the subset of the rational numbers, $D_S = \{ a/b : \gcd(a, b) = 1 \text{ and } b = 1 \text{ or is a product of primes in } S \}$. (Recall Problem 8.1.4.) Show that every ideal in D_S is principal. If p is a prime in D_S describe $D_S/\langle p \rangle$ in familiar terms.

8.2.2   In $\mathbb{Z}[\sqrt{2}] = \{ a + b\sqrt{2} : a, b \in \mathbb{Z} \}$ let $J = \langle 2 \rangle$. Write out a multiplication table for R/J. Find a maximal ideal in R/J.

8.2.3 Complete this sentence: "If (0) is a prime ideal of a commutative ring R then R is ---."

8.2.4 Show that in a commutative ring with identity every maximal ideal is prime.

8.2.5 Show that if every ideal of a ring is principal, then every ideal of every homomorphic image is principal.

8.2.6 Show that every ideal of the ring R,

$$R = \left\{ \begin{pmatrix} a & 0 \\ 0 & b \end{pmatrix} : a, b \in \mathbb{Z} \right\}$$

is principal. (This ring was introduced in Problem 7.2.8.) Find the prime and the maximal ideals in this ring and show that not every prime ideal is maximal.

8.2.7 Prove or disprove: If R and S are commutative rings in which every ideal is principal, then every ideal of $R \oplus S$ is principal.

8.2.8 Let R be an integral domain in which every ideal is principal. Show that a prime ideal is either $\{0\}$, R, or a maximal ideal of R.

8.2.9 Let R be a ring with an identity. Use the maximal principal 2.2.6 to show that every ideal, except R itself is contained in a maximal ideal.

8.3 INTRODUCTION TO FIELDS

One of the main themes of this book is the structure of fields, their subfields, their extensions, and their automorphisms. Recall that a field $\langle F, +, \cdot \rangle$ is a commutative ring with identity in which every nonzero element has a multiplicative inverse. In brief:

$$\langle F, + \rangle \text{ is an abelian group with } 0 \text{ as its identity}$$

and

$$\langle F - \{0\}, \cdot \rangle = \langle F^\times, \cdot \rangle \text{ is an abelian group with } 1 \text{ as its identity}$$

satisfying the distributive laws

$$a(b + c) = ab + ac \quad \text{and} \quad (a + b)c = ac + bc$$

for all a, b, and $c \in F$.

Each nonzero element is a unit (has a multiplicative inverse) and so a field has only two ideals, the zero ideal, $\{0\} = \langle 0 \rangle$, and the field, $F = \langle 1 \rangle = \langle a \rangle$ if $a \neq 0$.

This means in turn that a homomorphism of a field into a ring is either one-to-one (kernel $= \langle 0 \rangle$) or maps every element to 0 (kernel $= F$). Thus we shall not look for the kind of structure provided by the study of homomorphic images of groups; there are no interesting "factor fields."

However, there are often many isomorphisms. In the case where these are automorphisms, that is, isomorphisms of the field onto itself, we find that they form a group and the structure of the group will tell us a great deal about the structure of the subfields of a field. All this in later chapters.

The first important characteristic (pun intended!) to determine about a field is its characteristic (as a ring). There are two possibilities, either 0 or a prime p. Here is the relevant theorem.

8.3.1 Theorem.   Characteristic of a Field

The characteristic of a field F is either 0 or p, p a prime.

Proof.   From Definition 7.2.10 we know that the characteristic of a field as a ring is either 0 or n. If n were composite, say $n = rs$, then

$$0 = n \cdot 1 = rs \cdot 1 = (r \cdot 1)(s \cdot 1)$$

and since neither $r \cdot 1$ nor $s \cdot 1 = 0$ this would mean that the field has zero divisors, contrary to assumption.

Look ahead to Theorem 8.3.8. The two possibilities for the characteristic continue to distinguish a fundamental property of the field, namely, its smallest subfield, but we are ahead of our story.

Usually a field has many subfields and there are the usual definitions and introductory theorems which we now mention.

8.3.2 Definition.   Subfield

A nonempty subset S of a field $\langle F, +, \cdot \rangle$ is a subfield if $\langle S, +, \cdot \rangle$ is a field.
     A subfield S is called *proper* if $S \neq F$.

Thus a subfield is a subring that contains a^{-1} whenever it contains a, if a is not equal to 0, of course. This is the usual test that needs to be applied to a subring to verify that it is a subfield.

Next we introduce two notions related to that of a "subfield" which play a significant role in the theory of fields.

8.3.3 Definition.   Field Extensions and Embeddings

If S is a subfield of F, we say that F is an extension of S. If S is a field and F is a field containing a subfield S^* which is isomorphic to S, then we also say that S is embedded in F.

As examples we can observe that the field $\mathbb{Q}$ of rational numbers is embedded in the field $\mathbb{R}$ of real numbers which is embedded in the field $\mathbb{C}$ of complex numbers. Here is another interesting example.

8.3.4 Example.   A Field with Four Elements

We shall construct a field with four elements and show that it is unique, up to isomorphism. (In Chapter 10 we give a complete determination of all finite fields.)

First, if there is a field with four elements, what must be its characteristic? Well, the order of the additive group is 4. And we know that its characteristic is the additive order of 1 and hence must divide 4. Also it must be a prime.

Thus the characteristic of a field with four elements is 2. In turn this means that the order of each element divides 2 since

$$x + x = 2x = (2 \cdot 1) \cdot x = 0 \cdot x = 0$$

and so the additive group must be the four-group.

We can also easily determine the multiplicative group. Its order is 3 and so the multiplicative group is the cyclic group of order 3. In particular it must be that $x^3 = 1$ if $x \neq 0$ and

$$x^4 = x$$

for all x in a field of four elements.

Now we can write the addition and multiplication tables. Let u be any element not equal to 0 or 1. Then we may construct these tables:

Addition

0	1	u	$(1 + u)$
1	0	$(1 + u)$	u
u	$(1 + u)$	0	1
$(1 + u)$	u	1	0

Multiplication

1	u	u^2
u	u^2	1
u^2	1	u

From a comparison of the tables we conclude that

$$u^2 = 1 + u$$

and that the elements must be $\{0, 1, u, u^2 = 1 + u\}$. It is remarkable that the distributive laws hold for these tables. They do, and to show that there is a field with four elements all one needs to do is to check the distributive laws. This is a tedious chore, although not so horrible in this case. Instead we prefer to give another model for this field.

Pause first to realize that the tables above were constructed simply from the assumption that a field of four elements existed. Thus any field of four elements must have these tables; we conclude that any two fields consisting of four elements are isomorphic.

Now here is another construction. Let F consist of the following matrices with entries from $\mathbb{Z}_2$:

$$F = \left\{ \begin{pmatrix} a & b \\ b & a + b \end{pmatrix} : a, b \in \mathbb{Z}_2 \right\}$$

Verify that these matrices are closed under matrix addition and multiplication. Since addition and multiplication of matrices satisfy the ring properties we need only see that the nonzero matrices have multiplicative inverses. If

$$u = \begin{pmatrix} 0 & 1 \\ 1 & 1 \end{pmatrix} \quad \text{then} \quad u^2 = u + I = \begin{pmatrix} 1 & 0 \\ 1 & 1 \end{pmatrix}$$

and $u^3 = I$.

Notice that the diagonal matrices

$$\begin{pmatrix} a & 0 \\ 0 & a \end{pmatrix}$$

form a subfield isomorphic to $\mathbb{Z}_2$. Thus we can say that $\mathbb{Z}_2$ is embedded in this field. Equivalently, F is an extension of $\mathbb{Z}_2$.

8.3.5 Theorem. Lattice of Subfields

The subfields of a field form a complete lattice under set inclusion.

Proof. Clearly F itself is a subfield and is the top element of this partially ordered set of subfields under set inclusion. We have already seen that the set intersection of a collection of rings is a subring: we need only show that if $a \neq 0$ then a^{-1} belongs to this intersection whenever a does. But since all the rings in the collection are sub*fields*, it follows that a^{-1} belongs to each subfield in the collection and so to the set intersection.

Thus we have verified the criterion of Theorem 2.2.13, and the proof is complete.

What is the bottom element of this lattice? It will be the set intersection of all the subfields of a field. Well, it is not $\{0\}$ since that is not a field (any field contains at least two elements). It turns out that this bottom subfield is isomorphic either to the field of rational numbers $\mathbb{Q}$ or to $\mathbb{Z}_p$ for some prime p. It is not difficult to prove this but we do need to recognize how $\mathbb{Q}$ is constructed from the integers $\mathbb{Z}$. This construction can be carried out on any integral domain, and many interesting fields are constructed in this fashion. The next theorem gives this construction.

8.3.6 Theorem. Construction of the Field of Fractions

An integral domain $\langle D, +, \cdot \rangle$ may be embedded in a field in the following manner: Let

$$M = \{(a, b): a, b \in D, b \neq 0\}$$

Define an equivalence relation $\sim$ on M by

$$(a, b) \sim (c, d) \qquad \text{if and only if} \quad ad = bc$$

Let F^* denote the set of equivalence classes under $\sim$. Denote by the "fraction" a/b the equivalence class containing (a, b). Define addition and multiplication on F^* by

$$a/b + c/d = (ad + bc)/bd$$
$$(a/b)(c/d) = (ac)/(bd)$$

Then F^* is a field and the set

$$D^* = \{a/1: a \in D\}$$

is isomorphic to D.

Moreover F^* is the smallest field in which D can be embedded in the following sense: If E is a field containing D then E is an extension of F^*. In particular if D is contained in no proper subfield of E then E and F^* are isomorphic.

Proof. There are many details for this proof. None are difficult and we shall leave most of them as exercises. We do need to be careful to check off the things that have to be proved.

First, we must verify that $\sim$ is an equivalence relation on M.

Second, we must prove that the definitions for the sum and product of fractions a/b and c/d do not depend on the equivalence class representatives chosen.

Third, we must prove that the field axioms are satisfied.

Fourth, we must prove that the map

$$a \to a/1$$

is an isomorphism.

Finally, we must show the minimal nature of F^*. To do this let us suppose that E is a field containing D. We first verify that the mapping from the field of fractions F^* into E via

$$a/b \to ab^{-1}$$

is an isomorphism. As usual we must first verify that this map is well defined, that is, that it does not depend on the equivalence class representative chosen to determine the image of a/b. Then we must verify that the ring homomorphism properties hold. Now the kernel does not contain 1 since

$$1/1 \to 1 \neq 0$$

and hence the mapping must be one-to-one. Thus its image is a field containing D and contained in E. This means that E is an extension of F^*. If in addition E has no proper subfields containing D then E is the range of the map and so E is isomorphic to F^*.

We leave the details as an exercise. The reason that the algebraic details are familiar is that they are just the rules of algebra that we have long been using in dealing with the rational numbers. Indeed, this is the way the rational numbers $\mathbb{Q}$ are constructed from the integers.

Now we can determine the nature of the bottom element in the lattice of all subfields of a field.

8.3.7 Definition.   Prime Subfield

The prime subfield of a field is the intersection of all of its subfields.

8.3.8 Theorem.   Characterization of the Prime Subfield of a Field

Let F be a field. The prime subfield of F is the smallest subfield containing 1. There are two possibilities:

If char$(F) = 0$ the prime subfield is isomorphic to $\mathbb{Q}$.

If char$(F) = p$, p a prime, the prime subfield is isomorphic to $\mathbb{Z}_p$.

Proof. From Theorem 7.2.9 and its proof we know the image of $\mathbb{Z}$ under the ring homomorphism Φ from $\mathbb{Z}$ into F defined by

$$\Phi : n \to n \cdot 1$$

is isomorphic to either $\mathbb{Z}$ (in case char(F) = 0) or to $\mathbb{Z}_p$. Also we see that the image contains 1 and hence the image is contained in every subfield of F.

In the case that char(F) = p, p a prime, we know that $\mathbb{Z}_p$ is a field and thus the homomorphic image of $\mathbb{Z}$ is both a subfield of F and contained in all subfields of F. Hence $\mathbb{Z}_p$ is isomorphic to the smallest field containing $1 \in F$.

If char(F) = 0 the range of the homomorphism is isomorphic to $\mathbb{Z}$, which is not a field. But the range consists of elements that must be *in* the smallest field containing 1 and so we just need to see what elements must be added to create a subfield of F. These are the elements b^{-1} for all $b \neq 0$ in the range and then, for closure's sake, ab^{-1}. Finally there are sums of these elements and their inverses to check.

Alternatively we can extend the map Φ to a map from $\mathbb{Q}$, viewed as the field of fractions of $\mathbb{Z}$, into F via

$$a/b \to (a \cdot 1)(b \cdot 1)^{-1}$$

The details that this is a ring homomorphism are by now familiar. The mapping must be one-to-one since not every element of $\mathbb{Q}$ is mapped to zero. Thus F will contain a subfield isomorphic to $\mathbb{Q}$, and by the minimal nature of the field of fractions it follows that this image of $\mathbb{Q}$ in F must be the smallest subfield containing 1.

8.3.9 Theorem.   Fixed Field of an Automorphism

Let F be a field and let σ be an automorphism of F. The set of elements fixed by σ form a subfield, called the *fixed field* of σ. This subfield necessarily contains the prime subfield of F.

Proof. The closure properties are routine. Problem 8.3.9 asks for details. As the last theorem of this section we record for future reference a fact that we shall use often, especially in the case that the ring is a field.

8.3.10 Theorem.   Ring Automorphisms Form a Group

Let R be a ring. The set of automorphisms of R form a group G under composition. If S is a subset of R, the automorphisms fixing each element of S form a subgroup of G.

Proof. It is routine to verify that the composition of automorphisms is again an automorphism of R. We know that composition of functions, hence also of automorphisms, is associative. The identity function is an automorphism and is the identity element of the group of automorphisms.

Perhaps the only fact that is not immediate is the existence of inverse automorphisms. So, let σ be an automorphism of R,

$$\sigma: r \to \sigma(r)$$

We define the inverse of σ, denoted here by τ, the only way we can:

$$\tau(s) = r \quad \text{if and only if} \quad s = \sigma(r) \tag{1}$$

Because σ is a one-to-one function from R onto R, condition (1) does define a one-to-one function from R onto R. (You might want to check out once more that (1) means that $\tau(\sigma(s)) = s$ and $\tau(\sigma(r)) = r$ for all s and r in R.) What we need to prove is that τ obeys the homomorphism condition, that is, that τ preserves addition and multiplication. We ask

$$\tau(a + b) \stackrel{?}{=} \tau(a) + \tau(b) \quad \text{and} \quad \tau(ab) \stackrel{?}{=} \tau(a)\tau(b)$$

From (1) these questions are equivalent to

$$\sigma(\tau(a) + \tau(b)) \stackrel{?}{=} a + b \quad \text{and} \quad \sigma(\tau(a)\tau(b)) \stackrel{?}{=} ab$$

Now we calculate, using the fact that σ is a homomorphism:

$$\sigma(\tau(a) + \tau(b)) = \sigma(\tau(a)) + \sigma(\tau(b)) = a + b$$

and

$$\sigma(\tau(a) \cdot \tau(b)) = \sigma(\tau(a)) \cdot \sigma(\tau(b)) = a \cdot b$$

Finally it is easy to see that the set of automorphisms fixing each element of a subset S is closed under composition. And if σ fixes each element of S so does its inverse σ^{-1}. Hence the automorphisms fixing each element of S form a subgroup.

EXERCISE SET 8.3

8.3.1 Prove that the direct sum of two fields is never a field.

8.3.2 Complete the details of the field of fractions theorem 8.3.6.

8.3.3 Let F be the set of 2×2 matrices with rational coefficients of the form

$$\begin{pmatrix} a & b \\ 2b & a \end{pmatrix}$$

Show that F is a field under the usual operations of matrix addition and multiplication. Show that the rational numbers $\mathbb{Q}$ are embedded in F and find the prime subfield of F. Show that F is isomorphic to

$$\mathbb{Q}(\sqrt{2}) = \{r + s\sqrt{2} : r, s \in \mathbb{Q}\}$$

8.3.4 Find an automorphism of $\mathbb{Q}(\sqrt{2})$ that is not the identity.

8.3.5 Determine the fraction field of $\mathbb{Z}[\sqrt{2}]$. Determine the fraction field of D_S. (Refer to Problem 8.1.4 for the definition of D_S.)

8.3.6 Let F be a finite field, a field with m elements. Show that m must be a power of a prime. (*Hint*: Think additive groups!)

8.3.7 Let F be a finite field with p^n elements. Show that if S is a subfield then S has p^m elements where m divides n. *Hint*: Recall Problem 1.3.7. Now think multiplicative groups! We will later prove that there is exactly one subfield for each divisor of n.

8.3.8 Describe the smallest subfield of the field of real numbers containing $\sqrt{2}$ and $\sqrt{3}$. Show that it has five subfields. Are any two isomorphic?

8.3.9 Prove that if σ is an automorphism of a field F then $S = \{x: \sigma(x) = x\}$ is a subfield of F. Show that S always contains the prime subfield of F.

8.3.10 Let F be a field of characteristic p. Show that the mapping $\sigma: x \to \sigma(x) = x^p$ is an automorphism of F. Show that if F is a finite field then for all $a \in F$ the equation $x^p = a$ has a solution in F.

8.3.11 Let R be a ring with identity I and let σ be an automorphism of R. Show that $\sigma(I) = I$. Construct an example of a ring R with identity and a homomorphism Φ from R into R such that $\Phi(I) \neq I$.

Chapter 9

Polynomials

Polynomials are among the most familiar objects of study in beginning algebra and calculus courses. We study functions like

$$x \to f(x) = x^2 + x + 1$$

We determine its zeros (numbers r such that $f(r) = 0$), its points of local maximum and local minimum and its inflection points. We differentiate and integrate these functions. The symbol x becomes an old friend and takes on a life of its own.

In this text we are primarily interested in the ring properties of these functions. We make these functions into a ring by defining addition and multiplication of these functions in a manner that we call colloquially "pointwise":

$$(f + g)(x) = f(x) + g(x)$$
$$(f \cdot g)(x) = f(x)g(x)$$

This means that if

$$f(x) = a_0 + a_1 x + \cdots + a_k x^k + \cdots + a_n x^n$$

and

$$g(x) = b_0 + b_1 x + \cdots + b_k x^k + \cdots + b_m x^m$$

then we obtain

$$(f + g)(x) = (a_0 + b_0) + (a_1 + b_1)x + \cdots + (a_k + b_k)x^k + \cdots$$

by using the associative and commutative laws for addition and the distribu-

216

tive law to collect terms of equal exponent in x. We also obtain

$$(f \cdot g)(x) = c_0 + c_1 x + \cdots + c_k x^k + \cdots + c_{n+m} x^{n+m}$$

where

$$c_k = \sum_{0 \le h \le k} a_h b_{k-h} = \sum_{i+j=k} a_i b_j$$

by using associativity of addition and multiplication, distributivity, and the commutative law for addition. It is interesting and important to observe that the commutative law for multiplication is used only for computations involving x. The only commutativity we need for multiplication is

$$a_i x = x a_i \quad \text{and} \quad b_j x = x b_j$$

for all i's and j's. We do not need to know or assume that the a's commute with themselves or with the b's or that the b's commute with themselves or with the a's. This means that these ring operations for polynomial functions are valid even when the coefficients belong to a ring that may not be commutative as long as the elements substituted for x do commute with the coefficients.

What then is x? In this section we give substance to x itself and we construct new rings by adjoining the symbol x, now to be referred to as an "indeterminate." To do this reflect on how x is used in algebra and calculus in dealing with polynomials. There it is used primarily as a symbol to keep apart the various coefficients. In the next section we begin by defining polynomials without using x.

9.1 POLYNOMIAL RINGS

We begin with the definition of a polynomial and the ring of polynomials over a ring. The definition speaks of an addition and multiplication for this ring in anticipation of the next theorem, which establishes the right to use these terms.

9.1.1 Definition. Polynomials over a Ring

Let R be a ring with identity 1.

1. A polynomial over R is an infinite sequence

$$(a_0, a_1, \ldots, a_k, \ldots)$$

   in which each term $a_i \in R$ and in which *only a finite number* of terms are different from 0.
2. The zero polynomial is the sequence whose every term is 0. Denote this polynomial as 0.
3. The degree of a nonzero polynomial is the maximal index n for which $a_n \ne 0$. We write $\deg(f)$ for the degree of the polynomial f. The coefficient a_n is called the leading coefficient of f. If the leading

coefficient is 1 then f is called a monic polynomial. The zero polynomial is not assigned a degree.

4. The ring of polynomials over R is the set of all polynomials over R with an addition $(+)$ and a multiplication $(\cdot)$ defined by

$$(a_0, a_1, \ldots, a_k, \ldots) + (b_0, b_1, \ldots, b_k, \ldots)$$
$$= (a_0 + b_0, a_1 + b_1, \ldots, a_k + b_k, \ldots)$$
$$(a_0, a_1, \ldots, a_k, \ldots) \cdot (b_0, b_1, \ldots, b_k, \ldots) = (c_0, c_1, \ldots, c_k, \ldots)$$

where

$$c_k = \sum_{0 \le h \le k} a_h b_{k-h} = \sum_{i+j=k} a_i b_j$$

9.1.2 Theorem.   Ring Properties of Polynomials

Let R be a ring with identity 1. The operations in Definition 9.1.1 make the polynomials over R into a ring with identity. Moreover, if f and g are two nonzero polynomials then either

$$f + g = 0 \quad \text{or} \quad \deg(f+g) \le \max\{\deg(f), \deg(g)\} \tag{1}$$

and either

$$f \cdot g = 0 \quad \text{or} \quad \deg(f \cdot g) \le \deg(f) + \deg(g) \tag{2}$$

If the product of the leading coefficients of f and g is nonzero then

$$\deg(f \cdot g) = \deg(f) + \deg(g) \tag{3}$$

in particular equation (3) is true for monic polynomials in all rings or for any two nonzero polynomials if R is an integral domain.

Proof. We have to verify all the ring axioms. Most of the details are straightforward, and we shall just indicate how they go.

Begin by noting that the definition of addition and multiplication of polynomials produces other polynomials; that is, under the definition of $+$ and $\cdot$ only a finite number of terms in a sum or product are zero. More precisely, we verify the assertions about the degrees of the sum and product of polynomials. Having established that, we see that the set of all polynomials over R is closed under $+$ and $\cdot$.

The facts about the degree of a sum of two polynomials give no trouble because the definition of the sum is term-by-term. And now we can see that as far as addition goes the additive group structure of R is reflected in the addition of polynomials. In fact what we have just observed shows that the polynomials under the operation $+$ form a subgroup of the direct sum of an infinite number of copies of the additive group $\langle R, + \rangle$. Hence addition of polynomials is commutative and associative, and additive inverses exist with respect to the zero polynomial.

The facts about the degree of the product of polynomials are a little more delicate. Suppose that $\deg(f) = n$ and $\deg(g) = m$. First let us argue that the deg($f \cdot g$) is at most the sum of the two degrees. Consider the coefficient c_k in the product:

$$c_k = \sum_{i+j=k} a_i b_j$$

If $a_i b_j \neq 0$ then $a_i \neq 0$ and $b_j \neq 0$; thus $i \leq \deg(f) = n$ and $j \leq \deg(g) = m$ and so $k = i + j \leq n + m$. Hence $\deg(f \cdot g)$ is at most $n + m$. If $k = n + m$ it follows that $c_{n+m} = a_n b_m$ and so the degree of a product is the sum of the degrees provided the product of the leading coefficients is nonzero. In particular, the degree of the product of monic polynomials is always the sum of the degrees.

If R has divisors of zero then the product of nonzero polynomials may be the zero polynomial. For example, let $R = \mathbb{Z}_4$, the integers mod 4. Let $f = (0, 2, 0, 0, \dots)$. Compute f^2 to find that f^2 is the zero polynomial.

The ring axioms for multiplication and the distributivity may now be proved. The details are tedious; we don't recommend doing them and we shall omit them here!

There is an interesting and important corollary to these definitions and computations.

9.1.3 Corollary. Subring of Scalars

Let R be a ring with identity. The polynomials of degree zero together with the zero polynomial form a subring of the ring of polynomials over R which is isomorphic to R.

Proof.   Theorem 9.1.2 shows that the polynomials of degree 0 and the zero polynomial form a subring. Then it is easy to verify that the map from R into the ring of polynomials over R defined by

$$r \to (r, 0, 0, \dots)$$

is a ring isomorphism.

It is customary to identify the elements of R with the polynomials of degree zero. In this way we say that the ring R is embedded in its ring of polynomials. We also refer to these polynomials as *scalar* polynomials, anticipating some vector terminology.

Now about x. We need a lemma and a definition and then we shall forget about sequences. We can speak and write about polynomials in x knowing what that symbol stands for.

9.1.4 Lemma. The Indeterminate x and the Ring R[x]

Let R be a ring with identity. In the ring of polynomials over R let

$$x = (0, 1, 0, \dots) \quad \text{and} \quad x^0 = 1 = (1, 0, 0, \dots)$$

For positive integers k, x^k is the polynomial whose only nonzero coefficient has index k and value 1. For any polynomial of degree n,

$$f = (a_0, a_1, \ldots, a_k, \ldots) = \sum_{0 \le k \le n} a_k x^k$$

Proof. Once we have verified the form of x^k, the rest is a simple fact of pointwise addition. Since we have not developed a form for the product of k polynomials, the alleged form for x^k is perhaps most easily verified by a straightforward induction on k. We omit these details. But in any event note that x is a monic polynomial and so x^k is also; moreover degrees add so that we know x^k will have degree k. Perhaps you can "see" that all the preceding coefficients will be zero because nonzero products can arise only from the products of leading coefficients.

Remark. The use of a symbol x and the notation for polynomials

$$\sum_{0 \le k \le n} a_k x^k$$

is of course the ordinary way we write and think about polynomials. Our experience is to confuse them with polynomial functions. We hope that Definition 9.1.1 and Lemma 9.1.4 will point out the special nature of polynomials and the notational role of x. Common terminology encompasses this discussion by saying that "an indeterminate x has been adjoined to R." The intuition is that x is a new element and that a new and larger ring, containing R, has been created whose elements are the polynomials in x that appear in Lemma 9.1.4. To indicate the role of x as it appears in Lemma 9.1.4 we say that an indeterminate x has been adjoined to the ring R and we denote the ring of polynomials in x with coefficients in R by the symbol $R[x]$. The use of the brackets is standard convention which is almost universally followed. Later, in discussing fields we speak of $R(x)$ to mean something different. In any event Lemma 9.1.4 shows that we have constructed a new ring $R[x]$ that contains an isomorphic copy of the original ring R.

We could of course have used any symbol in place of x. Whatever symbol is used, the underlying construct is composed of infinite sequences in which only a finite number of terms are different from zero.

Because it is so important, we single out one of the results of Theorem 9.1.2.

9.1.5 Corollary.   If *R* Is an Integral Domain So Is *R*[*x*]

This is so because there are no zero divisors in R and hence for two nonzero polynomials $\deg(f \cdot g) = \deg(f) + \deg(g)$. In particular $f \cdot g$ cannot be the zero polynomial. Hence there are no zero divisors in $R[x]$. Also it is easy to see that multiplication in $R[x]$ is commutative if it is in R and that the scalar polynomial 1 is the identity of $R[x]$. As we said, we identify the elements of R with their corresponding scalar polynomials.

Notation.   If f is a polynomial in the indeterminate x, for emphasis we write

$$f(x) = \sum_{0 \le h \le k} a_h x^h$$

Now we want to elucidate the sometimes not so elementary notion of "substitution." Indeed, someone once said that anyone who could substitute correctly could earn a Ph.D. in mathematics. The intuitive idea is that we simply replace x with any element of a ring and proceed to calculate as before. Replacement sounds a little bit like dealing with a homomorphic image. And that is what is made precise in the next theorem. Now that we know a lot about ring homomorphisms, their kernels (ideals), and how to construct them (factor out an ideal) we can obtain a great deal of information from this idea. This next theorem is an extremely important one when we come to study field extensions and the ring $R[x]$ when R is a field.

9.1.6 Theorem.   Substitution for *x*

Let T be a ring with identity and let R be a subring of T that contains the identity. Let $u \in T$ be an element that commutes with all elements of R. Then the map from $R[x]$ to T given by

$$\sum_{0 \le k \le n} a_k x^k \rightarrow \sum_{0 \le k \le n} a_k u^k$$

is a homomorphism. If K is the kernel of this homomorphism then $K \cap R = \{0\}$. The range of this homomorphism, denoted $R[u]$, is the smallest subring of T containing R and $R[u] \cong R[x]/K$.

Remark.   This is called a "substitution" theorem because it tells us what it means to replace x by an element u and when it is safe to do so. It also tells that all the ring properties hold if x is replaced by u, as long as u commutes with all the elements of R. Thus any equalities between polynomials in x, like

$$x^n - 1 = (x - 1)(x^{n-1} + x^{n-2} + \cdots + x + 1)$$

hold when x is replaced by u. Suppose, for example, that $R = \mathbb{Q}$ and T is the ring of 2×2 matrices with elements in $\mathbb{Q}$. The equality above holds when any 2×2 matrix is substituted for x.

Proof.   The conditions for a homomorphism are readily verified. If we don't carry them out, at least check the following special computation to see why the commutativity of u with the elements of R is used:

$$ax \rightarrow au \quad \text{and} \quad bx \rightarrow bu$$

and so $(ax)(bx) \rightarrow (au)(bu)$. On the other hand, in $R[x]$, $(ax)(bx) = abx^2$ and so

$$(ax)(bx) = abx^2 \rightarrow abu^2$$

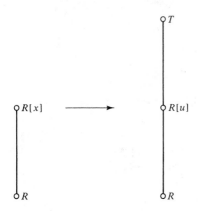

Figure 9.1 These rings are involved in the substitution theorem. The arrow shows the homomorphism that occurs when the element u is substituted for x.

and so it must be that $aubu = abu^2$. But this is so because we have assumed that b and u commute.

The general theory of ring homomorphisms that we investigated in Chapters 7 and 8 shows us that the kernel is an ideal and that the range of the homomorphism is isomorphic to $R[x]/K$. If $a \in R$ then a is a scalar polynomial and its image in the homomorphism is a. Hence a is in the kernel only if $a = 0$.

Notation.   As a result of this theorem we use the familiar functional notation to denote the substitution of u. In the homomorphism defined in the substitution theorem we write $f(u)$ for the image of $f(x)$.

9.1.7 Terminology.   Zero of a Polynomial

Given a polynomial $f(x)$ in $R[x]$, if T is a ring containing R and $u \in T$ commutes with the coefficients of T and $f(u) = 0$, then u is called a zero of $f(x)$. The element u may also be called a root of the equation $f(x) = 0$.

Much time and effort has been given to the question of how zeros of polynomials may be found. Indeed we shall spend some time on that question ourselves. And in the next section we study the construction of Kronecker to show that for any commutative ring R with identity and any polynomial in $R[x]$ whose leading coefficient is not a zero divisor in R there is a ring containing R in which the polynomial has a zero.

Now, to emphasize the distinction between polynomials and functions here is a definition and an example.

9.1.8 Definition.   Polynomial Function

Let R be a commutative ring. A function f from R into R is called a *polynomial function* if there is a polynomial $p(x) \in R[x]$ such that $f(u) = p(u)$ for every $u \in R$.

From this definition it follows that if R is a commutative ring with identity then every polynomial gives rise to a function from R into R. Indeed, if $f(x) \in R[x]$ then

$$f: u \rightarrow f(u)$$

is a function from R into R.

9.1.9 Example.   A Function That Is Not a Polynomial Function

Let $R = \mathbb{Z}_4$. Thus R has 4 elements and the number of functions from R into R is 4^4. This is so because for each of the four elements of the domain there are four choices for the range value. Thus there are only 4^4 functions. On the other hand, there exist an infinite number of polynomials. Obviously some polynomials must give rise to the same function. And it is possible that some functions cannot be represented by any polynomial.

First, although the polynomials x^2 and x^4 are different, the functions they represent are the same. The functions $x \rightarrow x^2$ and $x \rightarrow x^4$ are identical, for the next table confirms that $u^4 = u^2$ for all $u \in \mathbb{Z}_4$. Hence $u^5 = u^3$, $u^6 = u^2$, $u^7 = u^3$, and so on.

u	u^2	u^4
0	0	0
1	1	1
2	0	0
3	1	1

Second, we claim that no polynomial in $\mathbb{Z}_4[x]$ represents the function given explicitly by the following table:

u	$f(u)$
0	0
1	0
2	0
3	1

To begin with, we may assume the degree of the polynomial is at most 3. This is so because we have just shown that, *as functions on* $\mathbb{Z}_4$, there is no difference between $x \rightarrow x^2$ and $x \rightarrow x^4$. For this reason we may assume that if there were a polynomial representing this function it would have degree at most 3. Suppose $f(x) = a + bx + cx^2 + dx^3$. Then after substituting $x = 0$ the table information $f(0) = 0$ implies that $a = 0$. Similarly the table information $f(1) = 0$ implies that $b + c + d = 0$ and $f(3) = 3b + cx + 3d = 1$. Adding these last two equations in $\mathbb{Z}_4$ we obtain $2c = 1$, which has no solution in $\mathbb{Z}_4$. Hence this function is not a polynomial function.

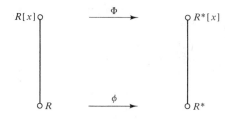

Figure 9.2 These rings are related in Theorem 9.1.10. A homomorphism ϕ from one ring to another may be extended to a homomorphism of their polynomial rings.

The next theorem will seem pretty obvious. It seems to be little more than a name and symbol-changing exercise. However, we shall have occasion to use it when $R^* = R$ and the homomorphism is an automorphism. Figure 9.2 shows the relationship of the rings involved.

9.1.10 Theorem.   Extensions of Homomorphisms and Zeros

Let R and R^* be rings and let

$$\phi: R \to R^*$$

be a ring homomorphism of R into R^*. The homomorphism ϕ can be extended to a homomorphism Φ of $R[x]$ into $R^*[x]$ with these properties:

1. If ϕ is one-to-one then so is Φ.
2. If $u \in R$ commutes with all the elements of R then $\phi(u)$ commutes with all the elements of $\phi(R)$.
3. If u is a zero of $f(x) \in R[x]$ and if $f^* = \Phi(f)$ then $\phi(u)$ is a zero of $f^*(x) \in R^*[x]$.

Proof.   We can define Φ as the natural map of sequences term-by-term. Thus

$$\Phi: (a_0, \ldots, a_k, \ldots) \to (\phi(a_0), \ldots, \phi(a_k), \ldots)$$

This mapping is readily seen to be an extension of the homomorphism ϕ in the sense that if f is a scalar polynomial then $\Phi(f) = \phi(f)$. Moreover, Φ is a homomorphism of $R[x]$ to $R^*[x]$.

A little care should be taken to prove that the extended mapping Φ preserves products of polynomials. This verification uses the fact that ϕ preserves both sums and products of elements in R. Here is a demonstration. Let c_k be the general term of the product of two polynomials following the notation of Theorem 9.1.2. Then

$$\phi(c_k) = \phi\left(\sum_{0 \le h \le k} a_h b_{k-h} \right) = \sum_{0 \le h \le k} \phi(a_h b_{k-h}) = \sum_{0 \le h \le k} \phi(a_h)\phi(b_{k-h})$$

It is easy to verify that if ϕ is one-to-one then so is its extension Φ.

Finally suppose that $u \in R$ is a zero of $f(x) \in R[x]$ and $f^* = \Phi(f)$. Say

$$f(x) = \sum_i a_i x^i \quad \text{and thus} \quad f^*(x) = \sum_i \phi(a_i) x^i$$

If $0 = f(u) = \sum_i a_i u^i$, then

$$0 = \phi(0) = \phi\left(\sum_i a_i u^i\right) = \sum_i \phi(a_i)\varphi(u)^i = f^*(\phi(u))$$

We shall derive the most useful properties of $R[x]$ in the case that R is a field. This is the subject of the next section. However, the following restricted division algorithm is useful in the more general context given here. For this result the ring need not be commutative or be assumed to be free of zero divisors. Look carefully at the proof; watch the multiplication!

9.1.11 Theorem. Division Algorithm in R[x]

Let R be a ring with identity. Let $f(x) \in R[x]$ be a polynomial of degree n

$$\sum_{0 \le k \le n} a_k x^k$$

whose leading coefficient a_n is a unit of R.

For any polynomial $g(x) \in R[x]$ there exist unique polynomials $q(x)$ and $r(x)$ such that

$$g(x) = f(x)q(x) + r(x) \tag{4}$$

where either $r(x) = 0$ or $\deg(r) < \deg(f)$.

Proof of Existence. First dispense with the case $g(x) = 0$. Simply choose $g(x) = r(x) = 0$.

We now proceed by induction on the degree of $g(x)$. We prove: For a fixed polynomial $f(x)$, whose leading coefficient a_n is a unit, if $\deg(g) = k$ then there exist polynomials q and r such that (4) holds. The crucial fact is that a_n, being a unit, has an inverse $(a_n)^{-1}$.

The induction hypothesis states that the division algorithm may be carried out on polynomials $g(x)$ of degree less than k. Suppose then that $\deg(g) = k$. Say that

$$g(x) = b_0 + \cdots + b_k x^k$$

Of course, if $\deg(f) > \deg(g) = k$ then choose $q(x) = 0$ and $r(x) = g(x)$ to obtain the result. This will always be the case if $k = 0$, which starts the induction, unless $n = k = 0$. But in this case f is a scalar polynomial that is a unit of R and hence divides all polynomials in $R[x]$. Now we handle the induction step.

Suppose that $\deg(g) = k \ge \deg(f)$. Find that

$$f(x)(a_n)^{-1}b_k x^{k-n} = c_0 + \cdots + b_k x^k$$

so that

$$h(x) = g(x) - f(x)(a_n)^{-1}b_k x^{k-n}$$

has degree that is less than k. Hence by induction there exist polynomials

$q(x)$ and $r(x)$ such that

$$h(x) = f(x)q(x) + r(x)$$

where either $r(x) = 0$ or $\deg(r) < \deg(f)$. Now equate the two expressions for $h(x)$, solve for $g(x)$, and collect terms

$$g(x) = f(x)(a_n)^{-1}b_k x^{k-n} + f(x)q(x) + r(x)$$
$$= f(x)\left((a_n)^{-1}b_k x^{k-n} + q(x)\right) + r(x)$$

and so the inductive step has been completed.

Proof of Uniqueness.   Suppose that

$$g(x) = f(x)q(x) + r(x) = f(x)q^*(x) + r^*(x)$$

where either $r(x)$ or $r^*(x)$ are zero or their degrees are less than n.
    By subtraction we have

$$0 = f(x)(q(x) - q^*(x)) + (r(x) - r^*(x))$$

or

$$f(x)(q(x) - q^*(x)) = r^*(x) - r(x)$$

Now the left-hand side is either zero or is a polynomial of degree greater than the degree of f since the leading coefficient of f is a unit. And a unit is not a zero divisor in R. But the polynomial on the right-hand side is either zero or had degree less than $\deg(f)$. Thus it must be that both sides are the zero polynomial. Hence

$$r(x) = r^*(x)$$

Now on the left-hand side, since the leading coefficient of f is a unit and not a zero divisor in R, it must be that $(q(x) - q^*(x))$ is thus the zero polynomial. Thus

$$q(x) = q^*(x)$$

A consequence of this result are the remainder and factor theorems.

9.1.12 Theorem.   Remainder Theorem

If R is a commutative ring with identity then for any $g(x) \in R[x]$ and $a \in R$

$$g(x) = (x - a)q(x) + g(a)$$

Proof.   Use the division algorithm with $f(x) = x - a$, a polynomial of degree 1, to obtain

$$g(x) = (x - a)q(x) + r$$

where the remainder either is 0 or has degree 0. In any event r is an element of R. Use the substitution theorem with $T = R$ and $u = a$ to find

$$g(a) = (a - a)q(a) + r$$

and hence $r = g(a)$.

9.1.13 Corollary. Factor Theorem

If R is a commutative ring with identity and $g(x) \in R[x]$ then $(x - a)$ divides $g(x)$ in $R[x]$ if and only if $g(a) = 0$, that is, if and only if a is a zero of $g(x)$.

Proof. Again use the division algorithm to divide $g(x)$ by $(x - a)$. Since the quotient $q(x)$ and the remainder r are uniquely determined, the result of the remainder theorem yields the factor theorem.

It is often desirable to consider polynomials in several indeterminates, just as we consider polynomial functions in several variables. Here are examples:

$$x^2 + xy + y^2 \text{ is a polynomial in } x \text{ and } y$$

$$z^2 x + 2(x + y)x - 4zy^3 \text{ is a polynomial in } x, \ y, \text{ and } z$$

The first may be considered as a polynomial in y with coefficients in $\mathbb{Z}[x]$. The "constant term" is x^2; the leading coefficient is 1.

The second may similarly be considered as a polynomial in z. Here the leading coefficient is $x \in \mathbb{Z}[x]$. The "constant" term is $2(x + y)x$. It is a polynomial in y with coefficients in $\mathbb{Z}[x]$. We say $z(x + y)x = 2x^2 + xy \in \mathbb{Z}[x, y]$.

There are several ways to introduce these polynomials. At this point it seems easiest to do this inductively.

9.1.14 Definition. Polynomial Ring in Several Indeterminates

Let R be a ring with identity. Define $R_0 = R$ and $R_1 = R_0[x]$. Proceeding inductively let $R_n = R_{n-1}[x]$ for $n = 2, 3, \dots$. The ring R_n is called the ring over R in n indeterminates.

Remark. Notice that R_n is, by Definition 9.1.1, a set of sequences of the form

$$(\hat{a}_0, \dots, \hat{a}_k, \dots)$$

where only a finite number of the $\hat{a}$'s are different from 0 and where each $\hat{a}_k$ is an element of R_{n-1}. As in the case of $R[x]$ it is more convenient to introduce n distinct new symbols $x_1, \dots, x_n$ and then express elements in R_n as sums of powers of these indeterminates. The next theorem shows that this can be done.

9.1.15 Theorem. Representation for Elements in $R[x_1, \dots, x_n]$

If R is a ring with identity and S is a ring over R in n indeterminates then there exists a chain of subrings

$$R_0 = R \subset R_1 \subset \cdots \subset R_n = S$$

where R_i is a ring in i indeterminates over R and moreover there exist distinct elements $x_1, \ldots, x_n$, each $x_i \in R_i$ such that each $f \in S$ is a sum of the form

$$f = \sum_{0 \le i_1 \le \cdots \le i_n \le n} a_{i_1 \cdots i_n} x_1^{i_1} \cdots x_n^{i_n} \qquad \text{where } a_{i_1 \cdots i_n} \in R$$

Proof. The proof is by induction on n. For $n = 1$ this is the content of Lemma 9.1.4.

Since S is a ring over R in n indeterminates there is by definition a ring, call it R_{n-1}, in $n - 1$ indeterminates so that $S = R_{n-1}[x]$.

Now every element in S has the form

$$f = \sum_{i_n} \hat{a}_{i_n} x_n^{i_n}$$

where $\hat{a}_{i_n} \in R_{n-1}$. We take

$$x_n = (\hat{0}, \hat{1}, \hat{0}, \ldots)$$

where $\hat{1}$ is the identity of R_{n-1} (which may be identified with the identity of R. Why?)

By induction R_{n-1} has a chain of subrings

$$R_0 \subset \cdots \subset R_{n-1}$$

and elements $x_i \in R_i$ such that every element $\hat{a}$ of R_{n-1} can be expressed in the form

$$\hat{a} = \sum_{1 \le i_1 \le \cdots \le i_{n-1} \le (n-1)} a_{i_1 \cdots i_{n-1}} x_1^{i_1} \cdots x_{n-1}^{i_{n-1}}$$

where

$$a_{i_1 \cdots i_{n-1}} \in R$$

Of course, these elements depend upon $\hat{a}$.

In particular let

$$\hat{a}_{i_n} = \sum_{1 \le i_1 \le \cdots \le i_{n-1} \le n-1} a_{i_1 \cdots i_{n-1} i_n} x_1^{i_1} \cdots x_{n-1}^{i_{n-1}}$$

where $a_{i_1 \cdots i_{n-1} i_n} \in R$. The dependence on the index i_n is reflected in the notation as the last subscript in the coefficient of

$$x_1^{i_1} \cdots x_{n-1}^{i_{n-1}}$$

Now multiply each coefficient $\hat{a}_{i_n}$ by $x_n^{i_n}$ and sum over the indices i_n. We have

$$f = \sum_{i_n} \hat{a}_{i_n} x_n^{i_n}$$

$$= \sum_{i_n} \left(\sum_{1 \le i_1 \le \cdots \le i_{n-1} \le n-1} a_{i_1 \cdots i_{n-1} i_n} x_1^{i_1} \cdots x_{n-1}^{i_{n-1}} \right) x_n^{i_n}$$

$$= \sum_{1 \le i_1 \le \cdots \le i_n \le n} a_{i_1 \cdots i_n} x_1^{i_1} \cdots x_{n-1}^{i_{n-1}} x_n^{i_n}$$

EXERCISE SET 9.1

9.1.1 Let $R = \mathbf{Z}_6$. Find a function from R into R that is not a polynomial function.

9.1.2 Let $R = \mathbf{Z}_2$. Show that every function from R into R is a polynomial function.

9.1.3 Let R be a commutative ring with identity 1. Show that the set of polynomials V in $R[x]$ defined by $V = \{ f(x) \in R[x]: f(a) = 0 \text{ for all } a \in R \}$ form an ideal of $R[x]$. Describe the residue classes of $R[x]/V$ in terms of the polynomial functions from R into R.

9.1.4 Show that if R is a ring and every function from R into R is a polynomial function then R is a division ring.

9.1.5 Show that $\langle 2 \rangle$ is a prime ideal in $\mathbf{Z}[x]$. Show that $\langle 2 \rangle$ is not a maximal ideal in $\mathbf{Z}[x]$.

9.1.6 Let R be a commutative ring with identity and let $G = \{ \sigma_1, \ldots, \sigma_n \}$ be a finite group of automorphisms of R. Show that $T = \{ r \in R: (r)\sigma = r \text{ for all } \sigma \in G \}$ is a subring of R. Show that for any $a \in R$

$$(x - \sigma_1(a))(x - \sigma_2(a)) \cdots (x - \sigma_n(a)) \in T[x]$$

9.1.7 Prove that multiplication of polynomials is associative by an induction on τ, the sum of the number of nonzero coefficients in polynomials $f(x)$, $g(x)$, $h(x)$. Use the distributive law.

9.1.8 For A, B in QUAT, the ring of quaternions (see Examples 7.1.2, part 12) find $q(x)$ and $r(x)$ in QUAT$[x]$ so that $x^2 - A = (x - B)q(x) + r(x)$ where $r(x) = 0$ or $\deg(r) = 0$.

9.1.9 Show that in QUAT

$$\begin{pmatrix} -2 & 0 \\ 0 & -2 \end{pmatrix}$$

has an infinite number of square roots of the form

$$\begin{pmatrix} i & -\beta \\ \beta & -i \end{pmatrix}$$

9.1.10 Let $R\{x\}$ denote the set of polynomials in the indeterminate x with addition defined in its usual manner. However, let multiplication be defined:

$$\left(\sum_k a_k x^k \right)\left(\sum_k b_k x^k \right) = \sum_k a_k b_k x^k$$

Assume that R is a ring with identity. Is $R\{x\}$ a ring? Which, if any, of the results of this chapter hold in $R\{x\}$?

9.1.11 Let R be a ring. Let F denote the set of all functions from R into R. Let addition be defined pointwise. Let multiplication be defined by composition:

$$f \Diamond g(x) = f(g(x))$$

Is $\langle F, +, \Diamond \rangle$ a ring?

9.1.12 Show that if R is a commutative ring with identity the mapping of $R[x]$ into $R[x]$ defined by $f(x) \to f(x + b)$ is an automorphism of $R[x]$.

Place a sufficient condition on a so that

$$f(x) \to f(ax + b)$$

is an automorphism of $R[x]$.

Show that

$$f(x) \to f(x^2)$$

is never an automorphism of $R[x]$. When is this mapping one-to-one?

9.2 POLYNOMIALS OVER A FIELD

The ring of polynomials $F[x]$ where F is a field has great utility for us. The fact that the underlying ring of coefficients is a field gives us lots of special properties. For example, we have full commutativity of all coefficients. There are no zero divisors, so we know that $F[x]$ will be an integral domain. More than that, each nonzero polynomial is assigned a degree and

$$\deg(f(x)g(x)) = \deg(f(x)) + \deg(g(x))$$

Because every nonzero element of a field is a unit we can strengthen the division algorithm (Theorem 9.1.11):

For every nonzero polynomial $f(x) \in F[x]$, and for every polynomial $g(x) \in F[x]$ there are unique polynomials $q(x)$ and $r(x)$ in $F[x]$ such that

$$g(x) = f(x)q(x) + r(x) \quad \text{where} \quad r(x) = 0 \text{ or } \deg(r) < \deg(f) \quad (1)$$

These two facts enable us to prove much more about $F[x]$. The things we are interested in are collected in the next theorem. Please note how many of these facts resemble the basic arithmetic facts about the integers. For example, in the division algorithm the degree of a polynomial plays the same role as the absolute value of an integer. And for both absolute value and degree it is true that

$$\text{If } f \text{ and } g \text{ are integers} \quad \text{and} \quad f \mid g \quad \text{then} \quad |f| \le |g|$$

$$\text{If } f \text{ and } g \text{ are polynomials} \quad \text{and} \quad f \mid g \quad \text{then} \quad \deg(f) \le \deg(g) \quad (2)$$

If you wonder whether there is some pervasive underlying principle at work, the answer is "Yes!" and in the next section we disclose that principle and use it to define a wider class of integral domains. We use property (2) instead of the stronger fact that the degree of the product of two polynomials is the sum of their degrees.

9.2.1 Theorem.   Arithmetic in $F[x]$

Let F be a field.

1. The units of $F[x]$ are the polynomials of degree 0.
2. Every nonzero polynomial has an associate that is a monic polynomial.
3. If $f(x) \mid g(x)$ in $F[x]$ then f is either a unit, an associate of g or $\deg(f) < \deg(g)$.

4. All pairs of nonzero polynomials $f(x)$, $g(x) \in F[x]$ possess a greatest common divisor. The $\gcd(f, g)$ can be computed by an euclidean algorithm and there exist polynomials $u(x)$ and $v(x)$ such that $\gcd(f, g) = f(x)u(x) + g(x)v(x)$.
5. Every irreducible polynomial in $F[x]$ is a prime element.

Proof of 1. Suppose $f(x)$ is a unit. This means that there is another polynomial $g(x)$ such that

$$f(x)g(x) = 1$$

and thus $\deg(fg) = 0$. But from (2) it follows that $\deg(fg) \geq \deg(f)$ and hence $0 \geq \deg(f)$. Since it is always true that $0 \leq \deg(f)$, it follows that $\deg(f) = 0$ and so f is a nonzero scalar polynomial. Conversely, if $\deg(f) = 0$ then f is a nonzero scalar polynomial and so is a nonzero element of F; hence a unit in F and so is a unit of $F[x]$ also.

Proof of 2. Suppose that c is the leading coefficient of the nonzero polynomial $f(x)$. Then the polynomial $c^{-1}f(x)$ is monic.

Proof of 3. If $f(x) \mid g(x)$, say, $g(x) = f(x)d(x)$ then it follows from (2) that $\deg(g) \geq \deg(f)$. Suppose $\deg(g) = \deg(f)$. We ask if $g(x) \mid f(x)$. If so, then because $F[x]$ is an integral domain, it follows that $f(x)$ and $g(x)$ are associates in $F[x]$. If $g(x) \nmid f(x)$ then from (1)

$$f(x) = g(x)h(x) + r(x) \quad \text{where} \quad \deg(r) < \deg(g)$$

and so substituting for $g(x)$,

$$f(x) = f(x)d(x)h(x) + r(x)$$

and we may conclude that $f(x) \mid r(x)$ and so $\deg(f) \leq \deg(r)$. But now we have a contradiction for $\deg(r) < \deg(g)$ and $\deg(g) = \deg(f)$ so that we would be able to infer that $\deg(f) < \deg(f)$. Tilt! Thus it must be that f and g are associates in $F[x]$. Query: Where did the possibility that $f(x)$ be a unit enter the proof?

Proof of 4. The proof is patterned after our discussion of the euclidean algorithm for integers. Formally we prove by induction on the $\deg(f(x))$ that if $f(x)$ is a nonzero polynomial in $F[x]$ then for any other polynomial $g(x)$, $\gcd(f, g)$ exists and there are polynomials $u(x)$ and $v(x)$ such that $\gcd(f, g) = f(x)u(x) + g(x)v(x)$.

If $\deg(f) = 0$, then f is a scalar polynomial, hence a unit in $F[x]$ and so divides $g(x)$. Hence $f = \gcd(f, g) = 1 \cdot f(x) + 0 \cdot g(x)$.

Now the induction hypothesis: Suppose that if $\deg(t(x)) < k$ then for all polynomials $g(x) \in F[x]$, $\gcd(t, g)$ exists and there are polynomials $n(x)$ and $m(x)$, depending on $g(x)$, such that $\gcd(t, g) = t(x)n(x) + g(x)m(x)$. So now let $\deg(f(x)) = k$. And let a polynomial $g(x)$ be given. First perform the division algorithm and so find polynomials $q(x)$ and $r(x)$ such that

$$g(x) = f(x)q(x) + r(x) \quad \text{where} \quad r(x) = 0 \quad \text{or} \quad \deg(r) < \deg(f)$$

If $r(x) = 0$ then $f \mid g$ and so $\gcd(f, g) = f$ and there is nothing to prove. If $f \nmid g$ then, as in Lemma 1.3.4, it follows easily that any divisor of f and g is a divisor of f and r and conversely. Hence they have the same set of common divisors. Now we can apply the induction hypothesis to find that $\gcd(r(x), g(x))$ exists and that there are polynomials $n(x)$ and $m(x)$ such that

$$\gcd(r, g) = r(x)n(x) + g(x)m(x)$$

In particular, $\gcd(f, g)$ exists and $\gcd(f, g) = \gcd(r, g)$. Now substitute for $r(x)$ to obtain

$$\gcd(f, g) = (g(x) - f(x)q(x))n(x) + g(x)m(x)$$
$$= f(x)(-q(x)n(x)) + g(x)(n(x) + m(x))$$

And this completes the proof. The euclidean algorithm is just the iteration of this inductive step until a remainder is obtained that does divide. It may be programmed for computation in exactly the same way as the one for the integers.

Proof of 5. Suppose that $p(x)$ is an irreducible polynomial. Remember that means that its only divisors are units and associates of $p(x)$. Suppose that $p(x) \mid f(x)g(x)$. We are to show that $p \mid f$ or that $p \mid g$. Suppose that $p \nmid f$. Since the only divisors of $p(x)$ are units and associates of $p(x)$ it follows that $\gcd(p, f) = 1$. Hence by 3 there are polynomials $u(x)$ and $v(x)$ such that

$$1 = p(x)u(x) + f(x)v(x)$$

and so, multiplying through by $g(x)$

$$g(x) = p(x)u(x)g(x) + f(x)v(x)g(x)$$

and since $p \mid f \cdot g$ it follows that $p(x)$ divides the right-hand side of the last equation and thus the left-hand side also; so $p(x)$ divides $g(x)$.

Remark. It is not easy to determine whether a particular polynomial is irreducible in $F[x]$. We shall have some special results that are not exhaustive but are helpful. The first thing to realize now is that any polynomial of degree 1 is irreducible in $F[x]$. If $\deg(f) = 1$ and $f(x) = g(x)h(x)$ from 3 we see that g is an associate of f or that $\deg(g) < \text{def}(f) = 1$; in this case $\deg(g) = 0$ and g is a unit.

Remark. The fact that $F[x]$ is an integral domain with a function, the degree function, from $F[x] - \{0\}$ to the nonnegative integers satisfying equations (1) and (2) enables us to prove that every irreducible is a prime, which leads to the unique factorization theorem in $F[x]$. In the next section we shall see a deeper reason why this is so.

9.2.2 Theorem.   Unique Factorization in $F[x]$

If F is a field then every nonzero polynomial that is not a unit can be expressed as a product

$$f(x) = up_1(x) \cdots p_n(x)$$

where u is a unit and each $p_i(x)$ is irreducible. If we prescribe, as we may, that each p_i is a monic polynomial then the factorization is unique except for the order of the irreducible factors.

Proof. The proof of this theorem is almost word for word the same as the proof we have given for the unique factorization in the integers $\mathbb{Z}$. The difference is that here we induct on the degree of the polynomial rather than the absolute value of the integer. (Actually we had the luxury in discussing $\mathbb{Z}$ of dealing only with positive integers.) And in $\mathbb{Z}$ there were only two units, 1 and -1. In $F[x]$ any scalar polynomial, except zero, is a unit. We do simplify things a little by insisting that the irreducible factors be monic and using a leading unit, u, to adjust the equality.

Here is a sketch of the proof of the existence. First, we consider the case that $\deg f(x) = 1$. Then $f(x) = ax + b = a(x + ba^{-1})$ is a factorization into a unit and a monic irreducible polynomial. Now we proceed by induction. If $f(x)$ is irreducible itself then we may factor out the leading coefficient and obtain the desired form. If f is not irreducible, say, $f(x) = g(x)h(x)$ where neither g nor h is a unit or associate, then $\deg(f) > \deg(g)$ and $\deg(f) > \deg(h)$. Then by the induction hypothesis these polynomials may be factored into products of irreducibles. In a similar way the proof of the uniqueness follows the uniqueness part of Theorem 1.3.10.

Another similarity with $\mathbb{Z}$ is the fact that every ideal in $F[x]$ is principal. A glance back at Theorem 1.3.15, the corresponding theorem in $\mathbb{Z}$, shows how the proof should go here.

9.2.3 Theorem. $F[x]$ Is a Principal Ideal Domain

If F is a field, every ideal of $F[x]$ is principal.

Proof. Let A be an ideal of $F[x]$. If $A = \{0\}$ then $A = \langle 0 \rangle$. Otherwise let $m(x)$ be a nonzero polynomial of least degree in A. We claim $A = \langle m(x) \rangle$. Clearly $A \supseteq \langle m(x) \rangle$. Conversely, let $f(x)$ be any other polynomial in A. We claim $m \mid f$, for if not then from equation (1),

$$f(x) = m(x)q(x) + r(x) \quad \text{and} \quad \deg(r) < \deg(m)$$

However, $r(x) = f(x) - m(x)q(x)$ and so belongs to A since both $f(x)$ and $m(x)q(x)$ do. But $r(x)$ has smaller degree than $m(x)$, contrary to the choice of $m(x)$.

9.2.4 Theorem. A Sufficient Condition for $f(x)$ to Be Reducible

If $f(x) \in F[x]$ has a zero in F and $\deg(f) \geq 2$ then $f(x)$ is reducible.

Proof. Apply the factor theorem.

9.2.5 Theorem. Reducibility Test for Quadratics and Cubics

A polynomial of degree 2 or 3 is reducible in $F[x]$ if and only if it has a zero in F.

Proof. Suppose $f(x)$ is reducible; say,

$$f(x) = g(x)h(x)$$

where neither factor is a unit in $F[x]$. Since

$$\deg(f) = \deg(g) + \deg(h) \leq 3$$

and $\deg(g)$ and $\deg(h)$ are positive integers, one of them must equal one. That is, f must have a factor of degree 1. We may assume that it is monic and so by the factor theorem find a zero of $f(x)$.

Caution. The dependence on degree 2 or 3 is crucial. The quartic $x^4 + x^2 + 1$ has no zero in $\mathbb{Q}$ but is reducible in $\mathbb{Q}[x]$:

$$x^4 + x^2 + 1 = x^4 + 2x^2 + 1 - x^2 = (x^2 + 1)^2 - x^2$$

$$= (x^2 + 1 - x)(x^2 + 1 + x)$$

In the next section we see that there is an algorithm for factoring polynomials in $F[x]$ as the product of irreducible ones in case F is a finite field or the field of rational numbers. The methods are not particularly elegant.

The next important observation is that reducibility-irreducibility of a particular polynomial depends heavily on the field. Thus the quadratic

$$x^2 - 2 = (x - \sqrt{2})(x + \sqrt{2})$$

is irreducible in $\mathbb{Q}[x]$ but is reducible in $\mathbb{Q}(\sqrt{2})[x]$.

On the other hand, the greatest common divisor of two polynomials does not change when going from one field to an extension of it.

9.2.6 Theorem. Preservation of the gcd

Let F be a subfield of a field E. Let $f(x)$ and $g(x)$ be two polynomials in $F[x]$. If $d(x)$ is their greatest common divisor in $F[x]$ then $d(x)$ is their greatest common divisor in $E[x]$.

Proof. In $F[x]$ we know that

$$\gcd(f, g) = f(x)u(x) + g(x)v(x) \tag{3}$$

and this holds in $E[x]$ since $F[x]$ is embedded in $E[x]$.

On the other hand, if $c(x) \in E[x]$ were a common divisor of $f(x)$ and $g(x)$ in $E[x]$ then equation (3) shows that $c(x)$ would also divide $\gcd(f, g)$. Hence there can be no change in the greatest common divisor.

In a strictly logical development we would now state and prove Kronecker's extension theorem. In this text this extremely important theorem is postponed to the end of this section because the notation and details of the proof are more complicated than the next few interesting applications of its main result. The part of Kronecker's theorem that is needed now is the following. Given a field F and a polynomial $f(x) \in F[x]$, there is a field $E \supseteq F$ in which $f(x)$ has a zero. We use this result freely in the next several theorems and present Kronecker's theorem as the climax of this section. The proof we give of Kronecker's theorem constructs the field as a homomorphic image of $F[x]$. If you wish, you may proceed to Theorem 9.2.16 now, but do return for these applications!

The first application we make is a test to decide whether two polynomials in $F[x]$ have a common zero in some extension field E containing F. The decision is achieved through an application of the euclidean algorithm carried out, not in $E[x]$, but in $F[x]$.

9.2.7 Theorem.　Test for a Common Zero

Let F be a field and let $f(x)$ and $g(x)$ be two polynomials in $F[x]$. Then f and g have a common zero in an extension E of f if and only if

$$\gcd(f, g) \neq 1 \text{ in } F[x]$$

Proof. Suppose first that $\gcd (f, g) \neq 1$. Then f and g have a common factor in $F[x]$ whose degree is larger than 0. Let E be a field containing a zero of the factor. In E this zero is a common root of $f(x)$ and $g(x)$.

Conversely, suppose that f and g have a common zero in an extension field E of F. In $F[x]$ we cannot have $\gcd(f, g) = 1$, for then

$$1 = f(x)u(x) + g(x)v(x) \tag{4}$$

for polynomials u and v in $F[x]$. Since $F[x] \subseteq E[x]$ Equation (4) holds in $E[x]$. But then the substitution of the common zero for x in (4) results in the classic contradiction that $1 = 0$.

Here is another theorem of special importance in $F[x]$ when F is a field. It is false if F is not commutative; remember that in the division ring of QUAT, the quaterions, we found (see Problem 9.1.9) an infinite number of square roots of a certain element.

9.2.8 Theorem.　Number of Zeros of a Polynomial

Let F be a field, let $f(x) \in F[x]$, and let E be an extension of F. The number of distinct zeros of $f(x)$ in E is at most the degree of $f(x)$.

Proof.   If a and b are zeros of $f(x)$ in E then, by the factor theorem in $E[x]$,

$$(x - a) \mid f(x) \quad \text{and} \quad (x - b) \mid f(x)$$

Further, if a and b are different elements of E then $(x - a)$ and $(x - b)$ are relatively prime in $E[x]$ and it follows just as in $\mathbb{Z}$ that

$$(x - a)(x - b) \mid f(x)$$

This argument extends to as many distinct zeros of f as there are in E. If

$$a_1, a_2, \ldots, a_m$$

are distinct zeros of $f(x)$ we know that

$$(x - a_i) \mid f(x)$$

Since these m factors are pairwise coprime it follows that

$$\prod_{1 \leq i \leq m} (x - a_i) \text{ divides } f(x)$$

and hence the degree of the product, which is m, must be less than or equal to $\deg(f)$.

Remark.   This theorem is often used in its contrapositive form: If a polynomial has degree at most n and has more than n zeros then it must be the zero polynomial. We can use this in the proof of the next theorem.

9.2.9 Theorem.   Lagrange Interpolation Formula

Let F be a field and let

$$p_1, p_2, \ldots, p_n$$

be n distinct elements of F and let

$$a_1, a_2, \ldots, a_n$$

by any n elements in F.

There exists a unique polynomial of degree at most $n - 1$ such that for $i = 1, \ldots, n$,

$$f(p_i) = a_i$$

Proof.   The existence is achieved by exhibiting the polynomial. Here it is:

$$f(x) = \sum_{1 \leq i \leq n} a_i \frac{(x - p_1) \cdots (x - p_{i-1})(x - p_{i+1}) \cdots (x - p_n)}{(p_i - p_1) \cdots (p_i - p_{i-1})(p_i - p_{i+1}) \cdots (p_i - p_n)}$$

To verify that this f has the requisite properties note that each summand has degree $n - 1$ and so the whole polynomial has degree at most $n - 1$.

Next when substituting p_i we see that every summand but the ith is zero. On the other hand, the fraction in the ith term is 1 and hence the ith summand is a_i.

The uniqueness comes about as follows. If there were two polynomials of degree at most $n - 1$ agreeing at n places then the difference of the two polynomials would have zeros at these distinct n places but would have degree at most $n - 1$ and thus, by Theorem 9.2.8, be the zero polynomial.

Remark. The existence proof gives no room for motivation; how did Lagrange figure it out? We might imagine that he knew it was easy to get a polynomial that is zero at, say, $p_2, \ldots, p_n$ and not at p_1. One such polynomial is

$$g_1(x) = (x - p_2) \cdots (x - p_n)$$

Adjusting that polynomial to have the value a_1 at p_1 is easy too; simply multiply by the right fix-up factor. Thus

$$h_1(x) = \frac{a_1}{g_1(p_1)} g_1(x)$$

fills the bill. So it would be easy to construct n polynomials, $h_1(x), \ldots, h_n(x)$, such that $h_i(x)$ is zero at p_j if $j \neq i$ and $h_i(p_i) = a_i$.

The neat trick is that the desired polynomial $f(x)$ is just the sum of the n polynomials h_i:

$$f(x) = h_1(x) + \cdots + h_i(x) + \cdots + h_n(x)$$

For example

$$f(p_1) = h_1(p_1) + 0 + \cdots + 0 = a_1$$

Next we develop a test to determine whether or not a polynomial in $F[x]$ has multiple zeros. The test is based on a formal derivative. First some definitions.

9.2.10 Definition.   Multiple Zero

Let F be an integral domain and let E be a field containing F as a subring. Let $f(x) \in F[x]$. An element $u \in E$ is said to be a zero of $f(x)$ of multiplicity k if

$$(x - u)^k \mid f(x) \text{ in } E[x]$$

but

$$(x - u)^{k+1} \nmid f(x)$$

in $E[x]$.

If u is a zero of multiplicity greater than 1 then it is called a multiple zero. Zeros of multiplicity 1 are called simple zeros.

Note that if $f(x)$ has a zero in a field E then its multiplicity cannot increase in any extension field. This is due to the uniqueness of the quotient and remainder in the division algorithm. If

$$(x - u)^{k+1} \nmid f(x) \text{ in } E[x]$$

then

$$f(x) = (x - u)^{k+1} g(x) + r(x)$$

where now $r(x)$ is not zero and so has degree less than $(k + 1)$. This equation continues to hold over any extension field of E and hence $(x - u)^{k+1}$ cannot divide $f(x)$ ever.

Next we establish a simple test that can be performed in $F[x]$ to decide whether or not a polynomial $f(x) \in F[x]$ has a multiple root in any field containing F. For this we need the following definition.

9.2.11 Definition.   Formal Derivative of a Polynomial

Let F be a commutative ring with identity and let

$$f(x) = \sum_{0 \le h \le n} a_h x^h$$

then the derivative of f, in symbols $f' = f'(x)$, is defined

$$f'(x) = \sum_{1 \le h \le n} h a_h x^{h-1}$$

9.2.12 Theorem.   Algebraic Properties of Derivatives

For all polynomials in $R[x]$

$$(f + g)' = f' + g'$$
$$(fg)' = f'g + fg'$$

In particular

$$\left[(x - a)^k\right]' = k(x - a)^{k-1}$$

Expressions with the exponent 0 should always be interpreted as 1. The proofs of this theorem are routine, and we shall omit the details.

The test for multiple zeros now follows:

9.2.13 Theorem.   Multiple Zero Test

Let F be a field and let $f(x) \in F[x]$. The polynomial $f(x)$ has a zero of multiplicity $k > 1$ in a field extension of F if and only if

$$\gcd(f, f') \ne 1$$

Proof.   Suppose that $f(x)$ has a zero of multiplicity greater than 1 in a field E containing F; say

$$f(x) = (x - u)^k g(x) \in E[x]$$

Then computing f' we obtain

$$f'(x) = k(x - u)^{k-1} g(x) + (x - u)^k g'(x)$$

and we see that $(x - u)$ is a factor of both f and f' and so the $\gcd(f, f')$ must be divisible by $(x - u)^{k-1}$ in fact.

Conversely, suppose that

$$d = \gcd(f, f')$$

is a polynomial of degree greater than 1. Now let E be an extension of F containing a zero of $d(x)$ and hence of $f(x)$ and $f'(x)$. In E let $f(x)$ have multiplicity k; $k \geq 1$. Thus

$$f(x) = (x - u)^k g(x)$$

in $E[x]$ and $(x - u) \nmid g(x)$ or else $(x - u)^{k+1}$ divides $f(x)$. By inspection of the derivative $f'(x)$ above, if $k = 1$, then $(x - u)$ cannot divide $f'(x)$ without dividing $g(x)$, which it does not and hence $k \geq 2$.

9.2.14 Corollary.   Sufficient Condition for Simple Zeros

If F is a field of characteristic zero then an irreducible polynomial has only simple zeros.

Proof.   Let $p(x)$ be the irreducible polynomial. Without loss of generality we may assume that $p(x)$ is monic. Then its derivative $p'(x)$ has leading coefficient equal to $\deg(p)$ and $\deg(p') = \deg(p) - 1$. In any event it is not the zero polynomial and so cannot have a nonunit factor in common with an irreducible polynomial of higher degree. Hence $\gcd(p, p') = 1$.

Now we come to a very important idea in the search for the zeros of a polynomial. The great German mathematician, Leopold Kronecker (1823–1891), gave a construction which, given a field F and a polynomial in $F[x]$, produced a field E containing F and a zero of the given polynomial. The method is in fact quite simple; its concept lies close to the substitution theorem 9.1.6 and we shall revert to that situation to give the main details. Refer to Figure 9.3 for a diagram to show the subrings and ideals involved in the proof.

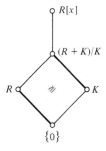

Figure 9.3 This lattice shows the relationships of the rings and ideals involved in Theorem 9.7.15. The diamond isomorphism for rings shows that $(R + K)/K$. a subring of $R[x]/K$, is isomorphic to R. In Part 2 of the theorem, K is chosen to be the principal ideal $\langle f(x) \rangle$. The theorem shows that the residue class $\alpha = x + \langle f(x) \rangle$ is a zero of $f(x)$ in $R[\alpha] = R[x]/K$.

9.2.15 Theorem.   Existence of Zeros

1. Let R be a commutative ring with identity and K be an ideal of $R[x]$ such that $K \cap R = \{0\}$. The factor ring $R[x]/K$ contains a subring, $(R + K)/K$, which is isomorphic to R. Moreover, if $\alpha = x + K$ then $R[x]/K = R[\alpha]$.
2. Let $f(x) \in R[x]$ be a nonscalar polynomial whose leading coefficient is not a zero divisor in R. Choose K to be the principal ideal $\langle f(x) \rangle$. Then $\alpha = x + K$ is a zero of $f(x)$ in $R[x]/K$.

Proof of 1.   Suppose that K is an ideal of $R[x]$ such that $K \cap R = \{0\}$. The homomorphism theorem for rings (7.2.13) shows that the ring $(R + K)/K$ is a subring of $R[x]/K$. The diamond lemma for rings (8.2.1) tells us that $(R + K)/K \cong R$. Now we apply the substitution theorem (9.1.16). Let $T = R[x]/K$ and let $u = \alpha$ be the residue class $x + R$. The homomorphism defined in the substitution theorem (9.1.6) sends

$$x \rightarrow \alpha = x + K$$

is thus the natural homomorphism of $R[x]$ onto $R[x]/K$. The kernel is K and the range of the homomorphism, identified in Theorem 9.1.6 as $R[\alpha]$, is $R[x]/K$.

Proof of 2.   Let $f(x)$ be a nonscalar polynomial belonging to $R[x]$. Apply part 1 by choosing K to be the principal ideal $\langle f(x) \rangle$. First we must verify that $K \cap R = \{0\}$. Well, all the elements of $\langle f(x) \rangle$ are multiples of $f(x)$. Because $\deg(f) \geq 1$, and because the leading coefficient of f is not a zero divisor in R, any multiple of $f(x)$ that is not itself zero will have degree at least that of $f(x)$ itself and hence cannot belong to R.

But now by part 1, $R[\alpha] = R[x]/K$ and $R[\alpha]$ contains an isomorphic copy of R. This means R can be thought of as a subring of $R[\alpha]$. To show that $\alpha = x + K$ is a zero of $f(x)$, calculate as residue classes mod K,

$$f(\alpha) = f(x + K) = f(x) + K = K$$

since $f(x) \in K$. But, in $R[x]/K$, K is the zero element. Thus $f(\alpha) = 0$ in $R[x]/K$. Thus a ring has been constructed that contains R as a subring and has a zero of the original polynomial, $f(x)$.

The next theorem gives the special features of the situation when R is a field and the polynomial is irreducible.

9.2.16 Theorem.   Kronecker's Extension Theorem

Let F be a field and $p(x)$ be an irreducible polynomial in $F[x]$. There exists a field E containing F as a subfield in which $p(x)$ has a zero.

Proof.   We claim that a field isomorphic to $F[x]/\langle p(x) \rangle$ will satisfy the theorem. First we do know from the preceding theorem that $F[x]/\langle p(x) \rangle$ will contain a subfield isomorphic to F and will contain a zero of $p(x)$. What we need to establish is that $F[x]/\langle p(x)\rangle$ is a field. We give two arguments for that important result. The first constructs the inverse; the second gives a more elegant structural reason.

First Argument. We already know that $F[x]/\langle p(x)\rangle$ has all the ring properties except possibly the existence of an inverse for each nonzero element. So, suppose that an element of $F[x]/\langle p(x)\rangle$ is nonzero. That means it is a residue class $f(x) + \langle p(x)\rangle$ such that $f(x) \notin \langle p(x)\rangle$. This means that $p(x) \nmid f(x)$ and therefore $\gcd(p(x), f(x)) = 1$. Hence

$$1 = f(x)u(x) + p(x)v(x) \tag{5}$$

for suitable $u(x)$ and $v(x)$. Because $p(x)v(x) \in \langle p(x)\rangle$ we see that in $F[x]/\langle p(x)\rangle$ this equation becomes

$$1 + \langle p(x)\rangle = f(x)u(x) + \langle p(x)\rangle$$

and hence that

$$(f(x) + \langle p(x)\rangle)(u(x) + \langle p(x)\rangle) = 1 + \langle p(x)\rangle$$

Thus the multiplicative inverse of $f(x) + \langle p(x)\rangle$ is $u(x) + \langle p(x)\rangle$.

If we write, as we did in Theorem 9.2.15, $\alpha = x + \langle p(x)\rangle$ then we may record our result as "The multiplicative inverse of $f(\alpha)$ is $u(\alpha)$ where $1 = f(x)u(x) + p(x)v(x)$ in $F[x]$."

There is another way to see this. Recall from Theorem 9.2.15 that if $\alpha = x + \langle p(x)\rangle$ then the mapping of $F[x]$ into $F[x]/\langle p(x)\rangle$ defined by $x \to \alpha$ is a ring homomorphism and since equation (5) holds in $F[x]$ it must be the case that

$$1 = f(\alpha)u(\alpha)$$

in $F[x]/\langle p(x)\rangle$.

Second Argument. We know from Theorem 8.2.9 that to show that $F[x]/\langle p(x)\rangle$ is a field all we have to do is to show that $\langle p(x)\rangle$ is a maximal ideal in $F[x]$. To show this we may suppose that H is an ideal such that $\langle p(x)\rangle \subset H \subseteq F[x]$. And suppose that $f(x) \in H$ but $f(x) \notin \langle p(x)\rangle$. Arguing as before, we find that equation (5) holds. Thus $1 \in H$ and so $H = F[x]$.

After all this we have that $F[x]/\langle p(x)\rangle$ is a field containing a zero of $p(x)$. The field $F[x]/\langle p(x)\rangle$ has an isomorphic copy of F, namely,

$$(F + \langle p(x)\rangle)/\langle p(x)\rangle = \{u + \langle p(x)\rangle : u \in F\}$$

Usually we simply say that we identify F with this subfield, set $E = F[x]/\langle p(x)\rangle$, and proclaim "Q.E.D." If we were being extremely rigorous, almost pedantic, we would describe a set theoretic construction whereby the elements of $(F + \langle p(x)\rangle)/\langle p(x)\rangle$ are replaced by the elements of F according to the isomorphism. Then new operations of addition and multiplication would be defined which would be isomorphic to the operation of $F[x]/\langle p(x)\rangle$. We shall never be this formal and simply use the euphemism "identify the elements of F with $(F + \langle p(x)\rangle)/\langle p(x)\rangle$."

9.2.17 Example.   The Field $\mathbb{Z}_3[x]/\langle x^2 + 1\rangle$

We denote the elements of $\mathbb{Z}_3$ as $0, 1, -1$. The polynomial $x^2 + 1$ is irreducible in $\mathbb{Z}_3$ because no element of $\mathbb{Z}_3$ is a zero of it. Every residue class of $\mathbb{Z}_3[x]/\langle x^2 + 1\rangle$ has a representative of the form $(a + bx)$ since every polynomial in $\mathbb{Z}_3[x]$ may be divided by $x^2 + 1$ with a remainder of 0 or of degree at most 1. So let

$$\alpha = x + \langle x^2 + 1\rangle$$

and thus each element of $\mathbb{Z}_3[x]$ may be expressed as $a + b\alpha$ where $a, b \in \mathbb{Z}_3$. Thus this field has nine elements. One of the exercises at the end of this section asks you to build the addition and multiplication tables of the field $\mathbb{Z}_3[x]/\langle x^2 + 1\rangle$.

The most important calculation is that $\alpha^2 = -1$. The equivalent form, $\alpha^2 + 1 = 0$, is true since the Kronecker theorem guarantees that α is a zero of $x^2 + 1$. But let us see that directly in terms of the congruence classes:

$$\alpha^2 = \left(x + \langle x^2 + 1\rangle\right)^2 = x^2 + \langle x^2 + 1\rangle$$
$$= -1 + x^2 + 1 + \langle x^2 + 1\rangle = -1 + \langle x^2 + 1\rangle$$

Now we can calculate more freely. For example,

$$(1 + \alpha)^2 = 1 + 2\alpha + \alpha^2 = 1 - \alpha - 1 = -\alpha$$

and

$$(1 + \alpha)(1 - \alpha) = 1 - \alpha^2 = 1 - (-1) = 2 = -1$$

To find, for example, the inverse of $1 + \alpha$, we know that $\gcd((1 + x), (1 + x^2)) = 1$ and an application of the euclidean algorithm in $\mathbb{Z}_3[x]$ (a one-step process in this example) gives

$$1 = (1 + x)(-1 + x) - (1 + x^2)$$

so that, substituting $x = \alpha$, this equation shows that $(1 + \alpha)^{-1} = \alpha - 1$.

EXERCISE SET 9.2

9.2.1 Show that if F is a finite field then every function from F into F is a polynomial function. * Show that if R is a ring with identity and every function from R into R is a polynomial function then R is a finite field.

9.2.2 Let $\mathbb{Q}$ be the field of rational numbers. Find the gcd of these polynomials over $\mathbb{Q}$:

$$x^4 + x^3 + 2x^2 + 3x + 1 \quad \text{and} \quad x^4 + x^3 - 2x^2 - x + 1$$

9.2.3 Solve the congruence $(x^2 + 1)f(x) \equiv 1 \pmod{x^3 + 1}$ over $\mathbb{Q}$.

9.2.4 Solve the congruence, if possible,

$$(x^4 + x^3 + x^2 + 1)f(x) \equiv (x^2 + 1) \pmod{x^3 + 1}$$

over the finite field $\mathbb{Z}_2$.

9.2.5 Let $\mathbb{R}$ be the field of real numbers and $\mathbb{C}$ be the field of complex numbers. Prove or disprove: $\mathbb{R}[x]/\langle x^2 + 1 \rangle$ is isomorphic to $\mathbb{C}$.

9.2.6 Show that $p(x) = x^2 + x + 1$ is irreducible over $\mathbb{Z}_2$. Let $u = x + \langle p(x) \rangle$. List the elements of $\mathbb{Z}_2(x)/\langle p(x) \rangle$ in the form $au + b$ where $a, b \in \mathbb{Z}_2$. Express u^{-1} in this form.

9.2.7 Let F be a field and let $g(x) \in F[x]$. In the field of quotients of $F[x]$ show that there exist polynomials $f_h(x)$ and $p_h(x)$ such that

$$\frac{1}{g(x)} = \frac{f_1(x)}{p_1(x)} + \cdots + \frac{f_t(x)}{p_t(x)}$$

where $\deg(f_h) < \deg(g_h)$ and each p_h is a power of a polynomial irreducible in $F[x]$.

9.2.8 Use Problem 9.2.7 to find the indefinite integral of

$$\frac{1}{(x + 1)(x^2 + x + 1)(x^2 - 2x + 1)}$$

If you should use the method of "undetermined coefficients," show why this will work over any field.

9.2.9 Let F be a field, let $f(x) \in F[x]$, and let $a, b \in F$. What can you say about the remainder $r(x)$ in $f(x) = (x - a)(x - b)q(x) + r(x)$? What happens if $a = b$? (Don't take a limit as $b \to a$!)

9.2.10 Consider the following four polynomials to be in $\mathbb{Q}[x]$. Of the six possible pairs of polynomials chosen from these polynomials, which have common zeros?

$$(x - 1), \qquad (x^3 + 2x^2 + 2x + 1)$$
$$(x^3 + x^2 + x + 1), \qquad (x^2 - x + 1)$$

9.2.11 Which of these polynomials in $\mathbb{Q}[x]$ have multiple zeros in some extension field?

$$(x^4 - x^3 + 2x^2 - x + 1) \quad \text{and} \quad (x^4 - 2x^3 + 2x^2 - 2x + 1)$$

9.2.12 Let F be a field of characteristic p, p a prime. Suppose that $p(x)$ is both irreducible in $F[x]$ and has multiple zeros. Determine the form of $p(x)$. Determine the possible multiplicities of the zeros.

9.2.13 Let R be an integral domain and let F be its field of fractions. We know that in general $R[x]$ is also an integral domain. Compare the field of fractions of $R[x]$ and the field of fractions of $F[x]$.

9.2.14 In $\mathbb{Z}_2[x]$ find $\gcd(x^3 + 1, x^5 + 1) = d(x)$ and express it in the form $d(x) = r(x)(x^3 + 1) + s(x)(x^5 + 1)$.

9.2.15 In $\mathbb{Z}_4[x]$ find a unit that is not a polynomial of degree 0.

9.3 EUCLIDEAN DOMAINS AND PRINCIPAL IDEAL DOMAINS

In this section we study the effect on an integral domain of having a "degree-like" function. We shall see that two simple properties guarantee a euclidean algorithm. In turn that guarantees that every ideal is principal and in turn that guarantees that the unique factorization theorem holds. The implication for the unique factorization property is then studied in more detail in Section 9.4.

9.3.1 Definition. Euclidean Domain

A *euclidean domain* is an integral domain D for which there is a function σ from the nonzero elements of D into the nonnegative integers such that the following two properties hold:

1. For all nonzero $a, b \in D$, $\sigma(a) \le \sigma(ab)$.
2. If a is not zero then for all $b \in D$ there exist $q, r \in D$ such that

$$b = aq + r$$

   and either $r = 0$ or $\sigma(r) < \sigma(a)$.

A function that satisfies these conditions is called a *size* function on D.

Remark.   Property 1 may be restated to say that if $a \mid c$ then $\sigma(a) \le \sigma(c)$. Note too, that there is no requirement that the q and r of property 2 be unique.

We have seen two examples of euclidean domains: The integers $\mathbb{Z}$ and $F[x]$ when F is a field. For $\mathbb{Z}$, the size function is absolute value. For $F[x]$ the size function is the degree of the polynomial. There is another interesting example of a euclidean domain. It is called the gaussian integers. These are the complex numbers $\{a + bi: a, b \in \mathbb{Z}\}$. This domain is discussed in Section 12.3. If you are interested now you already have all the tools necessary to read that section.

Now we begin with the first consequences of the size function.

9.3.2 Theorem. Elementary Properties of Euclidean Domains

Let D be a euclidean domain with a size function σ.

1. The units of d are elements whose σ-values are minimal. In D, u is a unit if and only if $\sigma(u) \le \sigma(a)$ for all nonzero elements $a \in D$.
2. Every ideal J of D is a principal ideal. Indeed, $J = \langle a \rangle$ where $\sigma(a)$ is minimal among all nonzero elements in J.

Proof.   Begin by noting that the range of the size function σ is a subset of the nonnegative integers. By the well-ordering property of $\mathbb{Z}$ we know that there is an element $u \in D$ such that

$$\sigma(u) \le \sigma(a)$$

for all nonzero a in D. Now we prove 1.

First, let u be a unit. Since u divides all elements a, it follows from Property 1 that $\sigma(u) \le \sigma(a)$. Conversely, suppose that $\sigma(u)$ is minimal and let $a \in D$. Now property 2 implies that

$$a = uq + r$$

where either $r = 0$ or $\sigma(r) < \sigma(u)$. Now since u was chosen so that $\sigma(u)$ is minimal, the second alternative cannot hold. So u divides a. In particular, $u \mid 1$ and so is a unit.

Now to prove 2. The same argument that we used above will show that every ideal is principal. Let J be an ideal of D. If $J = \{0\}$, then $J = \langle 0 \rangle$ and so is principal. Thus we may assume that J has some nonzero elements. The range of σ for the nonzero elements in J is again a subset of the nonnegative integers and we may apply the well-ordering principle again to find a nonzero element $a \in J$ such that $\sigma(a) \le \sigma(t)$ for all $t \in J$. Now apply property 2. There exist elements q and r in D such that

$$t = aq + r$$

where either $r = 0$ or $\sigma(r) < \sigma(a)$. In either case, since a and t belong to J, it follows that

$$r = t - aq$$

also belongs to J. Thus by the choice of a as an element having the property that $\sigma(a)$ is minimal in J, it cannot be that

$$\sigma(r) < \sigma(a)$$

thus $r = 0$ and a divides t. Thus

$$J = \langle a \rangle$$

This is the important result:

Every ideal of a euclidean domain is principal.

Much follows from this fact, and to emphasize this we prove several things using only the assumption that every ideal in an integral domain is principal.

9.3.3 Definition.   A Principal Ideal Domain (PID)

A *principal ideal domain* is an integral domain in which every ideal is principal.

9.3.4 Theorem.   Properties of a PID

Let D be a PID. Then

1. Every irreducible element is a prime.
2. Every pair of nonzero elements has a greatest common divisor. If $d = \gcd(a, b)$ then there exist elements $r, s \in D$ such that

$$d = ra + sb$$

3. There is no infinite ascending chain of ideals in D. That is, if

$$J_1 \subseteq J_2 \subseteq \cdots \subseteq J_n \subseteq \cdots$$

is an ascending chain of ideals then there is an integer k such that

$$J_k = J_{k+1} = \cdots = J_m$$

for all $m \geq k$.

Proof of 1. Suppose that p is an irreducible in D and that p divides ab. We want to show that $p \mid a$ or $p \mid b$. Consider the ideal

$$\langle p, a \rangle = \{ rp + sa : r, s \in D \}$$

This ideal is principal, say,

$$\langle p, a \rangle = \langle c \rangle$$

In particular p is a multiple of c and a is a multiple of c. But p is irreducible; therefore, either c is an associate of p or c is a unit. If c is an associate, then p divides c, and since c divides a, then p divides a. On the other hand, if c is a unit, say, $cd = 1$, then $1 \in \langle c \rangle = \langle p, a \rangle$ and so

$$1 = rp + sa$$

for some r and s in D. But then

$$b = rpb + sab$$

and since p divides rpb and sab, it follows that p divides b.

Proof of 2. Given $a, b \in D$. Consider the ideal

$$\langle a, b \rangle = \{ ra + sb : r, s \in D \}$$

It is principal, say, $\langle a, b \rangle = \langle d \rangle$. In particular

$$d = ra + sb \qquad \text{for some } r, s \in D$$

Thus any common divisor of a and b divides d. On the other hand, since a and b are in $\langle d \rangle$, d divides a and b. Hence $d = \gcd(a, b)$

Proof of 3. Suppose we have an ascending chain of ideals

$$J_1 \subseteq \cdots \subseteq J_n \subseteq \cdots$$

Form the set union of all the J's; call it U. We claim that U is an ideal of D.
Check the conditions. First suppose that $u, v \in U$. Then

$$u \in J_s \quad \text{and} \quad v \in J_t$$

for some s and t. Suppose that $s \leq t$. Then since $J_s \subseteq J_t$ we may conclude that both u and $v \in J_t$ and so $u - v \in J_t \subseteq U$.

The swallow-up condition is even easier to verify. Suppose that $u \in U$ and $r \in D$, then $ur \in U$. So, because U is an ideal and every ideal is principal in D it follows that

$$U = \langle g \rangle$$

for some $g \in U$. But because U is the set union of ideals it must be the case that for some k, $g \in J_k$, in which case

$$\langle g \rangle \subseteq J_k \subseteq U = \langle g \rangle$$

and thus $U = J_k$. But this means that if $m \geq k$ then

$$U = J_k \subseteq J_m \subseteq U$$

and so $J_m = U$.

9.3.5 Theorem.   Unique Factorization in a PID

Let D be a PID. Every element can be written as a product of a finite number of primes; this product is unique up to unit factors and the order in which the primes occur.

Proof.   The tricky part is the existence of the representation. Its uniqueness follows in exactly the same way the uniqueness of the factorization in $\mathbb{Z}$ follows.

Suppose the theorem were false, that is, that there is an element a that cannot be written as the product of a finite number of primes. We will see that this assumption leads to the construction of an infinite ascending chain of ideals. Begin by noting that a is not a prime and so

$$a = a_1 b_1$$

where neither factor is a unit. This means that $a \nmid a_1$ and $\langle a \rangle \subset \langle a_1 \rangle$. Similarly $\langle a \rangle \subset \langle b_1 \rangle$.

Now not both a_1 and b_1 can be written as the product of a finite number of primes; else putting both factorizations together would give a factorization of a. Assume that a_1 cannot be written as the product of a finite number of primes. We start a chain:

$$\langle a \rangle \subset \langle a_1 \rangle$$

where $\langle a \rangle \neq \langle a_1 \rangle$

This is the crucial step: Replace a with a_1. Repeat the argument to find an element a_2 such that

$$\langle a \rangle \subset \langle a_1 \rangle \subset \langle a_2 \rangle$$

and such that a_2 cannot be written as the product of a finite number of primes. Now continue. This iteration cannot be stopped. In this way, we could construct an infinite ascending chain of ideals. But infinite ascending chains of ideals do not exist in PIDs. This contradiction means that every element must be expressible as the product of a finite number of primes.

We omit the details of the uniqueness of the factorization which is based on an induction on the number of primes in the factorization.

Remark.   While these theorems tell us about the arithmetic of PIDs (and hence of euclidean domains) they do not give us an algorithm to find, say, the gcd of two elements. There do exist PIDs in which no algorithm exists for this determination. There do exist PIDs that are not euclidean domains. An elementary example is given by Oscar A. Cámpoli, A Principal Ideal Domain That Is Not a Euclidean Domain, *American Mathematics Monthly*, vol. 95, pp.

868–871 (1988). An article by T. Motzkin also gives a lot of information about euclidean domains. (See T. Motzkin, "The Euclidean Algorithm," *Bulletin of the American Mathematical Society*, vol. 55, pp. 907–908, 1949.)

Therefore, the following result, which is the whole point of studying euclidean domains, is of interest and utility. You can be expected to carry out the algorithm in various euclidean domains. It would be nice in each case to modify the program you have already written in response to a problem of Exercise 1.3.14.

9.3.6 Theorem.   Euclidean Algorithm in Euclidean Domains

Let D be a euclidean domain with size function σ. Let a and b be elements of D; a not equal to zero. Then there exists a sequence

$$a = r_0, b = r_1, r_2, \ldots, r_n, r_{n+1} = 0$$

such that for $i \leq j \leq n$

$$\sigma(r_i) < \sigma(r_j)$$

$$r_n = \gcd(a, b)$$

There exist elements u_h, v_h in D such that for $0 \leq h \leq n$

$$r_h = au_h + bv_h \tag{1}$$

in particular

$$\gcd(a, b) = r_n = au_n + bv_n$$

Proof.   Define $u_0 = 1$, $v_0 = 0$; $u_1 = 0$, $v_1 = 1$, and for $h \geq 2$ define r_h to be the remainder obtained from the division algorithm upon dividing r_{h-2} by r_{h-1}. Thus

$$r_{h-2} = r_{h-1}q_h + r_h$$

where either $r_h = 0$ (in which case $h = n + 1$, $\gcd(a, b) = r_{h-1}$ and the process stops) or $\sigma(r_h) < \sigma(r_{h-1})$.

If $r_h \neq 0$ note that

$$\gcd(r_{h-2}, r_{h-1}) = \gcd(r_{h-1}, r_h)$$

since any common divisor of (r_{h-2}, r_{h-1}) is a common divisor of (r_{h-1}, r_h).

Verify that equation (1) holds for $h = 0$ and $h = 1$. Thus, inductively,

$$r_h = r_{h-2} - r_{h-1}q_h$$

$$= (au_{h-2} + bv_{h-2}) - (au_{h-1} + bv_{h-1})q_h$$

$$= a(u_{h-2} - u_{h-1}q_h) + b(v_{h-2} - v_{h-1}q_h)$$

$$= au_h + bv_h$$

EXERCISE SET 9.3

9.3.1 Show that if D is a euclidean domain with size function σ then τ defined by $\tau(a) = \sigma(a) - \sigma(1)$ is also a size function on D. Thus show that without loss of generality we may require of a size function that $\sigma(1) = 0$.

9.3.2 Let D be a PID. Show that any two nonzero elements have a least common multiple in D.

9.3.3 Find a size function for the subring D_S of $\mathbf{Q}$. This ring was defined in Exercise 8.1.4.

9.3.4 Show that if D is a PID and p is a prime then $\langle p \rangle$ is a maximal ideal of D.

9.3.5 Let D be a PID and let n be a nonzero element of D. Define a subset of $\bar{D} = D/\langle n \rangle$ by

$$D_n^\times = \{ \bar{a} \in \bar{D} : \gcd(a, n) = 1$$

Show that $D_n^\times$ is a multiplicative group. Show that if $\gcd(n, m) = 1$ then $D_{nm}^\times \simeq D_n^\times \times D_m^\times$.

9.4 $\mathbf{Z}[x]$ AND OTHER UNIQUE FACTORIZATION DOMAINS

Clearly it is important to know as much as we can about the factorization of polynomials in $\mathbf{Z}[x]$. Since $\mathbf{Z}$ is not a field we cannot apply directly the information of the preceding section and yet since $\mathbf{Q}$ is the field of fractions of $\mathbf{Z}$ we do know something about $\mathbf{Q}[x]$. It is reasonable to see how much information we can obtain about $\mathbf{Z}[x]$ from $\mathbf{Q}[x]$. In this section we learn how to relate questions of zeros of polynomials and factorizations of polynomials in $\mathbf{Z}[x]$ to those of $\mathbf{Q}[x]$ and conversely. In particular we learn that every polynomial in $\mathbf{Z}[x]$ can be factored in essentially only one way as the product of irreducible polynomials in $\mathbf{Z}[x]$.

All would be wonderful if $\mathbf{Z}[x]$ were a euclidean ring or even a PID, but it is not. Fortunately it turns out that what we do use is not the euclidean nature of $\mathbf{Z}$ but only the fact that every integer has a unique factorization into a product of irreducibles and that every irreducible is a prime.

Let us begin by establishing a couple of elementary things about $\mathbf{Z}[x]$.

9.4.1 Example. $\mathbf{Z}[x]$ Is Not a PID

To begin with, the units of $\mathbf{Z}[x]$ are just the units of $\mathbf{Z}$. In fact that is true more generally:

If D is an integral domain then the units of $D[x]$ are the units of D.

This is so because if

$$1 = f(x)g(x)$$

then the degree argument:

$$0 = \deg(1) = \deg(f) + \deg(g)$$

implies that

$$\deg(f) = \deg(g) = 0$$

and so f and g are scalar polynomials, hence units themselves in D. (This argument requires that D be an integral domain to ensure that the degree of a product of polynomials is the sum of the degrees.)

Now we can give an example of an ideal in $\mathbb{Z}[x]$ that is not principal. Consider

$$\langle 2, x \rangle = \{2f(x) + xg(x) : f, g \in Z[x]\}$$

Verify easily that this set is an ideal. Can it be principal? Is there a polynomial $d(x)$ such that

$$\langle d(x) \rangle = \langle 2, x \rangle$$

If there were, the polynomial $d(x)$ would divide 2 and x. Since it would divide 2, $d(x)$ would be a scalar polynomial, and hence be a unit (1 or -1) or an associate (2 or -2). (By the way, this proves that 2 is an irreducible in $\mathbb{Z}[x]$.) But $d(x)$ would also divide x in $\mathbb{Z}[x]$. Now 2 does not divide x in $\mathbb{Z}[x]$ since every polynomial multiple of 2 has even coefficients but x is a monic polynomial. Thus the only candidate for $d(x)$ is 1 or -1. However, for no polynomials f and $g \in \mathbb{Z}[x]$ does

$$1 = 2f(x) + xg(x)$$

since every coefficient of $2f(x)$ is even while every nonzero term of $xg(x)$ has degree at least 1.

Now the general setting.

9.4.2 Definition.   Unique Factorization Domains

An integral domain D is called a *unique factorization domain* (UFD) if every nonzero element is either a unit or can be expressed as a product of irreducibles in only one way, except for unit factors and the order of the irreducibles in the product. Specifically if $x \in D$ and $x \neq 0$ then either x is a unit or there exist irreducibles $p_1, \ldots, p_n$ such that

$$x = p_1 \cdots p_n$$

If

$$x = p_1 \cdots p_n = q_1 \cdots q_m$$

are two factorizations of x into a product of irreducibles, then $n = m$ and upon proper numbering each p_i is an associate of q_i.

It is of course true that every PID, hence every euclidean domain, is a UFD.

One of the major results of this section is to prove that if D is a UFD then so is $D[x]$. Thus by induction so is $D[x_1, \ldots, x_n]$.

Another major result is that, apart from factors of degree 0, a polynomial is reducible in $D[x]$ if and only if it is reducible in $F[x]$ where F is the field of fractions of D.

Before we can prove these results we need some more elementary results about UFDs.

9.4.3 Theorem.   Arithmetic in UFDs

Let D be a UFD.

1. Every irreducible is a prime.
2. Any two nonzero elements in D have a gcd in D.
3. If $d = \gcd(a, b)$ then $1 = \gcd(a/d, b/d)$.
4. If $d \mid ab$ in D and $\gcd(d, a) = 1$ then $d \mid b$.

Proof.   All these facts are verified by studying the factorization of the elements into irreducibles. To prove that an irreducible is a prime, suppose that p, an irreducible, divides a product ab but does not divide a. Suppose

$$ab = pr$$

Compare factorizations of the left-hand side and the right-hand side. Since they must agree and since p is an irreducible it occurs as a factor on the left-hand side. Since $p \nmid a$, it must be that p occurs in the factorization of b, that is, that $p \mid b$.

To compute the $\gcd(a, b)$ write

$$a = up_1^{e_1}p_2^{e_2} \cdots p_n^{e_n} \quad \text{and} \quad b = vp_1^{f_1}p_2^{f_2} \cdots p_n^{f_n}$$

where u and v are units and each p_i is an irreducible. We have simply collected the distinct irreducibles that divide either a or b into an exponential form. To achieve the same set of irreducibles as divisors of a and b we permit the exponents e_i and f_j to be 0. To account for possible associates of the irreducibles we have introduced unit factors u and v. Of course, these units do not affect the gcd. Now let

$$d = p_1^{m_1}p_2^{m_2} \cdots p_n^{m_n}$$

where the exponent $m_i = \min(e_i, f_i)$ for $i = 1, \ldots, n$. By definition, d divides both a and b. On the other hand, the factorization into irreducibles of any divisor of both a and b must have the exponent on p_i that is less than or equal to m_i. Hence

$$d = \gcd(a, b)$$

Property 3 follows in a similar way. Property 4 again is proved by an examination of the factorizations of ab and d.

Remark.   Properties 2 and 3 may be extended to several elements:

2'. Any finite set of nonzero elements in D has a gcd in D.
3'. If $d = \gcd(a_1, \ldots, a_n)$ then $1 = \gcd(a_1/d, \ldots, a_n/d)$.

However, we cannot claim, as in a euclidean domain, that if $d = \gcd(a, b)$ there are elements $u, v \in D$ such that $d = au + bv$. Thus in $\mathbb{Z}[x]$ we have seen that $1 = \gcd(2, x)$ but we cannot find polynomials $u(x)$ and $v(x)$ such that

$$1 = 2u(x) + xv(x)$$

As an example of an application of the above rules we prove a generalization of the well-known test for a rational zero of a polynomial with integer coefficients.

9.4.4 Theorem.   Test for Fractional Zeros in a Field of Fractions

Let D be a UFD. Let F be its field of fractions.
   If the polynomial $f(x) = a_0 + \cdots + a_n x^n \in D[x]$ has a zero $u/v \in F$ where $\gcd(u, v) = 1$ then $u \mid a_0$ and $v \mid a_n$.

Proof.   Substitute u/v in $f(x)$ and clear denominators by multiplying through by v^n. We obtain

$$0 = v^n f\left(\frac{u}{v}\right) = a_0 v^n + a_1 u v^{n-1} + \cdots + a_{n-1} u^{n-1} v + a_n u^n$$

Thus, since u divides 0 and every term of the right-hand side except possibly $a_0 v^n$, it follows that

$$u \mid a_0 v^n$$

as well. But since u and v are coprime we conclude from Theorem 9.4.3, part 4 that $u \mid a_0$. Similarly $v \mid a_n$.

   Now we turn our attention to the relation between irreducibles in $D[x]$ and $F[x]$ where D is a UFD and F is its field of fractions. One annoying complication we face is that polynomials in $D[x]$ may have scalar factors that are not units. In F these scalars become units and so play no role in the factorization in $F[x]$. Thus

$$2x^2 + 2 = 2(x^2 + 1)$$

is reducible in $\mathbb{Z}[x]$ while it is irreducible in $\mathbb{Q}[x]$ because the factor 2 is a unit of $\mathbb{Q}$. We factor out common scalar factors and deal with what remains. Here is the definition.

9.4.5 Definition.   Primitive Polynomial

Let D be a UFD. A polynomial $f(x) \in D[x]$ is called *primitive* if the greatest common divisor of all its coefficients is 1.

9.4.6 Lemma.   Form for Polynomials in $D[x]$

Let D be a UFD. Every polynomial $f(x) \in D[x]$ can be expressed in the form

$$f(x) = cg(x)$$

where $c \in D$ and $g(x)$ is a primitive polynomial. Moreover if

$$cg(x) = dh(x) \tag{1}$$

where c and d are elements of D and g and h are primitive polynomials in $D[x]$ then c and d are associates in D and $g(x)$ and $h(x)$ are associates in $D[x]$.

Proof.   The existence is proved by simply factoring out the greatest common divisor of all the coefficients of $f(x)$. By the natural extension of Theorem 9.4.3, part 3 to any finite number of terms, it follows that the resulting polynomial is primitive.

To prove uniqueness, suppose that equation (1) holds. From $f(x) = cg(x)$ we find that the greatest common divisor of the coefficients of $f(x)$ is c (remember $g(x)$ is primitive) and from $f(x) = dh(x)$ we find that the greatest common divisor is d. Hence c and d are associates, say, $c = du$, u a unit. Substitute in (1) and then cancel d to obtain

$$ug(x) = h(x)$$

and so $g(x)$ and $h(x)$ are associates in $D[x]$.

Now we state and prove the key lemma in this development, one due to Gauss and traditionally referred to as the "Lemma of Gauss."

9.4.7 Lemma of Gauss

The product of primitive polynomials is primitive.

Proof.   Let $f(x)$ and $g(x)$ be primitive polynomials of $D[x]$ where D is UFD. We are to show that $f(x)g(x)$ is primitive. This means that we have to examine the coefficients of the product expressed in terms of the coefficients of f and g. The argument is a classic one.

We do need some notation. Let

$$f(x) = a_0 + a_1 x + \cdots + a_n x^n$$

Let

$$g(x) = b_0 + b_1 x + \cdots + b_m x^m$$

Suppose there were an irreducible p that divided all the coefficients of $f(x)g(x)$. Then it would divide the constant term $a_0 b_0$ and hence either a_0 or b_0. Suppose it divides a_0. Since both f and g are primitive, p cannot divide all the a's or all the b's.

Let h be the least index such that $p \nmid a_h$.

Let k be the least index such that $p \nmid b_k$.

Thus either $k = 0$ or $b \mid b_0, \ldots, b_{k-1}$. Now examine the coefficient of x^{h+k}. It is

$$(a_0 b_{h+k} + \cdots + a_{h-1} b_{k+1}) + a_h b_k + (a_{h+1} b_{k-1} + \cdots + a_{h+k} b_0)$$

Now p divides the first expression enclosed in parentheses because p divides all the factors a_i in it. Similarly p divides the second expression enclosed in parentheses because p divides all the factors b_j in it. But p does not divide the middle summand $a_h b_k$. Thus p cannot divide all the coefficients of the product $f(x)g(x)$.

9.4.8 Theorem. Factorization in $D[x]$

Let D be a UFD and let F be its field of fractions. Let $f(x) \in D[x]$ be a primitive polynomial of degree greater than or equal to 1. Then $f(x)$ is reducible in $D[x]$ if and only if it is reducible in $F[x]$.

Proof. Since D may be regarded as a subring of F we may also regard $D[x]$ as a subring of $F[x]$.

Now suppose that $f(x)$ is reducible in $D[x]$, say

$$f(x) = g(x)h(x)$$

Since f is primitive and $\deg(f) \geq 1$ it must be that $\deg(g)$ and $\deg(h)$ are at least 1. But then, since $D[x]$ is a subring of $F[x]$ this equation holds in $F[x]$ as well and $f(x)$ is reducible in $F[x]$.

Conversely, suppose that in $F[x]$ we find

$$f(x) = g(x)h(x)$$

where $\deg(f)$ and $\deg(g)$ are both greater than or equal to 1 and the coefficients of g and h lie in F, the field of fractions of D. In particular we replace all the fractions that are coefficients by equivalent ones that have a common denominator and then we factor out that denominator and so have in $F[x]$

$$f(x) = g(x)h(x) = \frac{1}{d} g_1(x)h_1(x)$$

where now g_1 and h_1 are in $D[x]$.

Next factor out from g_1 and h_1 the greatest common divisor of the coefficients of each polynomial and thus obtain

$$f(x) = \frac{e}{d} g_2(x)h_2(x) = \frac{u}{v} g_2(x)h_2(x)$$

where g_2 and h_2 are primitive polynomials of $D[x]$ and for good measure we have reduced e/d to u/v where $\gcd(u, v) = 1$. Now from the right side of the above equations we know all the coefficients are elements of D. Thus v must divide all the coefficients of the product $g_2(x)h_2(x)$. But since g_2 and h_2 are primitive, the lemma of Gauss says that the product $g_2(x)h_2(x)$ is primitive. Thus v can't divide all the coefficients of g_2h_2 unless v is a unit. Thus

$$f(x) = wug_2(x)h_2(x) \tag{2}$$

where w is a unit of D. Now since $f(x)$ is a primitive polynomial the greatest common divisor of all its coefficients is a unit. On the other hand, the greatest common divisor of all the coefficients on the right is u (times a unit). Since these gcds are equal, we conclude that u is also a unit of D. Hence equation (2) is an equality in $D[x]$ and so f is reducible in $D[x]$.

9.4.9 Example.   Factorization in Q[x]

Is $x^2 + 10x + 6$ reducible in $\mathbb{Q}[x]$? If it is reducible in $\mathbb{Q}[x]$ then it is reducible in $\mathbb{Z}[x]$. If

$$x^2 + 10x + 6 = (x - a)(x - b)$$

in $\mathbb{Z}[x]$ then a and b are integers whose product is 6 and whose sum is 10. No such integers exist and so no factorization occurs in $\mathbb{Z}[x]$ and hence, by the factorization theorem, no factorization occurs even when rational numbers are permitted for a and b.

There is an important corollary to this theorem.

9.4.10 Corollary.   Irreducibles Are Primes in D[x]

If D is a UFD, then an irreducible polynomial is a prime in $D[x]$.

Proof.   Let $p(x)$ be the irreducible polynomial. In $D[x]$ an irreducible polynomial must be primitive, else the greatest common divisor of all the coefficients would factor out and $p(x)$ would be reducible.

If $\deg(p) = 0$ then $p(x) = p$ is an irreducible of D and hence a prime in D. We must also show that it is prime in $D[x]$. Suppose that $p \mid f(x)g(x)$. Write

$$f(x) = af_1(x) \quad \text{and} \quad g(x) = bg_1(x)$$

where a and b are in D and f_1 and g_1 are primitive polynomials and hence so is their product. Thus the only way that each coefficient of $f(x)g(x)$ can be divisible by p is for $p \mid ab$ and hence $p \mid a$ or $p \mid b$. But this means that $p \mid f(x)$ or $p \mid g(x)$, respectively.

Now suppose $\deg(p) \geq 1$. As before, let $p(x) \mid f(x)g(x)$. We are to show that $p(x) \mid f(x)$ or $p(x) \mid g(x)$. Consider this situation in $F[x]$ where F is the field of fractions of D. The polynomial $p(x)$ is still irreducible in $F[x]$ by the preceding theorem, and hence in $F[x]$ we may suppose that $p(x) \mid f(x)$, or that

$$f(x) = p(x)k(x) \quad \text{for some } k(x) \in F[x]$$

By now familiar arguments we may write

$$k(x) = \frac{u}{v} k_1(x)$$

where $k_1(x)$ is a primitive polynomial in $D[x]$ and $\gcd(u, v) = 1$. Hence

$$f(x) = \frac{u}{v} p(x) k_1(x) \tag{3}$$

Since both p and k_1 are primitive polynomials, so is their product and thus, since $f(x)$ has coefficients in D, every coefficient of $p(x)k_1(x)$ must be divisible by v. Therefore, v is a unit of D, and thus $u/v \in D$ and so equation (3) shows that $p(x)$ divides $f(x)$ in $D[x]$.

9.4.11 Theorem. $D[x]$ Is a UFD If D Is a UFD

If D is a UFD, then so is $D[x]$.

Proof. Let $f(x)$ be a nonzero polynomial in $D[x]$. We are to show that f can be expressed as a product of irreducibles in essentially one way. The proof of both the existence and uniqueness is by induction on the degree of f. Begin with $\deg(f) = 0$.

If the $\deg(f) = 0$ then $f \in D$. If f were reducible in $D[x]$, say,

$$f = g(x)h(x)$$

then $\deg(g) = \deg(h) = 0$ and so f would also be reducible in D. In other words, an irreducible in D remains irreducible in $D[x]$. Now in D both the existence of a decomposition of f into a product of irreducibles in D, and hence in $D[x]$, as well as the uniqueness of this product holds.

Now the argument can proceed just like the argument for the factorization of integers into prime factors. Review Theorem 1.3.10. We now know that an irreducible polynomial is a prime in $D[x]$ and so the only change needed in the wording is that the induction is on the degree of the polynomial rather than its character as a positive integer. We omit the details.

Next we give a very useful sufficient condition for a polynomial in $D[x]$ to be irreducible in $D[x]$, or equivalently in $F[x]$ where F is the field of fractions of D.

9.4.12 Theorem.   Eisenstein's Irreducibility Criterion

Let D be a UFD and let F be its field of fractions. Let $f(x) \in D[x]$:

$$f(x) = a_0 + a_1 x + \cdots + a_n x^n$$

If there exists a prime dividing all coefficients but the highest and whose square does not divide the constant coefficient, that is, if there is a prime

$p \in D$ such that

$$p \mid a_0, \ldots, p \mid a_{n-1}$$

and

$$p \nmid a_n \quad \text{and} \quad p^2 \nmid a_0$$

then $f(x)$ is irreducible in $F[x]$.

Proof. First we may assume that f is primitive. If not factor out the greatest common divisor of the coefficients of f:

$$f = af_1(x)$$

where $a \in D$ and f_1 is primitive in $D[x]$. Since $p \nmid a_n$ it follows that $p \nmid a$. Now we can verify that the same condition holds on the coefficients of f_1 with respect to the prime p that holds on $f(x)$. So if we can prove that $f_1(x)$ is irreducible in $F[x]$ it follows that $f(x)$ is also since they are associates in $F[x]$. Thus for this proof we further assume that $f(x)$ is primitive.

Next we invoke Theorem 9.4.8 to conclude that it suffices to show that $f(x)$ is irreducible in $D[x]$. Suppose to the contrary that in $D[x]$

$$f(x) = g(x)h(x)$$
$$= (b_0 + b_1(x) + \cdots + b_r x^r)(c_0 + c_1 x + \cdots + c_s x^s)$$

Examine the coefficients of the product with respect to divisibility by p. First

$$p \mid a_0 = b_0 c_0$$

and hence $p \mid b_0$ or $p \mid c_0$. Without loss of generality we can assume that $p \mid b_0$. But then $p \nmid c_0$ since $p^2 \nmid a_0$.

Second, p cannot divide all the coefficients b_k of $g(x)$ since $p \nmid a_n$. Let m be the least index such that $p \nmid b_m$. Thus

$$p \mid b_0, \ldots, p \mid b_{m-1}, \quad \text{and} \quad p \nmid b_m$$

It is also true that $m \le r = \deg(g) < \deg(f) = n$. Finally consider

$$a_m = b_0 c_m + b_1 c_{m-1} + \cdots + b_{m-1} c_1 + b_m c_0$$

Since $m < n$, we find that $p \mid a_m$. Moreover p divides $b_0, \ldots, b_{m-1}$. It follows that $p \mid b_m c_0$. But p divides neither factor, a contradiction.

Hence $f(x)$ is irreducible in $D[x]$.

9.4.13 Examples. Application of Eisenstein's Irreducibility Criterion

1. $x^4 + 9x^2 + 6$ is irreducible in $\mathbb{Q}[x]$. Find that the prime integer 3 divides all coefficients but the leading one and 9 does not divide the constant coefficient. Apply Eisenstein.

2. For any prime p

$$f(x) = \frac{x^p - 1}{x - 1} = x^{p-1} + x^{p-2} + \cdots + x + 1$$

   is irreducible in $\mathbb{Q}[x]$.

Èisenstein's criterion does not apply directly, but there is a standard device that makes it applicable. Consider

$$g(x) = f(x + 1) = \frac{(x + 1)^P - 1}{(x + 1) - 1} = \frac{(x + 1)^P - 1}{x}$$

$$= x^{P-1} + px^{P-2} + \cdots + \binom{p}{k}x^{k-1} + \cdots + \binom{p}{2}x + p$$

Now Eisenstein's criterion applies directly with the prime p to show that $g(x)$ is irreducible in $\mathbb{Q}[x]$. But then so is $f(x)$ for if

$$f(x) = g(x)h(x)$$

then

$$g(x) = f(x + 1) = g(x + 1)h(x + 1)$$

by the substitution theorem.

3. $x^4 + xy^2 + y$ is irreducible in $\mathbb{Q}[x, y]$. Note that $\mathbb{Q}[x, y] = \mathbb{Q}[y][x]$ and that y is a prime (irreducible) of $\mathbb{Q}[y]$. Now apply Eisenstein's criterion.

4. Let $F = \mathbb{Z}_2[y]$ where y is an indeterminate over $\mathbb{Z}_2$. Consider

$$x^2 + y$$

in $F[x]$. The element y is a prime of F and so by Eisenstein's criterion, $x^2 + y$ is irreducible in $F[x]$.

EXERCISE SET 9.4

9.4.1 Find all the irreducible polynomials of degree ≤ 4 over $\mathbb{Z}_2$.

9.4.2 Write an algorithm suitable for a computer program to list all the irreducible polynomials of degree $\leq n$ over $\mathbb{Z}_2$. * Write the program. (Your list may be shortened by the observation in Problem 9.4.3 and omitting the reverse of polynomials already listed.)

9.4.3 Suppose that F is a UFD and that $f(x)$ has degree n in $F[x]$. Show that if $f(0) \neq 0$ then $x^n f(1/x)$ is a polynomial in $F[x]$. This polynomial is called the reverse of f. Why? Show that $f(x)$ is reducible if and only if its reverse is reducible.

9.4.4 Let F be a finite field having q elements. Let $p(x)$ be irreducible in $F[x]$. How many elements are in the field $F[x]/\langle p(x)\rangle$?

9.4.5 For the finite field $\mathbb{Z}_2$ and for each irreducible polynomial $p(x)$ of degree 4 over $\mathbb{Z}_2$ find the multiplicative order of $u = x + \langle p(x)\rangle$ in $\mathbb{Z}_2[x]/\langle p(x)\rangle$. (The grunge is not so terrible, but you may want to write a program and use it.)

9.4.6 Formulate an algorithm suitable for adaption to a computer program to determine all the irreducible factors of a polynomial in $\mathbb{Z}[x]$. Here's the idea: If $f(x) \mid g(x)$ in $\mathbb{Z}[x]$ then, for every integer a, $f(a) \mid g(a)$ in $\mathbb{Z}$. Thus given $g(x)$ we can build a table of the possible factors of $g(a)$ and thus a table of the possible values of $f(a)$ for any factor of $g(x)$. If we want to find the possible factors f of degree k, we should choose $k + 1$ values for a, determine all the possible integer factors of $g(a)$ at these $k + 1$ places, and for each choice use the Lagrange interpolation formula to determine the corresponding candidate for a factor $f(x)$ of $g(x)$. Yes, it is a long process; we might even say, "lagrungian"! Use your algorithm to test the irreducibility of $x^5 + x + 1$.

9.4.7 Let $f(x) \in \mathbf{Z}[x]$. Show that if $f(x)$ is irreducible in $\mathbf{Z}_p[x]$ for some prime p, then $f(x)$ is irreducible in $\mathbf{Z}[x]$. Use this method to show that $x^4 + 3x + 1$ is irreducible in $\mathbf{Q}[x]$.

9.4.8 Use a device like the one used in Example 9.4.13, part 2 to show that $x^6 + x^3 + 1$ is irreducible in $\mathbf{Z}[x]$.

Chapter **10**

Field Extensions

10.1 VECTOR SPACES

Vectors are often introduced in calculus courses. They often appear as two-dimensional constructs to describe displacements in the plane. Later they are extended to space. The algebraic rules that cover addition, subtraction, and scalar multiplication of vectors are developed and are used to solve problems in geometry. Later vector valued functions of a single variable, usually thought of as time, are introduced. These functions can be differentiated and integrated and are used to express various laws of motion. Inherent in much of this development is the underlying geometry of euclidean space. For example, a dot or scalar product is introduced to characterize in vector terms the geometric notion of the angle between two vectors, and hence of perpendicularity.

In this section we review the algebraic aspect of vectors. The generality we seek is to permit any *field* to serve as the field of scalars. This generality poses no problem algebraically, but geometric terms no longer have the same meaning and significance. So beware.

In Chapter 16 we extend these algebraic notions even further. There we consider a further generalization of the vector space concept. In these constructs the scalars come from a ring, not necessarily a field. For both vector spaces and modules the elementary algebraic notions have a straightforward development. The terms and the theorems and the strategies reflect their background in the basic algebraic constructs we have been studying. For example, we shall study subspaces of a vector space and homomorphisms (called linear transformations in this context) of one vector space to another. We begin with the definition.

10.1.1 Definition.   Vector Space over a Field

A *vector space* over a field F is an abelian group $\langle V, + \rangle$ together with a scalar operation, that is, a function from $F \times V$ into V, which will be denoted by juxtaposition and called scalar multiplication. This means that for all $a \in F$ and all $v \in V$, $av \in V$. We say that av is a scalar multiple of v.

Moreover the following axioms hold:

1. Scalar multiplication distributes over addition in F. For all $a, b \in F$ and $v \in V$

$$(a + b)v = av + bv$$

2. Scalar multiplication distributes over addition in V. For all $a \in F$ and $v, w \in V$

$$a(v + w) = av + aw$$

3. Scalar multiplication is associative over multiplication in F. For all $a, b \in F$ and $v \in V$

$$(ab)v = a(b(v))$$

4. The identity of F is an identity for scalar multiplication. For all $v \in V$

$$1v = v$$

As the terminology suggests, the elements of V are commonly called vectors, unless they come from some other set and have other well-accepted nomenclature. The elements of F are called scalars. Our notation does not distinguish between the addition operation in F and the addition operation of vectors in the abelian group. And, please note that no multiplication of vectors is assumed.

Perhaps the simplest vector space consists of the zero vector: $\langle V, + \rangle = \{0: 0 + 0 = 0\}$ and $a0 = 0$ for all $a \in F$. We call this the *null* space.

10.1.2 Example.   Vector Space of *n*-tuples over *F*

Another simple vector space to construct over any field F is the set of n-tuples of elements of F. Addition and scalar multiplication are performed component by component. Here are the details: Let n be a positive integer. Let

$$V_n(F) = F \times F \times \cdots \times F = \{(a_1, \ldots, a_n): a_i \in F\}$$

Define vector addition and scalar multiplication by

$$(a_1, \ldots, a_n) + (b_1, \ldots, b_n) = (a_1 + b_1, \ldots, a_n + b_n)$$

$$s(a_1, \ldots, a_n) = (sa_1, \ldots, sa_n)$$

It is routine to show that the vector space axioms hold. We always denote this vector space by $V_n(F)$ or by V_n if the field F is understood in context.

We run the risk of a slight confusion in notation. We shall, as everyone does, use 0 for both the zero element of the underlying field F and the zero vector, the zero of the additive group $\langle V, + \rangle$.

The elementary consequences of these axioms are similar to those for rings. Scalar multiplication acts almost like a ring multiplication. The following observations about the rules involving the zero of the field F and the 0 vector are "ringlike."

The scalar product of $0 \in F$ and $v \in V$ is the zero vector:

$$0v = 0$$

This is so because $0v$ is an additive idempotent in $\langle V, + \rangle$:

$$0v + 0v = (0 + 0)v = 0v$$

For the same reason any scalar times the zero vector is the zero vector:

$$s0 = 0$$

for all $s \in F$. Similarly $(-1)v = -v$ since

$$0 = 0v = (1 + (-1))v = 1v + (-1)v = v + (-1)v$$

For us the following class of examples is extremely important.

10.1.3 Example.   A Field Extension as a Vector Space

Let F be a field and let K be an extension of F; that is, F is a subfield of K. It is important to view K as a vector space over F. The addition of vectors (elements of K) is just field addition and scalar multiplication is just field multiplication. The axioms are valid because they are field axioms holding in K with F as a subfield. Specific examples of this situation are provided by:

1. The real numbers $\mathbb{R}$ are a vector space over $\mathbb{Q}$.
2. The field $\mathbb{Q}(\sqrt{2})$ is a vector space over $\mathbb{Q}$.
3. The complex numbers $\mathbb{C}$ are a vector space over $\mathbb{R}$.
4. The complex numbers $\mathbb{C}$ are a vector space over $\mathbb{Q}$.

Remark.   The last two vector spaces are different because the underlying field of scalars is different.

More generally the following construction will be used frequently.

5. Let F be a field and let $p(x)$ be a polynomial in $F[x]$. The residue class ring $F[x]/\langle p(x) \rangle$ is a vector space over F if we identify F with the scalar polynomials; that is the reason polynomials of degree 0 are called "scalar polynomials." Remember that if $p(x)$ is irreducible in $F[x]$ then $\langle p(x) \rangle$ is a maximal ideal so that $F[x]/\langle p(x) \rangle$ is a field extension of F.

Remark. There is no requirement that the multiplication of field elements be commutative. Therefore, all of Example 10.1.3 goes through in the same fashion if we assume that K, the overfield, is merely a ring with identity. Thus

6. The quaternions, QUAT, of Examples 7.1.2 are a vector space over the field $\mathbb{R}$ if we define the scalar product

$$r\begin{pmatrix} \alpha & -\beta \\ \bar{\beta} & \bar{\alpha} \end{pmatrix} = \begin{pmatrix} r\alpha & -r\beta \\ r\bar{\beta} & r\bar{\alpha} \end{pmatrix}$$

This is possible because $\overline{r\alpha} = r\bar{\alpha}$ since $r \in \mathbb{R}$.

10.1.4 Definition.   Subspaces

A subspace of a vector space V over a field F is a subgroup $\langle S, + \rangle$ of $\langle V, + \rangle$ which is a vector space when the scalar multiplication is restricted to S.

In other words an additive subgroup $S \subseteq V$ is a subspace over F if for all $a \in F$ and $s \in S$, $as \in S$. The axioms then are automatically satisfied.

10.1.5 Example.   Span of a Subset of Vectors

Let X be a nonempty subset of vectors in V, a vector space over F. By the span of X we mean the set

$$\langle X \rangle = \{ u \in V \colon u = a_1 x_1 + \cdots + a_m x_m; \ a_i \in F; \ x_i \in X; \ m = 1, \ldots \}$$

It is not difficult to show that the span of X is closed under vector addition and scalar multiplication and so is a subspace of V.

Just as with abelian groups we shall be concerned with subspaces and meets and joins of subspaces. Under set inclusion they form a lattice; the proof is essentially the same as for subgroups; the scalar multiplication causes no complication.

10.1.6 Theorem.   Lattice of Subspaces

The set of subspaces of a vector space over a field F is a complete lattice under set inclusion. The lattice meet is set intersection and the join of two subspaces S and T,

$$S \vee T = \{ s + t \colon s \in S \text{ and } t \in T \}$$

is the abelian group join of the subspaces as subgroups.

The *modular* law holds: For subspaces R, S, T of V

$$\text{If } S \subseteq T \quad \text{then} \quad T \cap (S \vee R) = S \vee (T \cap R)$$

(Recall Exercises 3.2.19, 4.3.10, and 7.2.12.)

Just as we did for groups and rings we may speak of a direct sum of vector spaces.

10.1.7 Definition. Direct Sum of Vector Spaces

A vector space V is the direct sum of subspaces S and T provided
$$V = S \vee T \quad \text{and} \quad \{0\} = S \cap T$$

In this case we can prove that each element of V can be written in one and only one way in the form
$$s + t \quad \text{where } s \in S \text{ and } t \in T$$
In this case V is isomorphic to the set of ordered pairs
$$\{(s, t): s \in S, t \in T\}$$
where all the operations are done componentwise, including scalar multiplication.

Much of this has already been proved since V is an abelian group. What is new here is the role of the scalar multiplication. But as you can see, it causes no complications.

The most important concepts for a vector space over a field F are those of linear dependence, basis, and dimension.

10.1.8 Definition.   Dependence and Independence of a Set of Vectors

A nonempty set of vectors S in a vector space V over a field F is said to be linearly *dependent* over F if there is a finite subset of S, say, $s_1, \ldots, s_n$, and scalars $a_1, \ldots, a_n$ *not all zero* such that
$$a_1 s_1 + \cdots + a_n s_n = 0 \tag{1}$$

A nonempty set is called *independent* if it is not dependent. Note that any set containing the zero vector is dependent. Also, any set consisting of one nonzero vector is independent.

The condition that a set be independent is often expressed by saying that for any nonempty subset of vectors in the set the condition
$$a_1 s_1 + \cdots + a_n s_n = 0 \quad \text{implies} \quad a_1 = \cdots = a_n = 0$$

The condition that a set be dependent is equivalent to saying that one of the vectors in the set can be expressed as a linear combination of the other vectors in the set and so is in their span. For example, from equation (1), if $a_1 \neq 0$ then
$$s_1 = -\left(a_1^{-1} a_2 s_2 + \cdots + a_1^{-1} a_n s_n\right)$$
In particular, if $\{v, w\}$ is dependent, then either $w = 0$ or $v = aw$ for some scalar a.

10.1.9 Definition.   Basis

A *basis* for a vector space over F is a set of vectors that span V over F and are independent over F.

Thus if B is a basis for V over F every $v \in V$ may be expressed in one and only one way as a *finite* sum of scalar multiples of the elements in the basis. That is, given $v \in V$ there exist vectors $b_1, \ldots, b_n$ in B and scalars $v_1, \ldots, v_n$ in F such that

$$v = v_1 b_1 + \cdots + v_n b_n$$

Moreover, if

$$v_1 b_1 + \cdots + v_n b_n = w_1 b_1 + \cdots + w_n b_n$$

then $v_i = w_i$ for all i.

Terminology.   If we write the basis B as a sequence of vectors

$$B = \{b_1, b_2, \ldots, b_n \ldots\}$$

then the sequence of coefficients

$$v_1, v_2, \ldots v_n, \ldots$$

are called the *coordinates* of v in the B-basis for V.

10.1.10 Examples of Bases

1. *Standard Basis for $V_n(F)$.* The standard basis for the space of n-tuples of elements from F, $V_n(F)$, is

$$U_1 = (1, 0, 0, \ldots, 0)$$
$$U_2 = (0, 1, 0, \ldots, 0)$$
$$U_n = (0, 0, 0, \ldots, 1)$$

The calculation

$$(a_1, a_2, \ldots, a_n) = a_1 U_1 + a_2 U_2 + \cdots + a_n U_n$$

shows that the U's span V_n and indeed a linear combination of the U's such as the right side of the last equation is the zero vector only if all the coefficients a_i are zero.

2. *$\mathbb{C}$ as a Vector Space over $\mathbb{R}$.* The complex numbers have $\{1, i\}$ as a basis. Certainly $\{1, i\}$ span the complex numbers; one of the familiar representations for complex numbers is $a + bi$ where $a, b \in \mathbb{R}$. The set $\{1, i\}$ is independent over $\mathbb{R}$ since neither 1 nor i is a *real* multiple of the other.

10.1.11 Theorem.   Existence of a Basis

Every vector space over a field except the null space has a basis.

Proof.   Let the field of scalars be denoted by F and the vector space by V.

If we further assume that V could be spanned by a finite number of vectors then the proof of the existence of a basis can follow from the Steinitz replacement theorem 10.1.12. However, if we wish to prove the result more generally we must make use of an additional property from set theory. While a full discussion of this is beyond the scope of this text, there is a proof of this theorem that involves a typical application of the maximal principle. You may find this proof interesting in its own right even though we have no immediate use for this theorem in its full generality.

The idea behind this proof is to "choose" as large a set of independent vectors as we can in V and to show that this set forms a basis. The subtlety lies in the method used to make the choice. In this proof the choice is that of a maximal element in the set of all independent sets of vectors in V. Here are the details:

Let P be the set whose elements are independent sets of vectors in V. Partially order P by set inclusion. P is not empty since $\{v\}$ is in P if v is any nonzero vector in V. We wish to show that P has a maximal element (which is thus an independent set of vectors from V) and to show that this maximal element is a basis.

To do this we invoke the maximal principal 2.2.6. To use the maximal principle, we must first show that any chain $\hat{C}$ of elements in P, that is, any chain of independent sets from V, has an upper bound in P.

To do this let S be the set union of all the sets that are elements of $\hat{C}$. Thus

$$S = \cup \hat{C} = \{v: v \in \text{ some (independent) set in } \hat{C}\}$$

We claim that S is in P and is an upper bound for $\hat{C}$. If we can show that S is in P, then it surely is an upper bound for $\hat{C}$ since it contains all the vectors belonging to any set that is an element of $\hat{C}$. The point then is to show that the set of vectors in S is independent. From the definition of an independent set, to do this we need only show that any finite subset of vectors in S is independent.

Suppose that $T = \{v_1, \ldots, v_n\}$ is a finite subset of S. Thus each v_i belongs to some set C_i in the chain $\hat{C}$. Since $\hat{C}$ is a chain one of the C_i must contain all the others. (It is at this point that the finite nature of the set T is used. Were T not finite, there would not necessarily be one set among the C_i which contained all the others.) So suppose that M is one of the C_i's that contains all the others. Thus $T \subset M$. And since M is in P it is, by definition, independent and hence any nonempty subset of it, T in particular, is independent. Thus S is an independent set in V.

Finally we show that S is a basis. By construction and the proof above we know that S is independent. We need to show that the span of S is V.

Let $v \in V$. If $v \in S$ then v is in the span of S. If not, consider the set $\{v, S\}$. We claim this set is dependent. If it were independent, it would belong to P and properly contain S, contradicting the choice of S as a maximal element of P. This means that there are scalars

$$c_0, c_1, \ldots, c_m \in F$$

not all zero and vectors

$$v_1, \ldots, v_m \in S$$

such that

$$c_0 v + c_1 v_1 + \cdots + c_m v_m = 0 \qquad (2)$$

Now if $c_0 = 0$ then $\{v_1, \ldots, v_m\}$ would be dependent, contrary to the fact that S is an independent set. Since $c_0 \neq 0$ we can solve equation (2) for v; and thus v is in the span of S.

What is important for us, because we deal mostly with vector spaces in which there is a finite basis, is the following result of E. Steinitz (1871–1928).

10.1.12 Theorem.   Steinitz Replacement Theorem

Let V be a vector space over a field F. Let W be a nonempty set of vectors and let S be the subspace they span.

If $\{v_1, \ldots, v_n\}$ is an independent set of n vectors in S then there exists a subset of vectors

$$\{w_1, \ldots, w_n\} \subseteq W$$

such that the set obtained by replacing the w's by the v's,

$$(W - \{w_1, \ldots, w_n\}) \cup \{v_1, \ldots, v_n\}$$

span S. In particular $|W| \geq n$.

The proof is by induction on n. We omit the details.

From this theorem come many important results:

1. If a subspace is spanned by an independent set of n vectors, every set of vectors that span the subspace contains at least n vectors.
2. If a subspace has a finite basis, every basis has the same number of elements. (Actually this is true for infinite cardinals as well, but that result is more difficult to prove and uses techniques in set theory beyond the scope of this text.)
3. If T is a finite independent set of vectors in a subspace, there is a set of vectors R such that $T \cup R$ is a basis for the subspace. We say that T has been extended to a basis.
4. If T is a finite set of vectors, there is a subset $S \subseteq T$ such that S is a basis for the span of T.

These results make it possible to define a dimension for a subspace.

10.1.13 Definition.   Dimension

If V is a vector space over a field F, the *dimension* of V over F is the (cardinal) number of vectors in a basis for V over F.

The language is not pedantic. The same set of elements may be considered as a vector space over different fields. Its "dimension" depends on the field of scalars. Here's an example:

As we've seen, the complex numbers $\mathbb{C}$ are a vector space of dimension 2 over the field of real numbers $\mathbb{R}$.

The complex numbers are also a vector space over the rationals $\mathbb{Q}$, but the dimension of $\mathbb{C}$ over $\mathbb{Q}$ is infinite.

We shall show that the dimension of $\mathbb{C}$ over $\mathbb{Q}$ cannot be finite by exhibiting, for each integer n, a set of n complex numbers that is independent over $\mathbb{Q}$. The set we use is the first n powers of the positive nth root of 2. Let

$$a = \sqrt[n]{2}$$

We claim that

$$1, a, a^2, \ldots, a^{n-1}$$

are independent over $\mathbb{Q}$. Indeed, if there were nonzero rationals $c_0, \ldots, c_{n-1}$ such that

$$c_0 + c_1 a + \cdots + c_{n-1}a^{n-1} = 0$$

then a would be a zero of a polynomial of degree less than n in $\mathbb{Q}[x]$. But we know that

$$x^n - 2$$

is irreducible in $\mathbb{Q}[x]$ and so a is not a zero of a polynomial of degree less than n.

Thus the dimension of $\mathbb{C}$ over $\mathbb{Q}$ could not be finite, for we can exhibit as large a finite set of independent elements as we please by choosing n as large as we please.

This argument does not show that $\mathbb{C}$ contains an infinite set that is independent over $\mathbb{R}$. Can you show that the positive nth roots of 2:

$$\sqrt{2}, \sqrt[3]{2}, \sqrt[4]{2}, \ldots, \sqrt[n]{2}, \ldots$$

are linearly independent over $\mathbb{Q}$? However, they do not form a basis for $\mathbb{C}$ over $\mathbb{Q}$.

Here is another example of how the underlying field plays an important role in "dimension." Let

$$K = \left\{ a + b\sqrt[4]{2} + c\sqrt{2} + d\sqrt[4]{8} : a, b, c, d \in \mathbb{Q} \right\}$$

K is a field; in fact $K = \mathbb{Q}[\sqrt[4]{2}]$. We know from Kronecker's theorem 9.2.16 and the substitution theorem 9.1.6 that $K \cong \mathbb{Q}[x]/\langle x^4 - 2 \rangle$ since the fourth root of 2 is a zero of the irreducible polynomial $x^4 - 2$. Now K contains the subfield F,

$$F = \mathbb{Q}(\sqrt{2}) = \left\{ a + c\sqrt{2} \right\} : a, c \in \mathbb{Q} \right\}$$

which too is a field whose prime subfield is $\mathbb{Q}$. It is not difficult to show that K has dimension 4 over $\mathbb{Q}$, in fact

$$\left\{ 1, \sqrt[4]{2}, \sqrt{2}, \sqrt[4]{8} \right\}$$

is a basis for K over $\mathbb{Q}$. On the other hand, K has dimension 2 over F since

$$\sqrt[4]{8} = \sqrt{2}\sqrt[4]{2}$$

is a linear relation *over F* and thus

$$\left\{ 1, \sqrt[4]{2} \right\}$$

is a basis for K over F.

One of the important corollaries we obtain from a dimension argument is the number of elements in a finite field. Recall that a finite field must have characteristic p, p a prime, and have $\mathbb{Z}_p$ as its prime subfield.

10.1.14 Theorem.   The Number of Elements in a Finite Field

A finite field has p^n elements where p is the characteristic of F and n is the dimension of F as a vector space over $\mathbb{Z}_p$.

Proof.   Let the dimension of F over its prime subfield $\mathbb{Z}_p$ be n and let

$$\left\{ x_1, \ldots, x_n \right\}$$

be a basis for F over the prime subfield. Then each element of F has a unique representation as

$$a_1 x_1 + \cdots + a_n x_n$$

where the coefficients a_i lie in the prime subfield, $\mathbb{Z}_p$. There are p elements in $\mathbb{Z}_p$. Thus there are p choices for each a_i and so there are p^n elements in F.

Another consequence of the notion of a basis is the following representation theorem.

10.1.15 Theorem.   Isomorphism of Vector Spaces of Equal Dimension

Every vector space V over F of dimension n is isomorphic to $V_n(F)$.

Proof.   Choose a basis $\{ v_1, v_2, \ldots, v_n \}$ for V and now verify that the mapping

$$a_1 v_1 + \cdots + a_n v_n \rightarrow \left(a_1, \ldots, a_n \right)$$

is an isomorphism. You must prove that the mapping is one-to-one and that both vector addition and scalar multiplication are preserved.

Note that many isomorphisms are possible since the choice of the basis, or even the ordering of the basis elements implicit in the proof, is quite arbitrary.

The next theorem gives us a useful formula for computing dimensions of subspaces in a vector space.

10.1.16 Theorem. Dimension Formula

Let S and T be finite dimensional subspaces of a vector space V over a field F. Then

$$\dim(S) + \dim(T) = \dim(S \vee T) + \dim(S \cap T)$$

Proof.   Let
$$\{a_1, \ldots, a_r\}$$
be a basis for $S \cap T$. Extend this set of independent vectors in two different ways; to a basis for S and to a basis for T. Thus let
$$\{a_1, \ldots, a_r, b_1, \ldots, b_s\}$$
be a basis for S and let
$$\{a_1, \ldots, a_r, c_1, \ldots, c_t\}$$
be a basis for T.

Now prove that the set union of these two bases is a basis for $S \vee T$. Then a count shows
$$\dim(S \vee T) = r + s + t = (r + s) + (r + t) - r$$
$$= \dim(S) + \dim(T) - \dim(S \cap T)$$
We omit the details of this argument.

10.1.17 Example.   Systems of Linear Equations

An important application of these results is the analysis of the solutions of a system of m linear equations in n unknowns with coefficients in a field F. Consider

$$\begin{aligned}
a_{11}x_1 + a_{12}x_2 + \cdots + a_{1n}x_n &= b_1 \\
a_{21}x_1 + a_{22}x_2 + \cdots + a_{2n}x_n &= b_2 \\
&\cdots \\
a_{m1}x_1 + a_{m2}x_2 + \cdots + a_{mn}x_n &= b_m
\end{aligned} \tag{3}$$

and the special so-called homogeneous case:

$$\begin{aligned}
a_{11}x_1 + a_{12}x_2 + \cdots + a_{1n}x_n &= 0 \\
a_{21}x_1 + a_{22}x_2 + \cdots + a_{2n}x_n &= 0 \\
&\cdots \\
a_{m1}x_1 + a_{m2}x_2 + \cdots + a_{mn}x_n &= 0
\end{aligned} \tag{4}$$

There are at least two important ways to look at these systems of equations. One is to consider the columns of coefficients as vectors in the space V_m of m-tuples over F. Let the columns be denoted
$$C_1, \ldots, C_n, B$$
where
$$C_j = (a_{1j}, a_{2j}, \ldots, a_{mj})$$
or more graphically as

$$C_j = \begin{pmatrix} a_{1j} \\ a_{2j} \\ \vdots \\ a_{mj} \end{pmatrix}$$

and thus equation (3) can be rewritten as

$$x_1 C_1 + \cdots + x_n C_n = B$$

Thus the system of equations (3) has a solution if and only if

$$B \in \text{span}\langle C_1, \ldots, C_n \rangle$$

hence, if and only if

$$\dim \langle C_1, \ldots, C_n \rangle = \dim \langle C_1, \ldots, C_n, B \rangle$$

An immediate consequence of this formulation of the problem arises with $n = m$. We state it in its classical classroom terminology: If the number of unknowns (n) equals the number of equations (m) and if the column vectors are linearly independent, then there is always a unique solution to the system (3). This is so because in this case the column vectors form a basis for V_n and so every $B \in V_n$ can be expressed in terms of this basis in one and only one way.

A second important way to think of these systems is to consider the solutions $(x_1, \ldots, x_n)$ as vectors in V_n, the space of n-tuples over F. Thus we can see that the difference of any two solution vectors of (3) yields a solution of (4). Moreover, the set of solutions of (4) form a subspace of V_n. If X_0 is one solution of (3) then every solution of (3) has the form

$$X_0 + Z$$

where Z is a solution of (4). Thus every solution of (3) is generated from X_0 and a basis of solutions of (4). Expressed in the language of groups, the solutions of (3) form a coset of the solution space of (4).

It is not surprising that there is an intimate relationship between the space spanned by the columns C_j $(j = 1, \ldots, n)$ and the space of solutions of (4).

Of course, (4) always has the trivial solution in which all x's $= 0$. On the other hand, if n, the number of unknowns, exceeds m, the number of equations, there will always be nontrivial solutions. This is so because if $n > m$, the number of column vectors will exceed the dimension of the space to which they belong and so the column vectors will necessarily be dependent.

Terminology. We call the span $\langle C_1, \ldots, C_n \rangle$ the *column space* of the homogeneous system (4).

10.1.18 Theorem. Solutions of a System of Linear Equations

The solution space of a homogeneous system of m linear equations in n unknowns and its column space satisfy the following relationship:

$$n = \dim(V_n) = \dim(\text{solution space}) + \dim(\text{column space})$$

Remark. Let the system of equations be given in the form (4). Thus n is the number of columns and so the column space has dimension at most n. On the other hand, the space spanned by the columns is a subspace of V_m while the space of solutions is a subspace of V_n, so the result may seem surprising.

Proof.   Begin by selecting a basis for the solution space. Let $\{K_1, \ldots, K_u\}$ be a basis for the solution space. If $u = n$ then every n-tuple is a solution and it follows that every coefficient a_{ij} is zero, in which case the theorem holds. Now assume that $u < n$ and extend the basis for the solution space to a basis for V_n. Suppose that $\{K_1, \ldots, K_u, K_{u+1}, \ldots, K_n\}$ is a basis for V_n. Denote each of the n vectors K_v as follows:

$$K_v = (x_{v1}, x_{v2}, \ldots, x_{vn}) \qquad (v = 1, \ldots, n)$$

For v in the range 1 to u, K_v is a solution for (4) and so

$$\Sigma_h x_{vh} C_h = 0 \qquad \text{for } 1 \leq v \leq u \tag{5}$$

Next, for each v, $(u < v \leq n)$ define

$$D_v = \Sigma_h x_{vh} C_h = x_{v1} C_1 + x_{v2} C_2 + \cdots + x_{vn} C_n \tag{6}$$

By its form each D_v is in the column space. We now prove that

$$\text{Span}\langle C_1, \ldots, C_n \rangle = \text{span}\langle D_{u+1}, \ldots D_n \rangle$$

and that the set $\{D_{u+1}, \ldots, D_n\}$ is linearly independent.

First we show that the spans are the same. To do this it suffices to show, for each $i = 1, \ldots, n$, that C_i can be expressed in terms of the D's. Now fix an index i.

To express C_i in terms of the D's we note that because the K's are a basis for V_n we can express the standard basis element U_i (recall Example 10.1.10) in terms of the K's. Thus

$$U_i = (0, \ldots, 1, \ldots, 0) = \sum_{v=1}^{n} s_v K_v$$

and so, if $j \neq i$,

$$0 = \sum_{v=1}^{n} s_v x_{vj} \tag{7}$$

and for $j = i$,

$$1 = \sum_{v=1}^{n} s_v x_{vi} \tag{8}$$

Now, trivially, we may write $C_i = 0C_1 + \cdots + 1C_i + \cdots + 0C_n$ and if we substitute equations (7) and (8) for each corresponding 0 and 1 we can express C_i not so trivially:

$$C_i = \sum_{j=1}^{n} \left(\sum_{v=1}^{n} s_v x_{vj} \right) C_j = \sum_{v=1}^{n} s_v \left(\sum_{j=1}^{n} x_{vj} C_j \right)$$

Now consider the terms $\Sigma_{v=1}^{n} x_{vj} C_j$. For $j = 1, \ldots, u$, equation (5) says these terms are zero. From equation (6), the last terms, for $j = u + 1, \ldots, n$ are D_v.

Hence

$$C_i = \sum_{v > u}^{n} s_v D_v$$

where of course the coefficients s_v depend upon i but our simplified notation does not reflect that.

Next we must show that the D's are independent. Suppose

$$0 = \sum_{v > u}^{n} t_v D_v$$

Substituting for the D's we can write

$$0 = \sum_{v > 1}^{n} t_v \left(\sum_{j=1}^{n} x_{vj} C_j \right) = \sum_{j=1}^{n} \left(\sum_{v > u}^{n} t_v x_{vj} \right) C_j$$

Because the C's are independent we know that for $j = 1, \ldots, n$, each coefficient

$$\sum_{v > u}^{n} t_v x_{vj} = 0$$

and hence the solution space contains

$$T = \left(\sum_{v > u}^{n} t_v x_{v1}, \sum_{v > u}^{n} t_v x_{v2}, \ldots, \sum_{v > u}^{n} t_v x_{vn} \right) = \sum_{v > u}^{n} t_v K_v$$

Thus $T \in$ (solution space) $\cap \langle K_{u+1}, \ldots, K_n \rangle = 0$. Thus $\sum_{v > u}^{n} t_v K_v = 0$, and from the independence of the K's it follows that each $t_v = 0$. Thus the D's must be independent.

The intimate connection between the row and column space of a matrix is shown by the following theorem. First we pause to define a matrix and its transpose.

10.1.19 Definition.   Matrix and Transposed Matrix

A $(m \times n)$ *matrix* over a ring R is a rectangular array of mn elements of R arranged in m rows and n columns.

We often denote the element in the ith row and jth column by the pair of subscripts ij. If the size of the matrix is understood, we simply indicate the element in the ith row and jth column generically. Thus

$$A = \begin{bmatrix} a_{11} & a_{12} & \cdots & a_{1n} \\ a_{21} & a_{22} & \cdots & a_{2n} \\ \cdot & & \cdots & \cdot \\ a_{m1} & a_{m2} & \cdots & a_{mn} \end{bmatrix} = [a_{ij}]$$

The *transpose* of the $(m \times n)$ matrix A, denoted A^T, is the $(n \times m)$ matrix whose rows are the columns of A and whose columns are the rows of A. Thus the transform A^T can be formed from A by flipping it across the main diagonal which runs from upper left to lower right. The element in the ith row and jth column of A^T is a_{ji}. Thus

$$A^T = \left[b_{ij} \right] \qquad \text{where } b_{ij} = a_{ji}$$

In more detail

$$A^T = \begin{bmatrix} a_{11} & a_{21} & \cdots & a_{m1} \\ a_{12} & a_{22} & \cdots & a_{m2} \\ \cdot & & \cdots & \cdot \\ a_{1n} & a_{2n} & \cdots & a_{mn} \end{bmatrix}$$

10.1.20 Theorem.   Row Rank Equals Column Rank

Let A be a matrix whose elements lie in a field.

$$A = \begin{bmatrix} a_{11} & a_{12} & \cdots & a_{1n} \\ a_{21} & a_{22} & \cdots & a_{2n} \\ \cdot & & \cdots & \cdot \\ a_{m1} & a_{m2} & \cdots & a_{mn} \end{bmatrix}$$

Let Col_A be the space spanned by the column vectors

$$C_1 = (a_{11}, a_{21}, \ldots, a_{m1})$$
$$C_2 = (a_{12}, a_{22}, \ldots, a_{m2})$$
$$C_n = (a_{1n}, a_{2n}, \ldots, a_{mn})$$

and let Row_A be the space spanned by the row vectors,

$$R_1 = (a_{11}, a_{12}, \ldots, a_{1n})$$
$$R_2 = (a_{21}, a_{22}, \ldots, a_{2n})$$
$$R_m = (a_{m1}, a_{m2}, \ldots, a_{mn})$$

Then $\dim(\text{Col}_A) = \dim(\text{Row}_A)$. This number is called the rank of A.

Proof.   We shall prove that for any matrix A

$$\dim(\text{Col}_A) \leq \dim(\text{Row}_A)$$

Then for a particular matrix M and its transpose we obtain

$$\dim(\text{Col}_M) \leq \dim(\text{Row}_M) \quad \text{and} \quad \dim(\text{Col}_{M^T}) \leq \dim(\text{Row}_{M^T})$$

Now since the column space of M^T is the row space of M and the row space of M^T is the column space of M we also conclude

$$\dim(\text{Row}_M) \leq \dim(\text{Col}_M)$$

and hence that the dimensions of these two spaces are equal.

Suppose first that the m rows of the $(m \times n)$ matrix A are linearly independent. Then $\dim(\text{Row}_A) = m$. However, since the column space is a subspace of V_m it follows that

$$\dim(\text{Col}_A) \leq m = \dim(\text{Row}_A)$$

The argument in general is only slightly more difficult. We need one construction and one key observation.

Suppose that $\dim(\text{Row}_A) = k < m$. We may choose a subset of k vectors from the rows that are linearly independent and by a reordering if necessary we may suppose that these are the first k rows:

$$R_1, \ldots, R_k$$

A reordering of the rows makes no change in row space and only permutes the coordinates of the column space, hence makes no change in the dimension of the column space.

Thus the rows $R_{k+1}, \ldots, R_n$ are linear combinations of $R_1, \ldots, R_k$.

Now for the construction. Consider the $(k \times n)$ matrix consisting of the first rows of A. Call it A^*.

$$A^* = \begin{bmatrix} a_{11} & a_{12} & \cdots & a_{1n} \\ a_{21} & a_{22} & \cdots & a_{2n} \\ \cdot & & \cdots & \cdot \\ a_{k1} & a_{k2} & \cdots & a_{kn} \end{bmatrix}$$

Because the remaining rows of A are linear combinations of these, for each h in the range $k < h \le m$ there are constants r_{ih} such that

$$R_h = \sum_{1 \le i \le k} r_{ih} R_i$$

in particular for each coordinate a_{hj} of R_h,

$$a_{hj} = \sum_{1 \le i \le k} r_{ih} a_{ij} \tag{9}$$

Of course, the row spaces of A and A^* are equal so that $\dim(\text{Row}_A) = \dim(\text{Row}_{A^*})$. Now let the columns of A^* be $C_1^*, C_2^*, \ldots, C_m^*$. As before, it is true that

$$k = \dim(\text{Row}_{A^*}) \ge \dim(\text{Col}_{A^*}) = \dim(\langle C_1^*, \ldots, C_m^* \rangle)$$

Now for the key observation: We claim that $\dim(\langle C_1, \ldots, C_m \rangle) = \dim(\langle C_1^*, \ldots, C_m^* \rangle)$. Then we will have proved

$$\dim(\text{Col}_A) = \dim(\text{Col}_{A^*}) \le \dim(\text{Row}_{A^*}) = \dim(\text{Row}_A)$$

To verify our claim it suffices to show that any linear dependence relation

$$\sum_j x_j C_j = 0$$

among the C's is also a linear dependence relation among the C^*'s:

$$\sum_j x_j C_j^* = 0$$

and conversely. This means that an independent set in Col_A corresponds to an independent set in Col_{A^*} and conversely.

First note that any linear dependence among the C's remains a linear dependence among the C^*'s since only some coordinates have been chopped from the C's to form the C^*'s. Thus

$$\dim(\langle C_1, \ldots, C_m \rangle) \geq \dim(\langle C_1^*, \ldots, C_m^* \rangle)$$

To demonstrate the converse, suppose that

$$\sum_j x_j C_j^* = 0$$

Thus at the hth coordinate $(1 \leq h \leq k)$ we find

$$0 = \sum_j x_j a_{hj}$$

Now using (9) we conclude that for $k < h \leq m$,

$$\sum_j x_j a_{hj} = \sum_j x_j \left(\sum_{1 \leq i \leq k} r_{ih} a_{hj} \right) = \sum_{1 \leq i \leq k} r_{ih} \left(\sum_j x_j a_{ij} \right) = \sum_{1 \leq i \leq k} r_{ih} 0 = 0$$

and thus $\sum_h x_j C_j = 0$.

10.1.21 Definition.   Rank of a Matrix

The *rank* of a matrix whose elements lie in a field is the dimension of the row (equivalently the dimension of the column) space.

We may restate the result on the solutions of a homogeneous system of equations (4) as:

The number of linear independent solutions of (4) is the number of unknowns minus the rank of the coefficient matrix.

We conclude this section with some standard terminology.

10.1.22 Definition.   Singular and Nonsingular Matrices

An $n \times n$ matrix $[a_{ij}]$ with elements in a field F is said to be *nonsingular* if its rank is n and *singular* otherwise.

A result of this definition and Theorem 10.1.18 is that if $n = m$ then the system (4) has only the trivial (all zero) solution if and only if the coefficient matrix is nonsingular.

EXERCISE SET 10.1

10.1.1   Prove the Steinitz replacement theorem.

10.1.2   Prove the four listed consequences of the Steinitz replacement theorem.

10.1.3   Which of the following are vector spaces? Of those that are, find a basis. Take vector addition and scalar multiplication as the usual ones *except where noted*.
   1. The ring $F[x]$ over F where F is a field.
   2. The set of all sequences $(a_0, \ldots, a_i, \ldots)$ where the a's are elements of a field.

3. Same as the above except that only a finite number of terms a_i differ from zero.

4. Same as 2 except that at least one of the a's is zero.

5. The set of ordered pairs (a, b) from a field F where scalar multiplication is defined $u(a, b) = (0, ub)$.

6. The complex numbers over $\mathbb{R}$ where addition is as usual but where scalar multiplication is defined:

$$r(a + bi) = ra - rbi \qquad \text{for } r \in \mathbb{R}$$

10.1.4 Prove or disprove each of the following for finite dimensional vector spaces:
1. If S is a subspace of T then $\dim(S) \leq \dim(R)$.
2. If $\dim(S) < \dim(T)$ then $S \neq T$.
3. If $S \subset T$ and $S \neq T$ then $\dim(S) < \dim(T)$.

10.1.5 Let F be a finite field with p elements (p a prime). How many different bases are there for $V_n(F)$?
How many nonsingular $n \times n$ matrices are there with entries in F?

10.1.6 Show that the following row transformations do not change the row space of a matrix with entries in a field:
R1: Permute the rows.
R2: Multiply any row by a nonzero scalar.
R3: Replace row i by a scalar times row j ($i \neq j$).
What is the effect of R1, R2, and R3 on the column space of the matrix?

10.1.7 Show that by using the row transformations of Exercise 10.1.6 an $n \times m$ matrix of rank k can be transformed into one whose last $n - k$ rows are all 0's and whose ith row has an initial string of zeros and the first nonzero entry is a 1 in column $\sigma(i)$ ($\sigma(i) \geq i$) and so looks like:

$$(0, \ldots, 0, 1, r_{i\sigma(i)+1}, \ldots)$$

and where the σ's are increasing:

$$1 \leq \sigma(1) < \sigma(2) < \cdots < \sigma(k)$$

and where each column $\sigma(i)$ has zeros except in row i. This form is called the "row reduced echelon" form of the matrix. If A is a matrix of coefficients of a system (4), then the solution of the row reduced form for A is easy!

10.1.8 Find all solutions of

$$3x + 2y + 3z = 7$$
$$-x + 2y - z = -5$$
$$x + y + z = 2$$

10.1.9 What is the condition that integers a, b, and c must satisfy in order that

$$2w - x + y - 3z = a$$
$$w + x - y = b$$
$$4w + x - y - 3z = c$$

can be solved in integers?

10.1.10 Let S be a subspace of dimension k in V_n over $\mathbb{Z}_3$. Count the number of subspaces T such that

$$S \cap T = \{0\} \quad \text{and} \quad S \vee T = V_n$$

(Remember $S \vee T$ is the least subspace containing both S and T.)

10.2 CLASSIFICATIONS OF FIELD EXTENSIONS

In the next several sections we study the relation between two fields E and F when $E \supseteq F$. In this case we say that F is a subfield of E or that E is an extension of F.

Recall from Section 8.3 that E and F have prime subfields. Since $E \supseteq F$, the prime subfield of F is automatically the prime subfield of E and so E and F have the same characteristic. We also know that the subfields of E containing F form a complete lattice.

One of the most useful parameters to be associated with a field extension $E \supseteq F$ is the dimension of E as a vector space over F. As you recall, this situation was anticipated in Example 10.1.3. We now give a special name to its dimension over F.

10.2.1 Definition. Degree of E over F

The *degree* of a field extension E over F is the dimension of E as a vector space over F.

Notation.

$$[E : F] = \text{degree of } E \text{ over } F = \dim_F(E)$$

We say that an extension E over F is a finite extension or an infinite extension according as the degree of E over F is finite or infinite.

We begin with two very important examples.

10.2.2 Theorem. An Infinite Extension

If F is a field then the field of fractions of $F[x]$ is an extension of F of infinite degree.

Proof. We know that $F[x]$, the ring of polynomials in the indeterminate x, is an integral domain. We construct its field of fractions and denote it as $F(x)$.

Each element in this field of fractions can be represented as the quotient of two polynomials. We have

$$F(x) = \{ f(x)/g(x) \colon g(x) \neq \text{zero polynomial}\}$$

as this field. Note the use of the *round* parentheses in the notation as opposed to the brackets used when designating the ring of polynomials.

An important property of this field is that it contains an isomorphic copy of F. The field of fractions $F(x)$ contains a copy of the generating integral domain $F[x]$. In turn $F[x]$ contains a copy of the field F through the polynomials of degree 0. We identify this field with F and say that $F(x)$ is an extension of F.

The degree $[F(x):F]$ is infinite. To see this we show that the set of powers of the indeterminate x,

$$\{1, x, x^2, \ldots\}$$

is an independent set in $F(x)$, although it is not a basis for $F(x)$ over F. We argue as follows:

Certainly no two distinct powers are equal as polynomials and so the set is infinite. Moreover no finite subset of them is dependent since no combination of different powers of x with coefficients from F can be the zero polynomial unless each coefficient is zero. Thus $F(x)$ can have no finite basis over F.

On the other hand, the powers of x cannot be a basis for $F(x)$ over F. Indeed the sub*space* of $F(x)$ spanned by the powers of x is just the set of polynomials in x. Thus the fraction $1/x$ is not contained in the span of these powers of x.

The field $F(x)$ is commonly referred to as the field of rational functions or rational expressions over F. Indeed, in calculus courses a lot of time is spent differentiating and integrating these functions.

A second seminal example is provided by the homomorphic image of the polynomial ring $F[x]$ when a maximal ideal is factored out. Because $F[x]$ is a principal ideal domain such a maximal ideal has the form $\langle p(x)\rangle$ where $p(x)$ is an irreducible polynomial in $F[x]$.

10.2.3 Theorem.   A Finite Extension

Let F be a field and let $p(x)$ be an irreducible polynomial in $F[x]$. Then the degree of the field

$$F[x]/\langle p(x)\rangle = F(u)$$

over F where

$$u = x + \langle p(x)\rangle$$

is the degree of the irreducible polynomial $p(x)$:

$$[F(u):F] = \deg(p(x))$$

Proof.   We have already established in Theorem 9.2.16 that

$$F[x]/\langle p(x)\rangle$$

is a field in which $p(x)$ has a root. This root is the residue class

$$u = x + \langle p(x)\rangle$$

Remember that this field contains an isomorphic copy of F defined by the correspondence

$$a \rightarrow a + \langle p(x) \rangle$$

between elements $a \in F$ and residue classes $a + \langle p(x) \rangle$. Through this isomorphism the elements of F are identified with elements of $F(u)$.

Now we want to show that the dimension of $F(u)$ over F is the degree of $p(x)$. Let $n = \deg(p(x))$. We show in particular that

$$\{1, u, \ldots, u^{n-1}\}$$

is a basis for $F(u)$ over F.

First of all, any element of $F(u)$ is the image of some polynomial $g(x) \in F[x]$ and so has the form

$$g(u) = g(x) + \langle p(x) \rangle$$

Moreover the polynomial can be chosen to have degree at most $n - 1$. This follows from the division algorithm: For any $g(x) \in F[x]$ we have

$$g(x) = p(x)q(x) + r(x)$$

where $r(x) = 0$ or $\deg(r) < \deg(p)$ and in either case,

$$g(x) \in r(x) + \langle p(x) \rangle$$

and so

$$g(u) = r(u)$$

Thus the powers of U:

$$1, u, u^2, \ldots, u^{n-1}$$

span $F(u)$. Now we claim that this set is independent over F. Suppose that a linear combination of the u's equals 0:

$$a_0 + a_1 u + \cdots + a_{n-1} u^{n-1} = 0$$

Thus the image of the corresponding polynomial

$$t(x) = a_0 + a_1 x + \cdots + a_{n-1} x^{n-1}$$

belongs to the kernel $\langle p(x) \rangle$ and hence is divisible by $p(x)$. But the $\deg(p(x)) > \deg(t(x))$. The only way for this to happen is for $t(x)$ to be the zero polynomial; hence all the coefficients a_i are zero. This completes the proof that the first n powers of u form a basis and so the degree of this extension is

$$[F[x]/\langle p(x) \rangle : F] = n = \deg(p(x))$$

Next we want to distinguish between two types of elements in a field extension, $E \supseteq F$. The technique is similar to that used for determining the characteristic of a ring used in Theorem 7.2.9.

10.2.4 Definition.   Algebraic and Transcendental Elements

Let E be an extension of F. Let $a \in E$ and let Φ be the homomorphism of $F[x]$ into E defined by the substitution

$$\Phi: f(x) \rightarrow f(a)$$

The element a is called *transcendental* over F if the kernel of Φ is $\{0\}$. Otherwise the element a is called *algebraic* over F. If a is algebraic over F then the kernel of Φ is $\langle p(x) \rangle$ where $p(x)$ must necessarily be an irreducible polynomial in $F[x]$.

Some of the assertions of the definition require justification. To begin, the substitution theorem (9.1.6) shows that the mapping mentioned in the definition is a homomorphism. Since the image is a subring of a field, it is thus an integral domain. By Theorem 8.2.6 this means that the kernel is a prime ideal. Since $F[x]$ is a PID, the kernel has the form $\langle p(x) \rangle$ and is a prime ideal if and only if $p(x) = 0$ or $p(x)$ is an irreducible polynomial.

The terminology is suggestive. If an element is an *algebraic* element over a field F then it is a zero of some polynomial in $F[x]$. Determining the zeros of polynomials is a central theme of much of classical algebra. If this is not the case, the element "transcends" the scope of algebra; in particular a transcendental element cannot be expressed or computed from the elements of F by algebraic means alone.

As examples we see that x, in the field $F(x)$ of Theorem 10.2.2, is a transcendental element over F while u, in the field $F(u)$ of Theorem 10.2.3, is an algebraic element over F.

It is also important to realize that every element of F is algebraic over F since $a \in F$ is the zero of the first-degree polynomial $x - a \in F[x]$. We summarize these important concepts and examples in the next theorem.

10.2.5 Definition.   The Subfield F(a)

Let E be a field extension of the field F. If $a \in E$ let $F(a)$ denote the smallest subfield of E containing F and a. Thus

$$F(a) = \cap \{ K: a \in K \text{ and } E \supseteq K \supseteq F \}$$

10.2.6 Theorem.   The Degree of [F(a):F]

Let E be an extension of F. Let $a \in E$.

1. If a is transcendental over F then $F(a) \cong F(x)$ and $[F(a):F]$ is infinite.
2. If a is algebraic over F then $F(a) \cong F[x]/\langle p(x) \rangle$ where $\langle p(x) \rangle$ is the kernel of the homomorphism of $F[x]$ into E defined by $f(x) \rightarrow f(a)$, the element a is a zero of the irreducible polynomial $p(x)$, and $[F(a):F] = \deg(p(x))$.

Proof. Any subfield of E containing F and a must, by closure, contain all the polynomials in a. Thus we are led to consider the homomorphism of $F[x]$ into E defined by

$$f(x) \to f(a)$$

and its kernel. If the kernel is $\{0\}$ then a is transcendental over F by definition and E contains an isomorphic copy of $F[x]$, and hence an isomorphic copy of its field of quotients $F(x)$ by extending the isomorphism as follows:

$$f(x)/g(x) \to f(a)/g(a)$$

Note that because the kernel is zero, $g(a) \neq 0$ for every nonzero polynomial.

On the other hand, if the kernel is $\langle p(x) \rangle$ where $p(x)$ is an irreducible polynomial then a is algebraic by definition and the image of $F[x]$, which contains a, the image of x, in E is isomorphic to

$$F[x]/\langle p(x) \rangle$$

10.2.7 Definition.   Algebraic and Transcendental Extensions

An extension E of a field F is called *algebraic* over F if every element of E is algebraic over F. Otherwise it is called *transcendental* over F.

Note that an algebraic extension over F contains only elements that are algebraic over F. A transcendental extension of F may contain elements that are algebraic over F. For example, let $K = \mathbf{Q}(\sqrt{2})$ and let $E = K(x)$. Then E must be transcendental over $\mathbf{Q}$ but it contains $\sqrt{2}$, which is algebraic over $\mathbf{Q}$.

10.2.8 Theorem.   Finite Degree Implies Algebraic Extension

Let E be a finite extension of F. Then E is an algebraic extension of F.

Proof. Suppose that $[E : F] = n$. Let a be any element of E. We have that the $(n + 1)$ elements

$$1, a, a^2, \ldots, a^n$$

are linearly *dependent* over F. Hence there exist coefficients, *not all zero*, such that

$$c_0 \cdot 1 + c_1 a + c_2 a^2 + \cdots + c_n a^n = 0$$

and so the nonzero polynomial

$$g(x) = c_0 + c_1 x + \cdots + c_n x^n$$

lies in the kernel of the mapping of $F[x]$ into E defined by

$$f(x) \to f(a)$$

and so the kernel is not zero. Hence a is algebraic over F.

10.2.9 Theorem.   Extensions of Degree 2

Let E be an extension of degree 2 over a field F whose characteristic is not 2. Then $E = F(\sqrt{d})$ where $d \in F$.

Proof.   Start with a basis. Pick 1 as a basis element and let r be any element in E but not in F. The set $\{1, r\}$ is independent over F; otherwise $r \in F$. Since $[E : F] = 2$, it is a basis.

On the other hand, $\{1, r, r^2\}$ are dependent over F and so for coefficients a, b, and $c \in F$, $a \neq 0$, we have

$$ar^2 + br + c = 0$$

so that r is a zero of the polynomial $ax^2 + bx + c$. By completing the square we see that

$$a\left(r + \frac{b}{2a}\right)^2 = (b^2 - 4ac)/2a$$

(this is where we use the fact that the characteristic of F is not 2) and so

$$r = \left[-b + \sqrt{b^2 - 4ac}\,\right]/2$$

The number inside the square root symbol is called the discriminant of the quadratic:

$$d = b^2 - 4ac$$

and

$$E = F(r) = F(\sqrt{d})$$

since r and $\sqrt{d}$ can be expressed in terms of each other with coefficients in F. See Problem 10.2.12 for conditions when

$$F(\sqrt{u}) = F(\sqrt{v})$$

There is another classification of field extensions. Many and indeed most of those we study in this text have the property that there is some one element a in E such that $E = F(a)$. This was the case in the examples of Theorems 10.2.2 and 10.2.3 by the very nature of their construction.

10.2.10 Definition.   Simple Extensions

An extension E of a field F is called simple if there is an element $a \in E$ such that $E = F(a)$. We say that E is obtained by adjoining the element a to F. In this case E is the smallest subfield of E containing a and F. It is also useful to consider the smallest subfield of E containing F and a subset S of elements from E.

10.2.11 Definition.   Extensions Generated by a Set of Elements

Let E be an extension of F. Let S be a set of elements in E. Denote by $F(S)$ the smallest subfield of E containing F and the subset S. Thus

$$F(S) = \bigcap_{S \subseteq K \subseteq E} K$$

where the intersection is taken over all *subfields* of E containing F and the subset S.

When $S = \{s\}$ we write $F(s)$. The field $F(s)$ is a *simple* extension of F.

Please distinguish between brackets, [,], which are used to denote a polynomial ring and parentheses, (,), which are used to denote field extensions of F.

10.2.12 Example.   The Fields $\mathbb{Q}(\sqrt{2}\,)$, $\mathbb{Q}(\sqrt{3}\,)$, $\mathbb{Q}(\sqrt{2}\,)$, $\sqrt{3}\,)$

Let $E = \mathbb{R}$ be the field of real numbers and let $F = \mathbb{Q}$ be the field of rational numbers. Then

$$\mathbb{Q}(\sqrt{2}\,) = \{a + b\sqrt{2} : a, b \in \mathbb{Q}\}$$
$$\mathbb{Q}(\sqrt{3}\,) = \{a + b\sqrt{3} : a, b \in \mathbb{Q}\}$$
$$\mathbb{Q}(\sqrt{2}\,,\sqrt{3}\,) = \{a + b\sqrt{2} + c\sqrt{3} + d\sqrt{6} : a, b, c, d, \in \mathbb{Q})$$

We omit the verification that these fields have the elements specified above.

It is interesting that $\mathbb{Q}(\sqrt{2}\,,\sqrt{3}\,)$ is a simple extension. Here is a sketch of how you might verify that

$$\mathbb{Q}(\sqrt{2}\,,\sqrt{3}\,) = \mathbb{Q}(\sqrt{2} + \sqrt{3}\,)$$

Let

$$u = \sqrt{2} + \sqrt{3}$$

Show that $\sqrt{2} \in \mathbb{Q}(u)$. One way to do this is to compute u^2 and show that

$$\sqrt{6} = (u^2 - 5)/2$$

Then observe that

$$u\sqrt{6} = 2\sqrt{3} + 3\sqrt{2}$$

Now eliminate $\sqrt{3}$ between this last expression and the one for u.

10.2.13 Lemma.   Successive Adjunction

Let E be an extension of F. Let $a, b \in F$.

$$F(a, b) = F(a)(b) = F(b)(a)$$

In general if R and T are subsets of E then

$$F(R \cup T) = F(R)(T)$$

Proof. The proof is simply a matter of understanding how the notation is being used: If E is an extension of K and S is a subset of E then $K(S)$ is the least subfield of E containing K.

Thus $F(R)$ is the least subfield of E containing F and R. Then $F(R)(T)$ is the least subfield of E containing the subfield $F(R)$ and the subset T. In particular $F(R)(T)$ must contain both the sets R and T and since it is a field must contain $F(R \cup T)$.

Conversely, $F(R \cup T)$ contains R and so $F(R \cup T) \supset F(R)$. But $F(R \cup T) \supset T$ so $F(R \cup T) \supset F(R)(T)$. Thus

$$F(R \cup T) = F(R)(T)$$

The next theorem is a very useful one and is analogous to the theorem of Lagrange in groups.

10.2.14 Theorem. **Multiplication of Degrees**

Let E be a finite extension of a field K. Let K be a finite extension of a field F. Then E is a finite extension of F and

$$[E : F] = [E : K][K : F]$$

Proof. It suffices to exhibit a basis for E over F and count its elements. Let a basis for E over K be

$$a_1, a_2, \ldots, a_n$$

and let a basis for K over F be

$$b_1, b_2, \ldots, b_m$$

We shall show that nm elements

$$a_i b_j$$

form a basis for E over F.

First we show these elements span E over F. Let u be any element of E. Regarding u as an element of E over K there are coefficients in K, $c_1, \ldots, c_n$, such that

$$u = c_1 a_1 + \cdots + c_n a_n$$

Since each c_i is in K, regarding K as a vector space over F we have, for each i, coefficients $d_{1i}, \ldots, d_{mi}$ in F such that

$$c_i = d_{1i} b_1 + \cdots + d_{mi} b_m$$

hence

$$c_i a_i = d_{1i} b_1 a_i + \cdots + d_{mi} b_m a_i$$

Now plugging all this into the expression for u we see that u can be written as a linear combination of the elements $a_i b_j$, which belong to E, with coefficients d_{ij} from F. Thus these nm elements span E over F.

Now to show the independence of the $a_i b_j$ over F. Suppose that a linear combination

$$u = d_{11} b_1 a_1 + \cdots + d_{ji} b_j a_i + \cdots + d_{mn} b_m a_n = 0$$

where the d's belong to F. For each i, combine and factor out the terms involving a_i. This will give back

$$c_i a_i = (d_{1_i} b_1 + \cdots + d_{m_i} b_m) a_i$$

Note that this coefficient of a_i belongs to K. Because the a's are independent over K, each $c_i = 0$. But then, the independence of the b's over F requires that all the d's are zero. Hence the mn elements $b_j a_i$ are independent over F.

As an example of this theorem recall Example 10.2.12. You can see that we have that

$$\left[\mathbb{Q}(\sqrt{2}, \sqrt{3}) : \mathbb{Q} \right] = \left[\mathbb{Q}(\sqrt{2}, \sqrt{3}) : \mathbb{Q}(\sqrt{2}) \right] \left[\mathbb{Q}(\sqrt{2}) : \mathbb{Q} \right]$$

This is a very useful theorem. One of the immediate consequences is that if E is a finite extension of K and K is a finite extension of F then E is a finite extension of F.

Another immediate application is that if E, K, F are fields and $E \supseteq K \supseteq F$ and E is a finite extension of F then the degree of K over F divides the degree of E over F. We use this to show that certain fields cannot be subfields of other fields. As a specific example consider the field of real numbers $\mathbb{R}$ as an extension of $\mathbb{Q}$. An element of $\mathbb{R}$ is

$$\alpha = \sqrt[3]{2}$$

so that α is a cube root of 2 and is a zero of the irreducible polynomial $x^3 - 2$. Thus $E \supseteq \mathbb{Q}(\alpha)$ and moreover

$$\mathbb{Q}(\alpha) \cong \mathbb{Q}[x]/\langle (x^3 - 2) \rangle$$

is an extension of degree 3 over $\mathbb{Q}$. It follows, for example, that $\mathbb{Q}(\alpha)$ cannot contain $\sqrt{2}$. If it did, then $\mathbb{Q}(\alpha)$ would also contain $\mathbb{Q}(\sqrt{2})$ as a subfield. However, $\mathbb{Q}(\sqrt{2})$ is an extension of degree 2 over $\mathbb{Q}$ and since $2 \nmid 3$ it cannot be a subfield of $\mathbb{Q}(\alpha)$.

Now we use this multiplicative property to obtain another important result.

10.2.15 Theorem. Algebraic Elements Form a Field

Let E be an extension of F. The set of algebraic elements in E over F form a subfield of E containing F.

In particular, if $a, b \in E$ are algebraic over F then $F(a, b)$ is a finite extension of F containing $a + b$, ab, a^{-1} (if $a \neq 0$).

Proof. Since a is algebraic over F we have

$$F(a) \text{ isomorphic to } F[x]/\langle p(x) \rangle$$

where $p(x)$ is an irreducible polynomial such that $p(a) = 0$. Similarly $F(b)$ is isomorphic to

$$F[x]/\langle q(x) \rangle$$

where $q(x)$ is an irreducible polynomial such that $q(b) = 0$. Now $q(x) \in F(a)[x]$ but it need not be irreducible over $F(a)$. In any event there will be an irreducible factor $q_1(x)$ of $q(x)$ in $F(a)[x]$ such that $q_1(b) = 0$ and so

$$F(a)(b) = F(a, b) \text{ is isomorphic to } F(a)[x]/\langle q_1(x) \rangle$$

Now it follows that $F(a, b)$ is a finite extension of F since

$$[F(a, b): F] = [F(a, b): F(a)][F(a): F]$$

and each factor is finite. Thus $F(a, b)$ is an algebraic extension of F.

10.2.16 Theorem.   Transitivity of Algebraic Extensions

An algebraic extension of a field that is itself an algebraic extension of a field F is algebraic over F.

Proof.   Let E be an algebraic extension of K and let K be an algebraic extension of F. We claim that E is an algebraic extension of F. To show this, let $a \in E$. Since E is algebraic over K there is a polynomial in $K[x]$ of which a is a root, say, a is a root of

$$c_0 + c_1 x + \cdots + c_n x^n$$

where the c's are in K. By applying the previous theorem n times we have that

$$F(c_0, c_1, \ldots, c_n) = K$$

is a finite extension of F. But now we have that $K(a)$ is a finite extension of K, since $K(a)$ is a finite extension of F and so a is algebraic over F.

Remark.   With apologies to Gertrude Stein, "An algebraic extension of an algebraic extension is an algebraic extension."

EXERCISE SET 10.2

10.2.1   i. Show that $x^4 + x + 1$ is irreducible over $\mathbb{Z}_2$. Let

$$u = x + \langle x^4 + x + 1 \rangle$$

ii. List the 16 elements of $E = \mathbb{Z}_2[x]/\langle x^4 + x + 1 \rangle$ as polynomials in u. Find all the zeros of this polynomial in E. *Hint*: If b is a zero of $p(x)$ then so is b^2. Prove the hint; it's true for any polynomial in $\mathbb{Z}_2[x]$.

iii. Show that the multiplicative group of E is cyclic by exhibiting a generator.

10.2.2   i Show that if E is an extension of prime degree over F then E has no subfields except itself and F.

ii. Let an extension E of F contain subfields H and K each of which contains F. Show that if $[E: K]$ is relatively prime to $[E: H]$ then the least subfield of E containing both K and H is E.

iii. Show that if $[K: F]$ is relatively prime to $[H: F]$ then $H \cap K = F$.

10.2.3   Show that if an extension E over F contains subfields H and K each of degree 2 over F then either $H = K$ or E contains a subfield of degree 4 over F.

10.2.4   **i.** Let E be an extension of $\mathbb{Q}$ that contains α, a zero of $x^3 - 2$. Show that $[\mathbb{Q}(\alpha): \mathbb{Q}] = 3$. Show that $x^2 + x + 1$ is irreducible over $\mathbb{Q}(a)$.

   ii. Suppose further that E contains ω, a primitive cube root of unity; that is, ω is a zero of $x^2 + x + 1$. Determine $[\mathbb{Q}(\alpha, \omega): \mathbb{Q}]$. Show that if a subfield of E contains two of the three zeros of $x^3 - 2$ then it contains ω and so all three cube roots of 2. Show that $\mathbb{Q}(\alpha, \omega)$ contains $\mathbb{Q}(\alpha)$, $\mathbb{Q}(\omega\alpha)$, and $\mathbb{Q}(\omega^2\alpha)$ and that these three fields are distinct and isomorphic to each other.

10.2.5   Find all the subfields of $\mathbb{Q}(\sqrt[3]{2}, \omega)$ where ω is a primitive cube root of unity; a zero of $x^2 + x + 1$ over $\mathbb{Q}$. (This problem anticipates Example 10.3.7. See how much you can do!)

10.2.6   Show that $f(x) = 2x^3 + 9x + 6$ is irreducible over $\mathbb{Q}$. In a field E containing $\mathbb{Q}$ and a zero, θ, of $f(x)$, express the multiplicative inverse of $1 + \theta$ in the form $a + b\theta + c\theta^2$ where a, b, and c are in $\mathbb{Q}$.

10.2.7   Let E be an extension of a field F. Let $a \in E$. Regarding E as a vector space over F show that the subspace spanned by $\{1, a, a^2, \dots\}$ is a field if and only if a is algebraic over F.

10.2.8   Let R be a ring and let E and F be fields such that $E \supset R \supset F$ are subrings. If E is an algebraic extension of F prove that R is a field.

10.2.9   Let $f(x)$ and $g(x)$ be irreducible over F and have degrees that are relatively prime. Suppose that $g(x)$ has a zero, call it α, in an extension E of F. Show that f is irreducible over $F(\alpha)$.

10.2.10   A field E is called *algebraically closed* if E has no proper algebraic extensions. Let E be an algebraic extension of F. Prove that the following are equivalent:
   1. E is algebraically closed.
   2. Every irreducible polynomial in $E[x]$ is linear.
   3. Every polynomial in $F[x]$ has all its zeros in E.

10.2.11   Let $\mathbb{R}$ be the real numbers. Let V be a vector space of dimension 3 over $\mathbb{R}$. Show that it is impossible to define a multiplication on V so that V becomes a field. (Use the fact that the complex numbers are algebraically closed over $\mathbb{R}$.)

10.2.12   Let E be an extension of degree 2 over F, a field of characteristic not 2. Give necessary and sufficient conditions on u and v in F such that

$$E = F(\sqrt{u}) = F(\sqrt{v})$$

10.3 SPLITTING FIELDS

In this section, given a field F and a polynomial $f(x) \in F[x]$, we construct an extension field having all the zeros of $f(x)$. Then we prove a very strong isomorphism theorem for these constructions. This theorem forms the basis for all the structure theorems about subfields of algebraic extensions, particularly their Galois theory discussed in Chapter 13.

10.3.1 Definition.   Splitting Field

Let F be a field and let $f(x) \in F[x]$. The polynomial $f(x)$ is said to *split* in an extension $E \supseteq F$ if $f(x)$ can be factored as a product of linear factors in

$E[x]$, that is, in $E[x]$

$$f(x) = (x - a_1)(x - a_2)\ldots(x - a_n)$$

Equivalently, $f(x)$ splits in E if E contains all the zeros of $f(x)$.

A subfield K of E that contains F is called a *splitting field for $f(x)$ over F* if $f(x)$ splits in K and in no proper subfield of K containing F.

10.3.2 Example. The Splitting Field of $(x^2 - 2)(x^2 - 3)$ over $\mathbb{Q}$

Consider

$$f(x) = x^4 - 5x^2 + 6 = (x^2 - 2)(x^2 - 3)$$

in $\mathbb{Q}[x]$. Then $f(x)$ splits in $\mathbb{R}$ and $K = \mathbb{Q}(\sqrt{2}, \sqrt{3})$ is a splitting field for $f(x)$ over $\mathbb{Q}$.

10.3.3 Theorem. Existence of a Splitting Field

If F is a field then for every polynomial $f(x) \in F[x]$ there is a splitting field for $f(x)$ over $F[x]$.

If $f(x)$ splits in $E \supseteq F$ and has zeros, $a_1, \ldots, a_n$, then

$$F(a_1, \ldots, a_n)$$

is the splitting field of $f(x)$ over F.

The degree $[F(a_1, \ldots, a_n): F]$ is at most $m!$ where $m = \deg(f)$.

Proof. The construction of a splitting field is an iteration of the basic technique. Let $p(x) \in F[x]$ be an irreducible factor of $f(x)$. Construct

$$F_1 = F[x]/\langle p(x)\rangle$$

We know that $[F_1 : F] = \deg(p)$ and $\deg(p) \leq \deg(f) = m$. We know that $p(x)$, and hence $f(x)$, has at least one zero, call it α_1, in F_1 and so in F_1

$$f(x) = (x - \alpha_1)f_1(x)$$

Now consider $f_1(x)$ in $F_1[x]$. Iterate!

Formally proceed by induction on the degree of $f(x)$. To begin, suppose the degree of $f(x)$ is 1, that is,

$$f(x) = x - a$$

and so $a \in F$; thus F itself is the splitting field of $f(x)$.

By induction, then, we can construct a splitting field D of $f_1(x)$ over F_1 and by induction its degree over F_1 is at most $(m - 1)!$ since the $\deg(f_1) \leq (m - 1)$. Thus

$$[D: F] = [D: F_1][F_1 : F] \leq m(m - 1)! = m!$$

We can easily see that $f(x)$ splits in D since all its zeros lie in this field.

A splitting field $K \supseteq F$ must contain all the zeros of $f(x)$, since a is a zero of $f(x)$ if and only if $(x - a)$ is a factor of $f(x)$. Thus

$$K \supseteq F(a_1, \ldots, a_n)$$

if the a's are zeros of $f(x)$. Conversely, since the a's are in $F(a_1, \ldots, a_n)$ and if these are all the zeros of $f(x)$ in E then $f(x)$ splits in $F(a_1, \ldots, a_n)$ and so $F(a_1, \ldots, a_n) \supseteq K$.

The next theorem and its corollary are central to classical algebra.

10.3.4 Theorem. Extension of an Isomorphism to an Isomorphism of Splitting Fields

Let $\phi: F \to F^*$ be an isomorphism of a field F onto a field F^*. Let $\phi(a) = a^*$ for $a \in F$.

1. The isomorphism ϕ can be extended to an isomorphism ϕ of $F[x]$ onto $F^*[x]$ by defining

$$\phi\left(\sum_{i=1}^{i=n} a_i x^i\right) = \sum_{i=1}^{i=n} \phi(a_i) x^i = \sum_{i=1}^{i=n} a_i^* x^i = f^*(x)$$

2. If E and E^* are splitting fields for $f(x)$ and $f^*(x)$ over F and F^*, respectively, then the isomorphism ϕ can be extended to an isomorphism Φ from E onto E^*. Moreover,
3. If α is a zero of an irreducible factor $p(x)$ of $f(x)$ in $F[x]$ and α^* is any zero of the corresponding irreducible factor $p^*(x) = \phi(p(x))$ of $f^*(x)$ in $F^*[x]$ then the extension may be constructed so that $\Phi(\alpha) = \alpha^*$.

Proof.   The proof of 1 is routine, and we omit the details.

We prove 2 by induction on the degree of the polynomial $f(x)$ and in the first step establish 3. Note that 1 ensures that the degrees of $f(x)$ and $f^*(x)$ are the same. Note too that since ϕ is an isomorphism of $F[x]$, $\phi(p(x))$ is irreducible in $F^*[x]$ if and only if $f(x)$ is irreducible in $F[x]$.

If the degree of $f(x)$ is 1 then $f(x) = (x - a_1)$ splits in F; $f^*(x) = (x - a_1^*)$ splits in F^* and nothing needs to be proved.

Let α be a zero of an irreducible factor $p(x)$ of $f(x)$ and let α^* be a zero of the corresponding irreducible polynomial $p^*(x)$. Now we know that

$$F(\alpha) \cong F[x]/\langle p(x) \rangle \quad \text{and} \quad F^*[x]/\langle p^*(x) \rangle \cong F(\alpha^*)$$

and in the first isomorphism, $\alpha \to x + \langle p(x) \rangle$; in the second $x + \langle p^*(x) \rangle \to \alpha^*$. If we can show that

$$F[x]/\langle p(x) \rangle \cong F^*[x]/\langle p^*(x) \rangle \text{ with } x + \langle p(x) \rangle \to x + \langle p^*(x) \rangle$$

then the composition of these three isomorphisms shows that ϕ can be extended to an isomorphism ϕ_1 between

$$F(\alpha) \text{ and } F^*(\alpha^*)$$

in which

$$\alpha \rightarrow \alpha^*$$

The details are thus completed by showing that

$$g(x) + \langle p(x) \rangle \rightarrow g^*(x) + \langle p^*(x) \rangle$$

is an isomorphism from $F[x]/\langle p(x) \rangle$ onto $F^*[x]/\langle p^*(x) \rangle$. This is the result of Theorem 7.2.13 by interpreting the results of that theorem in the case that the homomorphism is an isomorphism. In the application we need here the isomorphism can be verified easily by checking the requisite conditions for an isomorphism.

Next we simply have to apply the induction hypotheses to the polynomial

$$f_1(x) = \frac{f(x)}{(x - \alpha)}$$

which is in $F(\alpha)[x]$ and to the extended isomorphism ϕ_1. We have $\deg(f_1) < \deg(f)$ and the extended isomorphism sends

$$\frac{f(x)}{(x - \alpha)} \rightarrow \frac{f^*(x)}{(x - \alpha^*)}$$

Moreover, E is a splitting field of $f_1(x)$ over F; first of all, $f_1(x)$ splits in E and second, any subfield of E containing F_1 and all the zeros of f_1 also contains all the zeros of $f(x)$; hence it would be a splitting field of $f(x)$ and thus contain E. Thus induction shows that ϕ_1 can be extended to an isomorphism Φ of E onto E^*. Since ϕ_1 is an extension of ϕ in which $\phi_1(\alpha) = \alpha_1$, we see that Φ satisfies all the conditions of the theorem.

This theorem is an extremely important one. For one thing it establishes (Corollary 10.3.5) that a splitting field is essentially unique. In mathematical jargon, we say that the splitting field is unique up to isomorphism. Second, it provides us with lots of automorphisms of the splitting field. In fact, we'll get from Corollary 10.3.6 a whole group of automorphisms. The structure of this group plays a central role in the determination of the subfields of a splitting field and whether the zeros of a polynomial can be expressed in terms of radicals.

10.3.5 Corollary. Uniqueness of the Splitting Field of $f(x)$ over F

If E_1 and E_2 are splitting fields of $f(x)$ over the field F then E_1 and E_2 are isomorphic by an isomorphism fixing each element of F.

Proof. In Theorem 10.3.4 choose ϕ to be the identity isomorphism. The isomorphism Φ of the theorem provides the isomorphism guaranteed by the corollary.

10.3.6 Corollary.   Existence of Automorphisms of a Splitting Field

Let E be a splitting field of an irreducible polynomial $f(x) \in F[x]$ over F. Let α and β be distinct zeros of $f(x)$ in E. Then there is an automorphism Φ of E that fixes each element of F and sends α into β:

$$\Phi(\alpha) = \beta \quad \text{and} \quad \Phi(u) = u \qquad \text{for each } u \in F$$

The order of the group of automorphisms of E that fix each element of F is at least the number of distinct zeros of the irreducible polynomial f.

Proof.   Take $F = F^*$ and ϕ = identity and $\alpha^* = \beta$ in Theorem 10.3.5. Let α be a particular zero of $f(x)$. For each zero β ($\alpha = \beta$ is O.K.) of $f(x)$ we know there is an automorphism of E fixing F and sending α into β. Hence there are at least as many automorphisms of the splitting field fixing F as there are distinct zeros of f.

10.3.7 Example.   The Splitting Field of $x^3 - 2$ over $\mathbb{Q}$

The zeros of $x^3 - 2$ in the field of complex numbers $\mathbb{C}$ are

$$\sqrt[3]{2}, \ \omega\sqrt[3]{2}, \ \omega^2\sqrt[3]{2}$$

where ω is a zero of $(x^3 - 1)/(x - 1) = x^2 + x + 1$ and so ω is a complex number such that $\omega^3 = 1$. There is no need to identify which complex number it is.

First we claim that $K = \mathbb{Q}(\sqrt[3]{2}, \omega)$ is the splitting field of $x^3 - 2$ over $\mathbb{Q}$ contained in $\mathbb{C}$. To see this, argue that K contains all the zeros of $x^3 - 2$ and conversely if any field contains all the zeros then it contains $\sqrt[3]{2}$, of course, and

$$\omega = \frac{\left(\omega\sqrt[3]{2}\right)}{\sqrt[3]{2}}$$

Thus any field containing all the zeros of $x^3 - 2$ contains K.

Now we want to determine the degree of K over $\mathbb{Q}$. Since $K \supset \mathbb{Q}(\sqrt[3]{2})$ we know that 3 divides $[K : \mathbb{Q}]$. On the other hand, we claim that $x^2 + x + 1$ is irreducible over $\mathbb{Q}(\sqrt[3]{2})$.

In this special case we can see that $\mathbb{Q}(\sqrt[3]{2})$ is a subfield of the real numbers and ω is complex and hence is not contained in $\mathbb{Q}(\sqrt[3]{2})$, so $x^2 + x + 1$ is irreducible over $\mathbb{Q}(\sqrt[3]{2})$.

More generally (see Problem 10.3.3) we can argue as follows: Suppose that $x^2 + x + 1$ is reducible over $\mathbb{Q}(\sqrt[3]{2})$; thus $x^2 + x + 1$ has a zero ω in $\mathbb{Q}(\sqrt[3]{2})$ and so

$$\mathbb{Q}\left(\sqrt[3]{2}\right) \supset \mathbb{Q}(\omega) \supset \mathbb{Q}$$

and thus $[\mathbf{Q}(\omega): \mathbf{Q}]$ divides $[\mathbf{Q}(\sqrt[3]{2}: \mathbf{Q}]$ but

$$[\mathbf{Q}(\omega): \mathbf{Q}] = 2 \quad \text{and} \quad \left[\mathbf{Q}(\sqrt[3]{2}: \mathbf{Q}\right] = 3$$

but $2 + 3$. Tilt!

Thus we have that

$$\left[\mathbf{Q}(\sqrt[3]{2}, \omega): \mathbf{Q}\right] = \left[\mathbf{Q}(\sqrt[3]{2}, \omega): \mathbf{Q}(\sqrt[3]{2})\right]\left[\mathbf{Q}(\sqrt[3]{2}): \mathbf{Q}\right] = 2 \cdot 3 = 6$$

We can also determine all the subfields of $\mathbf{Q}(\sqrt[3]{2}, \omega)$. First we have the three subfields of degree 3, namely,

$$\mathbf{Q}(\sqrt[3]{2}), \mathbf{Q}(\omega\sqrt[3]{2}), \mathbf{Q}(\omega^2\sqrt[3]{2})$$

There can be no other subfield of degree 3 over $\mathbf{Q}$. If there were such a subfield, call it H, then as the general argument in the preceding paragraph shows, $x^3 - 2$ would either be irreducible over H and hence the splitting field would contain $H(\sqrt[3]{2})$ a field of degree 9 over $\mathbf{Q}$ or H would have a zero of $x^3 - 2$ and hence coincide with one of the three known subfields of degree 3 over $\mathbf{Q}$. Neither alternative is possible.

A similar argument, citing the results of Problem 10.3.4, shows that no other subfield of degree 2 over $\mathbf{Q}$ except $\mathbf{Q}(\omega)$ can exist. These subfields are illustrated in Figure 10.1.

We may determine the automorphisms of E that fix each element of $\mathbf{Q}$. First there are two automorphisms of $\mathbf{Q}(\omega)$; the identity and complex conjugation that sends ω into ω^2. Indeed this automorphism is given to us by Theorem 10.3.4 applied to the irreducible polynomial $x^2 + x + 1$ over $\mathbf{Q}$.

Now we want to apply Theorem 10.3.4 to the polynomial $x^3 - 2$ over $F = \mathbf{Q}(\omega)$. As in Theorem 10.3.4 we may take ϕ to be either the identity automorphism of F or complex conjugation. In each case, because $x^3 - 2$ is

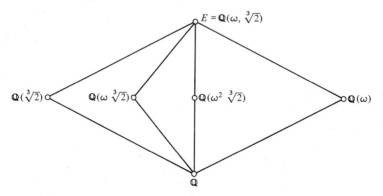

Figure 10.1 The lattice of subfields of E, the splitting field of $x^3 - 2$, over $\mathbf{Q}$. There are six automorphisms of E over $\mathbf{Q}$, one for each of the six permutations of the three zeros of $x^3 - 2$.

irreducible over F it follows that we may extend either of these automorphisms and, as we choose, map

$$\sqrt[3]{2} \rightarrow \sqrt[3]{2}, \omega\sqrt[3]{2}, \quad \text{or} \quad \omega^2\sqrt[3]{2}$$

Since there are two choices for the automorphism of F to be extended and three choices for the image of $\sqrt[3]{2}$, it follows that there are six automorphisms of E over $\mathbb{Q}$.

There can be no more for

$$E = \mathbb{Q}\left(\sqrt[3]{2}, \omega\sqrt[3]{2}, \omega^2\sqrt[3]{2}\right)$$

so that any automorphism is completely determined when we know the images of these zeros of $x^3 - 2$. We also know that any automorphism of E must take a zero of $x^3 - 2$ into another zero. Hence there can be at most as many automorphisms as there are permutations of the zeros. In this case this means that there are at most six, and we've found six. That's all there is!

That may be all the automorphisms, but there is a very important idea just below the surface of this example that we should pause to savor. Remember that the automorphisms of E form a group (Theorem 8.3.10). What group is it?

Our remarks above have suggested that each automorphism of E results in a permutation of the zeros of $x^3 - 2$. Said mathematically, there is an isomorphism from this group of automorphisms into the group $\mathbb{S}_3$ of permutations on the three zeros. It is an isomorphism because if an automorphism fixes all zeros of a polynomial $f(x) \in F[x]$ and all the elements of F then it fixes the splitting field of $f(x)$ over F. In this case the isomorphism must map the group of automorphisms onto all of $\mathbb{S}_3$. Here are two automorphisms of E that generate the group of automorphisms of E over $\mathbb{Q}$.

$$A: \omega \rightarrow \omega \quad \text{and} \quad \sqrt[3]{2} \rightarrow \omega\sqrt[3]{2}$$

and

$$B: \omega \rightarrow \omega^2 \quad \text{and} \quad \sqrt[3]{2} \rightarrow \sqrt[3]{2}$$

Check the following computations to determine the effect of A and B on the zeros of $x^3 - 2$:

$$A\left(\omega\sqrt[3]{2}\right) = A(\omega)A\left(\sqrt[3]{2}\right) = \omega\omega\sqrt[3]{2} = \omega^2\sqrt[3]{2}$$

and

$$A\left(\omega^2\sqrt[3]{2}\right) = A(\omega^2)A\left(\sqrt[3]{2}\right) = \omega^2\omega\sqrt[3]{2} = \sqrt[3]{2}$$

so that the effect of A on the zeros of $x^3 - 2$ is the 3-cycle

$$\left(\sqrt[3]{2}, \omega\sqrt[3]{2}, \omega^2\sqrt[3]{2}\right)$$

Now verify that the effect of B on the zeros of $x^3 - 2$ is the 2-cycle

$$\left(\omega\sqrt[3]{2}, \omega^2\sqrt[3]{2}\right)$$

and thus that automorphisms A and B generate $\mathbb{S}_3$.

EXERCISE SET 10.3

In these exercises you are asked to "find" or "describe" a field extension $E \supseteq F$. These words mean that you should give as much information as you can; certainly determine the degree of the extension and a minimal set S such that $E = F(S)$.

10.3.1 Let F be any field. Describe the splitting field of a quadratic polynomial over F.

10.3.2 Find necessary and sufficient conditions on $a, b \in F$ such that the fields $F(\sqrt{a})$ and $F(\sqrt{b})$ are isomorphic. Show that if both fields lie in a common extension then isomorphism implies equality.

10.3.3 Let $E \supseteq F$. Let $p(x)$ be an irreducible polynomial of degree 2 or 3 in $F[x]$ whose degree does not divide $[E : F]$. Show that $p(x)$ remains irreducible in $E[x]$.

10.3.4 Let $E \supset F$ and suppose that the characteristic of F is not 2. Suppose that E contains two distinct subfields K_1 and K_2 each of degree 2 over F. Show that E contains K, a subfield of degree 4 over F; indeed $K = K_1 \vee K_2$.

10.3.5 For each n, $1 < n < 11$, find the splitting field of $x^n - 1$ over $\mathbb{Q}$. In each case find a zero α of an irreducible factor of $x^n - 1$ such that the splitting field is $\mathbb{Q}(\alpha)$. *Hint.* If α is a zero of $x^n - 1$ so is α^k.

10.3.6 Let E be the splitting field of $p(x) = x^4 - 2x^2 - 2$ over $\mathbb{Q}$.
   1 Find the degree $[E : F]$.
   2 Let α be a zero of $p(x)$. Find $c \in \mathbb{Q}$ so that $\mathbb{Q}(\alpha)$ contains $\mathbb{Q}(\sqrt{c})$.
   3 Let β be a zero of $p(x)$ distinct from α and $-\alpha$. Find $[\mathbb{Q}(\alpha\beta): \mathbb{Q}]$.
   4 Theorem 10.3.4 guarantees an automorphism of E sending $\sqrt{c}$ into $-\sqrt{c}$. What are the possible images for α under such an automorphism?
   5 Find all the subfields $E \supset K \supset \mathbb{Q}$ such that $[K : \mathbb{Q}] = 2$.
   6 What is the group of automorphisms of E?

10.3.7 Find the splitting field of $x^4 - bx^2 + 1$ over $\mathbb{Q}$ where b is a positive rational number.

10.3.8 Let E be an extension of $\mathbb{Q}$ of degree 6. What are the possibilities for the lattice of subfields of E?

10.4 FINITE FIELDS

In this section we consider the special topic of finite fields and how to construct them. Finite fields are important for many applications in combinatorics, the design of statistical experiments, and algebraic coding theory. Then we take up cyclotomic fields, fields extending the rational numbers that are generated by roots of unity.

We know something of finite fields. We know, for example, that if F is finite its characteristic is a prime p and its prime subfield must be $\mathbb{Z}_p$. Then, because it certainly has finite degree over $\mathbb{Z}_p$, it must have, as a vector space over its prime subfield, p^n elements where n is the degree of the field over $\mathbb{Z}_p$. It is also true that its multiplicative group must be cyclic. Here is a stronger statement of some interest.

10.4.1 Lemma. Finite Multiplicative Subgroups of a Field

If G is a finite subgroup of the multiplicative group of a field then G is cyclic. In particular $\mathbb{Z}_p^\times$ is cyclic if p is a prime.

Proof. Since we are in a field there are at most n solutions to $x^n = 1$. Hence the criterion of Theorem 5.3.6 holds and so G is cyclic. But the most important question is unanswered. Do finite fields exist?

We get some clue about where to look by noting that the multiplicative group of a finite field with p^n elements has order $p^n - 1$ and so for all $x \neq 0$ we have

$$x^{p^n - 1} = 1$$

and thus, including 0, we can say that every element is a zero of

$$x^{p^n} - x$$

This suggests that we consider the splitting field of this polynomial over $\mathbb{Z}_p$. We do, and we obtain the existence and uniqueness of finite fields.

10.4.2 Theorem. Existence and Uniqueness of Finite Fields

The splitting field of

$$x^{p^n} - x$$

over $\mathbb{Z}_p$ is a finite field with p^n elements. Conversely any field with p^n elements is the splitting field of $x^{p^n} - x$ over $\mathbb{Z}_p$.

Notation. The finite field with p^n elements is often called, in honor of Evariste Galois (1811–1832), the Galois field of p^n elements and is denoted

$$GF(p^n)$$

Proof. The first thing to note is that we know the splitting field exists and has finite degree over $\mathbb{Z}_p$. Thus the splitting field will have only a finite number of elements. What we don't know yet is that it has p^n elements.

Second, we establish that the polynomial $x^{p^n} - x$ has n distinct zeros. To do this simply apply the criterion of Theorem 9.2.14. The derivative of $x^{p^n} - x$ over $\mathbb{Z}_p$ is

$$p^n x^{(p^n - 1)} - 1 = -1$$

and thus has no zeros! So $x^{p^n} - x$ and its derivative have no common zero.

Now, owing to the miraculous properties of characteristic p, it turns out that the zeros of $x^{p^n} - x$ are closed under addition, multiplication, and inversion. Thus the zeros of $x^{p^n} - x$ will form the splitting field. We establish these closure properties by proving a more general result.

10.4.3 Lemma.   Automorphisms of a Finite Field

If F is a finite field of characteristic p then the mapping

$$\sigma: a \longrightarrow a^p$$

is an automorphism of F fixing each element of $\mathbb{Z}_p$.

Proof.   To check the homomorphism condition we must verify that

$$(a + b)^p = a^p + b^p$$

and

$$(ab)^p = a^p b^p$$

The second of these is immediate because multiplication in F is commutative.

The first is a consequence of the binomial theorem and an elementary property of the binomial coefficients. Compute:

$$(a + b)^p = a^0 + \left(\sum_{1 \leq 1 < p} \binom{p}{k} a^{p-k} b^k \right) + b^p$$

Now the binomial coefficients within the terms of Σ are zero in F because as integers, p divides the numerator $p!$ but not the denominator $k!(p - k)!$ if $1 \leq 1 < p$. Thus in F

$$\sigma(a + b) = (a + b)^p = a^p + b^p = \sigma(a) + \sigma(b)$$

The mapping σ is one-to-one because the kernel is 0; if $a^p = 0$ then $a = 0$. The mapping σ is onto because the splitting field is finite and for finite sets, one-to-one mappings are onto. Thus σ is an automorphism of the splitting field. Moreover the homomorphism maps a zero of $x^{p^n} - x$ into another zero because

$$0 = \sigma(0) = \sigma(a^{p^n} - a) = \sigma(a^{p^n}) - \sigma(a) = (\sigma(a))^{p^n} - \sigma(a)$$

Now it is easy to show the closure properties. The nth iterate of σ is

$$\sigma^k(a) = a^{p^k}$$

and is an automorphism and thus, for example,

$$(a + b)^{p^k} = a^{p^k} + b^{p^k}$$

Similarly, the products of zeros are zeros and the inverse of a zero is a zero.

The result of all this is simply that the zeros of $x^{p^n} - x$ constitute a subfield of F, and owing to the minimal nature of the splitting field, it follows that F is this set of zeros. In particular, $|F| = p^n$ elements, all of which are zeros of $x^{p^n} - x$. Thus F contains the splitting field of this polynomial over $\mathbb{Z}_p$. But, as we've argued, the splitting field has p^n elements and so must be equal to F. Now the general theory of splitting fields shows that any two fields with p^n elements must be isomorphic. Finally σ fixes 1 and hence all elements of $\mathbb{Z}_p$.

Now we prove that there are irreducible polynomials of degree n over $\mathbb{Z}_p$ and that the finite field of p^n elements is

$$\mathbb{Z}_p[x]/\langle p(x)\rangle$$

where $p(x)$ is one of these polynomials.

10.4.4 Theorem.   Construction of $GF(p^n)$

If ω is a generator of the multiplicative group of $GF(p^n)$ then ω is the zero of an irreducible polynomial $t(x)$ of degree n in $\mathbb{Z}_p[x]$ and

$$GF(p^n) \text{ and } \mathbb{Z}_p[x]/\langle t(x)\rangle \text{ are isomorphic}$$

It follows that any polynomial $q(x)$ that is irreducible of degree n over $\mathbb{Z}_p$ generates $GF(p^n)$ as

$$\mathbb{Z}_p[x]/\langle q(x)\rangle$$

Proof.   First, $GF(p^n) = \mathbb{Z}_p(\omega)$ since ω generates $GF(p^n)$ as a multiplicative group (except, of course, for 0). Since $GF(p^n)$ has finite degree over $\mathbb{Z}_p$ it follows that ω is algebraic over $\mathbb{Z}_p$ and so it is the zero of some polynomial $t(x)$ that is irreducible over $\mathbb{Z}_p$. Thus $GF(p^n)$ must contain a subfield isomorphic to $\mathbb{Z}_p[x]/\langle t(x)\rangle$. By Theorem 9.2.16 this field is isomorphic to $\mathbb{Z}_p(\omega) = GF(p^n)$. In particular they have the same degree over $\mathbb{Z}_p$. By Theorem 10.1.14 since $|GF(p^n)| = p^n$, $[GF(p^n) : \mathbb{Z}_p] = n$. On the other hand, by Theorem 10.2.6 this degree is equal to $\deg(t(x))$.

Finally if $q(x)$ is any polynomial of degree n and irreducible over $\mathbb{Z}_p$ then

$$\mathbb{Z}_p[x]/\langle q(x)\rangle$$

has degree n over $\mathbb{Z}_p$ and hence is a finite field with p^n elements, hence isomorphic to $GF(p^n)$.

We can count the number of monic polynomials whose zeros are generators for the multiplicative group. We know that there are $\varphi(p^n - 1)$ generators of the multiplicative group and any irreducible polynomial of degree n has n of these as zeros. Hence

$$\frac{\varphi(p^n - 1)}{n}$$

is the number of polynomials irreducible over $\mathbb{Z}_p$ having zeros that generate $GF(p^n)$. This gives an alternate proof that $n|\varphi(p^n - 1)$; see Problem 3.4.20.

Now we want to obtain a formula for the total number of polynomials of degree n that are irreducible over $\mathbb{Z}_p$. The first step is this simple number theoretic fact.

10.4.5 Lemma.   A Divisibility Property in $\mathbb{Z}$

If a is a positive integer greater than 1 and n and m are positive integers, then

$$(a^n - 1)|(a^m - 1) \qquad \text{if and only if } n|m$$

Proof. If $n|m$ then the divisibility is the result of summing the geometric series:

$$1 + a^{m/n} + \cdots + \left(a^{m/n}\right)^{(n-1)}$$

Conversely write $m = nq + r$ where $0 \le r$ and then

$$a^m - 1 = a^{nq+r} - 1 = a^r(a^{nq} - 1) + (a^r - 1)$$

Now since $(a^n - 1)$ divides $(a^m - 1)$ and $(a^{nq} - 1)$ it follows that $(a^n - 1)$ divides $(a^r - 1)$, which, considering the range of r, is impossible unless $r = 0$. Thus $n|m$.

The second step in this count is to determine the subfields of $GF(p^n)$. This is an important result in its own right.

10.4.6 Theorem. Subfields of $GF(p^n)$

The subfields of $GF(p^n)$ are in one-to-one correspondence with the divisors of n. For each divisor d of n there is precisely one subfield $GF(p^d)$.

Proof. We use the fact that the multiplicative group is cyclic of finite order and so has exactly one subgroup, which of course is cyclic, for each order dividing the order of the group. Thus, for any k there can be at most one subfield with k nonzero elements because there is at most one subgroup of order k of the multiplicative group. The number of elements in a subfield is p^d and thus, arguing on the orders of the multiplicative subgroups, we have

$$p^d - 1 | p^n - 1$$

By the lemma this means that $d|n$.

Conversely, if $d|n$ then $(p^d - 1)|(p^n - 1)$ and so there is a subgroup of the multiplicative group of $GF(p^n)$ of that order. The $p^d - 1$ element of this subgroup are zeros of

$$x^{p^d} - x$$

and thus are all the zeros of this polynomial. Hence they form a subfield isomorphic to $GF(p^d)$.

Now we can determine the irreducible factors of $x^{p^n} - x$ in $\mathbb{Z}_p[x]$.

10.4.7 Theorem. Factorization in $\mathbb{Z}_p[x]$

Over $\mathbb{Z}_p$

$$x^{p^n} - x$$

is the product of all the monic irreducible polynomials of degree d where $d|n$.

If $W_p(d)$ denotes the number of monic irreducible polynomials of degree d over $\mathbb{Z}_p$ then the factorization gives this count of degrees:

$$p^n = \sum_{d|n} dW_p(d) \tag{1}$$

Proof. Certainly $x^{p^n} - x$ is the product of irreducibles which, by adjusting unit factor, we may take to be monic. Suppose $q(x)$ divides $x^{p^n} - x$ and is monic and irreducible of degree d. Since $x^{p^n} - x$ has all its zeros in $GF(p^n)$ the unique factorization theorem guarantees that all the zeros of $q(x)$ are in $GF(p^n)$; let λ be one of them. Moreover we have (Theorem 10.2.6) that

$$\mathbb{Z}_p[x]/\langle q(x)\rangle \text{ and } \mathbb{Z}_p(\lambda) \text{ are isomorphic}$$

Now we know the subfield has degree $d(= \deg(q(x)))$ over $\mathbb{Z}_p$ and thus from the preceeding theorem, $d|n$.

On the other hand, if $q(x)$ is an irreducible polynomial over $\mathbb{Z}_p$ and $d|n$ then $\mathbb{Z}_p[x]/\langle q(x)\rangle$ is a finite field of degree d over n; by the uniqueness we know this field is isomorphic to $GF(p^d)$ and by the subfields theorem we know, since $d|n$, that this field is a subfield of $GF(p^n)$. Thus $q(x)$ will have a zero (and hence all of its zeros) in $GF(p^n)$. But this means that the factorization of $x^{p^n} - x$ into monic linear factors contains the factorization of $q(x)$ into monic linear factors. But this means that

$$q(x)|x^{p^n} - x$$

A comparison of the degrees of $x^{p^n} - x$ and its factorization gives the relationship between p^n and the number of monic irreducible polynomials of degree d which is expressed in equation (1). We want a better formula for $W_p(n)$, but before we develop one here are a few special cases of interest.

10.4.8 Examples.   Special Cases of $W_p(n)$

1. For all primes p we have

$$W_p(1) = p$$

since every monic polynomial $x + a$ is irreducible and there are p choices for a since there are p elements in $\mathbb{Z}_p$.

2. If q is a prime then

$$W_p(q) = (p^q - p)/q$$

since in this case formula (1) says

$$p^q = qW_p(q) + W_p(1)$$

and we can solve for $W_p(q)$.

3. If $n = qt$ is the product of two distinct primes

$$W_p(qt) = (p^{qt} - p^q - p^t + p)/qt$$

since in this case formula (1) says

$$p^{qt} = qtW_p(qt) + qW_p(q) + tW_p(t) + W_p(1)$$

and so we can solve for $W_p(qt)$.

4. If $n = q^u$ is a power of a prime then

$$W_p(q^u) = (p^{q^u} - p^{q^{u-1}})/q^u$$

In this case formula (1) says

$$p^{q^u} = q^u W_p(q^u) + \sum_{d|q^{u-1}} dW_p(d)$$

since any divisor of q^u is a divisor of q^{u-1}. But formula (1) also says

$$p^{q^{u-1}} = \sum_{d|q^{u-1}} dW_p(d)$$

and so

$$p^{q^u} = q^u W_p(q^u) + p^{q^{u-1}}$$

Now solve for $W_p(q^u)$.

We can generalize these examples as follows:

10.4.9 Theorem.   The Number of Monic Polynomials Irreducible over $\mathbb{Z}_p$

Let the distinct prime factors of n be $q_1, \ldots, q_k$. Then the number of monic irreducible polynomials with coefficients in $\mathbb{Z}_p$ is

$$W_p(n)$$

$$= \frac{1}{n}\left(p^n - \sum_i p^{n/q_i} + \sum_{i<j} p^{n/q_i q_j} + \cdots \right.$$

$$\left. + (-1)^t \sum_{i_1 < \cdots < i_t} p^{n/q_{i_1} \cdots q_{i_t}} + \cdots \right) \qquad (2)$$

Proof.   The proof begins with equation (1):

$$p^n = \sum_{d|n} dW_p(d)$$

and so

$$W_p(n) = \frac{1}{n}\left\{ p^n - \sum_{\substack{d|n \\ d<n}} dW_p(d) \right\}$$

where now the summation extends over divisors of n which are less than n.

To get a further simplification we argue as follows: If $d < n$ and $d|n$ then d must divide one or more of

$$\frac{n}{q_1}, \ldots, \frac{n}{q_k}$$

Hence if u is a specific divisor of n, $u < n$, the term $uW_p(u)$ will appear in one or more expressions

$$\sum_{d|n/q_i} dW_p(d)$$

and by (1) each of these sums equals p^{n/q_i}.

So all the terms $dW_p(d)$ will appear in

$$\sum_i \left(\sum_{d|n/q_i} dW_p(d) \right) = \sum_i p^{n/q_i}$$

However, if the divisor u divides, say, n/q_1 and n/q_2, then it appears in *two* of these sums and so we must compensate. If this happens, u must also be a divisor of n/q_1q_2. To compensate for this we must subtract off all terms $dW_p(d)$ where $d|n/q_1q_2$. And indeed for all $i < j$, the terms $dW_p(n)$ where $d|n/q_iq_j$.

Systematic compensation for the inclusion and exclusion of the proper terms $dW_p(d)$ leads to

$$\sum_{\substack{d|n \\ d<n}} dW_p(n) = \sum_{1 \le i_1 < i_2 < \cdots < i_t \le k} (-1)^{t-1} \left(\sum_{d|n/q_{i_1}q_{i_2}\cdots q_{i_t}} dW_p(d) \right)$$

$$= \sum_{1 \le i_1 < i_2 < \cdots < i_t \le k} (-1)^{t-1} p^{n/q_1q_2\cdots q_t}$$

And this in turn gives the formula of the theorem.

This formula can be rewritten in a neater way using the Möbius μ-function. This is a handy function to have in studying number theory and combinatorial theory. Here is its definition.

10.4.10 Definition. Möbius μ-Function

Define a function from the nonnegative integers into $(-1, 0, 1)$ by

$$\mu(0) = 0 \qquad \mu(1) = 1$$

$\mu(n) = 0$ of m is divisible by the square of a prime

$\mu(n) = (-1)^k$ of m is the product of k distinct primes

Then we have

$$W_p(n) = \frac{1}{n} \sum_{d|n} \mu\left(\frac{n}{d}\right) p^d \tag{3}$$

Exercises 10.4.1 and 10.4.2 develop formula (3) directly by using the Möbius μ-function to solve the relationship in Theorem 10.4.9 for $W_p(n)$ directly.

Our final theorem determines the group of automorphisms of a finite field.

10.4.11 Theorem. Automorphisms of $GF(p^n)$

The group of automorphisms of $GF(p^n)$ is a cyclic group of order n generated by

$$\sigma: a \longrightarrow a^p \tag{4}$$

Proof. We have already seen that σ is an automorphism of $GF(p^n)$. Furthermore we have for $1 \le k \le m$ that

$$\sigma^k(a) = a^{p^k}$$

and is the identity if and only if the equation

$$x^{p^n} = x$$

has p^n solutions, hence if and only if $k = n$. Thus there are at least n automorphisms of $GF(p_n)$.

On the other hand let ω generate the multiplicative group of $GF(p^n)$. Let $q(x)$ be the irreducible polynomial of degree n in $\mathbb{Z}_p[x]$ of which ω is a zero. Any automorphism σ of $GF(p^n)$ fixes 1 and hence the prime subfield $\mathbb{Z}_p$ and so σ will also fix $q(x)$. It follows from Theorem 9.1.10 that the image of ω is also a zero of $q(x)$. This means that there are at most $n = \deg(q)$ possibilities for the image $\sigma(\omega)$. But since $\mathbb{Z}_p(\omega) = GF(p^n)$ it follows that once $\sigma(\omega)$ is determined, every image of $GF(p^n)$ under σ is determined. Hence there can be at most as many automorphisms as there are possibilities for $\sigma(\omega)$; that is, there are at most n automorphisms of $GF(p^n)$.

Combining the arguments of the last two paragraphs we see that the group of automorphisms of $GF(p^n)$ is a cyclic group of order n and generated by the automorphism defined in (4).

Here is a table of irreducible monic polynomials of low degree over $\mathbb{Z}_2$, $\mathbb{Z}_3$, and $\mathbb{Z}_5$. You may find this information useful in the next exercise set.

MONIC IRREDUCIBLE POLYNOMIALS OF LOW DEGREE OVER $\mathbb{Z}_p$ ($p = 2, 3,$ AND 5)

In this table $f(x) = x^n + a_{n-1}x^{n-1} + \cdots + a_1x + a_0$ is represented by the n-tuple $(a_{n-1}, a_{n-2}, \ldots, a_1, a_0)$. If $f(x)$ appears in the table, its reverse, $(1/a_0)x^nf(1/x)$, is not listed.

Over $\mathbb{Z}_2$:
Degree 1: $(0), (1)$
Degree 2: $(1,1)$
Degree 3: $(0,1,1)$
Degree 4: $(0,0,1,1), (1,1,1,1)$
Degree 5: $(0,0,1,0,1), (0,1,1,1,1), (1,0,1,1,1)$
Degree 6: $(0,0,0,0,1,1), (0,0,0,1,0,1), (0,1,0,1,1,1), (0,1,1,0,1,1), (1,0,0,1,1,1)$
Degree 7: $(0,0,0,0,0,1,1), (0,0,0,1,0,0,1), (0,0,0,1,1,1,1), (0,0,1,1,1,0,1),$
$\quad\quad\quad (0,1,0,0,1,1,1), (0,1,0,1,0,1,1), (0,1,1,1,1,1,1), (1,0,0,1,0,1,1), (1,1,0,1,1,1,1)$

Over $\mathbb{Z}_3$:
Degree 1: $(0), (1), (2)$
Degree 2: $(0,1), (1,2)$
Degree: $\quad (0,2,1), (0,2,2), (1,1,2), (1,2,1)$
Degree 4: $(0,0,1,2), (0,0,2,2), (0,1,0,2), (0,1,1,1), (0,1,2,1)$
$\quad\quad\quad (1,0,2,1), (1,1,1,1), (1,1,2,2), (2,1,1,2), (2,1,2,1)$

Over $\mathbb{Z}_5$:
Degree 1: $(0), (1), (2), (3), (4)$
Degree 2: $(0,2), (1,1), (1,2), (2,3), (2,4), (4,1)$
Degree 3: $(0,1,1), (0,1,4), (0,2,1), (0,2,4), (0,3,2), (0,3,3), (0,4,2)$
$\quad\quad\quad (0,4,3), (1,1,3), (1,1,4), (1,3,1),(1,3,4), (1,4,1), (1,4,3)$
$\quad\quad\quad (2,1,3), (2,1,4), (2,2,3), (3,1,2), (3,2,3), (3,4,1)$

EXERCISE SET 10.4

10.4.1 Prove the Möbius inversion formula. Let f and F be two functions from the positive integers $\mathbb{Z}^+$ into $\mathbb{Z}^+$ related by

$$f(m) = \sum_{d|m} F(d)$$

Show that

$$F(m) = \sum_{d|m} \mu\left(\frac{m}{d}\right)f(d) = \sum_{d|m} \mu(d)f\left(\frac{m}{d}\right)$$

where μ is the Möbius μ-function.

10.4.2 Use Problem 10.4.1 and the basic formula (1) of Theorem 10.4.7 to derive the formula of Theorem 10.4.9 directly.

10.4.3 Use the table of irreducible polynomials of $GF(2)$ to verify that $x^{16} - x$ is the product of all the irreducible polynomials of degree 1, 2, and 4.

10.4.4 Which of the two polynomials of degree 4 over $GF(2)$ has as its zeros generators of the multiplicative group of $GF(16)$? Express the zero of the other polynomial as a power of a generator.

10.4.5 Note that both $x^3 + 2x + 1$ and $x^3 + 2x + 2$ are irreducible over $GF(3)$. Theory says that

$$GF(3)/\langle x^3 + 2x + 1\rangle \quad \text{and} \quad GF(3)/\langle x^3 + 2x + 2\rangle$$

are isomorphic to $GF(9)$. Exhibit an isomorphism directly between the two fields.

10.4.6 Show that there is a one-to-one correspondence between the subgroups of the group of automorphisms of $GF(p^n)$ and the subfields of $GF(p^n)$ given by

$$H \rightarrow \{u \in GF(p^n): \sigma(u) = u \text{ for all } \sigma \in H\}$$

where H is a subgroup of the group of automorphisms.

10.4.7 Let $a \in GF(p)$. Show that

$$x^{p^n} - x + na$$

is always divisible by

$$x^p - x + a$$

over $GF(p)$.

10.4.8 Show that if p is a prime and $a \neq 0$ in $GF(p)$ then

$$x^p - x + a$$

is always irreducible over $GF(p)$.

10.4.9 Let a and b be elements in $GF(2^n)$. If n is odd show that $a^2 + ab + b^2 = 0$ implies $a = b = 0$. What if n is even?

10.4.10 Let K be a field of characteristic $p > 0$ and let $f(x) \in K[x]$. Show that

$$f(x) = f(x + 1) \qquad \text{if and only if} \quad f(x) = h(x^p - x)$$

for some $h(x) \in K[x]$.

10.4.11 Let $F = GF(p)(y)$ where y is an indeterminate. Show that $x^2 + y$ is irreducible in $F[x]$. Show that $x^2 + y$ has multiple zeros. Determine the splitting field of $x^2 + y$ over F and the group of automorphisms fixing each element of F.

10.4.12 Write an algorithm suitable for computer programming to determine and list the monic irreducible polynomials of degree 1 to degree n over $GF(p)$.

10.4.13 Write an algorithm suitable for computer programming to determine the multiplicative order of an element α in $GF(p^n)$. Assume that you are given the irreducible polynomial of which α is a zero.

For the next two problems recall definition 3.1.12 of a latin square. Here we denote a $q \times q$ latin square on the symbols S,

$$S = \{s_1, \ldots, s_q\}$$

as a matrix $[m_{i,j}]$ whose entries are elements of S. The matrix is a latin square if each row and column is a permutation of the elements of S.

Two latin squares

$$A = [a_{ij}] \quad \text{and} \quad B = [b_{ij}]$$

are said to be *orthogonal* provided each ordered pair of the cartesian product $S \times S$ occurs in the form (a_{ij}, b_{ij}). Thus the squares

$$\begin{bmatrix} 1 & 2 & 3 \\ 3 & 1 & 2 \\ 2 & 3 & 1 \end{bmatrix} \qquad \begin{bmatrix} 1 & 2 & 3 \\ 2 & 3 & 1 \\ 3 & 1 & 2 \end{bmatrix}$$

are orthogonal.

10.4.14 Show that if $A_1, \ldots, A_m$ are m latin squares on S such that any two are orthogonal then $m \leq q - 1$.

10.4.15 Show that if q is a power of a prime, $q = p^n$ then there exists a collection of $n - 1$ latin squares such that any two are orthogonal. For S take the elements of the finite field $GF(q)$, say, $S = \{s_1 = 0, \ s_2 = 1, \ldots, s_q\}$. *Hint:* For each $0 \neq \alpha \in GF(q)$ define a square A_α whose entry in row i and column j is $x_i \alpha + x_j$. It is an unsolved problem to show that if there exists a collection of $q - 1$ orthogonal latin squares of size q then q is a power of a prime.

PART **TWO**

SELECTED TOPICS

Chapter **11**

Topics in Groups

In this chapter we complete our introduction to group theory. The most important theorems are the three Sylow theorems which give information about the number of subgroups of order p^k where k is the highest power of a prime p dividing the order of a group. These theorems help us to determine all the groups of small order. Along the way we discuss yet another construction for groups that generalizes the notion of a direct product. In this way we are able to give constructions for most groups of small order. The prerequisites for this chapter are Chapters 1 through 5 of Part One.

11.1 CONJUGACY AND THE CLASS EQUATION OF A GROUP

In this section we introduce what is called the "class equation" of a group. Our first application is to give a second proof of Cauchy's theorem: If a prime p divides the order of a group then the group contains an element of order p.

In the next section we use the ideas introduced here to establish the Sylow theorems.

11.1.1 Definition. Conjugacy

Let G be a group and let S be a subgroup of G. Two subsets A and B of G are called *conjugate over* S whenever there exists an element $g \in S$ such that

$$B = g^{-1}Ag = \left\{ g^{-1}ag \colon a \in A \right\}$$

Notation.   Write

$$A \stackrel{S}{\approx} B$$

if A and B are conjugate sets over S.

11.1.2 Example.   Some Conjugate Sets in $\mathscr{A}_4$

Let G be the alternating group $\mathscr{A}_4$ with identity ι. Let S be the cyclic subgroup generated by a three-cycle: $S = \langle(1,2,3)\rangle$ and let $A = \{(2,3,4), ((1,2)(3,4))\}$. The sets conjugate to A over S are

$$\iota^{-1}A\iota = A$$
$$(1,3,2)\,A\,(1,2,3) = \{(1,4,3), ((1,4)(2,3))\}$$
$$(1,2,3)\,A\,(1,3,2) = \{(1,2,4), ((1,3)(2,4))\}$$

The term "conjugate" comes from the application of this concept to the study of the subfields of the splitting field E of an irreducible polynomial over a field F. Two subfields H and K of E containing F are conjugate over F if there is an automorphism σ of E fixing each element of F and sending H onto K. We shall have more to say about this situation in Section 13.1.

11.1.3 Lemma.   Properties of Conjugacy

Let G be a group and S be a subgroup of G.

1. If $A \stackrel{S}{\approx} B$, then A and B have the same cardinal number of elements.
2. Conjugacy over S is an equivalence relation on the subsets of G of equal cardinality.
3. If A and B are conjugate over S and A is a subgroup of G then B is a subgroup of G isomorphic to A.
4. The *normalizer of A in S* defined by

$$N_S(A) = \{x: x \in S \quad \text{and} \quad x^{-1}Ax = A\}$$

is a subgroup of S. Note

$$N_S(A) = S \cap N_G(A)$$

5. The number of subsets B of G conjugate to A over S is the index $[S: N_S(A)]$.

Proof.   The proof depends essentially on the observation that the mapping of G into G given by $x \to g^{-1}xg$ is an automorphism of G. Thus the cardinal number of elements of two conjugate sets is the same. If one set is a subgroup of G, so is the other; indeed it is an isomorphic subgroup. Thus we have parts 1 and 3.

The proof of part 2 is straightforward, and we omit the details. But note that to prove that if A is conjugate to B over S, then B is conjugate to A over S requires that if $g \in S$ so does g^{-1}; this is one place where the hypothesis that S be a *subgroup* of G is used. Indeed

$$B = g^{-1}Ag \qquad \text{if and only if} \quad gBg^{-1} = A$$

and since S is a subgroup we have $g \in S$ if and only if $g^{-1} \in S$. The closure of S under the group operation is used in proving that conjugacy over S is transitive.

The proof that the normalizer of A in S is a subgroup of S is routine, and we omit the details.

To prove the last part we show that there is a one-to-one correspondence between the right cosets $N_S(A)g$ and the conjugate sets $g^{-1}Ag$. It is easy to see that the following five conditions are equivalent for elements g and h in S:

$$N_S(A)g = N_S(A)h$$

$$h \in N_S(A)g$$

$$hg^{-1} \in N_S(A)$$

$$\left(hg^{-1}\right)^{-1}A\left(hg^{-1}\right) = A$$

$$h^{-1}Ah = g^{-1}Ag$$

Thus $h^{-1}Ah = g^{-1}Ag$ if and only if $N_S(a)g = N_S(A)h$. These equivalences establish the one-to-one correspondence between right cosets of the normalizer and conjugate sets of the same cardinality.

Remark. When A is a subgroup of S then

$$A \subseteq N_S(A)$$

in fact,

$$A \quad \text{is a *normal* subgroup of} \quad N_S(A)$$

and

$$N_S(A) \quad \text{is the *largest* subgroup of } S \text{ in which } A \text{ is normal}$$

11.1.4 Definition.   The Class Equation of a Group

Let G be a group. We specialize Lemma 11.1.3 to the case in which $S = G$ and the sets A are singleton sets $\{a\}$. Now

$$N_G(\{a\}) = N(a) = \{x \in G : x^{-1}ax = a\}$$

is the set of elements that commute with a. $N(a)$ is often called the *centralizer* of a. By Lemma 11.1.3 the number of distinct conjugates of a is the index

$$[G : N(a)] = |G|/|N(a)|$$

Let D be a set of distinct representatives, one from each of the conjugate classes. The class equation is

$$|G| = \sum_{a \in D} [G : N(a)]$$

Remark. Note that the identity is in a conjugacy class by itself, as is any element that commutes with all elements of the group, that is, any element that belongs to the center of G. Thus we may write the class equation in the form

$$|G| = |Z| + \sum_{\substack{a \in D \\ a \notin Z}} [G : N(a)]$$

where Z is the center of G.

Our first application is to give another proof of Cauchy's theorem. We begin with an easy case.

11.1.5 Lemma.   Cauchy's Theorem for Abelian Groups

An abelian group whose order is divisible by the prime p has an element of order p.

Proof.   Use induction on the order of the group. We can begin the induction with order 2. A group of prime order is cyclic and so the theorem holds. In general, let G denote the group, let its order be n, and let x be an element of the group, not the identity. Let the order of x be m. If p divides m then $x^{m/p}$ has order p.

If p does not divide m then m is not the order of the group. Consider the factor group $G^* = G/\langle x \rangle$ whose order is n/m, less than the order of G. Since p does not divide m, it must be that p divides n/m and so by induction G^* has an element h^* of order p. Let h be a preimage of h^* in G. Since the image of h^p is $h^{*p} = 1^*$, it follows that

$$h^p \in \langle x \rangle$$

We cannot say that $h^p =$ identity, but we can say that

$$h^p = x^t$$

for some t. Let the order of x^t be k. Of course, k divides m and thus p does not divide k. We have that

$$\left(h^k \right)^p = h^{pk} = x^{tk} = 1$$

and so the order of h^k is either p or 1. It cannot be one, for then, in the homomorphism from G to G^*, h^* would have order dividing k. But the order of h^* is p and $p \nmid k$. Tilt! Thus the proof of the lemma is complete.

11.1.6 Theorem.   Cauchy's Theorem for Finite Groups

If a prime p divides the order of a group then there is an element of order p in the group.

Proof.   We use induction on the order of G. For low orders the validity of the theorem is well known. Consider the class equation

$$|G| = |Z| + \sum_{\substack{a \in D \\ a \notin Z}} [G : N(a)]$$

where Z is the center of G.

Consider a term $[G: N(a)]$ where a does not belong to the center of G. In this case $N(a)$ is not G and so $|N(a)|$ is less than $|G|$. Now if p divides $|N(a)|$ then by induction $N(a)$, and hence G, has an element of order p.

Thus we can assume that for all elements a not in the center of G, p does not divide $|N(a)|$ and hence p does divide the index $[G: N(a)]$. Now, since p divides the left-hand side of the class equation and all terms in the summation, it follows that p divides the order of the center Z of G. But Z is an abelian group and hence, by the lemma, has an element of order p. Done!

11.1.7 Definition. *p*-Group

Let p be a prime. A p-group is a group in which the order of every element is a power of the prime p.

Remark.   The theorem of Lagrange implies that every group of order p^k (p a prime) is a p-group. The theorem of Cauchy implies that the order of a finite p-group is a power of p. Together these theorems imply that a finite group is a p-group if and only if its order is a power of the prime p. In the next set of exercises we develop the elementary properties of p-groups.

EXERCISE SET 11.1

11.1.1 Use the class equation to show that a finite p-group has a center consisting of more than the identity. Conclude that a finite p-group has a normal subgroup of order p.

11.1.2 Show that a finite p-group is *solvable*, that is, if G is a finite p-group of order p^k then there is a chain of subgroups
$$(\iota) \subset G_1 \subset \cdots \subset G_i \subset G_{i+1} \subset \cdots \subset G_k = G$$
such that each G_i is a normal subgroup of G_{i+1} and $[G_{i+1}: G_i] = p$.

11.1.3 Show that in a finite p-group every proper subgroup S is properly contained in its normalizer $N_G(S)$. Conclude that every maximal subgroup of a finite p-group is a normal subgroup.

11.1.4 Show that if G is a finite p-group and S is a subgroup then there is a chain of subgroups
$$S \subset H_1 \subset \cdots \subset H_m = G$$
where each H_i is a normal subgroup of H_{i+1} and $[H_{i+1}: H_i] = p$.

11.1.5 Show that a group of order p^2 is either cyclic or the direct product of two cyclic groups of order p. In this case how many subgroups of order p does the group have?

11.2 THE SYLOW THEOREMS

The three Sylow theorems stated on the next page give us important information about the structure of a finite group just from the arithmetical properties of its order. These properties are extremely useful in determining the precise structure of a group; in particular they enable us to determine all groups of low order. They do not help with p-groups. There are many proofs for these theorems; the ones given here seem direct, even natural, and were first presented to me by Helmut Wielandt. Here are the theorems.

11.2.1 Theorem.   The Sylow Theorem

Let p be a positive prime. Let G be a finite group of order $p^k n$ where $p \nmid n$. Then

1. G has a subgroup of order p^k. Such a subgroup is called a p-Sylow subgroup (p-SSG) of G.
2. Any subgroup that is a p-group is contained in some p-SSG of G. Any two p-SSGs are conjugate in G and therefore all subgroups of order p^k in G are isomorphic.
3. The number of p-SSGs in G is the index of the normalizer over G of a p-SSG and is congruent to 1 modulo p.

Remark.   In view of 2, a p-SSG of G is a normal subgroup of G if and only if there is only one p-SSG in G. In view of 3, the number of p-SSGs in G divides the order of G.

Proof of 1.   The proof is by induction on the order of the group. Clearly 1 is true for groups of small order. For a group G we examine the class equation

$$|G| = |Z| + \sum_{\substack{a \in D \\ a \notin Z}} [G \colon N(a)]$$

where Z is the center of G. Suppose that p does not divide some index $[G \colon N(a)]$ for some $a \notin Z$. Then it must be that p^k divides $|N(a)|$ which, because a is not in the center, is less than $|G|$. Thus, by induction it follows that the subgroup $N(a)$ and hence G has a subgroup of order p^k.

Otherwise it must be that p divides the order of the center Z. Let x be an element of order p in Z and consider the factor group

$$G/\langle x \rangle$$

of order $p^{k-1} n$. Again, by induction this group contains a subgroup of order p^{k-1}. Its preimage of G must be a subgroup of G whose order is

$$|x| \cdot p^{k-1} = p^k$$

Proof of 2.   Let S be a subgroup of order p^h and let P be a p-SSG of G. We shall prove that S is contained in some conjugate of P. Thus, when S is itself another p-SSG, whose order is p^k, the condition "contained in" implies equality, and so S will be a conjugate of P.

To execute the proof we consider the set of subgroups conjugate to P. By Lemma 11.1.3 there are $[G \colon N(P)]$ of these. Since the normalizer of P contains P, and since $[G \colon P] = n$, the index $[G \colon N(P)]$ is not divisible by p.

On the other hand, we may partition these subgroups conjugate to P into sets that are conjugate *over* S. And thus write

$$[G: N(P)] = \sum_{Q \in L} [S: N_S(Q)]$$

where L is a set of distinct representatives of subgroups conjugate over S and conjugate to P over G. We shall prove that $S \subseteq Q$ for some $Q \in L$.

Now p does not divide the left-hand side of the equation, so p cannot divide all the terms of the right-hand side. Thus for some p-SSG, Q, which is a conjugate of P, we must have that

$$[S: N_S(Q)]$$

is not divisible by p. But S is a p-group, so the only way p cannot divide this index is for

$$[S: N_S(Q)] = 1 \quad \text{and so} \quad S = N_S(Q)$$

Since $N_S(Q) = S \cap N_G(Q)$ we have

$$S = S \cap N_G(Q) \text{ or } S \subseteq N_G(Q)$$

Thus $N_G(Q)$ contains the p-groups S and Q. The subgroup Q is normal in $N_G(Q)$ and Q is conjugate to P. Now we claim that $S \subseteq Q$. In any event $S \vee Q = SQ$ is a subgroup of $N_G(Q)$ whose order is

$$\frac{|S| \, |Q|}{|S| \cap |Q|} = \frac{|S|}{|S| \cap |Q|} |Q| = p^e p^k$$

Unless $S \subseteq Q$, the exponent $e \geq 1$ and $S \vee Q$ is a p-subgroup of G whose order would be a power of p greater than k contrary to the choice of k.

Proof of 3. Begin by starting the proof of 2 in the case that S is another p-SSG and so obtain the equation

$$[G: N(P)] = \sum_{Q \in L} [S: N_S(Q)]$$

where L is a set of distinct representatives of subgroups conjugate over S and conjugate to P over G. Thus the left-hand side is the number of subgroups conjugate to P and is, by 2, equal to the number of all of the p-SSGs in G. On the right each term is divisible by p except in the cases in which

$$S = N_S(Q)$$

As in the proof of 2 this means that $S \subseteq Q$, in fact, $S = Q$, since both S and Q have order p^k. Thus this can happen exactly once and in this case $[S: N_S(Q)] = 1$. Hence the number of p-SSGs is congruent to 1 modulo p.

11.2.2 Example.   The Groups of Order 12

We use the Sylow theorems to determine the groups of order 12. For each prime divisor we determine the number of possible p-SSGs. For $p = 2$ we find that the number of 2-SSGs (each of order 4) has the form $1 + 2k$ and divides 12, hence is either 1 or 3.

Number of 2-SSG ↓ Either C_4 or V	Number of 3-SSG. Each is a C_3. →	
	1 3-SSG	4 3-SSG
1 2-SSG $\langle b \rangle = C_4$	$C_4 \times C_3 \simeq C_{12}$	No group. Only solution to $k^3 \equiv 1 \pmod 4$ is $k = 1$
$\langle b, c \rangle = V$	$C_2 \times C_2 \times C_3$	$a^3 = 1,\ b^2 = c^2 = 1$ $aba^{-1} = c,\ aba^{-1} = bc$ $\simeq \mathscr{A}_4$
3 2-SSG $\langle b \rangle = C_4$	$a^3 = 1,\ b^4 = 1$ $bab^{-1} = a^{-1}$	No group. Too many elements
$\langle b, c \rangle = V$	$a^3 = 1,\ b^2 = c^2 = 1$ $bab = a^{-1},\ cac = a^{-1}$ $\cong D_{12}$	No group. Too many elements

Figure 11.1 This table determines the five nonisomorphic groups of order 12. An analysis of the Sylow subgroups gives the possibilities of one or four 3-SSG, each of which must be isomorphic to C_3 and one or three 2-SSG. Either all are isomorphic to C_4 or all are isomorphic to the four-group V. Further computation gives the relations among the generators shown in the table and shows that some cases cannot occur.

Similarly for $p = 3$ the number of 3-SSGs (each of order 3) has the form $1 + 3k$ and divides 12. Hence there are either one or four 3-SSGs. The possibilities are listed in the box shown in Figure 11.1. The entry in the box shows the nature of the group of order 12.

There are either one or four 3-SSGs and each 3-SSG is cyclic of order 3.

There are either one or three 2-SSGs. But now since the order of a 2-SSG is 4 there are two choices, either C_4 or the four-group V. Because any two p-SSGs are isomorphic all are isomorphic either to V or to C_4.

If there are four 3-SSGs then because the intersection of any two of them is the identity, these four subgroups give rise to $4 \times 2 = 8$ elements and so there can be but one subgroup of order 4. So either a 3-SSG or a 2-SSG is normal. This information is very helpful in determining the possible groups of order 12.

Now to cases.

The easiest cases to dispose of are the ones in which there is just a single 2-SSG and a single 3-SSG, for then the group is a direct product of these two subgroups. This gives rise to

$$C_4 \times C_3 \cong C_{12}$$

and

$$V_4 \times C_3 \cong C_2 \times C_2 \times C_3$$

Next suppose that there is but one 3-SSG. Call it $C_3 = \langle a \rangle$. Consider the case that the 2-SSG is cyclic of order 4, say, generated by b. Since $\langle a \rangle$ is normal we have

$$bab^{-1} = a^k$$

By an argument once familiar (recall Problem 3.1.7)

$$b^h a b^{-h} = a^{k^h}$$

and so, since $b^4 = 1$, we have putting $h = 4$,

$$k^4 \equiv 1 \pmod 3$$

and since $k^3 \equiv k \pmod 3$ we have

$$k^2 \equiv 1 \pmod 3$$

and so $k = 1$ or -1. The alternative $k = 1$ means that a and b commute so that $\langle a, b \rangle \cong C_{12}$. The case $k = -1$ remains a definite possibility, and in fact Problem 5.29 shows that there is such a group. Alternatively, return to this point after the next section and use the semidirect product construction of Theorem 11.3.7.

Suppose next that the 2-SSGs are isomorphic to V and suppose that $V = \langle b, c \rangle$ is one of them. We have $bc = cb$ and $b^2 = c^2 = 1$.

As before we have

$$bab = a^k \quad \text{and} \quad cac = a^h$$

and as before $h = \pm 1$ and $k = \pm 1$. If both $h = k = 1$ then the resulting group is isomorphic to $V \times C_3$, a previously discussed case. In the cases

$$h = 1, k = -1$$

and

$$h = -1, k = 1$$

and

$$h = -1, k = -1$$

we find that one element of order 2 in V commutes with a, hence creating an element of order 6 in G. This element together with any other element of order 2 from V generates the dihedral group D_{12}.

Finally we consider the case in which there are but one 2-SSG and four 3-SSGs. Suppose the 2-SSG is cyclic of order 4, say $\langle b \rangle$. Let $\langle a \rangle$ be one of the elements of order 3. We have then

$$aba^{-1} = b^k$$

and conclude

$$b = b^{k^3}$$

or that

$$k^3 \equiv 1 \pmod 4$$

But this means that $k \equiv 1 \pmod 4$ and so $\langle a, b \rangle$ is again the direct product, isomorphic to C_{12}. Thus there is no group of order 12 with one cyclic subgroup of order 4 and four subgroups of order 3.

Finally then, suppose that there is but one subgroup $V = \langle b, c \rangle$ of order 4 and again let a be an element of order 3. By the normality of V we know

$$aba^{-1} \in V \quad \text{and} \quad aca^{-1} \in V$$

but we don't know which elements they are. If $axa^{-1} = a$ for all $x \in V$ then G is abelian and a direct product that we've excluded from this case. Hence we may suppose without loss of generality that

$$aba^{-1} = c$$

Now it is impossible for $aca^{-1} = b$, for if that were true then

$$a^2ca^{-2} = c$$

and so a^2 and hence $a^4 = a$ commutes with c, contrary to the assumption $aca^{-1} = b$. Thus it must be that

$$aca^{-1} = bc \quad \text{and} \quad \text{so } abca^{-1} = c$$

This happens in the alternating group $\mathscr{A}_4$; this case yields the alternating group $\mathscr{A}_4$. To recognize it in the form of Section 4.1 determine that these four elements have order 3:

$$a \quad bab \quad cac \quad bcabc$$

and they may be called the elements a, b, c, and d in Definition 4.1.3. You should check all the relations.

EXERCISE SET 11.2

11.2.1 Use an analysis of the possible Sylow subgroups to find the possible groups of order 18, 20, 28, 30, 44.

11.2.2 Use the Sylow theorems to show that every group of order 30 has a normal subgroup of order 15. Compare your solution with Exercise 6.3.8.

11.2.3 Show that there are two nonisomorphic groups of order 20 with the same Sylow subgroup structure.

*11.2.4 Determine the possible groups of order 24.

11.3 SEMIDIRECT PRODUCTS

In this section we give another method of constructing new groups from old ones. The method generalizes the direct product. It would be a good idea to review Chapter 5 on direct products. We begin, as in that section, with the notation of an *internal* semidirect product, or split extension, as it is often called.

11.3.1 Definition.   Internal Semidirect Product (Split Extension)

A group G is said to be the internal semidirect product of a normal subgroup A by a subgroup B (G is the split extension of A by B) if the following three conditions are met:

  i. A is a normal subgroup of G.
  ii. $G = BA = B \vee A$.
  iii. $B \cap A =$ identity.

Note that the difference between an internal direct and semidirect product is that only one subgroup is required to be normal in a semidirect product; both subgroups are required to be normal in a direct product. Thus we see that a direct product is a special case of a semidirect product.

Note that the complex BA is a subgroup because one of the subgroups, A, is normal in $B \vee A$. Moreover conditions ii and iii show that the order of G is the product of the order of B and A:

$$|G| = |B||A|$$

Groups that are semidirect products occur more frequently than direct products. Moreover we shall see that it is possible to construct semidirect products almost as easily as direct products.

To get the proper motivation for the definition of an external semidirect product we need to look closely at the implications of conditions i, ii, and iii. We establish these in the next lemma, but first here are some examples to show us how ubiquitous semidirect products are.

11.3.2 Examples.   Familiar Semidirect Products

As we've already noted, any direct product is a semidirect product. Any dihedral group D_{2n} is a semidirect product of the cyclic group of order n (this subgroup has index 2 in D_{2n}; hence it is normal) and an element of order 2 not in the cyclic subgroup.

The symmetric group $\mathbb{S}_n$ is a semidirect product of the alternating group $\mathscr{A}_n$ and a transposition. The quaternion group of order 8 is not a semidirect product except trivially. The alternating group $\mathscr{A}_4$ is the semidirect product of the four-group

$$V = \{\iota, (1,2)(3,4), (1,3)(2,4), (1,4)(2,3)\}$$

and any 3-cycle.

Thus almost every group we've studied so far is a semidirect product. Perhaps you have already observed that the two groups of order 6 are semidirect products of a cyclic group of order 3 and a cyclic group of order 2. Thus we shall not be able to prove very much in the way of isomorphisms of semidirect products. The interaction of group B and group A is what is important, as we find in the next lemma.

11.3.3 Lemma.   Properties of Split Extensions

Let G be the split extension of A by B. Then

1. Every element $g \in G$ can be written in one and only one way

$$g = ba \quad \text{where} \quad a \in A \text{ and } b \in B$$

2. G/A is isomorphic to B.

Proof.   Let $g \in G$. The existence of $a \in A$ and $b \in B$ such that $g = ba$ is precisely the import of ii in Definition 11.3.1. The uniqueness of the b and a is a consequence of iii. Indeed, if

$$ba = b^*a^*$$

then

$$(b^*)^{-1}b = a^*a^{-1} = x$$

and so $x \in B \cap A = \{1\}$. Thus we have

$$(b^*)^{-1}b = 1 \quad \text{so that} \quad b^* = b$$

and

$$a^*a^{-1} = 1 \quad \text{so that } a^* = a$$

It is easily shown that the map defined by

$$g \to b \quad \text{if} \quad g = ab \ (a \in A \text{ and } b \in B)$$

is a homomorphism of G onto B whose kernel is A. This uses the normality of A. The observation of what happens to preserve multiplication is important to us now. Suppose that

$$g^* = b^*a^* \quad \text{and} \quad g' = b'a'$$

where of course b^*, b' and a^*, a' belong to B and A, respectively. Then because A is a normal subgroup of G

$$g^*g' = b^*a^*b'a' = b^*b'(b')^{-1}a^*b'a' = b^*b'(b'^{-1}a^*b')a'$$

$$g^*g' = b^*b'a^{**}a'$$

where $a^{**} = b'^{-1}a^*b'$ is also in A. In a direct product the elements of A commute with the elements of B, so that *in a direct product* $a^{**} = a^*$, but this does not always hold in a semidirect product.

   The key observation here is that each element b of B sets up an automorphism of A by

$$a \to b^{-1}ab$$

and in fact the product g^*g' can be written

$$g^*g' = b^*b'a^{**}a'$$

where a^{**} is the image of a^* under the automorphism of A set up by b'.

Moreover, these automorphisms of A preserve the group operation of B. That is, if $b*$ sets up the automorphism of A,

$$a \overset{(1)}{\to} b*^{-1}ab*$$

and b' sets up the automorphism of A,

$$a \overset{(2)}{\to} b'^{-1}ab'$$

then the automorphism of A set up by the product $b*b'$ in B is the *composition* of the automorphisms set up by $b*$ and b', respectively. We verify this:

$$a \overset{(1)}{\to} b*^{-1}ab* \overset{(2)}{\to} b'^{-1}(b*^{-1}ab*)b' = (b*b')^{-1}ab*b'$$

Thus, if G is the internal semidirect product of A by B, there is a homomorphism from B to aut(G), the automorphism group of A. This is the final observation that we need to define an external semidirect product.

11.3.4 Definition.   External Semidirect Product of *A* by *B*

Let A and B be groups and suppose that ν is a homomorphism from B into the group of automorphisms of A. Then the semidirect product of A by B is the set of ordered pairs $B \times A$ under the operation

$$(b*, a*)(b', a') = (b*b', a*^{b'}a')$$

where $a*^{b'}$ denotes the image of $a*$ under the automorphism $(b')\nu$.
Thus the "only" difference between the semidirect product of ordered pairs and the direct product of ordered pairs is that the second member of the direct product is $a*a'$ while in the semidirect product it is $a*^{b'}a'$. Please note that the image of b under ν is written $(b)\nu$ as though ν were a permutation. Using this convention we can use the exponential notation a^b for the image of a under $(b)\nu$ rather than the more cumbersome $\nu(b)(a)$.

Notation.   Write

$$B \times_\nu A$$

for the semidirect product A by B where ν is the homomorphism of B into aut(A) used in the definition.
The essential properties of the external semidirect product are collected in the next theorem.

11.3.5 Theorem.   Properties of the Semidirect Product

1. $B \times_\nu A$ is a group under the operation in Definition 11.3.4.
2. $A \cong \hat{A} = \{(1, a): a \in A\}$. $B \cong \hat{B} = \{(b, 1): b \in B\}$.
3. $B \times_\nu A$ is a the split extension of $\hat{A}$ by $\hat{B}$.

Proof. Begin by observing that since ν is a homomorphism of B into aut(G) the identity of B must map onto the identity automorphism of A, which we now denote by ι, the Greek letter iota. Now it is easy to verify that $(1, 1)$ is the identity of $B \times_\nu A$

$$(b, a)(1, 1) = (b \cdot 1, a^\iota \cdot 1) = (b, a)$$

The inverse of (b, a) is $(b^{-1}, (a^{-1})^{b^{-1}})$. We verify that

$$(b, a)\left(b^{-1}, (a^{-1})^{b^{-1}}\right) = \left(bb^{-1}, a^{b^{-1}}(a^{-1})^{b^{-1}}\right) = (1, 1)$$

since under the automorphism corresponding to b^{-1}

$$a^{b^{-1}}(a^{-1})^{b^{-1}} = (a \cdot a^{-1})^{b^{-1}} = 1$$

The verification of the associative law takes a little more computation:

$$[(b, a)(b^*, a^*)](b', a') = (bb^*, a^{b^*}a^*)(b', a') = \left(bb^*b', (a^{b^*}a^*)^{b'}a'\right)$$

while

$$(b, a)[(b^*, a^*)(b', a')] = (b, a)(b^*b', a^{*b'}a') = (bb^*b', a^{b^*b'}a^{*b'}a')$$

thus we need only observe that under the automorphism $(b')\nu$ we have

$$(a^{b^*}a^*)^{b'} = (a^{b^*})^{b'}a^{*b'}$$

and since ν is a homomorphism

$$(b^*b')\nu = (b^*)\nu(b')\nu$$

hence

$$a^{b^*b'} = (a^{b^*})^{b'}$$

The proofs that $\hat{A}$ and $\hat{B}$ are isomorphic, respectively, to A and B are routine. The details depend upon the fact that $(1)\nu$ is the identity and $1^b = 1$ for all b.

Finally we must establish that $\hat{A}$ is normal in $B \times_\nu A$ and that $\hat{A} \cap \hat{B} = (1, 1)$ and that $B \times_\nu A = \hat{B}\hat{A}$. Only the normality is less than self-evident, and even that's easy. Verify that

$$(b, 1)^{-1}(1, a)(b, 1) = (1, a^*)$$

since the first component of a product is the product of the first components. This completes the proof.

One case arises so frequently that we give the details explicitly:

11.3.6 Example. Semidirect Product of Two Cyclic Groups

We want to construct a semidirect product of a cyclic group $C_n = \langle a \rangle$ by a cyclic group $C_m = \langle b \rangle$ of order m. To do this we need to recall the automorphisms of C_n. These are defined by

$$a \to a^k$$

where $\gcd(k, n) = 1$. Different choices for $k \pmod{n}$ lead to different automorphisms. [In fact the group of automorphisms of C_n, $\text{aut}(C_n)$ is isomorphic to the group of units mod n. We called this $\mathbb{Z}_n^{\times}$ in Example 3.1.4.]

The next question to resolve is the choice of the homomorphism ν from C_m to $\text{aut}(C_m)$. Since $C_m = \langle b \rangle$, ν is determined by the choice for the image of b under ν. If the automorphism $(b)\nu$ is

$$(b)\nu: a \to a^k$$

then it must follow that the automorphism that is the image of b^u is u iterations of the image of b; thus it is

$$(b^u)\nu: a \to a^{(k^u)}$$

In particular

$$(b^m)\nu: a \to a^{(k^m)}$$

Since b has order m it must be the case that the image of b^m is the identity automorphism, in other words, that

$$(b^m)\nu: a \to a^{(k^m)} = a$$

or that

$$k^m \equiv 1 \pmod{n}$$

Said equivalently, the order of k in $\mathbb{Z}_n^{\times}$ must divide m.

Thus to determine a semidirect product of C_n by C_m it suffices to choose k such that $\gcd(k, n) = 1$ and $k^m \equiv 1 \pmod{n}$.

If we take $C_m = \langle b \rangle$ and $C_n = \langle a \rangle$ and use multiplicative notation we have

$$C_m \times_k C_n = \{(b^u, a^x): u \in \mathbb{Z}_m \text{ and } x \in \mathbb{Z}_n\}$$

where

$$(b^u, a^x)(b^v, a^v) = (b^{u+v}, a^{xk^v + y})$$

since the automorphism corresponding to b^v is v iterations of the automorphism

$$a \to a^k$$

and so is

$$a \to a^{(k^v)}$$

and so the image of a^x under the automorphism corresponding to b^v is

$$(a^x)^{(k^v)} = a^{xk^v}$$

To simplify notation we can take the cyclic groups to be the additive groups mod m and n, respectively. If

$$C_m = \langle \mathbb{Z}_m, + \rangle \quad \text{and} \quad C_n = \langle \mathbb{Z}_n, + \rangle$$

and $\gcd(k, n) = 1$ and $k^m \equiv 1 \pmod{n}$ then we have a semidirect product

$$C_m \times_k C_n = \{(u, x): u \in \mathbb{Z}_m \text{ and } x \in \mathbb{Z}_n\}$$

where the group operation is defined by

$$(u, x) + (v, y) = (u + v, xk^v + y) \tag{1}$$

It may be instructive (see Exercise 11.3.1) to verify that the operation defined in (1) gives a group. It should be proved that the operation is independent of the congruence class representatives, u, x, v and y, chosen. The crucial detail is that

If $v \equiv w \pmod{m}$ then $(u, x)(v, y) = (u, x)(w, y)$

under definition (1) above.

We summarize this as a theorem.

11.3.7 Theorem.   The Group $\langle B, A: B^m = A^n = 1, AB = BA^k \rangle$

If $\gcd(k, n) = 1$ and $k^m \equiv 1 \pmod{n}$ then there is a group of order mn generated by an element B of order m, and element A of order n satisfying the relation

$$B^{-1}AB = A^k$$

In particular $\langle A \rangle$ is a normal subgroup of this group.

Proof.   Clearly the group is the semidirect product of C_n by C_m defined in the example above. We need only identify

$$B = (1, 0) \quad \text{and} \quad A = (0, 1)$$

and compute

$$A^h = (0, h)$$

which holds for all h. We can prove this easily by induction.

Finally compute that

$$AB = BA^k$$

As a specific example of all this consider the group that is the semidirect product of a cyclic group of order 3 by a cyclic group of order 4 with $k = -1$. This yields the group of order 12 promised by the preceding section.

Another frequent semidirect product arises when group $B = \langle b \rangle$ has order 2 and A is abelian. Then simply let b correspond to the automorphism of A in which each element is sent into its inverse. In fact this is how dihedral groups arise:

$$D_{2n} = C_2 \times_{(-1)} C_n$$

An important caveat in all this is that although different values of k give rise to different automorphisms of C_n, different values of k often give *isomorphic* semidirect products. So when determining the nonisomorphic groups of a given order be sure to check for possible isomorphism between different-appearing semidirect products!

EXERCISE SET 11.3

11.3.1 Give the details that formula (1) yields a group operation on the ordered pairs $\{(u, x): u \in \mathbb{Z}_m \text{ and } x \in \mathbb{Z}_n\}$.

11.3.2 Find an automorphism v of order 3 for the Klein four-group V and construct $C_3 \times_v V$ using it. Which group of order 12 is this?

11.3.3 Show that there is a nonabelian group of order 56 with a normal subgroup of order 8.

11.3.4 Construct a nonabelian group of order 55.

11.3.5 Determine all groups of order pq where p and q are primes.

11.3.6 Determine all groups whose order is less than 32 except for the groups of order 16 and 24.

11.3.7 Determine all groups of order 16 that are split extensions of smaller subgroups.

11.3.8 Determine all groups of order 24 that are split extensions of groups of smaller orders.

11.3.9 Determine all groups of order $4p$ where p is a prime greater than 4. Beware, the cases $p \equiv 1 \pmod 4$ and $p \equiv -1 \pmod 4$ are different!

11.3.10 Prove or disprove: If two groups have the same order and the same number of p-SSGs for all p, and if these p-SSGs are all isomorphic then the groups are isomorphic.

11.4 FINITE ABELIAN GROUPS

We now prove a fundamental result on finite abelian groups:

A finite abelian group is the direct product of cyclic groups.

This theorem is true under more general assumptions and there are more precise results (Theorems 16.3.2 and 16.4.4) that accompany this main decomposition result. However, the result quoted here is important and we can prove it with the tools at hand.

This theorem is analogous to the fact that an integer is the product of primes. The cyclic groups are the building blocks for abelian groups; the process is the direct product. Naturally we would like a uniqueness theorem to accompany this existence one. That requires more machinery, and we shall do it as an example of module theory in Section 16.4.

11.4.1 Lemma. (Problem 5.2.8) The Join of Normal Subgroups and Their Direct Product

Let G have normal subgroups R and S such that $rs = sr$ for all $r \in R$ and $s \in S$. Then $R \vee S$ is a homomorphic image of $R \times S$. The kernel is a subgroup of $R \times S$ isomorphic to $R \cap S$.

Proof. Verify that the mapping from $R \times S$ to the complex RS given by

$$(r, s) \rightarrow rs$$

is a homomorphism. The kernel consists of the set

$$\{(x, x^{-1}): x \in R \cap S\}$$

which is isomorphic to $R \cap S$ through the isomorphism

$$x \rightarrow (x, x^{-1})$$

Note that the condition that an element in R commute with an element in S implies that $R \cap S$ is abelian.

Next we need some precise information on the number and type of subgroups of the direct product of subgroups of prime power order. What we need is included in the next lemma.

11.4.2 Lemma. Subgroups in the Product of Cyclic Groups

Let p be a prime.

1. The direct product of two cyclic groups each of order p has $p + 1$ subgroups each of order p.
2. The direct product of two cyclic groups each of order p^2 has $p + 1$ subgroups of order p. Each subgroup of order p is contained in p subgroups each isomorphic to C_{p^2} and one subgroup isomorphic to $C_p \times C_p$.

Proof of 1. The direct product of two groups of order p has $p^2 - 1$ elements not equal to the identity. These must be in subgroups of order p that have only the identity in common. Each such subgroup contains $p - 1$ elements, not the identity; hence there are $(p^2 - 1)/(p - 1) = p + 1$ subgroups.

Proof of 2. To exhibit the structure of the direct product of two cyclic groups of order p^2 let each group be represented by $\langle \mathbb{Z}_{p^2}, + \rangle$. Thus a typical element is the pair (a, b) where a and b are taken modulo p^2. We want to count the elements of order p. Now an element will have order p if and only if

$$p(a, b) = (pa, pb) = (0, 0)$$

hence if and only if

$$a \equiv b \equiv 0 \,(\text{mod } p)$$

Integers modulo p^2 that satisfy this condition have the form kp where $0 \leq k < p$. There are thus p choices for a and p choices for b. Excluding the identity there are thus $p^2 - 1$ elements of order p. As before these elements must partition themselves into $p + 1$ subgroups of order p.

Now we want to determine the subgroups that contain a particular subgroup $\langle a, b \rangle$ of order p.

Let $\langle(a, b)\rangle$ be a typical subgroup of order p. Since $a \equiv 0 \pmod{p}$ we may write $a = pc$; similarly $b = pd$. Since $(a, b) \neq (0,0)$ it must be that p does not divide both c and d. Without loss of generality we assume that $p \nmid c$.

Now for $m = 0, 1, \ldots, (p - 1)$,

$$(a, b) = p(c, d + mp)$$

and so the element (a, b) belongs to the subgroups

$$\langle(c, d + mp)\rangle$$

We claim these p subgroups are distinct. Indeed, suppose that two subgroups were equal. Then we should have

$$u(c, d + mp) = (c, d + np)$$

and hence

$$uc \equiv c \pmod{p^2} \quad \text{and} \quad u(d + mp) \equiv d + np \pmod{p^2}$$

and thus, since $p \nmid c$ it follows that $u \equiv 1 \pmod{p^2}$ and so $m \equiv n \pmod{p}$.

The additional subgroup containing (a, b) is isomorphic to $C_p \times C_p$ and is the join of all the subgroups of order p.

Remark. The essential fact that we need from this lemma is that no factor group of $C_{p^2} \times C_{p^2}$ whose kernel has order p has a unique subgroup of order p.

Now we prove a stronger theorem than the statement of the decomposition theorem.

11.4.3 Theorem. Existence of a Direct Factor of an Abelian Group

Let A be a finite abelian group. Let p be a prime dividing the order of A and let $h \in A$ have maximal order among the elements whose orders are a power of p. Then either $A = \langle h \rangle$ or $\langle h \rangle$ is a direct factor of A.

Proof. The proof is by induction on the order of A. We know from our experience so far that this theorem holds for the abelian groups of order up to 12 and for all groups of prime order. We shall revert to multiplicative notation and denote by I the identity subgroup.

Suppose then that h has been chosen as in the hypothesis. We want to show that either $A = \langle h \rangle$ or there is a subgroup L of A such that

$$A = \langle h \rangle \times L$$

Let $|h| = p^e$ and let $\langle z \rangle$ be the subgroup of $\langle h \rangle$ of order p. We distinguish two cases. The lattice relations of the subgroup and factor groups are shown in Figure 11.2 on the next page.

Case 1. There is a subgroup W such that $W \cap \langle z \rangle = I$. This also implies that $\langle z \rangle \cap \langle h \rangle = I$ since the subgroups of $\langle h \rangle$ form a chain in which $\langle z \rangle$ covers I.

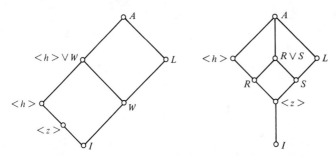

Figure 11.2 These lattices show the relationships of the subgroups and factor groups involved in the proof of Theorem 11.4.2. On the left is the diagram for case 1; on the right is the diagram for case 2.

We consider A/W. By the diamond lemma, $\langle h \rangle$ is isomorphic to $(\langle h \rangle \vee W)/W$ and so A/W contains an element (in fact it may be chosen as the coset hW) of maximal prime power order p^e.

By induction applied to A/W either $A/W = \langle hW \rangle$ or there exists a subgroup $\bar{L}$ in A/W such that

$$A/W = ((\langle h \rangle \vee W)/W) \times \bar{L}$$

In the former case, that is, when $A/W = \langle hW \rangle$, we find that

$$A/W = (\langle h \rangle \vee W)/W = \langle (hW) \rangle$$

and so

$$A = \langle h \rangle \vee W \quad \text{and} \quad \langle h \rangle \cap W = I$$

thus

$$A = \langle h \rangle \times W$$

and we are done.

In the latter case, since we may write

$$\bar{L} = L/W \quad \text{where} \quad A \supset L \supset W$$

we find that

$$A/W = ((\langle h \rangle \vee W)/W) \times (L/W)$$

Thus

$$(\langle h \rangle \vee W) \cap L = W \quad \text{and} \quad (\langle h \rangle \vee W) \vee L = A$$

But then it follows that

$$A = \langle h \rangle \times L$$

since

$$\langle h \rangle \vee L = \langle h \rangle \vee L \vee W = A$$

and

$$\langle h \rangle \cap L = I$$

since

$$\langle h \rangle \cap L \subset (\langle h \rangle \vee W) \cap L = W \quad \text{and} \quad \langle h \rangle \cap L \subset \langle h \rangle$$

and so $\langle h \rangle \cap L \subset W \cap \langle h \rangle = I$. This completes case 1.

Note that case 1 holds whenever the order of A is divisible by more than one prime, since W may be taken to be an element whose order is a prime not p.

Case 2. Every subgroup of A except I contains $\langle z \rangle$. We shall see that this means that A is cyclic. In any event, as we've just noted, in this case the order of A is a power of the prime p. If A has order p then the result holds. Therefore, we assume that the order of A is at least p^2. Also in this case the exponent $e \geq 2$. Otherwise every element is of order p and any element not in $\langle z \rangle$ generates a subgroup distinct from $\langle z \rangle$. This means that $\langle z \rangle$ is the unique subgroup of order p in A and is contained in every subgroup of A but I.

Now consider $A/\langle z \rangle$. First, the maximal prime power order of coset $x\langle z \rangle$ in $A/\langle z \rangle$ is p^{e-1}. This is so because every $\langle x \rangle$ contains $\langle z \rangle$ in A. Hence $x^{p^{e-1}} \in \langle z \rangle$. On the other hand, $h\langle z \rangle$ has order p^{e-1} in $A/\langle z \rangle$.

By induction applied to $A/\langle z \rangle$, either $A/\langle z \rangle = \langle h\langle z \rangle \rangle$ or for some L, $A \supset L \supset \langle z \rangle$,

$$A/\langle z \rangle = (\langle h \rangle / \langle z \rangle) \times (L/\langle z \rangle)$$

We now show the latter alternative cannot hold. The latter alternative implies that

$$\langle h \rangle \vee L = A \quad \text{and} \quad \langle h \rangle \cap L = \langle z \rangle$$

We know that $\langle h \rangle$ has a subgroup R of order p^2 that is of course cyclic. In addition L has order $\geq p^2$ and in fact must contain an element of order p^2, for otherwise L has more than $\langle z \rangle$ as a subgroup of order p. Let S be this cyclic subgroup of order p^2. Note that $R \cap S = \langle z \rangle$.

Now consider the subgroup $R \vee S$ of A. By Lemma 11.4.1 $R \vee S$ is a homomorphic image of $R \times S$ with a kernel of order $|R \cap S| = p$. But by Lemma 11.4.2 this factor group cannot have a unique subgroup of order p. On the other hand, as a subgroup of A, $R \vee S$ must have a unique subgroup of order p. This contradiction means that the other alternative:

$$A/\langle z \rangle = \langle h\langle z \rangle \rangle$$

holds. Compare the orders to find

$$|A/\langle z \rangle| = |A|/p = |h\langle z \rangle| = p^{e-1}$$

and so that

$$|A| = p^e = |h|$$

and thus

$$A = \langle h \rangle$$

This completes the proof of Theorem 11.4.3. By induction we now obtain the fundamental theorem as a corollary:

11.4.4 Corollary.   Fundamental Theorem for Finite Abelian Groups

A finite abelian group is the direct product of cyclic groups.

Proof. The proof is by induction on the order of the abelian group. We know the theorem is true for low values of n. By Theorem 11.4.3 a finite abelian group A is either cyclic or has a proper cyclic group as a direct factor. In this case, say $A = \langle h \rangle \times B$, where B has a lower order than A. By the induction hypothesis we conclude that B is the direct product of cyclic groups and substituting this product for B gives the result for A.

This theorem tells us that to construct all the abelian groups of a given order, say, n, we need only combine in direct products cyclic groups of orders dividing n. The only condition we must meet is that the product of the orders of these cyclic groups is n.

If we carry out this program slavishly we will construct many isomorphic abelian groups. For example, $C_4 \times C_3$ is a C_{12}. Thus we still must seek a method to determine when two abelian groups of the same order are isomorphic.

Now we return once more to the multiplicative groups $\mathbb{Z}_n^{\times}$ of integers relatively prime to n. When is this group cyclic?

11.4.5 Theorem.   When is $\mathbb{Z}_n^{\times}$ Cyclic?

The multiplicative group $\mathbb{Z}_n^{\times}$ is cyclic if and only if

$$n = 2, 4, p^e, 2p^e \tag{1}$$

where p is an odd prime.

If $n \geq 3$ then $\mathbb{Z}_{2^n}^{\times}$ is the direct product of a cyclic group of order 2^{n-2} and a cyclic group of order 2; in fact

$$\mathbb{Z}_{2^n}^{\times} = \langle [-1]_{2^n} \rangle \times \langle [x]_{2^n} \rangle$$

where the residue class $[x]_{2^n}$ has order 2^{n-2}.

Proof.   First we show that the only possibilities for $\mathbb{Z}_n^{\times}$ to be cyclic are those for which n is one of the values listed in the theorem or is a higher power of 2.

We begin by supposing that n is divisible by two different primes p and q. One of these must be odd, say that q is odd. Factor n

$$n = p^e q^f r \qquad \text{where} \quad \gcd(p, q, r) = 1$$

From Theorem 5.2.6 we have the direct product decomposition

$$\mathbb{Z}_{p^n}^{\times} \cong \mathbb{Z}_{p^e}^{\times} \times \mathbb{Z}_{q^f}^{\times} \times \mathbb{Z}_r^{\times}$$

The orders of the first two factors are $p^{e-1}(p-1)$ and $q^{f-1}(q-1)$, respectively. Each is divisible by 2 if p is odd or if $p = 2$ and $e \geq 2$. In these cases each of these factors contains at least one element of order 2. Thus in these cases $\mathbb{Z}_n^{\times}$ contains at least two elements of order 2 and so cannot be cyclic. This means that if $\mathbb{Z}_n^{\times}$ is to be cyclic n must have one of the forms listed in (1) or be a higher power of 2.

Now suppose that n has one of the forms listed in (1). We shall prove that $\mathbb{Z}_n^{\times}$ is cyclic. The verification is easy for $n = 2$ and $n = 4$. And if p is an odd prime,

$$\mathbb{Z}_{2p^n}^{\times} \cong \mathbb{Z}_2^{\times} \times \mathbb{Z}_{p^n}^{\times} \cong \mathbb{Z}_{p^n}^{\times}$$

To complete the proof we need to show that if p is an odd prime $\mathbb{Z}_{p^n}^\times$ is cyclic and also to determine the structure of $\mathbb{Z}_{2^n}^\times$ for $n \geq 3$.

This part of the proof relies on three rather technical lemmas concerning congruences. We begin by proving these lemmas.

11.4.6 Homomorphisms of $\mathbb{Z}_m^\times$

If n divides m then $\mathbb{Z}_n^\times$ is the homomorphic image of $\mathbb{Z}_m^\times$ via the mapping

$$\Phi: [x]_m \mapsto [x]_n$$

Proof. First, we observe that since n divides m,

$$x \equiv y \;(\text{mod } m) \quad \text{implies } x \equiv y \;(\text{mod } n)$$

This means that the definition of Φ does not depend on the residue class representative as it appears to; thus Φ is well defined.

Second, it is easy to verify that Φ preserves multiplication.

Third, since n divides m it follows that if $\gcd(x, m) = 1$ then $\gcd(x, n) = 1$ and so Φ maps $\mathbb{Z}_m^\times$ into $\mathbb{Z}_n^\times$. It is not so immediate that Φ is onto. This fact follows from an application of the Chinese remainder theorem 2.1.8(v). We may express

$$m = n n_1 m_1$$

where a prime p divides n_1 only if p divides n and $\gcd(n n_1, m_1) = 1$. Here are the details.

Let $[a] \in \mathbb{Z}_n^\times$ so that $\gcd(a, n) = 1$. From the Chinese remainder theorem we may find an x such that

$$x \equiv a \;(\text{mod } n n_1)$$

and

$$x \equiv 1 \;(\text{mod } m_1)$$

It follows that $x \equiv a \;(\text{mod } n)$ and so $\gcd(x, n) = 1$ since $\gcd(a, n) = 1$. From this we also have $\gcd(x, n n_1) = 1$. Next observe that $\gcd(x, m_1) = 1$ as well because the second congruence guarantees that any common divisor of x and m_1 is a divisor of 1. Finally, since $\gcd(n n_1, m_1) = 1$, we may conclude that $\gcd(x, m) = 1$. Thus

$$\Phi: [x]_m \mapsto [a]_n$$

and so we have proved that the homomorphism Φ is onto.

11.4.7 Two Related Congruences

If p is an odd prime, $a \geq 0$, and $n \geq 1$, or if $p = 2$, $a \geq 1$ and $n \geq 2$ then $x^{p^{a+1}} \equiv 1 \;(\text{mod } x^{p^{n+1}})$ implies $x^{p^a} \equiv 1 \;(\text{mod } x^{p^n})$.

Proof. The proof is by induction on n. Suppose first that p is an odd prime. We start the induction at $n = 1$. The hypothesis

$$x^{p^{a+1}} \equiv 1 \;(\text{mod } p^2)$$

implies

$$x^{p^{a+1}} \equiv 1 \pmod{p}$$

From this it follows that the order of $[x]_p$ is a power of p. But $\mathbb{Z}_p^{\times}$ has order $(p - 1)$ and since p does not divide $p - 1$, the only element whose order is a power of p is the identity, in which case the power is 0. Thus $x \equiv 1 \pmod{p}$ and thus

$$x^{p^a} \equiv 1 \pmod{p}$$

In the case $p = 2$ we start the induction at $n = 2$ so that $p^n = 2^2 = 4$ and $p^{n+1} = 2^3 = 8$. In these cases the element has order 1 or 2 in $\mathbb{Z}_8^{\times}$ so that if $a \geq 1$

$$x^{2^a} \equiv 1 \pmod{8} \quad \text{and hence} \quad x^{2^a} \equiv 1 \pmod{4}$$

Now for the induction step. Suppose the lemma holds for n. Remember, $n \geq 1$ if p is odd, and $n \geq 2$ if $p = 2$. Suppose that

$$x^{p^{a+1}} \equiv 1 \pmod{x^{p^{n+2}}}$$

then

$$x^{p^{a+1}} \equiv 1 \pmod{x^{p^{n+1}}}$$

and by induction

$$x^{p^a} \equiv 1 \pmod{x^{p^n}}$$

thus

$$x^{p^a} = 1 + up^n \tag{2}$$

and so

$$x^{p^{a+1}} = (1 + up^n)^p = 1 + pup^n + \sum_{k=2}^{p-1} \binom{p}{k}(up^n)^k + (up^n)^p \tag{3}$$

where the Σ term is empty if $p = 2$. We will now show that p^{n+2} divides all terms on the right-hand side of (3) except the first two.

For odd primes p and k such that $2 \leq k \leq (p - 1)$ it is true that p divides $\binom{p}{k}$ and so for k in this range the exponent on p is at least $nk + 1$. Thus

$$nk + 1 \geq 2n + 1$$

and because $n \geq 1$, $2n + 1 \geq n + 2$. Hence in this range

$$nk + 1 \geq n + 2$$

The exponent on p in the last term is pn. And,

$$\text{If } p \text{ is odd, } np \geq 3n \geq n + 2, \text{ since } n \geq 1$$

$$\text{If } p = 2, np = 2n \geq n + 2, \text{ since } n \geq 2$$

Thus, in all cases under consideration, if we treat (3) as a congruence mod p^{n+2} we have

$$x^{p^{a+1}} \equiv 1 + up^{n+1} \pmod{p^{n+2}}$$

But by hypothesis

$$x^{p^{a+1}} \equiv 1 \pmod{p^{n+2}}$$

Hence p must divide u and thus equation (2) becomes

$$x^{p^a} \equiv 1 \pmod{p^{n+1}}$$

11.4.8 Lifting Orders from Mod p^n to Mod p^{n+1}

Let p be an odd prime and $n \geq 1$ and $e \geq 1$, or let $p = 2$, $n \geq 2$ and $e \geq 2$. If $[x]_{p^n}$ has order p^e in $\mathbb{Z}_{p^n}^\times$ then $[x + p^n]_{p^{n+1}}$ has order p^{e+1} in $\mathbb{Z}_{p^{n+1}}^\times$.

Proof. We suppose that $x^{p^e} \equiv 1 \pmod{p^n}$. Since the order of $\mathbb{Z}_{p^n}^\times$ is $(p-1)p^{n-1}$ we may suppose without loss of generality that $e \leq n - 1$. Since x and $x + p^n$ are congruent modulo p^n we also have

$$(x + p^n)^{p^e} \equiv 1 \pmod{p^n}$$

and so

$$(x + p^n)^{p^e} = 1 + up^n$$

hence

$$(x + p^n)^{p^{e+1}} = (1 + up^n)^p = 1 + \sum_{k=1}^{p-1} \binom{p}{k}(up^n)^k + (up^n)^p$$

and by a calculation now familiar,

$$(x + p^n)^{p^{e+1}} \equiv 1 \pmod{p^{n+1}}$$

since, in the cases under consideration, we have

$$1 + nk \geq n + 1 \quad \text{and} \quad pn \geq n + 1$$

Thus the order of $[x + p^n]_{p^{n+1}}$ in $\mathbb{Z}_{p^{n+1}}^\times$ is at most p^{e+1}.
  Now suppose the order of $[x + p^n]_{p^{n+1}}$ is p^{a+1}. Thus

$$(x + p^n)^{p^{a+1}} \equiv 1 \pmod{p^{n+1}} \tag{4}$$

We know $a \leq e$. The congruence (4) yields

$$(x + p^n)^{p^{a+1}} \equiv 1 \pmod{p^n}$$

and so $a + 1 \geq e$; thus $a \geq e - 1$ and by hypothesis, $e - 1 \geq 0$ if p is odd, or $e - 1 \geq 1$ if $p = 2$. In either event the hypotheses of Lemma 11.4.7 concerning a and e are met and so from congruence (4) we may conclude

$$(x + p^n)^{p^a} \equiv 1 \pmod{p^n}$$

But since $[x]_{p^n}$ has order p^e in $\mathbb{Z}_{p^n}^\times$ it follows that $a \geq e$. Hence $e = a$, and we have shown that the order of $[x + p^n]_{p^{n+1}}$ is p^{a+1} in $\mathbb{Z}_{p^{n+1}}^\times$. Thus the proof of Lemma 11.4.8 is complete. We are ready to put these lemmas together.

Completion of the Proof of Theorem 11.4.5.   We consider the case that p is an odd prime. We proceed by induction on n to prove that $\mathbb{Z}_{p^n}^\times$ is cyclic. To begin the induction we know from Lemma 10.4.1 that $\mathbb{Z}_p^\times$ is cyclic. Here is the inductive step:

Suppose that $\mathbb{Z}_{p^n}^\times$ is cyclic of order $p^{n-1}(p-1)$ and thus has an element $[x]_{p^n}$ of order p^{n-1} and another of order $(p-1)$. Since $\gcd(p, p-1) = 1$ the product of these commuting elements has order $p^{n-1}(p-1)$ and so generates the group. The homomorphism of Lemma 11.4.6 shows that $\mathbb{Z}_{p^{n+1}}^\times$ has an element whose image has order $(p-1)$. This means in turn that $\mathbb{Z}_{p^{n+1}}^\times$ has an element of order $(p-1)$. Next, Lemma 11.4.8 shows that $\mathbb{Z}_{p^{n+1}}^\times$ has an element of order p^n. Hence $\mathbb{Z}_{p^{n+1}}^\times$ has an element of order $p^n(p-1)$ and so is cyclic.

Finally we consider the case $p = 2$. We verify that

$$\mathbb{Z}_8^\times = \langle[-1]_8\rangle \times \langle[3]_8\rangle$$

However, we must begin our induction on n with $n = 4$. Verify that

$$\mathbb{Z}_{16}^\times = \langle[-1]_{16}\rangle \times \langle[3]_{16}\rangle$$

Now for the induction step: Suppose that $n \geq 4$ and that

$$\mathbb{Z}_{2^n}^\times = \langle[-1]_{2^n}\rangle \times \langle[x]_{2^n}\rangle$$

where $[x]_{2^n}$ has order 2^{n-2}. Lemma 11.4.8 shows that $[x + 2^n]_{2^{n+1}}$ has order 2^{n-1} in $\mathbb{Z}_{2^{n+1}}^\times$. It remains only to show that $[-1]_{2^{n+1}}$ does not belong to the cyclic subgroup generated by $[x + 2^n]_{2^{n+1}}$. Since $[-1]_{2^{n+1}}$ has order 2, if it does belong to this subgroup it is the unique element of order 2 in this subgroup. This is

$$\left([x + 2^n]_{2^{n+1}}\right)^{2^{n-2}}$$

and so it must be that

$$\left([x + 2^n]_{2^{n+1}}\right)^{2^{n-2}} = [-1]_{2^{n+1}}$$

Now

$$(x + 2^n)^2 = x^2 + 2^{n+1}x + 2^{2n} \equiv x^2 \pmod{2^{n+1}}$$

so that

$$(x + 2^n)^{2^{n-2}} \equiv x^{2^{n-2}} \pmod{2^{n+1}}$$

and hence that

$$x^{2^{n-2}} \equiv -1 \pmod{2^{n+1}}$$

Thus

$$x^{2^{n-2}} \equiv -1 \pmod{2^n}$$

and so

$$\left([x]_{2^n}\right)^{2^{n-2}} = [-1]_{2^n}$$

But $[x]_{2^n}$ has order 2^{n-2}, and so we have the contradiction that

$$[1]_{2^n} = [-1]_{2^n}$$

This completes the proof of the theorem.

EXERCISE SET 11.4

11.4.1 Find all abelian groups whose order is at most 31.

11.4.2 Draw the lattice of subgroups of $C_q \times C_q$.

11.4.3 Show that there are as many subgroups of index p in $C_{p^2} \times C_{p^2}$ as there are subgroups of order p.
*Does this generalize?

11.5 A SIMPLE GROUP OF ORDER 168

In this section we give an example of a group of matrices that is representative of a whole class of finite simple groups. Remember, a simple group is one without any normal subgroups except the identity and the group itself. The most elementary simple groups are the cyclic groups of prime order. These are the only simple abelian groups. The most elementary class of nonabelian finite simple groups are the alternating groups $\mathscr{A}_n$ where $n \geq 5$. Theorem 6.2.12 establishes that these groups are simple.

The simple group of this section is an example of a class of groups called the projective special linear groups (PSL). In this section we describe this group as a homomorphic image of a group of 2×2 matrices over the finite field $\mathbb{Z}_7$.

First we need to recall a few facts about 2×2 matrices over any field F. If

$$A = \begin{bmatrix} a & b \\ c & d \end{bmatrix} \quad \text{and} \quad B = \begin{bmatrix} w & x \\ y & z \end{bmatrix}$$

then

$$AB = \begin{bmatrix} aw + by & ax + bz \\ cw + dy & cx + dz \end{bmatrix} \quad \text{and} \quad A + B = \begin{bmatrix} a + w & b + x \\ c + y & d + z \end{bmatrix}$$

and

$$rA = \begin{bmatrix} ra & rb \\ rc & rd \end{bmatrix}$$

$$\det(A) = ad - bc \quad \text{and} \quad \det(AB) = \det(A)\det(B)$$

If $\det(A) = \delta \neq 0$ then

$$A^{-1} = \begin{bmatrix} \dfrac{d}{\delta} & \dfrac{-b}{\delta} \\ \dfrac{-c}{\delta} & \dfrac{a}{\delta} \end{bmatrix}$$

Under addition these matrices form an abelian group. Multiplication is associative and both distributive laws hold. Multiplication is almost never commutative.

Under multiplication the matrices with nonzero determinants form a group, called the general linear group $GL(2,7)$. The matrices with determinant 1 form a subgroup, called the special linear group $SL(2,7)$. This group has a center Z, and it is the factor group $SL(2,7)/Z$, called the projective linear group $PSL(2,7)$, that we are interested in. We wish to prove that $PSL(2,7)$ is a simple group of order 168. In fact, it is the only simple group of that order, but we shall not prove this stronger uniqueness property. We begin by determining the order of $SL(2,7)$.

11.5.1 Lemma.   The Order of SL(2, 7)

The order of $SL(2,7)$ is 336.

Proof.   Consider a matrix in this group,

$$\begin{bmatrix} a & b \\ c & d \end{bmatrix} \qquad \text{where} \quad ad - bc = 1$$

To count the elements in $SL(2,7)$ we just need to count the number of solutions to the equation $ad - bc = 1$. Suppose first that $a \neq 0$. Then if b and c are arbitrary, d is determined by the equation

$$d = (1 + bc)/a$$

From $\mathbb{Z}_7$ there are six choices for a and seven for b and seven for c, a total of 6×7^2. If $a = 0$ then the choice of d is free while the choices for b and c are subject only to $bc = -1$. Thus if b is chosen to be nonzero then $c = -b^{-1}$. There are seven choices for d and six for b, giving 7×6 choices for a matrix of this type. In all there are $6 \times 7^2 + 6 \times 7 = 336$ elements in $SL(2,7)$.

11.5.2 Lemma.   The Center Z of SL(2, 7)

The center consists of those matrices such that $b = c = 0$ and $a = d$ and $a^2 = 1$. Exercise 11.5.1 asks for a proof of this fact. The equation $a^2 = 1$ has only solutions $a = 1$ and $a = -1$ in $\mathbb{Z}_7$, indeed in any field. Thus the center Z of $SL(2,7)$ is just $\{I, -I\}$ where I is the 2×2 identity matrix.

As an immediate corollary we compute the order of $PSL(2,7)$:

$$|PSL(2,7)| = |SL(2,7)|/|Z| = 336/2 = 168$$

Before we can begin our proof we need still one other fact about 2×2 matrices.

11.5.3 Definition.   Characteristic Polynomial of a Matrix

If A is a 2×2 matrix its *characteristic polynomial* $c_A(x)$ is defined to be the determinant $\det(A - Ix)$.

$$\text{If } A = \begin{bmatrix} a & b \\ c & d \end{bmatrix}$$

then

$$c_A(x) = \det\begin{bmatrix} a - x & b \\ c & d - x \end{bmatrix} = x^2 - (a + d)x + (ad - bc)$$

Notice that the constant term of $c_A(x)$ is $\det(A)$. Hence if $A \in SL(2, 7)$ then this constant term is 1. Also, the coefficient of the x term is the *negative* of the sum of the diagonal elements of the matrix. This sum is called the trace of the matrix. What is important for us is that conjugate elements in $SL(2, 7)$ have the same characteristic equation.

11.5.4 Lemma. Conjugate Elements Have Equal Characteristic Equations

Proof. This is so because a conjugate of A has the form SAS^{-1} and

$$SAS^{-1} - xI = SAS^{-1} - x\,SIS^{-1} = S(A - xI)S^{-1}$$

and so

$$\det(SAS^{-1} - xI) = \det S \det(A - xI)\det S^{-1} = \det(A - xI)$$

11.5.5 Theorem. PSL(2, 7) Is a Simple Group of Order 168

Strategy. The demonstration presented here that $PSL(2, 7)$ is simple is based on the fact that if an element belongs to a normal subgroup so do all the elements conjugate to it in the group. Thus a normal subgroup is the set union of certain conjugacy classes of elements in the group. So the first thing is to determine the size of the conjugacy classes in $PSL(2, 7)$. Then we see that no union of these classes can form a subgroup because the sum of the elements in any union of classes will not divide 168, unless of course the union is just the identity or all the conjugacy classes. This last part is not difficult.

Proof. First we shall determine the conjugacy classes in $SL(2, 7)$. Then we see how these classes coalesce when Z is factored out. Note that each element in the center Z is in a conjugacy class by itself. Because of this fact we can begin our classification of conjugacy classes by considering matrices with different characteristic polynomials. This tactic is enhanced with the observation that the matrix

$$A_t = \begin{bmatrix} 0 & 1 \\ -1 & t \end{bmatrix}$$

has the characteristic polynomial $x^2 - tx + 1$. (In the theory of matrices, the matrix A_t is called the companion matrix to the polynomial $x^2 - tx + 1$.)

There are seven possible values for t, and we next determine the size of the conjugacy classes of these seven matrices. From Lemma 11.1.3 the number of conjugates to an element g in a group is the index of the centralizer of g in the group. Thus we need to calculate the size of the centralizer of each of the matrices A_t in $SL(2,7)$. To this end let

$$S = \begin{bmatrix} a & b \\ c & d \end{bmatrix} \quad \text{where} \quad ad - bc = 1$$

and suppose that $SA_t = A_t S$. Compute

$$SA_t = \begin{bmatrix} a & b \\ c & d \end{bmatrix} \begin{bmatrix} 0 & 1 \\ -1 & t \end{bmatrix} = \begin{bmatrix} -b & a + bt \\ -d & c + dt \end{bmatrix}$$

$$A_t S = \begin{bmatrix} 0 & 1 \\ -1 & t \end{bmatrix} \begin{bmatrix} a & b \\ c & d \end{bmatrix} = \begin{bmatrix} c & d \\ -a + tc & -bt + td \end{bmatrix}$$

By equating entries we find that $c = -b$ and that $d = a + bt$. We may choose a and b freely as long as $ad - bc = 1$. Substituting the results of the preceding equalities we have

$$1 = a^2 + abt + b^2$$

Now for a fixed t and b the resulting quadratic in a is

$$a^2 + bta + (b^2 - 1)$$

which has two, one, or no solution in $\mathbb{Z}_7$ according as the discriminant

$$b^2 t^2 - 4(b^2 - 1) = b^2(t^2 - 4) + 4$$

has two, one, or no square roots in $\mathbb{Z}_7$. Thus for a choice of t we can try all $b \in \mathbb{Z}_7$ and count the total numbers of solutions. This number will be the order of the centralizer of A_t in $SL(2,7)$. The following table records the results of this step. The table is compressed because the results are the same for $\pm t$ and $\pm b$. It is helpful to compute the squares in $\mathbb{Z}_7$:

$$0^2 = 0, (\pm 1)^2 = 1, (\pm 2)^2 = 4, (\pm 3)^2 = 2$$

Thus 0 has one square root; 1, 4, and 2 each have two square roots; -1, -2, and 3 have no square roots in $\mathbb{Z}_7$. A typical entry in the following table is described more completely in the paragraph following the table.

THE VALUES OF $b^2(t^2 - 4) + 4$ AND THE INDEX OF THE CENTRALIZER OF A_t IN $SL(2,7)$

t	Matrix	$b = 0$	$b = \pm 1$	$b = \pm 2$	$b = \pm 3$	Number of solutions	Index of centralizer
0	A_0	4	0	2	3	8	42
± 1	A_1, A_{-1}	4	1	-1	-2	6	56
± 2	A_2, A_{-2}	4	4	4	4	14	24
± 3	A_3, A_{-3}	4	2	3	0	8	42

Here, for example, are the details of the computations for the last line. Here $t^2 = (\pm 3)^2 = 2$ in $\mathbb{Z}_7$ so that the discriminant is

$$b^2(2 - 4) + 4 = -2b^2 + 4$$

For $b = 0, \pm 1, \pm 2, \pm 3$, compute $b^2 = 0, 1, 4, 2$ and determine that the value of the discriminant $-2b^2 + 4$ is 4, 2, 3, 0, respectively. Thus in the column headed $b = 0$ the entry 4 shows that the discriminant is 4. This means that the quadratic in a has two solutions in $\mathbb{Z}_7$ for a and hence when $b = 0$, there are two matrices centralizing A_3 (there are also two other matrices centralizing

A_{-3}). Similarly the entry 2 in the column headed $b = \pm 1$ means that the discriminant $-2b^2 + 4 = 2$. This means that for each of $b = 1$ and $b = -1$ there are two solutions for the quadratic in a, and hence two matrices centralizing A_3, a total of four such matrices. (There are also four other matrices centralizing A_{-3}.) The entry of 3 in the column headed $b = \pm 2$ means that the quadratic has no solutions. Finally, in the column headed $b = \pm 3$ the entry 0 means that there is just one solution for a and hence just one matrix for each choice of b, $b = 3$ or $b = -3$ that centralizes A_3. Thus when $t = \pm 3$, there are a total $2 + 4 + 2 = 8$ matrices centralizing A_3. Similarly there are 8 matrices centralizing A_{-3}. The size of the conjugate class containing A_3 or A_{-3} is then $336/8 = 42$.

Up to this point we have found conjugacy classes of sizes

$$1, 1, 42, 56, 24, 24, 42, \text{ and } 42$$

Thus the total number of elements found is 288. Forty-eight elements are missing! It turns out that they are the conjugates to the matrices A_2^{-1} and A_{-2}^{-1}. We compute A_2^{-1}:

$$A_2^{-1} = \begin{bmatrix} 0 & 1 \\ -1 & 2 \end{bmatrix}^{-1} = \begin{bmatrix} 2 & -1 \\ 1 & 0 \end{bmatrix}$$

whose characteristic polynomial is $x^2 - 2x + 1$. Thus the only possible matrix listed to which A_2^{-1} could be conjugate is A_2 itself. However, A_2^{-1} is not conjugate to A_2 in $SL(2,7)$ as we now verify: Suppose

$$\begin{bmatrix} a & b \\ c & d \end{bmatrix}\begin{bmatrix} 0 & 1 \\ -1 & 2 \end{bmatrix} = \begin{bmatrix} 2 & -1 \\ 1 & 0 \end{bmatrix}\begin{bmatrix} a & b \\ c & d \end{bmatrix}$$

then $d = -a$ and $c = 2a + b$. However, the condition $ad - bc = 1$ requires that

$$1 = -a^2 - 2ab - b^2 = -(a + b)^2$$

but since -1 is not a square in $\mathbb{Z}_7$ there can be no solution. A similar computation shows that A_{-2}^{-1} is not conjugate to A_{-2}.

Next the centralizer of A_2^{-1} is easily computed by the techniques we have already used. These computations show the order of the centralizer to be 14 and so the size of its conjugacy class is 24. And the same result holds for A_{-2}^{-1}. We omit these details. Hence we obtain the following list of 11 class representatives and the number of elements in the corresponding conjugacy class.

Conjugacy classes of $SL(2,7)$			
Conjugacy class representative	Size of class	Conjugacy class representative	Size of class
I	1	$-I$	1
A_0	42		
A_1	56	A_{-1}	56
A_2	24	A_2^{-1}	24
A_{-2}	24	A_{-2}^{-1}	24
A_3	42	A_{-3}	42

Now we must see what happens to these conjugate classes under the homomorphism

$$SL(2,7) \to PSL(2,7) = SL(2,7)/Z$$

when the center Z is factored out. As usual if M is a matrix in $SL(2,7)$ we denote its image in $PSL(2,7)$ by $\bar{M}$.

Since $-I$ is in the center, the elements A and $-A$ have the same image in $PSL(2,7)$. This means that their conjugate classes in $SL(2,7)$ coalesce in $PSL(2,7)$. Furthermore, if $t = 0, 1$, or 3 then $-A_t$ is conjugate at A_{-t} in $SL(2,7)$, as we now pause to verify. Please check the computation

$$S_t(-A_t)S_t^{-1} = A_{-t}$$

where

$$S_0 = \begin{bmatrix} 2 & 3 \\ 3 & -2 \end{bmatrix} \quad S_1 = \begin{bmatrix} 3 & 3 \\ 3 & 1 \end{bmatrix} \quad \text{and} \quad S_3 = \begin{bmatrix} 2 & 2 \\ 2 & -1 \end{bmatrix}$$

Also, $-A_2$ is conjugate to $A_{-\frac{1}{2}}$ in $SL(2,7)$ because

$$\begin{bmatrix} 1 & -2 \\ 0 & 1 \end{bmatrix}(-A_2)\begin{bmatrix} 1 & 2 \\ 0 & 1 \end{bmatrix} = A_{-\frac{1}{2}}$$

Thus a tentative list of conjugate class representative in $PSL(2,7)$ is

$$I, \bar{A}_0, \bar{A}_1, \bar{A}_2, \bar{A}_{-2}, \bar{A}_3$$

What are the sizes of these classes? We must check the size of the centralizer of each of these in $PSL(2,7)$. Now $\bar{S}$ belongs to the centralizer of $\bar{A}$ in $PSL(2,7)$ if and only if

$$\bar{S}\bar{A}\bar{S}^{-1} = \bar{A}$$

and this holds if and only if SAS^{-1} is in the coset AZ of Z in $SL(2,7)$, and this holds if and only if

$$SAS^{-1} = A \text{ or } -A \text{ in } SL(2,7)$$

However, if $t \neq 0$, the characteristic polynomial for A_t is distinct from that for $-A_t$ and so, if $t \neq 0$,

$$\bar{S}\bar{A}\bar{S}^{-1} = \bar{A} \text{ in } PSL(2,7) \qquad \text{if and only if } SAS^{-1} = A \text{ in } SL(2,7)$$

This means that, if $t \neq 0$, the centralizer of $\bar{A}_t$ is the image of the centralizer of A_t. These centralizers of course contain Z and so the index of one of these centralizers in PSL(2,7) is the same as the index of the corresponding centralizer in $SL(2,7)$.

The situation for A_0 is a little different because there are matrices S in $SL(2,7)$ such that $SA_0S^{-1} = -A_0$. We count these solutions as before. Suppose

$$\begin{bmatrix} a & b \\ c & d \end{bmatrix}\begin{bmatrix} 0 & 1 \\ -1 & 0 \end{bmatrix} = \begin{bmatrix} 0 & -1 \\ 1 & 0 \end{bmatrix}\begin{bmatrix} a & b \\ c & d \end{bmatrix}$$

then $c = b$, $d = -a$ and $a^2 + b^2 = -1$. This last equation has solutions if and only if $b^2 = -3$ or $b^2 = 2$. For each of the four choices for b there are two values possible for a. Hence there are eight such matrices S.

Now the subgroup of $SL(2,7)$ mapped to the centralizer of $\overline{A}_0$ is

$$C_0 = \left\{ S: SA_0 S^{-1} = A_0 \quad \text{or} \quad SA_0 S^{-1} = -A_0 \right\}$$

and has order 16. (Remember we had already calculated the order of the centralizer of A_0 in $SL(2,7)$ to be 8 and we have just added 8 more matrices to the group C_0.) This means that the index of the $\overline{C}_0$ in $PSL(2,7)$ is 21.

Thus the table of the six conjugacy classes and their sizes in $PSL(2,7)$ is

Conjugacy class representative Size of class	I	$\overline{A}_0$	$\overline{A}_1$	$\overline{A}_2$	$\overline{A}_2^{-1}$	$\overline{A}_3$
	1	21	56	24	24	42

Now to complete the argument that $PSL(2,7)$ has no proper normal subgroup. Suppose that $\overline{N}$ were one. As we foretold, the argument proceeds by noting that $\overline{N}$ must be the disjoint union of conjugacy classes and finishes by showing that no sum of these classes can divide 168 unless the sum is 1 or 168.

The order of $\overline{N}$ must divide 168 and can be at most 84. On the other hand, $\overline{N}$ must contain I and at least one other conjugacy class. Hence the order of $\overline{N}$ satisfies $22 \le |\overline{N}| \le 84$.

Now the divisors of 168 between 22 and 84 are 24, 28, 42, 56, and 84. Since these numbers are all even $|\overline{N}|$ must be even. Moreover since I is in $\overline{N}$, it follows that $\overline{A}_0$ would have to be in $\overline{N}$, or else $|\overline{N}|$ would be odd.

On the other hand, suppose $\overline{A}_2$ is in $\overline{N}$. Then so is $\overline{A}_2^{-1}$ and so $\overline{N}$ has at least 70 ($= 22 + 48$) elements. Since 70 does not divide 168 we see that $\overline{N}$ must contain other classes as well. But then the order of $\overline{N}$ would exceed 84. Thus we can conclude that $\overline{N}$ does not contain either class of size 24.

Thus the number of elements in $\overline{N}$ has the form

$$22 + 42a + 48b + 56c$$

where a, b, and c are 0 or 1. It is easy to see that no such sum can equal 24, 28, 42, 56, or 84. This completes our proof.

EXERCISE SET 11.5

11.5.1 Show that the center of $GL(2,7)$ consists of all those matrices of the form

$$\begin{bmatrix} t & 0 \\ 0 & t \end{bmatrix} \quad \text{where } t \ne 0$$

Hint: Show that if $\begin{bmatrix} a & b \\ c & d \end{bmatrix}$ commutes with $\begin{bmatrix} 1 & 1 \\ 0 & 1 \end{bmatrix}$ severe restrictions are placed on a, b, c, and d.

11.5.2 Determine the center of $SL(2,7)$.

11.5.3 Determine the center of $SL(n,q)$.

11.5.4 Determine the groups $PSL(2,2)$ and $PSL(2,3)$.

11.5.5 Determine the orders of the seven matrices A_t in $SL(2,7)$ and $\bar{A}_t$ in $PSL(2,7)$. *Hint*: To avoid excessive matrix calculation prove and use the fact that a 2×2 matrix "satisfies its own characteristic polynomial"; that is, if $c(x) = x - tx + d$ is the characteristic polynomial for a matrix A then

$$A^2 - tA + dI = 0 \,(\text{the zero matrix})$$

Hence $A^2 = tA - dI$ and $A^3 = tA^2 - dA = (t^2 - d)A - dtI$, and so on.

11.5.6 Determine the Sylow subgroups of $PSL(2,7)$.

Chapter **12**

Topics in Fields

In this chapter we consider several applications of the basic theory of field extensions that we have already studied. The first sections take up some traditional topics that may seem disjoint at first glance, but it is the notion of "roots of unity" that binds them together. The section on the gaussian integers gives us still another example of a euclidean domain but one with interesting number theoretic applications. The final section on simple extensions and the "all or nothing" aspect of splitting fields are important for our development of Galois theory. The prerequite for this Chapter is chapter 10, except for Section 12.3, which requires only Section 9.3, and Theorem 12.2.7, which needs Section 11.1.

12.1 COMPASS AND STRAIGHTEDGE CONSTRUCTIONS

In this section we take a look at the classical "impossibility" results. We show that using only a compass and a straightedge it is impossible to trisect every angle or to double a cube. Later, after we have discussed more about the roots of unity, we can determine which regular polygons can be constructed by these means. These are the tools of the Greek geometers and these are problems they considered important.

Geometrically speaking we are concerned with points, lines, and circles drawn (constructed) on paper (in a plane) by means of a straightedge and a compass. We do not say "ruler" for that implies that we could measure distances or locate points of distance r, where r is any positive real number. And that is a concept the Greeks found either difficult or impossible.

The first hurdle we face is to formulate in an algebraic way the notions we hold intuitively about drawing lines and circles on paper. To do that we use as our plane the plane of complex numbers. A point will be a complex number, and even though we may use the geometric term "point" we always mean a complex number. Correspondingly, a line and a circle are defined in terms of complex numbers. Once this is done we find what it means algebraically to locate a point (complex number) as the intersection of two lines or of a line and a circle or two circles. To give away the answer, we find that starting with just the points 0 and 1, if we restrict ourselves to constructing new points at the intersections described, the points we obtain must lie in a field extension of $\mathbb{Q}$ whose degree over $\mathbb{Q}$ is a power of 2. Then the impossibility results follow by showing that, for example, if we could construct the cube root of 2 (necessary for constructing the edge of a cube whose volume is 2cubic units) we would have to find a point ω that is a zero of the polynomial $x^3 - 2$. Since this polynomial is irreducible over $\mathbb{Q}$, ω would have to belong to an extension of degree 3 over $\mathbb{Q}$. Since 3 does not divide a power of 2, ω is not constructible.

So let us begin to build this machinery. Remember that for complex numbers the norm of $\alpha = a + bi$ is denoted and defined

$$\|\alpha\| = \|a + bi\| = a^2 + b^2$$

The distance between two complex numbers α and β is denoted and defined by

$$|\alpha, \beta| = \sqrt{\|\alpha - \beta\|}$$

We write $|\alpha|$ for $|\alpha, 0|$. We also need to recall a little of the geometry of complex numbers. Each nonzero complex number has an argument, the least positive angle the line segment from 0 to α makes with the positive real axis. Two useful facts are that the argument of a product of complex numbers is the sum of the arguments of the numbers (mod 360°) and the norm of a product is the product of the norms of the numbers.

12.1.1 Definition. Points, Lines, Circles, and Constructible Numbers

A *point* is a complex number. A line through points α and β, $\alpha \neq \beta$ is denoted and defined

$$[\alpha, \beta] = \{\alpha + r(\beta - \alpha): r \in \mathbb{R}\}$$

A *circle* with center at α and radius $r \in \mathbb{R}$ ($r > 0$) is denoted and defined

$$C(\alpha, r) = \{z: \|z - \alpha\| = r^2\}$$

Now let $\mathcal{M}$ be a set of points, lines, or circles in the complex plane. A point, line, or circle S is said to be *1-step constructible* from $\mathcal{M}$ if

1. S is a point that is the intersection of two lines in $\mathcal{M}$, or the intersection of a line and a circle both in $\mathcal{M}$, or an intersection of two circles in $\mathcal{M}$.
2. S is a line $[\alpha, \beta]$ where $\alpha, \beta \in \mathcal{M}$.
3. S is a circle $C(\alpha, r)$ where $\alpha \in \mathcal{M}$, and r is the distance between two points in $\mathcal{M}$.

A point, line, or circle S is said to be *constructed* from $\mathcal{M}$ if there exists a finite sequence $S_0, S_1, \ldots, S_m = S$ such that S_0 is a 1-step construction from $\mathcal{M}$ and for $k > 0$, S_k is a 1-step construction from

$$\mathcal{M} \cup \{S_0, S_1, \ldots, S_{k-1}\}$$

To get things started, here are two elementary geometric constructions that are constructible in the sense of the definition above.

12.1.2 Lemma. Two Familiar Constructions

1. If α and β are distinct points then the line through α perpendicular to the line $[\alpha, \beta]$ is constructible from $\{\alpha, \beta\}$.
2. If α and β are distinct points and α' is a point not on $[\alpha, \beta]$ then the line through α' parallel to $[\alpha, \beta]$ is constructible from $\{\alpha, \beta, \alpha'\}$.

Proof of 1. We use a typical geometric construction and describe the sequence of 1-step constructions beginning with $\mathcal{M} = \{\alpha, \beta\}$. Construct $C(\alpha, |\alpha, \beta|)$ and let β' be the other point of intersection of this circle with the line $[\alpha, \beta]$. Then construct the circles $C(\beta, |\beta, \beta'|)$ and $C(\beta', |\beta, \beta'|)$. Let β'' be a point of intersection of these two circles. Finally then we may construct $[\alpha, \beta'']$.

Proof of 2. Construct the line $[\alpha, \alpha']$ and the circle $C(\alpha, |\alpha, \beta|)$. Choose γ to be the intersection of that line and circle so that $\gamma - \alpha$ is a positive real multiple of $\alpha' - \alpha$. Construct the circle $C(\alpha', |\alpha, \beta|)$ and construct γ' and the intersection of that circle and $[\alpha, \alpha']$, choosing the intersection so that $\gamma' - \alpha' = \gamma - \alpha$. Next construct $C(\gamma', |\gamma, \beta|)$ and select β' to be the intersection of this circle and $C(\alpha', |\alpha, \beta|)$ so that $\beta' - \alpha' = \beta - \alpha$. Construct the line $[\alpha', \beta']$. That's it!

12.1.3 Lemma. Rational Numbers Are Constructible from $\{0, 1\}$

If $a, b \in \mathbb{Q}$ then $z = a + bi$ is constructible from $\{0, 1\}$.

Proof. We sketch the argument and leave the details as an exercise. First prove by induction that the complex number $n = n + 0i$, $n \in \mathbb{Z}$ is constructible from $\{0, 1\}$. Second, use Lemma 12.1.2 to show that the imaginary axes and then the points mi, $m \in \mathbb{Z}$ are constructible from $\{0, 1\}$. Third, prove that any rational point r, $r \in \mathbb{Q}$ can be constructed from $\{0, 1\}$ and hence that any complex number $z = r + si$, where $r, s \in \mathbb{Q}$ is constructible from $\{0, 1\}$. It may help to do the following: Use Lemma 12.1.2 to construct a line from each point ki, $k = 1, 2, \ldots, n$ parallel to the line joining ni and 1. These lines intersect the real axis at the point k/n. Now complete the proof.

12.1.4 Theorem.   Field of Constructible Numbers

Let $\mathcal{M}$ be any set of complex numbers containing $\{0, 1\}$. The set of numbers constructible from $\mathcal{M}$ is a field. Moreover, if α is constructible from $\mathcal{M}$, so is its complex conjugate $\bar{\alpha}$ and its real and imaginary parts.

Proof.   In this proof please read "constructible from $\mathcal{M}$" for the shorter "constructible." Since we speak of subsets of $\mathbb{C}$, the complex numbers, we have only to show that the set of constructible numbers is closed under subtraction, multiplication, and multiplicative inversion. We do this by standard geometrical descriptions and omit the details of every stage of the 1-step constructions.

If α and β are constructible, then $\alpha - \beta$ is parallel to a diagonal of the parallelogram with vertices 0, α, $\alpha + \beta$, β. To construct the number $\alpha\beta$ recall that the argument of $\alpha\beta$, that is, the angle the line $[\alpha\beta, 0]$ makes with the positive real axis, is the sum of the arguments of α and β. Thus to construct a line on which $\alpha\beta$ lies it suffices to copy the argument of β on the line $[\alpha, 0]$ at 0. Then we have to locate $\alpha\beta$ on that line at a distance of $|\alpha\beta, 0|$. Since $\|\alpha\beta\| = \|\alpha\| \cdot \|\beta\|$ it suffices to show how to construct the product of two positive real numbers r and s and to take the square root of a positive real number and to construct the reciprocal of a positive number. These constructions are shown in Figures 12.1, 12.2, and 12.3.

To construct α^{-1} construct $C(0, 1/|\alpha|)$ and then find the intersection of this circle and the line whose angle with the positive real axis is the opposite of the argument of α. The complex conjugate $\bar{\alpha}$ is equal to $\|\alpha\|\alpha^{-1}$ and is thus on the line $[\alpha^{-1}, 0]$ at a distance of $|\alpha, 0|$ from the origin in the direction of α^{-1}.

Finally, the real part of α equals $(\alpha + \bar{\alpha})/2$ and the imaginary part of α equals $(\alpha - \bar{\alpha})/2$.

Now suppose we start with $\mathcal{M} = \mathbb{Q}$. We obtain a field of constructible numbers from $\mathbb{Q}$. Theorem 12.1.6 characterizes in algebraic terms the elements in this field. The crux of the matter is the following lemma.

12.1.5 Lemma.   Constructions and Subfields

If a subfield E of the complex numbers contains a set $\mathcal{M}$ and ω is 1-step constructible from $\mathcal{M}$ then either $\omega \in E$ or there exists $\delta \in E$ such that $\omega \in E(\sqrt{\delta})$.

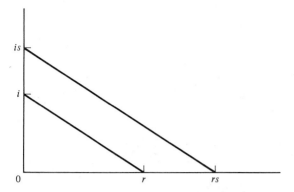

Figure 12.1 To construct the real number rs, construct the line parallel to the line $[i, r]$ through the number on the imaginary axis, is. The intersection of this line with the real axis occurs at the real number rs.

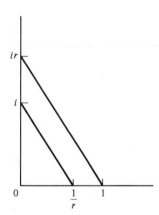

Figure 12.2 To construct $1/r$, the reciprocal of the real number r, construct the line parallel to $[ir, 1]$ through i. The intersection of this line with the real axis occurs at the real number $1/r$.

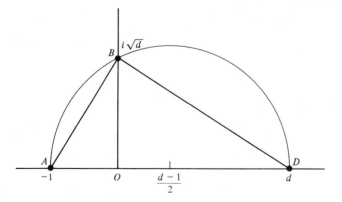

Figure 12.3 To construct a segment of length $\sqrt{d}$ we construct the number $i\sqrt{d}$. This number is the intersection of the imaginary axis and circle $C\{(d - 1)/2, (d + 1)/2\}$ shown as point B. Lines AB and BD are perpendicular and triangles OAB and OBD are similar.

Proof.   We simply have to examine how a point arises as a 1-step construction from $\mathcal{M}$. First, if ω is the intersection of two lines in $\mathcal{M}$ then we claim $\omega \in E$. For suppose that ω is on the line $[\alpha, \beta]$ and on the line $[\gamma, \delta]$. This means that there are real numbers r and s such that

$$\omega = \alpha + r(\beta - \alpha) = \gamma + s(\delta - \gamma)$$

By equating real and imaginary parts we have two equations in the unknowns r and s that have a unique solution since the lines are not parallel. Does this solution lie in E? Yes, because by assumption $\alpha, \beta, \gamma, \delta$ are in E, since their real and imaginary parts are also in E, and in solving linear equations only field operations are involved.

Second, suppose that ω is an intersection of a line $[\alpha, \beta]$ and a circle $C(\gamma, r)$ each in $\mathcal{M}$. Thus

$$\omega = \alpha + x(\beta - \alpha) \quad \text{and} \quad \|\omega - \gamma\| = r^2$$

where α, β, γ, and r^2 are members of E. The issue is whether these conditions force the real number x to belong either to E or to a quadratic extension of it. Substitute the expression for ω into $\|\omega - \gamma\| = r^2$; then let $\lambda = \alpha - \gamma$ and $\mu = \beta - \alpha$, so that we get the equation in x

$$\|\lambda + x\mu\| = r^2$$

which converts to

$$\mu\bar{\mu}x^2 + (\mu\bar{\lambda} + \mu\lambda)x + \lambda\bar{\lambda} - r^2 = 0$$

Thus the solution x is the zero of a quadratic polynomial whose coefficients belong to E and thus either $x \in E$ or $x \in E(\sqrt{d})$ where d is the discriminant of a quadratic polynomial.

Finally, suppose that ω is an intersection of two circles $C(\alpha, r)$ and $C(\beta, s)$. We want to show that the conditions

$$\|\omega - \alpha\| = r^2 \quad \text{and} \quad \|\omega - \beta\| = s^2$$

imply that ω belongs either to E or to a quadratic extension of E. Computations are simpler if we first translate one of the circles so that its center is the origin. Let $\gamma = \omega - \alpha$ and $\mu = \beta - \alpha$. Then the conditions become

$$\|\gamma\| = r^2 \quad \text{and} \quad \|\gamma - \mu\| = s^2$$

Expand the second of these equalities and substitute in the first relation to obtain

$$r^2 - \gamma\bar{\mu} - \bar{\gamma}\mu + \mu\bar{\mu} = s^2 \tag{1}$$

This implies that the real part of $\gamma\bar{\mu}$ is in E. Now set $\gamma = x + yi$ and $\mu = a + bi$ where x, y are real and a and b are in E because $\mu \in E$. The relation (1) yields

$$ax - by = c$$

where a, b, and c are all in the field E. (Note that if $a = b = c$ then the two circles are concentric and no solution exists.)

Thus we can solve for x in terms of y (if $a \neq 0$) or y in terms of x (if $b \neq 0$) and then substitute in the condition $\|\gamma\| = x^2 + y^2 = r^2$ to find that y (or x, depending on which substitution was made) is the zero of a quadratic polynomial. Determining the x (or y, depending on the substitution) gives us the result: Either $\gamma \in E$ or γ belongs to a quadratic extension of E. Hence the same is true for $\omega = \lambda + \alpha$.

By iterating this lemma we have the following theorem.

12.1.6 Theorem.   Necessary Condition for Constructibility

If ω is constructible from the rational numbers $\mathbb{Q}$ then there is a chain of fields:

$$\mathbb{Q} = E_0 \subset E_1 \subset \cdots \subset E_k$$

such that $\omega \in E_k$ and each $E_{i+1} = E_i(\sqrt{d_i})$ is an extension of degree 2 over E_i.

12.1.7 Corollary.   If ω Is Constructible from $\mathbb{Q}$ then $[\mathbb{Q}(\omega): \mathbb{Q}]$ Is a Power of 2

If ω is constructible from $\mathbb{Q}$ then ω is algebraic over $\mathbb{Q}$ and the degree of the monic irreducible polynomial in $\mathbb{Q}[x]$ of which ω is a zero is a power of 2.

Proof.   From the theorem we know that ω belongs to the top field E of a finite chain of fields each of which has degree 2 over the preceding one. Thus $[E : \mathbb{Q}]$ is a power of 2 by Theorem 10.2.14. By the same theorem the degree $[K : \mathbb{Q}]$ of any subfield $E \supseteq K \supseteq \mathbb{Q}$ must divide $[E : \mathbb{Q}]$ and so be a power of 2 also. But $\mathbb{Q}(\omega)$ is just such a subfield. And we know that $\mathbb{Q}(\omega) \cong \mathbb{Q}[x]/\langle p(x)\rangle$ where $p(x)$ is an irreducible polynomial of which ω is a zero and furthermore $\deg(p(x)) = [\mathbb{Q}(\omega): \mathbb{Q}]$, which is a power of 2.

It is not so surprising that the converse of this theorem is also true.

12.1.8 Theorem.   Sufficient Condition for Constructibility from $\{0, 1\}$

If there is a chain of fields $\mathbb{Q} = E_0 \subset E_1 \subset E_2 \subset \ldots \subset E_n$ where each extension has degree 2 over the preceding field then every member of E_n is constructible from $\{0, 1\}$.

Proof.   We know that the members of $\mathbb{Q}$ can be constructed from $\{0, 1\}$. We need only proceed up the chain one extension at a time. Thus suppose that all the members of E_{k-1} are constructible from $\mathbb{Q}$. Since the degree of E_k over E_{k-1} is 2 we may suppose (Theorem 10.2.9) that

$$E_k = E_{k-1}(\sqrt{\delta})$$

where $\delta \in E_{k-1}$ and is constructible from $\{0, 1\}$. Any member of E_k has the form $\alpha + \beta\sqrt{\delta}$ where α and β are members of E_{k-1}. Thus it suffices to show that $\sqrt{\delta}$ is constructible from E_{k-1}.

Remember that a square root of a complex number is the number whose argument is one-half that of the argument of δ and whose absolute value is $\sqrt{d}$ where $d = \|\delta\|$. By familiar compass and straightedge constructions (see Exercise 12.1.2) we can construct the angle bisector of the angle formed by line $[0, \delta]$ and the real axis. Thus it only remains to show that we can construct $\sqrt{d}$. This is indicated in Figure 12.3.

Now we can prove some of the classical impossibility results.

12.1.9 Theorem.   Cube Root of 2 and Angle Trisection

Not every angle can be trisected using only compass and straightedge.

Proof.   First we show that $c = \sqrt[3]{2}$ cannot be constructed from $\mathbb{Q}$. This is simply the result that c is a zero of the polynomial $x^3 - 2$ that is irreducible over $\mathbb{Q}$. Hence $[\mathbb{Q}(c): \mathbb{Q}] = 3$, and not a power of 2 as required by Corollary 12.1.7.

Next, we shall show that the trisector of a 60° angle cannot be constructed from $\mathbb{Q}$. First we construct ω, a sixth root of unity.

$$\omega = (1 + \sqrt{3}\,i)/2$$

We calculate that $\omega^6 = 1$ because the argument of ω is 60° and $\|\omega\| = 1$; hence the argument of ω^6 is $6 \times 60° = 360°$ and $\|\omega^6\| = \|\omega\|^6 = 1$. (Thus if we construct the points 1, ω, ω^2, ω^3, ω^4, and ω^5 we have constructed a regular hexagon.)

If we were able to construct the trisector of the angle formed by the real axis and the line $[0, \omega]$ then we would be able to construct the intersection of this trisector and the unit circle. Call this number ζ. Its argument is 20°. (Hence, in fact, $\zeta^{18} = 1$. And it is easy to show that 18 is the order of ζ in the multiplicative group of $\mathbb{C}$.) If we can construct ζ then we can construct its real part, which is $\cos 20°$. However, the trigonometric identity

$$\cos(3t) = 4\cos^3(t) - 3\cos(t)$$

with $t = 20°$ shows that $\cos 20°$ is a zero of

$$4x^3 - 3x - \frac{1}{2}$$

Clean this up a bit by substituting $x = y/2$. Then compute that $2\cos 20°$ is a zero of

$$y^3 - 3y - 1$$

Thus, if $\cos 20°$ were constructible from $\{0, 1\}$ then a zero of this polynomial would also be. But this polynomial is irreducible over $\mathbb{Q}$, and so a zero of it would have to belong to an extension of $\mathbb{Q}$ that is divisible by 3, contrary to the necessary condition of Corollary 12.1.7.

There is also a classical interest in determining those integers n for which there exists a construction from $\{0, 1\}$ for a regular polygon of n sides. We have just seen that we can do this for a regular hexagon. Exercise 12.1.5 shows that we can do it for $n = 5$ as well. The more general result requires that we know a little more about the nth roots of unity. This is so because it follows, just as it did in the argument about the trisection of 60°, that we can construct a regular polygon with n sides if and only if we can construct the nth root of unity,

$$\zeta = \cos\left(\frac{360°}{n}\right) + i\sin\left(\frac{360°}{n}\right)$$

EXERCISE SET 12.1

12.1.1 Complete the details of the proof of Lemma 12.1.3.

12.1.2 If α and β are constructible from $\{0, 1\}$ show that an angle bisector of the angle between the lines $[0, \alpha]$ and $[0, \beta]$ is constructible.

12.1.3 If a regular polygon can be constructed from $\{0, 1\}$ show that its center can be constructed from $\{0, 1\}$.

12.1.4 Show that if the edge of a cube whose volume is k^3 can be constructed from $\{0, 1\}$ then the edge of a cube whose volume is $2k^3$ cannot be constructed from $\{0, 1\}$.

12.1.5 Show that a regular pentagon can be constructed from $\{0, 1\}$. It may be helpful to proceed as follows: Let $\zeta = \cos 72° + i\sin 72°$. Show ζ is a fifth root of unity. The pentagon to be constructed has vertices 1, ζ, ζ^2, ζ^3, and ζ^4. Show it suffices to construct the length of a diagonal δ of the pentagon. Let $\delta = 1 - \zeta^2$ and compute $\delta\bar{\delta} = 2 - (\zeta^2 + \zeta^3)$. To find $\lambda = \zeta^2 + \zeta^3$, show that ζ is a zero of $(x^5 - 1)/(x - 1)$ and hence that

$$-1 = \zeta + \zeta^2 + \zeta^3 + \zeta^4$$

Then show that λ is a zero of $x^2 + x - 1$. Complete the details and construct with compass and straightedge a regular pentagon!

12.1.6 Let ζ be the fifth root of unity constructed in the previous exercise. Find $a \in \mathbb{Q}$ and $d \in \mathbb{Q}(\sqrt{a})$ so that

$$\mathbb{Q} \subset \mathbb{Q}(\sqrt{a}) \subset \mathbb{Q}(\sqrt{a}, \sqrt{d}) = \mathbb{Q}(\zeta)$$

Express ζ in terms of the basis $\{1, \sqrt{a}, \sqrt{d}, \sqrt{a}\sqrt{d}\}$ for $\mathbb{Q}(\zeta)$ over $\mathbb{Q}$.

12.2 ROOTS OF UNITY AND CYCLOTOMIC EXTENSIONS

In this section we discuss roots of unity in arbitrary fields.

12.2.1 Definition.   Roots of Unity

Let F be any field. Let n be a positive integer. An nth root of unity is an element $\zeta \in F$ such that $\zeta^n = 1$ for some positive integer n. An algebraic extension $E \supseteq F$ is called a cyclotomic extension of F if there is a root of unity ζ such that $E = F(\zeta)$.

As we saw in the previous section, we work with roots of unity when we construct regular polygons. Also every nonzero element of the finite field $GF(p^m)$ is a $(p^m - 1)$th root of unity.

In this section we are concerned with nth roots of unity where n is not divisible by the characteristic of the field F. This restriction is automatically satisfied if we deal with fields of characteristic zero, and indeed that is our major application.

Thus the setup for this section is that n is a positive integer and F is a field whose characteristic does not divide n. We let E denote the splitting field of $x^n - 1$ over F.

The first properties of the roots of unity are listed in the next theorem. They are all easy consequences of theorems we have already established. Please note that for these general consequences we make no use of the representation of roots of unity as complex numbers. The most important result of the section is to determine the irreducible polynomial whose roots are the "primitive" nth roots of unity over $\mathbf{Q}$.

As applications of these results we shall be able to give necessary and sufficient conditions on n for a regular polygon on n sides to be constructed from $\{0, 1\}$ by compass and straightedge. We shall also be able to give Witt's proof of Wedderburn's theorem that every finite division ring is in fact a field. Finally we obtain the structure of the group of automorphisms of the splitting field of $x^n - 1$ over $\mathbf{Q}$.

12.2.2 Theorem.   The Group of Roots of Unity

If F is a field whose characteristic does not divide n then

1. The polynomial $x^n - 1$ has n distinct zeros in its splitting field over F.
2. In any field extension of F in which $x^n - 1$ splits, the zeros form a cyclic multiplicative group of order n.
3. If ζ is a generator of this group then the set of generators is $\{\zeta^m : \gcd(n, m) = 1\}$.
4. The splitting field $E = F(\zeta)$ when ζ is a generator of the group of roots of unity.

Proof.   Property 1 is a direct result of the derivative test. Since $\operatorname{char}(F) \nmid n$, the derivative is $nx_{n-1} \neq 0$ and so has no zeros in common with $x^n - 1$.

For property 2 verify that the product of two nth roots of unity is again an nth root of unity. Since there are but a finite number of nth roots of unity and this set is closed under multiplication it follows that, in E, they form a subgroup of the multiplicative group of E. Thus they are a cyclic group (recall Theorem 10.4.1) of order n.

For property 3 recall that the order of ζ^m in the cyclic group generated by ζ is $n/\gcd(n, m)$. The generators of $\langle \zeta \rangle$ are the elements of order n and thus are the elements ζ^m where $\gcd(n, m) = 1$.

For property 4 observe that once an extension contains ζ it contains all the zeros of $x^n - 1$. Thus the splitting field is $E = F(\zeta) = F(1, \zeta, \zeta^2, \ldots, \zeta^{n-1})$.

12.2.3 Definition. Primitive Root of Unity

A generator of the group of nth roots of unity is called a *primitive* nth root of unity.

12.2.4 Definition. The nth Cyclotomic Polynomial

The nth cyclotomic polynomial is

$$\Phi_n(x) = \prod_{\gcd(n,\, m)=1} (x - \zeta^m)$$

It turns out that the cyclotomic polynomial $\Phi_n(x)$ has integer coefficients. More precisely $\Phi_n(x)$ has integer coefficients if $\operatorname{char}(F) = 0$; more generally Φ_n has coefficients in the prime subfield of F. And we even obtain a reasonably nice expression for it.

12.2.5 Theorem. Determination of a Cyclotomic Polynomial

Let F be a field whose characteristic does not divide n.

1. $x^n - 1 = \prod_{d \mid n} \Phi_d(x)$.
2. The cyclotomic polynomial $\Phi_n(x)$ has coefficients in the prime subfield of F. If $\operatorname{char}(F) = 0$ then $\Phi_n(x) \in \mathbb{Z}[x]$.
3. In terms of the Möbius μ-function

$$\Phi_n(x) = \prod_{d \mid n} \left(x^{\frac{n}{d}} - 1 \right)^{\mu(d)}$$

Proof. Let ζ be a primitive nth root of unity. For property 1 note that every zero of $x^n - 1$ has some order, say d, in the cyclic group of roots of unity in $F(\zeta)$. This order must divide the order of the group n. By definition this zero will be a zero of $\Phi_d(x)$.

Moreover, since each zero $\Phi_d(x)$ has order d, a root of unity can be the zero of only one cyclotomic polynomial.

Remark. Property 1 gives us another verification of the basic result on the Euler φ-function. The degree of $x^n - 1$ is n. The degree of $\Phi_d(x)$ is $\varphi(d)$. Equating the degrees of the left-hand side and left-hand side of the equation in 1 gives

$$n = \sum_{d \mid n} \varphi(d)$$

Remark. Property 1 also gives a clue to a better form for $\Phi_n(x)$. We can express this equation as

$$\Phi_n(x) = \frac{x^n - 1}{\displaystyle\prod_{\substack{d \mid n \\ d \neq 1}} \Phi_d(x)}$$

Then we can replace each of the denominator terms $\Phi_d(x)$, where now $d < n$, by similar expressions. In fact all we need to do is to divide out from $x^n - 1$ the primitive dth roots of unity where $d \mid n$. This algorithm is carefully computed in the expression of property 3.

We prove property 2 by induction on the integers n. We have $\Phi_1(x) = x - 1$ and $\Phi_2(x) = x + 1$, so 2 holds for $n = 1$ and $n = 2$.

By the induction hypothesis

$$\theta(x) = \prod_{\substack{d \mid n \\ d \neq 1}} \Phi_d(x)$$

has integer coefficients and is monic since each of the factors of $\theta(x)$ is monic. Hence

$$x^n - 1 = \Phi_n(x)\theta(x) \tag{1}$$

It remains to show that this forces $\Phi_n(x)$ to have integer coefficients. In any event we may apply the division algorithm (Theorem 9.1.11) to divide $x^n - 1$ by $\theta(x)$ in $\mathbb{Z}[x]$. From it we obtain unique polynomials $q(x)$ and $r(x)$ in $\mathbb{Z}[x]$ such that

$$x^n - 1 = \theta(x)q(x) + r(x) \tag{2}$$

where $q(x)$ and $r(x) \in \mathbb{Z}[x]$ and either $r = 0$ or $\deg(r(x)) < \deg(\theta(x))$. This equation still holds in $F[x]$ because either $\mathbb{Z}$ is a subring of F (in case $\mathrm{char}(F) = 0$) or $\mathbb{Z}$ maps homomorphically onto the prime subfield of F (in case $\mathrm{char}(F) = p$). So both equations (1) and (2) hold in $F[x]$. But now the uniqueness aspect of the division algorithm in $F[x]$ requires that $r(x) = 0$ and $q(x) = \Phi_n(x)$.

The formula of 3 is derived by an argument similar to the one we gave in counting the number of monic irreducible polynomials over a finite field. Here are some of the details of the argument in the present case.

Let n have $p_1, \ldots, p_k$ as its distinct prime divisors. Excepting n itself, any divisor of n is a divisor of one or more of

$$\frac{n}{p_1}, \frac{n}{p_2}, \ldots, \frac{n}{p_k}$$

Any divisor of n that divides two of these numbers is a divisor of one or more of

$$\frac{n}{p_1 p_2}, \ldots, \frac{n}{p_i p_j}, \ldots, \frac{n}{p_{k-1} p_k}$$

and so on. Thus

$$x^n - 1 = \prod_{d \mid n} \Phi d(x) = \Phi_n(x) \frac{\prod_{i=1}^{k}\left(\prod_{d \mid \frac{n}{p_i}} \Phi_d(x)\right) \prod_{i<j<k}\left(\prod_{d \mid \frac{n}{p_i p_j p_k}} \Phi_d(x)\right)\cdots}{\prod_{i<j}\left(\prod_{d \mid \frac{n}{p_i p_j}} \Phi_d(x)\right)\cdots}$$

(3)

To verify this you need only check to see that, after cancellation of multiple appearances in the numerator and denominator, each factor $\Phi_d(x)$ appears once, and in the numerator!

Now a factor such as $\prod_{d \mid n/p_i} \Phi_d(x) = x^{n/p_i} - 1$. Make these replacements and solve equation (3) for $\Phi_n(x)$. Obtain

$$\Phi_n(x) = \frac{(x^n - 1)\left(\prod_{i<j}\left(x^{\frac{n}{p_i p_j}} - 1\right)\right)\cdots}{\prod_{i=1}^{k}\left(x^{\frac{n}{p_i}} - 1\right)\prod_{i<j<k}\left(x^{\frac{n}{p_i p_j p_k}} - 1\right)\cdots}$$

This formula becomes more attractive using the Möbius μ-function.

$$\Phi_n(x) = \prod_{d \mid n}\left(x^{\frac{n}{d}} - 1\right)^{\mu(d)}$$

12.2.6 Examples. Special Cases of $\Phi_n(x)$

1. If p is a prime

$$\Phi_p(x) = (x^p - 1)/(x - 1) = x^{p-1} + x^{p-2} + \cdots + 1$$

2. $\Phi_8(x) = (x^8 - 1)/(x^4 - 1) = x^4 + 1$ since $\mu(8) = \mu(4) = 0$. Alternatively, dividing by $(x^4 - 1)$ eliminates the fourth roots of unity and leaves a polynomial whose degree is $\varphi(8) = 4$.

3. $\Phi_{12}(x) = (x^{12} - 1)(x^2 - 1)/(x^4 - 1)(x^6 - 1) = x^4 - x^2 - 1$ by evaluating the appropriate μ-values. Alternatively we must eliminate the primitive second, third, and sixth roots of unity. Hence we divide $(x^{12} - 1)$ by $(x^6 - 1)$. Then we must eliminate the primitive fourth roots of unity. Hence we must divide $(x^{12} - 1)$ by $(x^4 - 1)/(x^2 - 1)$.

Next we use this information about cyclotomic polynomials and the class equation for groups (11.1.4) to give Witt's classic proof of the theorem of Wedderburn on finite division rings.

12.2.7 Theorem.   Wedderburn's Theorem on Finite Division Rings

A finite division ring is a field.

Proof.   Recall that a division ring satisfies all the axioms of a field except that commutativity of multiplication is not assumed. Wedderburn's theorem tells us that if we assume that the division ring is finite then commutativity of multiplication follows from the other field axioms.

Let D be a finite division ring and let Z be its center:

$$Z = \{ z \in D: zd = dz \text{ for all } d \in D \}$$

It is routine to verify that Z is closed under addition and multiplication and of course multiplication is commutative inside Z by definition. Thus Z is a subring of D and is a finite field in its own right. Therefore, the number of elements in Z is a power of a prime. Let $|Z| = q = p^n$.

By the way, since D is finite any subset of D closed under multiplication and addition will be a subring, in fact a subdivision ring of D. If R is a subring of D then its nonzero elements form a multiplicative subgroup of the multiplicative group of D. As usual we denote this subgroup by $R^\times$.

Now D can be regarded as a vector space over Z where vector addition is addition in D and "scalar" multiplication is just multiplication in D by elements of Z. Remember, no multiplication of vectors, elements of D, is required for a vector space. Let m be the dimension of D as a vector space over its center Z and we conclude that the number of elements of D is a power of q, say q^m.

The order of $D^\times$ is $q^m - 1$. The order of $Z^\times$ is $q - 1$. We want to show that $D = Z$, or equivalently that $m = 1$. Suppose to the contrary that $m > 1$. We derive a contradiction using properties of the mth cyclotomic polynomial, $\Phi_m(x)$.

For any $a \in D$ let $N(a) = \{ x \in D: ax = xa \}$ be the centralizer of a in D. Verify that $N(a)$ is a subring of D, $N(a) \supseteq Z$, and that $N(a) = D$ if and only if $a \in Z$. If $a \neq 0$ then $N(a)^\times$ is also the normalizer of $\{a\}$ in $D^\times$.

Also, $N(a)$ is a subvector space of D over Z and has q^r elements for some positive integer r which of course depends on a. Hence $N(a)^\times$ is a subgroup of $D^\times$. We write $|N(a)^\times| = q^{r(a)} - 1$. Since $|N(a)^\times|$ divides $|D^\times|$, $(q^{r(a)} - 1)|(q^m - 1)$. By the old divisibility lemma (10.4.5) it follows that $r(a)|m$. (And remember, $r(a) = m$ if and only if $N(a) = D$, if and only if $a \in Z$.)

Now we write the class equation for $D^\times$:

$$q^m - 1 = (q - 1) + \sum_{a \in S} \frac{(q^m - 1)}{|N(a)^\times|}$$

where $(q - 1)$ is the number of elements in $Z^\times$, the center of $D^\times$, and S is a set of conjugacy class representatives for elements in $D^\times$ that are not in $Z^\times$.

So

$$q^m - 1 = (q - 1) + \sum_{a \in S} \frac{(q^m - 1)}{q^{r(a)} - 1} \tag{4}$$

and $r(a) \mid m$.

Now we invoke the cyclotomic polynomial over $\mathbb{Q}$. Let ζ be a primitive mth root of unity.

$$\Phi_m(x) = \prod_{\gcd(k, m) = 1} (x - \zeta^k)$$

Also $\Phi_m(x)$ is relatively prime to $x^r - 1$ if $r \mid m$ because no rth root of unity is a primitive mth root of unity if $r \mid m$. Hence

$$\Phi_m(x) \Big| \frac{x^m - 1}{x^{r(a)} - 1} \qquad \text{for all } a \in S$$

and by substitution

$$\Phi_m(q) \Big| \frac{q^m - 1}{q^{r(a)} - 1} \qquad \text{for all } a \in S$$

Hence we can conclude from the class equation (4) that $\Phi_m(q) \mid (q - 1)$ unless $m = 1$.

But now we estimate the absolute value of $\Phi_m(q)$ from its definition, regarding the mth roots of unity as complex numbers. Thus

$$|\Phi_m(q)| = \prod_{\gcd(k, m) = 1} |q - \zeta^k| > q - 1$$

because $q \in \mathbb{Z}$, $q \geq 2$ and if $m > k > 1$, $|q - \zeta^k| > q - 1$. Hence if $m > 1$ we would conclude that $|\Phi_m(q)| > q - 1$ contrary to the fact that $\Phi_m(q)$ divides $q - 1$. Tilt! Hence m must equal 1 and Z must equal D and D must be a finite field.

Now we turn to the important question of the irreducibility of $\Phi_n(x)$. It is remarkable that this polynomial is irreducible in $\mathbb{Q}[x]$ for all integers n. In fact this is one of the few classes of irreducible polynomials we can explicitly construct. However, this result is very field-dependent. For example, in $\mathbb{Z}_{13}[x]$,

$$\Phi_{12}(x) = x^4 - x^2 + 1 = (x - 2)(x + 2)(x^2 + 3)$$

12.2.8 Theorem.   Irreducibility of Cyclotomic Polynomials

The nth cyclotomic polynomial $\Phi_n(x)$ is irreducible in $\mathbb{Q}[x]$.

Proof.   Suppose that $\Phi_n(x)$ is reducible in $\mathbb{Q}[x]$. By Theorem 9.4.8 it is then also reducible in $\mathbb{Z}[x]$. Let

$$\Phi_n(x) = p(x)h(x)$$

in $\mathbb{Z}[x]$ where $p(x)$ is irreducible in $\mathbb{Z}[x]$. Without loss of generality we may assume that $p(x)$ is primitive, that is, that the greatest common divisor of all the coefficients of $p(x)$ is 1. Let ζ be a zero of $p(x)$ and thus a primitive nth root of unity. We shall show that if m is a prime not dividing n then $\tau = \zeta^m$ is also a zero of $p(x)$. Then, as we shall see at the end of the proof, it follows that all the nth primitive roots of unity are roots of $p(x)$. This means that $h(x) = 1$ and hence $\Phi_n(x)$ is irreducible in $\mathbb{Z}[x]$.

If τ is not a zero of $p(x)$, let $q(x)$ be the irreducible factor of $\Phi_n(x)$ of which τ is a zero. It must be the case that $\gcd(p(x), q(x)) = 1$ in $\mathbb{Z}[x]$ since each is irreducible and they are not associates since $q(x)$ has a zero that is not a zero of $p(x)$. Hence $p(x)q(x) \mid \Phi_n(x)$, in fact

$$x^n - 1 = p(x)q(x)u(x)$$

in $\mathbb{Z}[x]$. Moreover, we know that since $0 = q(\tau) = q(\zeta^m)$, ζ is a zero of $q(x^m)$. Thus $q(x^m)$ and $p(x)$ have a common zero and hence have a common factor in $\mathbb{Z}[x]$. But $p(x)$ is irreducible in $\mathbb{Z}[x]$ and this means that $p(x)$ must divide $q(x^m)$ in $\mathbb{Z}[x]$, say, that

$$q(x^m) = p(x)s(x)$$

in $\mathbb{Z}[x]$. Now we shift our attention to the finite field of m elements $\mathbb{Z}_m$. We know that $\mathbb{Z}_m$ is a homomorphic image of $\mathbb{Z}$ and hence $\mathbb{Z}_m[x]$ is a homomorphic image of $\mathbb{Z}[x]$.

Now consider the five polynomials

$$x^n - 1, \ p(x), \ q(x), \ s(x) \text{ and } q(x^m)$$

as polynomials in $\mathbb{Z}_m[x]$. There we have the wonderful identity

$$q(x^m) = q(x)^m$$

since we are doing things modulo m. Thus in $\mathbb{Z}_m[x]$

$$p(x)s(x) = q(x^m) = q(x)^m$$

This means that an irreducible factor of $p(x)$ in $\mathbb{Z}_m[x]$ is a factor of $q(x)$. Hence in some extension of $\mathbb{Z}_m$ a zero of $p(x)$ is also a zero of $q(x)$. But in $\mathbb{Z}[x]$, and hence also in $\mathbb{Z}_m[x]$,

$$x^n - 1 = p(x)q(x)u(x)$$

But this means that in some extension of $\mathbb{Z}_m$ the polynomial $x^n - 1$ has at least one multiple zero, the common zero of $p(x)$ and $q(x)$. But that can't be. In $\mathbb{Z}_m[x]$ since $m \nmid n$ the derivative of $x^n - 1$ is the nonzero polynomial nx^{n-1} that has no zero in common with $x^n - 1$. Tilt!

Thus it must be that τ is a zero of $p(x)$ as we alleged.

Now to complete the proof. Let ζ^k be any primitive nth root of unity. Express k as a product of primes, not necessarily distinct: $k = m_1 m_2 \ldots m_t$. By t applications of the argument we have just given we see that

$$\zeta, \zeta^{m_1}, \zeta^{m_1 m_2}, \ldots, \zeta^{m_1 m_2 \ldots m_t}$$

are all zeros of $p(x)$. In particular ζ^k is a zero of $p(x)$. Hence $p(x)$ and

$\Phi_n(x)$ have the same zeros and so must be associates in $\mathbb{Q}[x]$. Thus $\Phi_n(x)$ is irreducible in $\mathbb{Q}[x]$.

Remark. A classical conjecture, and one shared by most who first venture into the topic of cyclotomic polynomials, is that the only integers that appear in cyclotomic polynomials are 1, -1, and 0. That is true for $n \leq 104$ but not for 105. In fact the coefficients become arbitrarily large in absolute value for large n.

As a first application of the irreducibility of $\Phi_n(x)$ in $\mathbb{Z}[x]$ we can now state necessary and sufficient conditions on the positive integer n so that it is possible to construct a regular n-gon with only compass and straightedge. At this time we can prove the necessity of the conditions, but there is a gap in the proof of their sufficiency that we will have to wait to fill until we have had some Galois theory. But the discussion is useful to point out the need for more information.

12.2.9 Theorem.   Constructible Regular Polygons

A regular polygon of n sides ($n \geq 3$) may be constructed by compass and straightedge from $\{0,1\}$ if and only if its center and one vertex are constructible and

$$n = 2^k p_1 \ldots p_t$$

where $k \geq 0$ and the p's, if any, are distinct odd primes each of which has form

$$p_i = 2^{f_i} + 1$$

And in fact then each f_i is itself a power of 2; so

$$p_i = 2^{2^{h_i}} + 1$$

Proof. Without loss of generality we may assume that the polygon constructed or to be constructed has its center at 0 and has 1 as a vertex. We do this by the following steps. First suppose that a regular polygon has been constructed. Then its center is constructible (Exercise 12.1.3). Second we can translate the polygon so that its center is 0 by subtracting the center from each vertex. Hence each vertex has the same distance from the origin. This distance is constructible and hence we can normalize the distance to be 1 by dividing each vertex by the distance. Finally we can rotate the polygon so that one vertex is at 1 by multiplying all vertices by the conjugate $\bar{\alpha}$ of one of the vertices α. In this case the vertex with the least positive argument is an nth root of unity.

This process is reversible if the center and one vertex have been constructed. The crux of the matter is the number theoretic property of n. So the theorem comes down to the more restricted statement: "An nth root of unity is constructible if and only if $n = 2^k p_1 \ldots p_t$, where $k \geq 0$ and each p_i, if any, is an odd prime of the form described in the theorem."

We can prove the necessity of this condition. Suppose that ζ is an nth root of unity and is constructible from $\{0, 1\}$. Then ζ and indeed $\mathbb{Q}(\zeta)$ is contained in a field E_s for which there is a chain of quadratic extensions:

$$\mathbb{Q} = E_0 \subset E_1 \subset \ldots \subset E_s$$

and for each i, $[E_i : E_{i-1}] = 2$. But that means that $[E_s : \mathbb{Q}]$ is a power of 2 and hence, since $\mathbb{Q}(\zeta) \subseteq E_s$, $[\mathbb{Q}(\zeta): \mathbb{Q}]$ is also a power of 2. But the degree $[\mathbb{Q}(\zeta): \mathbb{Q}]$ is the degree of the irreducible polynomial in $\mathbb{Q}[x]$ of which ζ is a zero. This polynomial we have just proved is $\Phi_n(x)$ and so has degree $\varphi(n)$.

So let $n = 2^k p_1^{e_1} \ldots p_t^{e_t}$ be the prime factorization of n where the primes p_i, if any, are distinct odd primes. Since the Euler φ-function is multiplicative

$$\varphi(n) = \varphi(2^k)\varphi(p_1^{e_1})\ldots\varphi(p_t^{e_t}) = 2^{k-1}p_1^{e_1-1}(p_1 - 1)\ldots p_t^{e_t-1}(p_t - 1)$$

So for $\varphi(n)$ to be a power of 2 it must be that each $e_i = 1$ and each $p_i - 1$ is a power of 2.

Finally there is a little group theoretic argument to show that if an odd prime $p = 2^h + 1$ then h itself must be a power of 2. It goes like this: If $2^h + 1 = p$ then

$$2^h \equiv -1 \pmod{p} \quad \text{and} \quad \text{so } 2^{2h} \equiv 1 \pmod{p} \tag{5}$$

The order of the multiplicative group of the finite field $\mathbb{Z}_p$ is $p - 1$ and therefore a power of 2. The order of 2 in this group must divide the order of the group and hence is a power of 2. We claim that the order of $2 \in \mathbb{Z}_p$ is $2h$ and therefore h is a power of 2. Let the order of 2 be u. From (5) it follows that $2h \geq u$. Moreover, $2 \mid u$ since u is a power of 2 and $u \neq 1$. Hence $u/2$ is an integer and from $2^u \equiv 1 \pmod{p}$ we can conclude

$$2^{u/2} \equiv -1 \pmod{p} \quad \text{or} \quad 2^{u/2} \equiv 1 \pmod{p}$$

because in $\mathbb{Z}_p$ there are at most two square roots of 1. The second congruence above cannot hold since u is the order of 2. From the first congruence we know that p divides $2^{u/2} + 1$, but $p = 2^h + 1$ and hence $u/2 \geq h$. Thus $u = 2h$.

Now let's start on the sufficiency. We want to show that if n has the form given then ζ, a primitive nth root of unity is constructible from $\{0, 1\}$. To do this we need to find a chain of fields $\mathbb{Q} = E_0 \subset E_1 \subset \cdots \subset E_v$ where each degree $[E_i : E_{i-1}] = 2$ and $\mathbb{Q}(z) \subseteq E_v$.

We can conclude at this point that $[\mathbb{Q}(\zeta): \mathbb{Q}]$ is a power of 2. Here's the argument: If $n = 2^k p_1 \ldots p_t$ where the primes p_i, if any, are odd and have the form $2^h + 1$, then, as we've seen above, $\varphi(n) = 2^s$ for some s. And since $\Phi_n(x)$ is irreducible and has degree $\varphi(n)$ it follows that the degree

$$[\mathbb{Q}(\zeta): \mathbb{Q}] = 2^s$$

What we must do to complete the proof is to show that the necessary chain of subfields exists, ending at a field containing $\mathbb{Q}(\zeta)$. For that we will need to know something about the automorphisms of $\mathbb{Q}(\zeta)$. We do this next.

We can say a great deal about the group of automorphisms of the splitting field of a cyclotomic polynomial over a field F of characteristic 0. Theorem 12.2.2 tells us that this splitting field has the form $F(\zeta)$ where ζ is the zero of a factor of $\Phi_n(x)$ irreducible in $F[x]$. The following result serves as a good example of the general theorems on automorphisms of any splitting field.

12.2.10 Theorem. Automorphisms of Cyclotomic Extensions

Let E be a splitting field of $x^n - 1$ over a field F whose characteristic does not divide n. The group of automorphisms of E that fix each element of F is isomorphic to a subgroup of $\mathbb{Z}_n^\times$, the multiplicative group of units in $\mathbb{Z}_n$. The group is isomorphic to all of $\mathbb{Z}_n^\times$ if the cyclotomic polynomial is irreducible in $F[x]$.

Proof.   To begin, recall that the automorphisms of E that fix each element of F form a group (Theorem 8.3.10). Let $\mathcal{G}$ be that group.

From the nature of a primitive nth root of unity ζ, we know that $E = F(\zeta)$. Let σ be an automorphism of E that fixes each element of F. Since ζ generates the extension E, then its image, $\sigma(\zeta)$, determines the whole automorphism. And, as we've observed its image must be another zero of the same irreducible factor of $\Phi_n(x)$. More importantly, $\sigma(\zeta)$ must be another primitive nth root of unity. This primitive nth root of unity is a function of σ and so we write

$$\sigma(\zeta) = \zeta^{t(\sigma)}$$

and because $\sigma(\zeta)$ is a primitive nth root of unity it must be that $\gcd(n, t(\sigma)) = 1$. This suggests that the mapping T from $\mathcal{G}$ into $\mathbb{Z}_n^\times$ defined by

$$T\colon \sigma \to t(\sigma)$$

is an isomorphism.

First we must verify the "homomorphism" condition: If

$$T\colon \rho \to t(\rho)$$

that is, if $\rho(\zeta) = \zeta^{t(\rho)}$, then we must show that

$$T\colon \sigma\rho \to t(\sigma)t(\rho)$$

that is, that $(\sigma\rho)(\zeta) = \zeta^{t(\sigma)t(\rho)}$. So we must determine the action of the automorphism $\sigma\rho$ on ζ. Since products of automorphisms are their composition, we can compute

$$\sigma\rho(\zeta) = \sigma(\rho(\zeta)) = \sigma(\zeta^{t(\rho)}) = (\sigma(\zeta))^{t(\rho)} = \zeta^{t(\sigma)t(\rho)}$$

The third equality holds because σ is an automorphism of E. This verifies the "homomorphism" condition. It only remains to show that T is one-to-one. But by the definition σ is the identity if and only if $t(\sigma) = 1$. Hence the kernel of T is just the identity mapping.

Finally we must show that if $\Phi_n(x)$ is irreducible in $F[x]$ then the mapping T maps $\mathcal{G}$ onto $\mathbb{Z}_n^\times$. To do this we need to show that for each integer

m such that $\gcd(n, m) = 1$ there is an automorphism of E, fixing each element of F and sending ζ to ζ^m. But because $\Phi_n(x)$ is irreducible in $F[x]$, and because the splitting field $E = F(\zeta)$ it follows from Theorem 10.3.6 that such an automorphism exists.

Remark. In view of Theorem 11.4.5, we know that the group $\mathbb{Z}_n^{\times}$ is cyclic if $n = 2$ or 4 or if $n = p^k$ or $2p^k$ where p is an odd prime.

We conclude this section with another theorem about the automorphisms of a splitting field that is very much akin to the preceding one. And again this theorem has great utility in determining which polynomials have zeros that can be expressed in terms of radicals. Along the way this theorem provides a number of useful examples.

12.2.11 Theorem.   Automorphism Group of $x^n - a$

Let F be a field of characteristic 0 containing the nth roots of unity. Let $a \in F$ and let E be the splitting field of $x^n - a$ over F. The group of automorphisms of E that fix each element of F is cyclic.

Proof. We can't say much about the order of the group because we know nothing of the degree of E over F; $F = E$ is possible. However, we do know that in E we have one zero, call it α. Thus $\alpha^n = a$ in E. And we do know that we have a primitive nth root of unity; call it ζ in F, and thus in E as well. By inspection we can see that the n zeros of $x^n - a$ are

$$\alpha, \alpha\zeta, \ldots, \alpha\zeta^k, \ldots, \alpha\zeta^{n-1}$$

all of which lie in E. Thus any automorphism σ of E that fixes each element of F must map α onto another zero of $x^n - a$, say, $\alpha\zeta^k$. The exponent k depends on α, and to emphasize that we write

$$\sigma: \alpha \to \alpha\zeta^{k(\sigma)}$$

We claim that the mapping K defined by

$$K: \sigma \to k(\sigma)$$

is a group homomorphism from the group $\mathscr{G}$ of automorphisms of E fixing each element of F (and hence ζ) into the additive cyclic group $\langle \mathbb{Z}_n, + \rangle$.

Watch the verification of the "homomorphism" property to see how the fact that the automorphisms in $\mathscr{G}$ fix ζ is used. Let σ and τ belong to $\mathscr{G}$. Then

$$\sigma\tau(\alpha) = \sigma(\tau(\alpha)) = \sigma(\alpha\zeta^{k(\tau)}) = \sigma(\alpha)\sigma(\zeta^{k(\tau)}) = \alpha\zeta^{k(\sigma)}\zeta^{k(\tau)} = \alpha\zeta^{k(\sigma)+k(\tau)}$$

and this shows that

$$K: \sigma\tau \to k(\sigma\tau) = k(\sigma) + k(\tau)$$

and that verifies that K is a homomorphism from $\mathscr{G}$ into $\langle \mathbb{Z}_n, + \rangle$.

To show that K is an isomorphism we need to investigate its kernel. Suppose that K maps σ into the identity of $\langle \mathbb{Z}_n, + \rangle$ which is of course 0. This means that $k(\sigma) = 0$ and so

$$\sigma:\alpha \;\to\; \alpha\zeta^{k(\sigma)} = \alpha\zeta^0 = \alpha$$

and σ is the identity automorphism, the identity of $\mathscr{G}$. Thus K is one-to-one. Thus the image of $\mathscr{G}$ under K is isomorphic to a subgroup $\mathscr{H}$ of the cyclic group $\langle \mathbb{Z}_n, + \rangle$. Of course, $\mathscr{H}$ is also cyclic and its order divides n.

EXERCISE SET 12.2

12.2.1 Let $\Phi_n(x)$ be the nth cyclotomic polynomial in $\mathbb{Q}[x]$. Show that $\Phi_n(0) = 1$. Show that for $n > 1$, $\Phi_n(m) \equiv 1 \pmod{m}$ for all $m \in \mathbb{Z}$. Show that if ζ is a primitive nth root of unity over $\mathbb{Q}$ then

$$\prod_{\gcd(k,\,n)=1} \left(-\zeta^k\right) = 1$$

12.2.2 Prove the following relations on cyclotomic polynomials in $\mathbb{Q}[x]$:
1. $\Phi_n(x) = \Phi_d(x^m)$ where $n = dm$ and d is the product of all the distinct prime factors of n.
2. $\Phi_{2n}(x) = \Phi_n(-x)$ if n is odd.
3. $\Phi_{pn}(x) = \Phi_n(x^p)/\Phi_n(x)$ if p is a prime that does not divide n.

12.2.3 Show that if $r \mid n$ then $\Phi_n(x) \mid (x^n - 1)/(x^r - 1)$ in $\mathbb{Z}[x]$.

12.2.4 Show that the absolute value of one of the coefficients of $\Phi_{105}(x)$ over $\mathbb{Z}$ is 2.

12.2.5 (Migotti) Prove that if $n = p^a q^b$ where p and q are distinct primes then the coefficients of $\Phi_n(x)$ are either 1, -1, or 0.

12.2.6 Let ζ be a primitive tenth root of unity over $\mathbb{Q}$. Determine the monic irreducible polynomial in $\mathbb{Q}[x]$ of which ζ is a zero. Does $\mathbb{Q}(\zeta) = \mathbb{Q}(\zeta^2)$?

12.2.7 Determine the splitting fields over $\mathbb{Q}$ of the following polynomials:
1. $x^4 + x^2 + 1$
2. $x^4 + 1$
3. $x^6 + 1$
4. $x^6 + x^3 + 1$

12.2.8 Let ζ be a primitive fifth root of unity over $\mathbb{Q}$. Find $a \in \mathbb{Q}$ such that $\mathbb{Q} \subset \mathbb{Q}(\sqrt{a}) \subset \mathbb{Q}(\zeta)$.

12.2.9 Determine the splitting field E of $x^4 - 5$ over $\mathbb{Q}$. Determine the group of automorphisms of E. Find as many subfields of E as you can.

12.2.10 Determine the splitting field E of $x^5 - 3$ over $\mathbb{Q}$. Show that E contains ζ, a primitive fifth root of unity. Determine the group of automorphisms of E fixing each element of $\mathbb{Q}(\zeta)$. *Determine the group of automorphisms of E fixing each element of $\mathbb{Q}$. Determine as many subfields of E as you can.

12.2.11 Show that the following is a chain of extensions whose top field is $\mathbb{Q}(\zeta)$ where ζ is a primitive seventeenth root of unity:

$$\mathbb{Q} \subset \mathbb{Q}(\delta) \subset \mathbb{Q}(\gamma) \subset \mathbb{Q}(\beta) \subset \mathbb{Q}(\zeta)$$

and that each is an extension of degree 2 over the preceding field. Let

$$\delta = \zeta + \zeta^{-1} + \zeta^4 + \zeta^{-4} + \zeta^8 + \zeta^{-8} + \zeta^2 + \zeta^{-2}$$

$$\gamma = \zeta + \zeta^{-1} + \zeta^4 + \zeta^{-4}$$

$$\beta = \zeta + \zeta^{-1}$$

Show that $\mathbb{Q}(\delta) = \mathbb{Q}(\sqrt{17})$. Find the quadratic polynomial that each of δ, γ, β satisfies over the preceding field.

*Construct, using only compass and straightedge, a regular 17-gon.

12.3 THE GAUSSIAN INTEGERS

In this section we present an interesting example, the ring of gaussian integers. It is an example of a euclidean domain, but more than that, its properties provide a nice proof of the important result in number theory first stated by Fermat:

An odd prime is the sum of two squares if and only if it has the form $4k + 1$.

12.3.1 Definition.   The Gaussian Integers $\mathbb{Z}[i]$

The subring of the complex numbers,

$$\mathbb{Z}[i] = \{ a + bi \colon a, b \in Z \}$$

is called the ring of gaussian integers.

There is a geometric interpretation of the gaussian integers. Each gaussian integer represents a point in the complex plane with integer coordinates. Thus the gaussian integers form an orthogonal grid of points. The points of the grid form squares one unit on a side. See Figure 12.4.

From the complex numbers we borrow the notion of a complex conjugate and the norm of a complex number. Here is our notation and a definition of these quantities.

$$\text{If } \alpha = a + bi \quad \text{then} \quad \bar{\alpha} = a - bi$$

The norm of α is the square of its absolute value,

$$\|\alpha\| = a^2 + b^2 = \alpha\bar{\alpha}$$

In particular the norm of the difference of two complex numbers is the square of the distance between them. Note that if α is a gaussian integer then so is $\bar{\alpha}$ and its norm $\|\alpha\|$ is an integer. An important property satisfied by the norm is this multiplicative property

$$\|\alpha\beta\| = \|\alpha\| \, \|\beta\|$$

The algebraic properties of this ring are, first, that it is an integral domain since it is a subring of the complex numbers containing the identity 1. Second, as we shall show, every irreducible element is a prime, there is a euclidean algorithm, every ideal is a principal ideal, and every gaussian integer can be

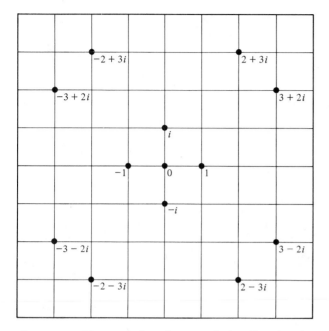

Figure 12.4 Here are plotted some of the Gaussian integers. Zero and the units $1, -1, i$, and $-i$ are plotted. The other eight numbers plotted are the associates of the number $2 + 3i$ and its complex conjugate $2 - 3i$. These numbers are Gaussian prime numbers because their norm is 13, a prime integer of the form $4k + 1$. Theorem 12.3.9 describes all the Gaussian primes.

written as the product of primes in $\mathbb{Z}[i]$ in one and only one way, except for unit factors and the order of the primes. We begin our investigation by determining the units of $\mathbb{Z}[i]$.

12.3.2 Lemma.   The Units of $\mathbb{Z}[i]$

The units of $\mathbb{Z}[i]$ are $\{1, -1, i, -i\}$. Thus the group of units is cyclic of order 4.

Proof.   Suppose that $\alpha = a + bi$ is a unit and so there is $\beta \in \mathbb{Z}[i]$ such that

$$\alpha\beta = 1$$

and so by conjugation

$$\bar{\alpha}\bar{\beta} = 1$$

Now multiplying we get

$$\alpha\bar{\alpha}\beta\bar{\beta} = 1$$

Since $\alpha\bar{\alpha}$ is a positive integer dividing 1, it follows that

$$\alpha\bar{\alpha} = a^2 + b^2 = 1$$

This means that either a or b, but not both, are zero. If $b = 0$ then $a = 1$ or -1; if $a = 0$ then $b = 1$ or -1; in this way the four units arise.

The most important fact for the gaussian integers is that there is a division algorithm. First we prove the following lemma.

12.3.3 Lemma. Proximity of Gaussian Integers

Let $\xi = x + yi$ be a complex number. There exists a gaussian integer $\tau = a + bi$ such that

$$\|\xi - \tau\| \le \frac{1}{2}$$

Proof. Here is a geometric argument. Consult Figure 12.4. Any complex number falls into one of the squares of the grid of gaussian integers. The distance from a point inside a square of the grid, or on the perimeter of the square, to the nearest vertex is at most half the diagonal distance of the square, which is $\sqrt{2}$. Hence the square of the distance from a point to the nearest vertex is at most

$$\left((\sqrt{2})/2\right)^2 = \frac{1}{2}$$

Here is an algebraic argument. Choose a to be the integer nearest x and choose b to be the integer closest to y. Thus

$$|x - a| \le \frac{1}{2} \quad \text{and} \quad |y - b| \le \frac{1}{2}$$

Thus we find

$$\|\xi - \tau\| = (x - a)^2 + (y - b)^2 \le \frac{1}{4} + \frac{1}{4} = \frac{1}{2}$$

12.3.4 Theorem. Division Algorithm in the Gaussian Integers

Let α be a nonzero gaussian integer. Then for every gaussian integer β there exist gaussian integers τ and ρ such that

$$\beta = \alpha\tau + \rho \quad \text{and} \quad \text{either } \rho = 0 \text{ or } \|\rho\| < \|\alpha\|$$

Proof. Let $\xi = \beta/\alpha$ and choose τ as in Lemma 12.3.3 so that $\|\xi - \tau\| < 1$. Then write

$$\beta = \alpha\tau + (\beta - \alpha\tau)$$

We claim that the norm of $\rho = \beta - \alpha\tau$ is less than $\|\alpha\|$. To see why, factor out α:

$$\rho = \alpha\left(\frac{\beta}{\alpha} - \tau\right) = \alpha(\xi - \tau)$$

So $\|\rho\| = \|\alpha\| \, \|\xi - \tau\| < \|\alpha\|$.

12.3.5 Corollary.   The Gaussian Integers Are Euclidean

The gaussian integers are a euclidean domain whose size function is the complex norm $\| \ \|$.

Proof. In view of the preceding theorem we only need to check the other requirement for the norm to be a size function: If α and β are nonzero, is $\|\alpha\| \leq \|\alpha\beta\|$? Well, in any event

$$\|\alpha\beta\| = \|\alpha\| \, \|\beta\|$$

Let $\beta = c + di$ where $c, d \in \mathbb{Z}$. Then $\|\beta\| = c^2 + d^2 \geq 1$ since not both c and d are zero. Thus

$$\|\alpha\beta\| = \|\alpha\| \, \|\beta\| \geq \|\alpha\|$$

Remark.   Now all good things about euclidean domains apply to the gaussian integers. Every ideal is principal, there is unique factorization into a product of primes, and so on.

Our interest in the gaussian integers is to establish the famous result of Fermat mentioned in the introduction of this section. This result will permit us to give a nice criterion for a gaussian integer to be a prime in that domain.

12.3.6 Lemma.   Properties of $\mathbb{Z}_p^\times$

Let p be an odd prime.

1. $\mathbb{Z}_p^\times$ contains one element of order 2.
2. If $p = 4k + 3$ then $\mathbb{Z}_p^\times$ contains no element of order 4.
3. If $p = 4k + 1$ then $\mathbb{Z}_p^\times$ contains two elements of order 4 and the congruence $y^2 + 1 \equiv 0 \pmod{p}$ is solvable.

Proof.   First a remark about this lemma and the proof that ensues. If we know that $\mathbb{Z}_n^\times$ is cyclic (Lemma 10.4.1) then we know that there is at most one subgroup of order 2 and one subgroup of order 4. Thus there is at most one element of order 2 and two elements of order 4. Since the order of the group is the even number $p - 1$, there is always a subgroup of order 2. There is a subgroup of order 4 if and only if $4 \mid (p - 1)$; which are conclusions 2 and 3.

Alternatively, the following proof does not rely on Lemma 10.4.1.

Since p is odd, $-1 \not\equiv 1 \pmod{p}$ and so -1 has order 2 in $\mathbb{Z}_p^\times$. Now suppose that x is any element of order 2 in $\mathbb{Z}_p^\times$. Thus

$$x^2 \equiv 1 \pmod{p}$$

and so

$$p \mid x^2 - 1 = (x - 1)(x + 1)$$

hence

$$p \mid x - 1 \quad \text{or} \quad p \mid x + 1$$

hence

$$x \equiv 1 \pmod{p} \quad \text{or} \quad x \equiv -1 \pmod{p}$$

Thus if x has order 2 in $\mathbb{Z}_n^\times$ then $x \equiv -1 \pmod{p}$.

If $p = 4k + 3$ then the order of $\mathbb{Z}_p^\times$ is $p - 1 = 4k + 2$, which is not divisible by 4. By Lagrange's theorem, the group cannot have an element of order 4.

Finally, suppose that $p = 4k + 1$ or that $4 \mid p - 1$. Consider the factor group

$$\overline{\mathbb{Z}} = \mathbb{Z}_p^\times / \langle -1 \rangle$$

whose order is $(p - 1)/2 = 2k$; another even number and so $\overline{\mathbb{Z}}$ has an element of order 2, call it $\bar{y}$. We have

$$\bar{y}^2 = I \quad \text{or} \quad y^2 \in \langle -1 \rangle$$

hence

$$y^2 \equiv 1 \pmod{p} \text{ or } y^2 \equiv -1 \pmod{p}$$

If the first alternative holds, then by 1, $y \equiv \pm 1 \pmod{p}$ and so $\bar{y} = I$, a contradiction.

Thus it must be that $y^2 \equiv -1 \pmod{p}$ and so y has order 4 in $\mathbb{Z}_p^\times$.

To show that y and $-y$ are the only elements of order 4 in $\mathbb{Z}_p^\times$, let x be any element of order 4. In particular x^2 has order 2 in $\mathbb{Z}_p^\times$ and so

$$x^2 \equiv -1 \equiv y^2 \pmod{p}$$

Hence

$$p \mid (x^2 - y^2) = (x - y)(x + y)$$

and so

$$p \mid (x - y) \quad \text{or} \quad p \mid (x + y) \quad \text{that is} \quad x \equiv y \text{ or } \equiv -y \pmod{p}$$

In the remainder of our discussion we must keep clear the distinction between primes in the integers $\mathbb{Z}$ and primes in the gaussian integers $\mathbb{Z}[i]$. We often refer to the first as "prime integers" and to the second as "gaussian primes."

12.3.7 Theorem (Fermat).   Primes That Are the Sum of Two Squares

A positive prime p can be written as the sum of two squares in $\mathbb{Z}$ if and only if $p = 2$ or $p = 4k + 1$.

Proof.   First observe that $2 = 1 + 1$. So from now on we assume that p is an odd prime. Now suppose that p can be written as the sum of two squares, say,

$$p = a^2 + b^2 \tag{1}$$

We shall show that $p = 4k + 1$. Suppose, to the contrary, that $p = 4k + 3$.

We may interpret equation (1) as a congruence modulo p:

$$a^2 + b^2 \equiv 0 \ (\text{mod } p)$$

Now neither a nor b can be divisible by p, for then equation (1) would imply that both were and moreover that $a^2 + b^2$ would be divisible by p^2. Hence in $\mathbb{Z}_p$ we find

$$\left(ab^{-1}\right)^2 = -1$$

But this means that ab^{-1} has order 4 in $\mathbb{Z}_p^\times$, a contradiction of Lemma 12.3.6, part 3.

Next suppose that $p = 4k + 1$. We shall show that it is possible to write p as the sum of two squares.

From Lemma 12.3.6, part 3 there is an integer y such that

$$y^2 + 1 \equiv 0 \ (\text{mod } p)$$

Thus $p \mid y^2 + 1$.

In the domain of gaussian integers, $\mathbb{Z}[i]$

$$y^2 + 1 = (y + i)(y - i)$$

If p were a prime in $\mathbb{Z}[i]$ then p would divide either $y + i$ or $y - i$. In addition p is an integer, and this means that p would divide the imaginary part of these factors; that is, p would divide 1, a contradiction.

Therefore, p must be composite as a gaussian integer; hence there is a factorization

$$p = \pi\rho$$

in $\mathbb{Z}[i]$ where neither π nor ρ is a unit. Multiplying p by its conjugate,

$$p^2 = \pi\bar{\pi}\rho\bar{\rho}$$

Now in $\mathbb{Z}$ this means that either

$$\|\pi\| = \|\rho\| = p$$

or one of π or ρ is a unit, contrary to the composite nature of p in $\mathbb{Z}[i]$.

Let $\pi = a + bi$ and compute

$$p = \|\pi\| = a^2 + b^2$$

One of the results of this theorem is that a positive prime integer of the form $4k + 1$ is not a gaussian prime since

$$\text{If } p = a^2 + b^2 \quad \text{then} \quad p = (a + bi)(a - bi)$$

We go on now to determine all the gaussian primes.

12.3.8 Lemma. Gaussian Primes and Their Norms

If π is a gaussian integer and $\|\pi\| = \pi\bar{\pi}$ is a prime integer then π is a gaussian prime. If π is a gaussian prime then

$$\|\pi\| = p \text{ or } p^2$$

where p is a prime integer. Moreover

If $\|\pi\| = p$ then $p = 2$ or has the form $4k + 1$.

If $\|\pi\| = p^2$ then p has the form $4k + 3$ and π is an associate of p in $\mathbf{Z}[i]$.

Proof. Suppose that the norm of π is a prime integer. Suppose that

$$\pi = \alpha\beta$$

then

$$\|\pi\| = \|\alpha\|\,\|\beta\|$$

Since $\|\pi\|$ is a prime it follows that either $\|\alpha\|$ or $\|\beta\|$ is 1 and hence that α or β is a unit. Thus π is a gaussian prime.

Now suppose that π is a gaussian prime. Consider the prime factorization of $\|\pi\|$ in the integers $\mathbf{Z}$. Since π divides $\|\pi\|$ in $\mathbf{Z}[i]$, π must divide one of the integer primes, call it p, in the integer prime factorization of $\|\pi\|$. Thus

$$p = \pi\alpha$$

in $\mathbf{Z}[i]$. But taking the norms

$$p^2 = \pi\bar{\pi}\alpha\bar{\alpha}$$

in $\mathbf{Z}$. Thus the integer prime factorization for $\|\pi\|$ is either

$$(1)\ p = \pi\bar{\pi} \quad \text{or} \quad (2)\ p^2 = \pi\bar{\pi}$$

In case 1 then

$$\pi\bar{\pi} = c^2 + d^2 = p$$

for integers c and d. If $p = 2$ then $c = d = 1$ and there is nothing more to say.

If p is an odd prime then it follows from Fermat's theorem 12.3.7 that $p = 4k + 1$.

Finally in case (2) we see that α is a unit and so

$$\pi = p\bar{\alpha}$$

thus π is an associate of p in $\mathbf{Z}[i]$.

It doesn't follow immediately that p has the form $4k + 3$. In any event 2 is not a gaussian prime since

$$2 = (1 + i)(1 - i)$$

and so $p \neq 2$.

Now, suppose that $p = 4k + 1$. Then, by Fermat's theorem p is the sum of two squares and so in $\mathbf{Z}[i]$

$$p = a^2 + b^2 = (a + bi)(a - bi)$$

so that p is not a prime.

Now we can summarize our arguments to determine the gaussian primes.

12.3.9 Theorem.   The Gaussian Primes

The primes in the gaussian integers $\mathbb{Z}[i]$ are:

1. The positive primes in $\mathbb{Z}$ of the form $4k + 3$ and their associates in $\mathbb{Z}[i]$.
2. Gaussian integers π whose norm is a prime in $\mathbb{Z}$, necessarily 2 or an odd prime of the form $4k + 1$.

Proof.   Lemma 12.3.8 shows that gaussian integers satisfying 2 are primes in $\mathbb{Z}[i]$.

If $p \in \mathbb{Z}$ is an odd prime of the form $4k + 3$ and if p is composite in $\mathbb{Z}[i]$ then, as we have seen, this would mean that p could be written as the sum of two squares and hence that $p \equiv 1 \pmod 4$, a contradiction.

Lemma 12.3.8 shows that 1 and 2 are only possibilities for primes in $\mathbb{Z}[i]$.

EXERCISE SET 12.3

12.3.1   Let 5 points be inside a square one unit on a side. Show that two of them must be at most $1/\sqrt{2}$ units apart.

12.3.2   In $\mathbb{Z}[i]$ let $\alpha = 1 + 2i$ and $\beta = 3 + 4i$. Show that α divides $\beta\bar{\beta}$. Find σ and ρ so that $\beta = \alpha\sigma + \rho$ where $\|\rho\| < \|\alpha\|$. Show that α is prime.

12.3.3   On a grid show all the gaussian primes such that $\|\pi\| \leq 100$. Remember, if π is a prime then so are its four associates and $\bar{\pi}$ and its four associates. When are the eight primes not distinct?

12.3.4   Under what conditions is $(\alpha, \bar{\alpha}) = 1$ in $\mathbb{Z}[i]$?

12.3.5   Determine the prime ideals in $\mathbb{Z}[i]$. Determine the maximal ideals in $\mathbb{Z}[i]$.

12.3.6   Describe the rings $\mathbb{Z}[i]/\langle 3 \rangle$, $\mathbb{Z}[i]/\langle 5 \rangle$, and $\mathbb{Z}[i]/\langle 2 + i \rangle$. Include in your description in each case the characteristic of the ring, the number of elements in the ring, and whether the ring is a field.

12.3.7   Let π be a prime in $\mathbb{Z}[i]$. Describe $\mathbb{Z}[i]/\langle \pi \rangle$. Your answer will depend on whether π is a prime integer.

12.3.8   Show that $D[\sqrt{-2}] = \{a + b\sqrt{-2} : a, b \in \mathbb{Z}\}$ is a euclidean domain.

12.3.9   Let $\omega = -1 + i\sqrt{3}/2$. Show that ω is a root of $x^3 - 1 = (x - 1)(x^2 + x + 1)$ and conclude that $\omega^3 = 1$ and $1 + \omega + \omega^2 = 0$. Let $\mathbb{Z}[\omega] = \{a + b\omega : a, b \in \mathbb{Z}\}$. Show that $\mathbb{Z}[\omega]$ is an integral domain. Find the units of $\mathbb{Z}[\omega]$. Prove a proximity theorem: For every complex number ζ there exists $\tau \in \mathbb{Z}[\omega]$ such that $\|\zeta - \tau\| < 1$. Show that $\mathbb{Z}[\omega]$ is a euclidean domain. Show that $1 - \omega$ is a prime in $\mathbb{Z}[\omega]$.
   Remark: This information can be used to prove a special case of Fermat's last theorem: There are no nonzero integers x and y such that $x^3 + y^3 = 3^3$. For more information see Goldstein, *Abstract Algebra: A First Course*, Prentice-Hall, 1973.

12.2.10   Show that $x^n - 2$ is irreducible over $\mathbb{Z}[i]$ for all positive integers n.

*12.2.11   Show that $x^4 + 1$ is irreducible over $\mathbb{Z}$ but is reducible over $\mathbb{Z}_p$ for every prime p.

12.4 TWO SPECIAL RESULTS OF FINITE EXTENSIONS

In this section we establish two theorems that have interest in their own right and are crucial to our development of Galois theory. The first guarantees that any finite extension E over a field of characteristic 0 is *simple*; that is, it is a result of the adjunction of a single element.

The second is, at first glance, a surprising property of splitting fields. It is the "all or none" property for the zeros of an irreducible polynomial. A splitting field of one polynomial contains either all the zeros of any other irreducible polynomial or none of them!

12.4.1 Lemma.   Existence of a Primitive Element

Let F be a field of characteristic 0 and E be an extension of F. If α and β are elements of E that are algebraic over F then

$$F(\alpha, \beta) = F(\gamma)$$

for some γ in E. Such an element is called a *primitive* element for $F(\alpha, \beta)$ over F.

Proof.   We shall show that $F(\alpha, \beta) = F(\alpha + u\beta)$ where u can be chosen to belong to F. This is the natural generalization of many of the examples we have seen; indeed

$$Q(\sqrt{2}, \sqrt{3}) = \mathbb{Q}(\sqrt{2} + \sqrt{3})$$

Here u could be chosen to be 1.

Let $p(x)$ be the irreducible polynomial satisfied by α and let $q(x)$ be the irreducible polynomial β satisfies; these polynomials belong to $F[x]$.

To create a field in which all of our constructions may take place, let L be a splitting field of $p(x)q(x)$ over E. In L let the roots of $p(x)$ be

$$\alpha = \alpha_1, \ldots, \alpha_n$$

and let the roots of $q(x)$ be

$$\beta = \beta_1, \ldots, \beta_m$$

Since F has characteristic zero and $q(x)$ is irreducible, $q(x)$ has no multiple roots. The lattice relation of these fields is shown in Figure 12.5.

Let u be chosen to be any element in F (even an integer) such that

$$u \neq \frac{\alpha_i - \alpha_1}{\beta_1 - \beta_j}$$

for all i, $1 \leq i \leq n$ and j, $1 < j \leq m$. This is possible because there are an infinite number of elements in F and we are excluding only a finite number of choices. In fact, we could choose u to be an integer. The reason for this choice will become apparent in a moment.

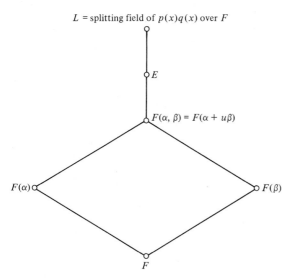

L = splitting field of $p(x)q(x)$ over F

E

$F(\alpha, \beta) = F(\alpha + u\beta)$

$F(\alpha)$

$F(\beta)$

F

Figure 12.5 The subfields constructed in the proof of Lemma 12.4.1.

Now we claim that $F(\alpha, \beta) = F(\alpha + u\beta)$. Clearly

$$F(\alpha, \beta) \supseteq F(\alpha + u\beta)$$

To prove the converse, it suffices to show that $\beta \in F(\alpha + u\beta)$, and we shall do this by showing that $(x - \beta)$ is the gcd of two polynomials that have coefficients in $F(\alpha + u\beta)$. These polynomials are

$$q(x) \quad \text{and} \quad p((a + u\beta) - ux)$$

Of course, the gcd of polynomials in this field has coefficients in this field, so if we show $(x - \beta)$ is the gcd, it follows that β is in this field. Here's how that argument goes:

In any event we know that β is a root of $q(x)$ and we substitute β into $p(\alpha + u\beta - ux)$ to find that

$$p(\alpha + u\beta - u\beta) = 0$$

so β must be a root of the greatest common divisor of these two polynomials. We claim that the greatest common divisor has no other root, for indeed if there were, it would have to be a root of $q(x)$ and hence be β_j for some $j \neq 1$. But it would also have to be a root of $p(\alpha + u\beta - ux)$ and thus $p(\alpha + u\beta - u\beta_j) = 0$. Thus for some i,

$$\alpha_j = \alpha + u\beta - u\beta_j$$

and hence

$$u = (\alpha_j - \alpha_1)/(\beta_1 - \beta_j)$$

contrary to the choice of u.

But this means that β is the only root of the gcd($q(x), p(\alpha + u\beta - ux)$). Moreover, β is not a multiple root. Hence

$$x - \beta = \gcd(q(x), p(\alpha + u\beta - ux)) \in F(\alpha + u\beta)[x]$$

hence $\beta \in F(\alpha + u\beta)$ and so $\alpha \in F(\alpha + u\beta)$ also.

Remark. The proof works perfectly well over any field F with an infinite number of elements provided the irreducible polynomial $q(x)$ does not have multiple roots.

12.4.2 Theorem.   Simple Extensions

Let E be a finite extension of a field F having characteristic zero. Then there exists $\tau \in E$ such that $E = F(\tau)$.

Proof.   Since $[E : F]$ is finite, E has a finite basis, say,

$$\alpha_1, \ldots, \alpha_n$$

over F. The result now follows by n applications of Lemma 12.4.1.

Terminology.   An extension E over a field F is said to be *simple* if there exists $\tau \in E$ such that $E = F(\tau)$. Note that a transcendental extension may be simple; however, we shall be chiefly concerned with algebraic extensions.

12.4.3 Theorem.   "All or None" Property of Splitting Fields

Let E be the splitting field of a polynomial $f(x) \in F[x]$ over F. Suppose that $p(x)$ is an *irreducible* polynomial in $F[x]$ that has a root in E. Then $p(x)$ splits in E.

Proof.   To create a field of reference, let L be the splitting field of $p(x)$ over E. Thus L contains all roots of $f(x)$ and $p(x)$.

Let α be a root of $p(x)$ in E and let β be a root of $p(x)$ in L. Thus E contains $F(\alpha)$ and we want to show that E also contains $F(\beta)$. In any event we know from Theorem 10.3.6 that there is an isomorphism φ,

$$F(\alpha) \cong F(\beta)$$

such that $\phi(\alpha) = \beta$ and we see that E is the splitting field of $f(x)$ over $F(\alpha)$ as well as over F. Let E' be the splitting field of $f(x)$ over $F(\beta)$ contained in L. We have $E' \supseteq E$ since E is the splitting field $f(x)$ over F. These subfield relationships are pictured in Figure 12.6.

Now apply the isomorphism theorem of splitting fields (Theorem 10.3.4) to see that the isomorphism ϕ may be extended to an isomorphism Φ of E and E'. In particular the degrees of E over $F(\alpha)$ and E' over $F(\beta)$ are equal,

L = splitting field of $p(x)$ over E

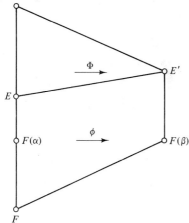

Figure 12.6 The subfields constructed in the proof of Theorem 12.4.3. The proof shows that $[E: F(\alpha)] = [E': F(\beta)]$.

and hence the degrees of E and E' over F are equal (to the product $[E: F(\alpha)][F(\alpha): F]$).

Now it follows that $E' = E$. Thus β belongs to E.

EXERCISE SET 12.4

12.4.1 Let E be the splitting field of $p(x) = x^4 - 2x^2 - 2$ over Q. Let the zeros of $p(x)$ in E be $\alpha, -\alpha, \beta, -\beta$. Thus $E = \mathbb{Q}(\alpha, \beta)$.
  1. Determine all rational numbers c such that $E = \mathbb{Q}(\alpha + c\beta)$.
  2. Find an irreducible polynomial $q(x) \in \mathbb{Q}[x]$ having $\alpha + \beta$ as a root.
  3. Determine the degree $[\mathbb{Q}(\alpha + \beta): Q]$.
  4. Show that $q(x)$ splits in E by exhibiting all its roots expressed in terms of α and β.

12.4.2 Show that if the characteristic of F is not 2 and $F(\sqrt{a}) \neq F(\sqrt{b})$ then $F(\sqrt{a}, \sqrt{b}) = F(\sqrt{a} + \sqrt{b})$.

Chapter 13

Galois Theory for Fields of Characteristic Zero

13.1 THE FUNDAMENTAL THEOREM OF GALOIS THEORY

By the term "Galois theory" we mean the study of the interrelation between the subfields of a field E and the subgroups of the group of automorphisms of E. When E is the splitting field of a polynomial $f(x)$ over a field F and $\mathbb{G}$ is the group of automorphisms of E that fixes each element of F, then there is a one-to-one correspondence between subgroups $\mathbb{H} \subseteq \mathbb{G}$ and intermediate fields $E \supseteq K \supseteq F$, given by

$$K = \{ u \in E \colon \sigma(u) = u \text{ for all } \sigma \in \mathbb{H} \}$$

It also turns out that if $[E : F]$ is finite then the order of $\mathbb{G}$ equals $[E : F]$.

This result has many important consequences. For one, it shows that if $[E : F]$ is finite then there are only a finite number of subfields K such that $E \supseteq K \supseteq F$. The correspondence gives a great deal of information about the subfields and ultimately yields the important criteria of Galois for a polynomial to have zeros that are expressible in terms of "radicals." Along the way it gives a method for deriving the formulas for the zeros of quadratic, cubic, and quartic polynomials as well as permitting the construction of a fifth-degree equation with integer coefficients whose zeros are not expressible in terms of radicals.

The development made here is restricted to the case that the characteristic of F is zero, thereby assuring that irreducible polynomials have only simple zeros and that E is a finite extension of F. There are many generalizations and extensions of Galois theory.

We begin with the initial definitions of $\mathbb{G}$ and introduce the notations and elementary ideas of the correspondence. The prerequisites are Chapters 1 to 10 and Sections 12.2 and 12.4.

13.1.1 Definition. The Group of an Extension and a Polynomial

Let E be a field extension of a field F. The automorphisms of E that fix each element of F form a group under composition called the group of E over F.

If E is the splitting field of $f(x)$ over F then this group is also called the group of $f(x)$ over F. In this case we may represent the group as a group of permutations on the zeros of $f(x)$.

To see this let the zeros of $f(x)$ be $\alpha_1, \ldots, \alpha_n$ so that

$$E = F(\alpha_1, \ldots, \alpha_n)$$

As we have shown (Theorem 9.1.10), any automorphism of E fixing each element of F must permute the α's since they are zeros of a polynomial in $F[x]$. Thus there is a mapping from the group of automorphisms into a group of permutations on the zeros. This mapping is in fact an isomorphism since the α's span E over F and thus any automorphism is completely determined by the images of the zeros. In particular, only the identity automorphism will fix all the zeros. We often determine an automorphism by giving the corresponding permutation of the zeros.

We will not give any further argument that the set of these automorphisms do form a group, actually a subgroup of the group (Theorem 8.3.10) of all automorphisms of E. It should be clear that the composition of automorphisms fixing an element again fixes that element. Also the inverse of an automorphism fixes all elements fixed by the automorphism. Next we associate a subfield with each subgroup of automorphisms.

13.1.2 Definition. The Fixed Field of a Group of Automorphisms

If $\mathbb{H}$ is a subgroup of the group of automorphisms of E over F then

$$K = \{ u \in E : u \text{ is fixed by each element of } \mathbb{H} \}$$

is a subfield of E containing F. We should probably supply some details, but it is easy to see that the set of elements fixed by $\mathbb{H}$ include F and are closed under addition, multiplication, and the formation of multiplicative inverses. The field K is called the *fixed field* of $\mathbb{H}$ and for emphasis on the dependence on $\mathbb{H}$ we write $K = K(\mathbb{H})$.

Remark.   If the subgroup $\mathbb{H}$ is the identity then K is all of E. If $\mathbb{H}$ is the whole group $\mathbb{G}$ then K certainly contains F but, unless E is a splitting field of some polynomial over F, K may contain more elements than F. We spend a good portion of this section proving this special property of splitting fields!

In view of these definitions we may set up the following correspondences. First, to each intermediate field $E \supseteq K \supseteq F$ we have the subgroup $\mathbb{H} = \mathbb{H}(K)$ of automorphisms of E fixing each element of K.

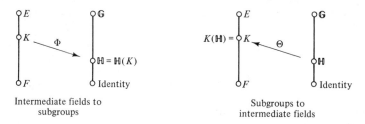

Figure 13.1 suggests these correspondences.

Figure 13.1 The Galois correspondence between subfields and subgroups

Second, to each subgroup $\mathbb{H}$ of the group of E over F we have the intermediate field $K = K(\mathbb{H})$ of elements fixed by all automorphisms of $\mathbb{H}$.

Figure 13.1 suggests these correspondences.

13.1.3 Theorem. Zeros of Polynomials and Automorphisms of a Field

Let σ be an automorphism of E over F. Let $u \in E$ be algebraic over F and let $f(x)$ be a polynomial in $F[x]$ of which u is a zero. Then:

1. The element $\sigma(u) \in E$ is also a zero of $f(x)$.
2. The order of the group of $F(u)$ over F is at most the degree of the *irreducible* polynomial in $F[x]$ of which u is a zero. In particular, if $[E : F]$ is finite, the order of the group of E over F is at most $[E : F]$.
3. If E is the splitting field of a polynomial over F then the group of E over F has order equal to the degree $[E : F]$.
4. If E is the splitting field of a polynomial over F then the group of E over F may be represented as a *transitive* permutation group.

Proof of 1. We have already established this result as part of Theorem 9.1.10, but the result is so important we repeat the details here. Write

$$f(x) = \sum_{i=0}^{i=n} c_i x^i$$

so that

$$0 = \sum_{i=0}^{i=n} c_i u^i$$

and hence, applying the automorphism σ,

$$0 = \sigma(0) = \sigma\left(\sum_{i=0}^{i=n} c_i x^i \right) = \sum_{i=0}^{i=n} \sigma(c_i u^i) = \sum_{i=0}^{i=n} \sigma(c_i)\sigma(u^i)$$

Now each $c_i \in F$ and is fixed by σ, and $\sigma(u^i) = (\sigma(u))^i$, hence

$$0 = \sum_{i=0}^{i=n} c_i(\sigma(u))^i = f(\sigma(u))$$

Proof of 2.   Note that the powers of u form a basis for $F(u)$ over F. Thus the image of each element of $F(u)$ under an automorphism of $F(u)$ is completely determined by the image of u. By 1, the image of u must be a zero of the irreducible polynomial for which u is a zero. Thus there are at most as many automorphisms as there are distinct zeros of the irreducible polynomial over F of which u is a zero. Since we have assumed char(F) = 0, this number is the degree $[F(u): F]$.

If E is a finite extension of F then by Theorem 12.4.2, $E = F(u)$ for some $u \in E$ and so there are at most $[E: F] = [F(u): F]$ automorphisms of E fixing F.

Proof of 3.   Now suppose E is the splitting field of a polynomial over F. We know (Theorem 12.4.2) that $E = F(u)$ for some $u \in E$. Let $p(x)$ be the irreducible polynomial in $F[x]$ of which u is a zero. From the "all or none" Theorem 12.4.3 we know that $p(x)$ splits in E. Also we know that for any two zeros u, v of the same irreducible polynomial over F, there is an isomorphism between the fields $F(u)$ and $F(v)$. Since $E = F(u)$ and $v \in E$ also we have $E = F(v)$ so the isomorphism is an automorphism. Thus there are as many possibilities for the image of u as there are zeros of $p(x)$, that is, as many as the degree of the polynomial, which is the degree $[E: F]$. We are guaranteed an automorphism for each of these possibilities by Theorem 10.3.4.

Proof of 4.   Part 3 of the proof gives us a little more information. We continue the notation. First, each automorphism may be recorded as a permutation on the zeros of $p(x)$. That is to say, the Galois group is isomorphic to a subgroup of the symmetric group on n symbols where n is the degree of the irreducible polynomial $p(x)$. The symbols are the zeros of $p(x)$. As we have just seen, for each pair of zeros u and v of $p(x)$ there is an automorphism of $E = F(u)$ taking u into v. Thus the subgroup of the symmetric group $\mathbb{S}_n$ to which the Galois group is isomorphic is *transitive* on the zeros of $p(x)$; that is, for each pair of symbols u, v there is a permutation of the subgroup taking u into v.

13.1.4 Definition.   Conjugate Elements and Conjugate Fields

If u and v are zeros of the same irreducible polynomial over F then u and v are said to be *conjugate* over F. The isomorphic fields $F(u)$ and $F(v)$ are said to be conjugate. Now we prove that three properties of a finite extension of F are equivalent. When an extension has any one, and hence all, of these properties we say it is a *normal extension* of F.

13.1.5 Theorem.   Normal Extension

Let F be a field of characteristic 0. The following three important properties of a finite extension E over F are equivalent:

1. If an irreducible polynomial $f(x) \in F[x]$ has one zero in E then $f(x)$ has all of its zeros in E.
2. E is the splitting field of a polynomial over F.
3. If $\mathbf{G}$ is the group of E over F then F is the fixed field of $\mathbf{G}$; that is, $F = \{ w \in E$ such that $\sigma(w) = w$ for all $\sigma \in \mathbf{G} \}$

Proof.   To prove that Property 1 implies Property 2, note that since E is a finite extension of F and char$(F) = 0$ it follows that $E = F(h)$ for some $h \in E$. Now Property 1 says that the irreducible polynomial in $F[x]$ having h as a zero splits in E. But since $E = F(h)$, that polynomial could split in no subfield of E; hence E is the splitting field of that polynomial.

We have already proved that 2 implies 1 in Theorem 12.4.3; thus 1 and 2 are equivalent. Now we prove that 1 and 2 imply 3. Let E be the splitting field of a polynomial $p(x) \in F[x]$.

By definition we know that each element of $\mathbf{G}$ fixes each element in F. Suppose that u lies in E but not in F. We will show that u is moved by some automorphism of E. Let $q(x)$ be the irreducible polynomial in $F[x]$ of which u is a zero. Since u is not in F we know the degree of $q(x) > 1$. Now by 1, $q(x)$ splits in E, so there is another zero v in E, distinct from u. Moreover, $F(u)$ and $F(v)$ are isomorphic by an isomorphism sending u into v. By the fundamental result (Theorem 10.3.4) on extending isomorphisms this isomorphism may be extended to an isomorphism of the splitting fields of $p(x)$ over $F(u)$ and $F(v)$, respectively. But since $E \supseteq F(u)$ and $E \supseteq F(v)$ it follows that E must be the splitting field of $p(x)$ over $F(u)$ and over $F(v)$. Thus the isomorphism of the splitting fields is an automorphism of E that sends u to v and fixes all elements of F. In other words, the only elements fixed by all automorphisms in the group of E over F are the elements of F.

Finally we prove that 3 implies 2. By Theorem 12.4.2 $E = F(u)$ for some $u \in E$. Now consider

$$f(x) = \sum_{\sigma \in \mathbf{G}} (x - \sigma(u))$$

We see that the coefficients of $f(x)$ are fixed by each $\sigma \in \mathbf{G}$ since such an automorphism only permutes the factors of $f(x)$. Thus by 3 these coefficients lie in F. Thus $f(x)$ splits in E and since $E = F(u)$, E must be the splitting field of $f(x)$ over F. Hence 2 holds. This completes the proof. Now we are ready to state and prove the fundamental theorem of Galois theory.

13.1.6 Definition.   Normal Extension and Galois Group

A finite extension $E \supseteq F$ is a *normal* extension of F if any one (and hence all) of the properties of Theorem 13.1.5 hold. Such extensions are also called

"Galois extensions." The group of automorphisms of E fixing F is called the *Galois group* of E over F. We often denote it by $\mathbb{G}(E/F)$.

13.1.7 The Fundamental Theorem of Galois Theory

Let F have characteristic 0 and let E be a normal extension of F. Let $\mathbb{G}$ be the Galois group of E over F.

1. There is a one-to-one correspondence between the subgroups of $\mathbb{G}$ and the intermediate fields K such that $E \supseteq K \supseteq F$ given by

$$\Phi: K \to \mathbb{H} = \mathbb{H}(K) = \text{the group of } E \text{ over } K$$

and

$$\Theta: \mathbb{H} \to K = K(\mathbb{H}) = \text{the fixed field of } \mathbb{H}$$

The group of E over $K(\mathbb{H})$ is $\mathbb{H}$ and the fixed field of $\mathbb{H}(K)$ is K.
2. The lattice of subgroups of $\mathbb{G}$ is the dual of the lattice of intermediate fields K, $E \supseteq K \supseteq F$.
3. An intermediate field K is a normal extension of F if and only if $\mathbb{H} = \mathbb{H}(K)$ is a normal subgroup of $\mathbb{G}$, in which case the group of K over F is isomorphic to $\mathbb{G}/\mathbb{H}$.

Proof of 1. We shall prove that

$$\Theta(\Phi(K)) = K$$

for all intermediate fields K and

$$\Phi(\Theta(\mathbb{H})) = \mathbb{H}$$

for all subgroups $\mathbb{H}$ of $\mathbb{G}$.

Notice that E is a normal extension of any intermediate field because it remains the splitting field of a polynomial over that field. Thus Theorem 13.1.5 applies to E and K and the group of E over K which is $\mathbb{H} = \Phi(K)$ by definition. Thus the fixed field of $\mathbb{H}$ is K and so

$$\Theta(\Phi(K)) = K$$

Thus Φ is one-to-one.

Now let $\mathbb{H}$ be any subgroup of $\mathbb{G}$. Let $K = \Theta(\mathbb{H})$ be the fixed field of $\mathbb{H}$. Consider the group of E over this intermediate field K. Call this group $\mathbb{H}'$. Since $\mathbb{H}$ fixes all the elements of $K = \Theta(\mathbb{H})$ by definition we have $\mathbb{H}' \supseteq \mathbb{H}$. We also have from Theorem 13.1.3, part 3, that the order of $\mathbb{H}'$ is the degree $[E : K]$. On the other hand, let $E = K(u)$ and consider the polynomial

$$f(x) = \prod_{\sigma \in \mathbb{H}} (x - \sigma(u))$$

This polynomial has coefficients in K because each coefficient is fixed by each automorphism of E in the group $\mathbb{H}$ and K is defined to be the elements of E fixed by $\mathbb{H}$. One of the zeros of $f(x)$ is u and so the degree $[E : K]$ is at most the order of $\mathbb{H}$; hence the order of $\mathbb{H}'$ is at most the order of $\mathbb{H}$. But since

$H' \supseteq H$ it follows that $H = H'$. Thus we have

$$\Phi(\Theta(H)) = H$$

From this it follows that Φ is onto and so it and its inverse map Θ are one-to-one correspondences of all intermediate fields to all subgroups.

Proof of 2. We need to show that the lattice join of two intermediate fields corresponds to the lattice meet of the corresponding subgroups and that the lattice meet of two intermediate subfields corresponds to the lattice join of two of the corresponding subgroups. Now this holds for any two lattices for which there is a one-to-one surjection Φ such that

$$K_1 \subseteq K_2 \quad \text{if and only if} \quad \Phi(K_1) \supseteq \Phi(K_2)$$

This result is not difficult to prove; here are the details.
In any event

$$K_1 \vee K_2 \supseteq K_1$$

and so

$$\Phi(K_1 \vee K_2) \subseteq \Phi(K_1)$$

and similarly,

$$\Phi(K_1 \vee K_2) \subseteq \Phi(K_2)$$

Thus

$$\Phi(K_1 \vee K_2) \subseteq \Phi(K_1) \cap \Phi(K_2)$$

On the other hand, suppose that H is any subgroup such that

$$H \subseteq \Phi(K_1) \quad \text{and} \quad H \subseteq \Phi(K_2)$$

then

$$K_3 = \Phi^{-1}(H) \supseteq K_1 \quad \text{and} \quad K_3 = \Phi^{-1}(H) \supseteq K_2$$

so that

$$K_3 \supseteq K_1 \vee K_2$$

and hence, applying Φ,

$$H \subseteq \Phi(K_1 \vee K_2)$$

Choosing $H = \Phi(K_1) \cap \Phi(K_2)$ we have

$$H = \Phi(K_1) \cap \Phi(K_2) \subseteq \Phi(K_1 \vee K_2) \subseteq \Phi(K_1) \cap \Phi(K_2)$$

thus

$$H = \Phi(K_1) \cap \Phi(K_2) = \Phi(K_1 \vee K_2)$$

A dual proof gives the other equality.

Proof of 3. Let us suppose that $E \supseteq K \supseteq F$ and the K is a normal extension of F. Let H be the group of E over K. We want to show that H is a normal subgroup of G. So let α be any automorphism in H and let σ be any automorphism in G. We want to show that

$$\sigma^{-1}\alpha\sigma \text{ is in } H$$

In view of the correspondence between subgroups and intermediate fields, to do this we need only show that $\sigma^{-1}\alpha\sigma$ fixes each element of K. If $v \in K$ then v is the zero of an irreducible polynomial over F. Since K is normal over F, that polynomial splits in K and so the other zeros of this polynomial lie in K too. But $\sigma(v)$ is one of those zeros (recall Theorem 13.1.3) and hence α fixes $\sigma(v)$. This means that

$$\alpha(\sigma(v)) = \sigma(v)$$

or that

$$\sigma^{-1}\alpha\sigma(v) = v$$

and hence $\sigma^{-1}\alpha\sigma$ fixes v, and thus any element of K. Hence $\sigma^{-1}\sigma\alpha$ lies in $\mathbb{H}$.

Conversely, let $\mathbb{H}$ be a normal subgroup of $\mathbb{G}$. We are to show that its corresponding intermediate fixed field is a normal extension of F. We will show that K is the splitting field of some polynomial $q(x)$ in $F[x]$. In any event we know that $K = F(u)$ for some $u \in K$. Let $q(x)$ be the monic irreducible polynomial over F of which u is a zero. We claim that $q(x)$ splits in K. In any event $q(x)$ splits in E and its zeros must all lie among the elements $\sigma(u)$ as σ ranges over the elements of $\mathbb{G}$. Thus to show that all the zeros of $q(x)$ lie in K it suffices, because of the one-to-one correspondence between intermediate fields and subgroups, to show that for each $\sigma \in \mathbb{G}$, $\sigma(u)$ is fixed by all $\alpha \in \mathbb{H}$. But

$$\alpha(\sigma(u)) = \sigma(u) \qquad \text{if and only if } \sigma^{-1}\alpha\sigma(u) = u$$

Since $\mathbb{H}$ is normal in $\mathbb{G}$, we have that σ^{-1} is in $\mathbb{H}$ and so fixes u, an element of K. It is important to note that these arguments show that if K is a normal extension of F then $\sigma(v) \in K$ for all $v \in K$.

Finally we want to show that if K is a normal extension of F then the group of K over F is isomorphic to the factor group $\mathbb{G}/\mathbb{H}$. To do this we want to define a homomorphism λ from $\mathbb{G}$ onto the group of K over F and verify that its kernel is $\mathbb{H}$.

We may choose λ to be what is called a *restriction* mapping. Take any $\sigma \in \mathbb{G}$. We have just shown that because K is normal over F if $v \in K$ then $\sigma(v) \in K$. Thus, if we restrict the domain of σ from E to K, we obtain an automorphism $\bar{\sigma}$ of K fixing each element of F. Thus the mapping

$$\lambda: \sigma \to \bar{\sigma}$$

is called the restriction mapping from $\mathbb{G}$, the group of all automorphisms of E fixing each element of F, to the group of automorphisms of K fixing each element of F. The mapping λ is easily seen to be a group homomorphism because the restriction of the composition of two automorphisms is the composition of the two restricted automorphisms.

To prove that λ maps $\mathbb{G}$ onto the group of automorphisms of K over F we prove that every automorphism of K over F arises as the restriction of some $\sigma \in \mathbb{G}$. Let $\bar{\sigma}$ be an automorphism of K over F. The larger field E still remains the splitting field of some polynomial over F and so, by the fundamental theorem on the extensions of isomorphism to splitting fields, $\bar{\sigma}$ may be extended to an automorphism σ of E over F. Since σ is an extension, its restriction back to K will be $\bar{\sigma}$. Thus the homomorphism λ is onto the group of K over F.

The kernel of this restriction homomorphism λ consists of those $\sigma \in G$ that act like the identity on K; but these are the automorphisms of E that fix K and thus are precisely the elements of $\mathbb{H}$. This completes the proof of the fundamental theorem of Galois theory. The mapping Φ of intermediate fields to subgroups or its inverse Θ is called a *Galois correspondence*.

13.1.8 Example.   The Group of $x^4 - 2$ over $\mathbb{Q}$

Let $F = \mathbb{Q}$ and let E be the splitting field of $x^4 - 2$ over $\mathbb{Q}$. In this example we determine the group of E over $\mathbb{Q}$ to be the dihedral group D_8 of order 8 and show explicitly the Galois correspondence of intermediate fields $\mathbb{Q} \subseteq K \subseteq E$ and subgroups of D_8.

We represent D_8 as permutations of the four zeros of $x^4 - 2$. All the information is entered on Figure 13.2. The numbering gives the Galois correspondence from subfield to subgroup and conversely.

The zeros of $x^4 - 2$ are

$$\sqrt[4]{2} \qquad -\sqrt[4]{2} \qquad i\sqrt[4]{2} \quad \text{and} \quad -i\sqrt[4]{2}$$

where i is a primitive fourth root of unity. Hereafter we call these zeros simply α, $-\alpha$, β, or $-\beta$, not only for convenience in writing but also to emphasize that all the algebraic information is in fact obtained from the polynomial, not from the fact that we have a more familiar notation for its zeros. In fact

$$x^4 - 2 = (x - \alpha)(x + \alpha)(x - \beta)(x + \beta) = (x^2 - \alpha^2)(x^2 - \beta^2)$$

so that by equating coefficients, $\alpha^2 + \beta^2 = 0$ and $\alpha^2\beta^2 = -2$. It follows that $\alpha^2/\beta^2 = -1$ and so α/β is a primitive fourth root of unity in E. We call this element i but insist that it is not necessarily the complex number!

1. We determine that $E = \mathbb{Q}(\alpha, \beta)$ and $[E : Q] = 8$. Here it is convenient to know that $\mathbb{Q}(\alpha)$ is a real field and thus does not contain β. To avoid this argument and use only information from the polynomial we could argue as follows to show that $\mathbb{Q}(\alpha)$ does not contain an element whose square is -1. First show that $\mathbb{Q}(\alpha^2)$ does not contain an element whose square is -1. Use the basis $\{1, \alpha^2\}$ for $\mathbb{Q}(\alpha^2)$ over $\mathbb{Q}$ and suppose that

$$-1 = (a + b\alpha^2)^2 = a^2 + 2b^2 + 2ab\alpha^2$$

where a and b belong to $\mathbb{Q}$. This means that $-1 = a^2 + 2b^2$ in $\mathbb{Q}$ which is not so. Next show that $\mathbb{Q}(\alpha)$ does not contain an element whose square is -1. Use the basis $\{1, \alpha\}$ for $\mathbb{Q}(\alpha)$ over $\mathbb{Q}(\alpha^2)$. Suppose that

$$-1 = (c + d\alpha)^2 = c^2 + d^2\alpha^2 + 2cd\alpha$$

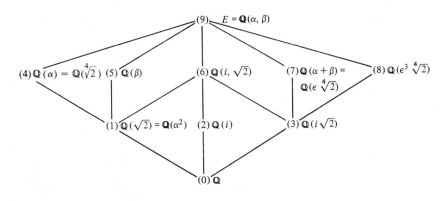

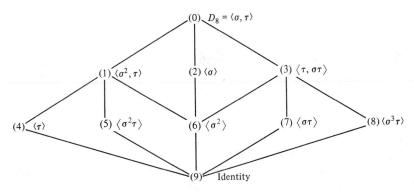

Figure 13.2 The Galois correspondence between subfields and the splitting field of $x^4 - 2$ and the subgroups of the group of automorphisms of E over $\mathbf{Q}$. The labeling of the subfields and the subgroups gives the correspondence. Note that the automorphisms for the generators for each subgroup fix the elements that generate the corresponding subfield. Here are the elements: $\alpha = \sqrt[4]{2}$, $\beta = i\sqrt[4]{2}$, $E = \mathbf{Q}(\alpha, \beta)$, $\mathbf{G} = D_8$, $\sigma = (\alpha, \beta, -\alpha, -\beta)$, and $\tau = (\beta, -\beta)$.

where c and d belong to $\mathbf{Q}(\alpha^2)$. This implies that the two equations

$$-1 = c^2 + d^2\alpha^2 \quad \text{and} \quad 0 = 2cd$$

have solutions in $\mathbf{Q}(\alpha^2)$. A little more argument shows that this leads to a contradiction.

2. Now $E = \mathbf{Q}(\alpha)(\beta)$ and so the identity automorphism of $\mathbf{Q}(\alpha)$ can be extended to an automorphism of E sending β into $-\beta$. Call this automorphism τ and we have

$$\tau = (\beta, -\beta)$$

Incidentally we see that this automorphism sends $i = \alpha/\beta$ into $-i = \alpha/\beta$ and hence τ is equivalent to complex conjugation.

3. $Q(\alpha)$ contains $Q(\alpha^2) = Q(\sqrt{2})$. The automorphism of this field sending α^2 into $-\alpha^2 = \beta^2$ can be extended to an automorphism of E sending α into β. This is so since $x^4 - 2$ splits into

$$(x^2 - \alpha^2)(x^2 - \beta^2)$$

over $Q(\alpha^2)$. The automorphism sending α^2 to β^2 sends $x^2 - \alpha^2$ to $x^2 - \beta^2$.

However, the theorem does not determine the image of β in this automorphism. There are two possibilities: β goes to either α or $-\alpha$. In fact both occur. Let σ denote the second possibility, say,

$$\sigma = (\alpha, \beta, -\alpha, -\beta)$$

then $\sigma\tau$, the composition of τ followed by σ, gives

$$\sigma\tau = (\alpha, \beta)(-\alpha, -\beta)$$

the other possibility. Similarly, $(\sigma\tau)\tau = \sigma$, so that either automorphism guarantees the existence of the other.

4. Now we find that the group generated by $\{\sigma, \tau\}$ is the dihedral group of order 8. Since the group of E over Q has order equal to the degree $[E : Q]$ we know that we have determined our group.

5. Now verify the Galois correspondence of the subgroup labeled n in Figure 13.1 to the intermediate field labeled n for $n = 0, 1, 2, 3, 4, 5, 6,$ and 9 by showing that the subgroups fix the primitive elements of the indicated intermediate fields.

6. The fields numbered 7 and 8 in Figure 13.1 come from the good observation that $\varepsilon = (1 + i)/\sqrt{2}$ is in E. Now ε is a primitive eighth root of unity. Thus the elements $\varepsilon^k \alpha$ are zeros of $x^4 + 2$ and belong to E. Alternatively argue as follows:

7. Find the field numbered 7 by observing what has been fixed by $\sigma\tau$. One such element is

$$u = \alpha + \beta$$

Find that $u^2 = 2i\sqrt{2}$ so that $u^4 = -8$. In fact $u(-i\sqrt{2}) = \varepsilon\alpha$.

EXERCISE SET 13.1

13.1.1 Determine the splitting field of $x^6 - 27$ over Q. Determine its Galois group and the correspondence between subfields and subgroups. Be sure to factor $x^6 - 27$ into irreducibles over Q.

13.1.2 Find the group of the cyclotomic polynomial $\Phi_n(x)$ over Q.

13.1.3 Find the group of $x^4 - 2x^2 - 2$ over Q. Find all the subfields of the splitting field. Give the correspondence guaranteed by the fundamental theorem. Determine the normal extensions and for each find an irreducible polynomial over Q whose zeros generate the extension.

13.1.4 For $n = 17$ find all subfields of $Q(\zeta)$ where ζ is an nth root of unity. Express each intermediate field not Q in the form $K(\sqrt{c})$, where c is an element in K

but not having a square zero in K. Thus explain the choices for the element β, γ, δ of Problem 12.2.11.

13.1.5 Find the splitting field E of $x^5 - 2$ over $\mathbb{Q}$ and all its subfields; find its Galois group and all of its subgroups. Give the Galois correspondence of subfields and subgroups. Determine the normal subgroups and normal subfields. Make legible lattice diagrams and indicate the Galois correspondence.

13.1.6 Let K be a normal extension of a field F with Galois group isomorphic to the direct product of n copies of the group of order 2. Prove that there exist elements $a_i \in F$, $d \in K$, $1 \leq i \leq n$, such that $K = F(d_1, \ldots, d_n)$ and $(d_i^2) = a_i$ for each i.

13.1.7 Let E be a finite normal extension of a field F and assume that the Galois group of the extension is abelian. Let $a \in K$ and let $f(x)$ be the monic irreducible polynomial over F of which a is a zero, and let b be a zero of $f(x)$. Prove that there exists $g(x) \in F[x]$ such that $b = g(a)$. (Do not assume that $K = F(a)$.)

13.1.8 Suppose that E is a normal extension of $\mathbb{Q}$ and that K is an intermediate field such that $[K : \mathbb{Q}] = m$. Suppose further that the Galois group of E over $\mathbb{Q}$ contains m automorphisms that map K into K and are distinct when restricted to K. That is, if α and β are two of these automorphisms then there exists $k \in K$ such that $\alpha(k) \neq \beta(k)$. Show that K is a normal extension of $\mathbb{Q}$.

13.1.9 Suppose that $f(x) = g(x)h(x) \in F[x]$ where F is a field of characteristic 0. Find a sufficient condition on the degree of g and h so that the Galois group of f over F is the direct product of the Galois groups of g and h over F.

Prove or disprove: If $g(x)$ and $h(x)$ have no common zero then the Galois group of f is the direct product of the Galois groups of g and h over F.

Determine the Galois groups of $(x^2 - 2)(x^3 - 3)$ and $(x^3 - 2)(x^3 - 3)$ over $\mathbb{Q}$.

13.2 THE FUNDAMENTAL THEOREM OF ALGEBRA

This is the classic theorem that says that the field of complex numbers is algebraically closed. There are many, many proofs of this fact. One of the most elegant ones uses a considerable amount of the analytic theory of complex numbers. Another interesting one is R.M. Redheffer, What! Another Note Just on the Fundamental Theorem of Algebra? *American Mathematical Monthly*, vol. 71, pp. 180–185, 1964. All proofs have an analytic part. In the version offered here, that is kept to a minimum and algebra becomes the workhorse.

13.2.1 Theorem.   The Fundamental Theorem of Algebra

Every polynomial with complex coefficients has a zero in the field of complex numbers.

Proof.   Let us begin by recasting the theorem in the notation and terminology we have developed. Let $\mathbb{R}$ denote the real numbers and let $\mathbb{C}$ denote the complex numbers. We take $\mathbb{C}$ to be a splitting field of $x^2 + 1$ over $\mathbb{R}$. If, as is the standard convention, we let i denote a zero of $x^2 + 1$ in $\mathbb{C}$ then $\mathbb{C} = \mathbb{R}(i)$.

The proof given here has two distinct parts, an analytic part and an algebraic part. The analytic part is the following lemma.

13.2.2 Lemma.   A Real Polynomial of Odd Degree Has a Real Zero

If $f(x) \in \mathbb{R}[x]$ has odd degree then there exists $u \in \mathbb{R}$ such that $f(u) = 0$.

Proof.   This is a standard theorem often presented in courses on calculus or real variables. It runs something like this:

Without loss of generality it may be assumed that

$$f(x) = x^n + c_1 x^{n-1} + \cdots + c_n$$

If n is odd then for large positive values of x (and it is not difficult to give bounds in terms of the coefficients c_i) $f(x)$ is positive while for large negative values of x, $f(x)$ is negative.

It is routine to show that as a function of the real variable x, $f(x)$ is a continuous function. While not exactly routine, it is a standard argument to show that continuous real-valued functions of a real variable have the intermediate value property; that is, if t is any value between $f(r)$ and $f(s)$ then for some u between r and s, $f(u) = t$. For polynomials this means that if $f(r)$ and $f(s)$ differ in sign then $f(u) = 0$ for some u between r and s.

Another property of the field of real numbers we need is that every positive real number has a real square root. But now the remainder of the argument can be algebraic.

13.2.3 Quadratic Polynomials in $\mathbb{C}[x]$ Have Zeros in $\mathbb{C}$

1. Every complex number has a square root.
2. If $\alpha(x) = x^2 + \beta x + \gamma \in \mathbb{C}[x]$ then $\alpha(x)$ splits in $\mathbb{C}[x]$

Proof.   Statement 2 can be reduced to the first by a completion of the square,

$$\alpha(x) = \left(x + \frac{\beta}{2}\right)^2 + \left(\gamma - \frac{\beta^2}{4}\right)$$

so that the zeros of $\alpha(x)$ are the square roots of $(\gamma - \beta^2/4)$ minus $\beta/2$.

The proof of statement 1 is essentially a computation to show that if

$$\alpha = a + bi$$

then a square root of α is $c + di$ where

$$d^2 = \frac{\sqrt{a^2 + b^2} - a}{2} \quad \text{and} \quad c = \frac{b}{2d}$$

Note that if $d = 0$ then α is real and so has a square root in $\mathbb{R}$ already.

The next lemma, the heart of the argument, is due to Gordan and appeared in H. Weber's treatise, *Kleines Lehrbuch der Algebra*, published in 1912.

13.2.4 Lemma.   Every Real Polynomial Has a Complex Zero

Every polynomial in $\mathbb{R}[x]$ has a zero in $\mathbb{C}$.

Proof. Let the degree of the polynomial be $n = 2^k q$ where q is odd. The proof is by induction on k, the exponent of 2. If $k = 0$ then the degree is odd and by Lemma 13.2.2, the polynomial has a real root, hence a root in $\mathbb{C}$.

Now suppose that $f(x)$ has degree $n = 2^k q$ where q is odd. Let S be a splitting field for f over $\mathbb{R}$ and let $\alpha_1, \ldots, \alpha_n$ be the zeros of $f(x)$, not necessarily distinct. For each integer $h \in \mathbb{Z}$ define a polynomial in $S[x]$.

$$g_n(x) = \prod_{i<j}\left(x - \left(\alpha_i + \alpha_j + h\alpha_i\alpha_j\right)\right)$$

There are two important things to note about each polynomial $g_n(x)$. The first is that its degree is

$$\binom{n}{2} = \frac{n(n-1)}{2} = 2^{k-1}q(n-1)$$

and that $q(n-1)$ is odd. The second thing to note is that $g_h(x)$ is fixed by any permutation of the zeros α_i of $f(x)$ and so it follows that $g_h(x) \in \mathbb{R}[x]$.

From these two facts we see that the induction hypothesis may be applied to each $g_n(x)$ and so we know that $\mathbb{C}$ contains at least one zero of each polynomial $g_h(x)$. There are an infinite number of h's but only a finite number of pairs (i, j) with $1 \leq i < j \leq n$ so that for some pair, say $(1, 2)$, and distinct integers h and k it must be that both

$$\beta_1 = \alpha_1 + \alpha_2 + h\alpha_1\alpha_2 \quad \text{and} \quad \beta_2 = \alpha_1 + \alpha_2 + k\alpha_1\alpha_2$$

belong to $\mathbb{C}$. Thus $\alpha_1\alpha_2 = (\beta_1 - \beta_2)/(h - k) \in \mathbb{C}$. But then $\alpha_1 + \alpha_2 \in \mathbb{C}$ also and so

$$x^2 - (\alpha_1 + \alpha_2)x + \alpha_1\alpha_2 \in \mathbb{C}[x]$$

Now Lemma 13.2.3 applies and it follows that zeros of this polynomial α_1 and α_2 lie in $\mathbb{C}$.

As an immediate extension of this result we can now prove Theorem 13.2.1 and we obtain even more by an easy induction.

13.2.5 Lemma.   Every Polynomial over $\mathbb{C}$ Splits in $\mathbb{C}$

Every polynomial on $\mathbb{C}[x]$ splits in $\mathbb{C}[x]$; that is, the field of complex numbers is algebraically closed.

Proof.   First we show that if $t(x) \in \mathbb{C}[x]$ then $t(x)$ has a zero α in $\mathbb{C}$ and so

$$t(x) = (x - \alpha)t_1(x) \in \mathbb{C}[x]$$

Thus a simple induction argument on the degree of polynomials will show that they split in $\mathbb{C}[x]$.

To show that $t(x)$ has a zero in $\mathbb{C}$, consider

$$f(x) = t(x)\bar{t}(x)$$

where $\bar{t}(x)$ denotes the polynomial in which every coefficient of $t(x)$ has been replaced by its complex conjugate. Note that $\beta \in \mathbb{C}$ is a zero of $t(x)$ if and only if $\bar{\beta}$ is a zero of $\bar{t}(x)$.

Now the polynomial $f(x) \in \mathbb{R}[x]$ since

$$f(x) = \bar{f}(x)$$

Thus we may apply Lemma 13.2.4 to show that $f(x)$ has a zero β in $\mathbb{C}$. Such a zero is either a zero of $t(x)$ or a zero of $\bar{t}(x)$. In the latter case $\bar{\beta}$ is a zero of $t(x)$.

13.3 THE DISCRIMINANT

It is not an easy problem to determine the Galois group of a polynomial, even a polynomial in $\mathbb{Z}[x]$. An even more difficult problem and one not completely solved is, given a group G, to determine whether there is a polynomial in $\mathbb{Z}[x]$ whose Galois group is G. An important first step in these directions is a test to decide whether, given a polynomial $f(x) \in F[x]$, its Galois group is a subgroup of an alternating group.

To be a bit more specific, let F be a field of characteristic zero. Let $f(x) \in F[x]$ have degree n. We want to decide whether or not the Galois group G of $f(x)$ over F is a subgroup of the alternating group $\mathscr{A}_n$. So let E be the splitting field of E over F. Let its zeros be $\alpha_1, \ldots, \alpha_n$. If $G \subseteq \mathscr{A}_n$ then we might seek some function of the zeros that is fixed by every even permutation of $\alpha_1, \ldots, \alpha_n$ and moved by every odd permutation. Since an odd permutation is a product of an odd number of transpositions and an even permutation is the product of an even number of transpositions we may begin our search by looking for a function of $\alpha_1, \ldots, \alpha_n$ that is moved by a single transposition but fixed by a composition of two transpositions. One attribute that flip-flops is the sign; suppose we could find a function, let's call it Δ, such that Δ goes to $-\Delta$ under every transposition. Then the composition of two (or an even number) of transpositions would send Δ into Δ. So consider $\alpha_i - \alpha_j$. The transposition (α_i, α_j) switches the sign of this difference of zeros. To deal with all the transpositions on $\alpha_1, \ldots, \alpha_n$ we are led to consider what is called the discriminant of $f(x)$.

13.3.1 Definition.   The Discriminant of a Polynomial

Let $f(x) \in F[x]$ and let $f(x) = c(x - \alpha_1) \cdots (x - \alpha_n)$ in a splitting field of $f(x)$ over F. Let

$$\Delta_f = \prod_{1 \le i < j \le n} (\alpha_i - \alpha_j)$$

The *discriminant* of $f(x)$ is Δ_f^2.

13.3.2 Theorem.   Role of the Square Root of the Discriminant

Let $f(x) = c(x - \alpha_1) \cdots (x - \alpha_n)$ in a splitting field of E over F. Let τ be an automorphism of E fixing each element of F. If τ effects a transposition on the symbols $\alpha_1, \ldots, \alpha_n$ then $\tau(\Delta_f) = -\Delta_f$. More generally, if τ effects an odd permutation of the zeros of $f(x)$ then $\tau(\Delta_f) = -\Delta_f$, if τ effects an even permutation of the zeros of $f(x)$ then $\tau(\Delta_f) = \Delta_f$. Thus Δ_f belongs to F if and only if the Galois group of $f(x)$ over F is a subgroup of the alternating group $\mathscr{A}_n$. All permutations fix the discriminant Δ_f^2.

Proof.   Suppose that τ effects the transposition (α_i, α_j). For brevity we write $\tau = (\alpha_i, \alpha_j)$. If a factor of Δ_f does not involve either α_i or α_j there is no change in the factor. Without loss of generality we may suppose that $i < j$. We consider the effect of τ in four cases involving factors with an α_i or an α_j.

Case 1.   Suppose $h < i < j$. The effect of τ is:

$$\tau: \alpha_h - \alpha_i \to \alpha_h - \alpha_j$$
$$\tau: \alpha_h - \alpha_j \to \alpha_h - \alpha_i$$

And so there is no change in the pair of factors $(\alpha_h - \alpha_i)(\alpha_h - \alpha_j)$ of Δ_f and so Δ_f is unchanged.

Case 2.   Suppose $i < j < h$. The effect of τ is:

$$\tau: \alpha_i - \alpha_h \to \alpha_j - \alpha_h$$
$$\tau: \alpha_j - \alpha_h \to \alpha_i - \alpha_h$$

And so as in case 1 there is no change in Δ_f.

Case 3.   Suppose $i < h < j$. The effect of τ is:

$$\tau: \alpha_i - \alpha_h \to \alpha_j - \alpha_h = -(\alpha_h - \alpha_j)$$
$$\tau: \alpha_h - \alpha_j \to \alpha_h - \alpha_i = -(\alpha_i - \alpha_h)$$

Again, the product of the two factors $(\alpha_i - \alpha_h)(\alpha_h - \alpha_j)$ is unchanged in Δ_f.

Case 4.   Suppose $h = i$ or $h = j$. Then $\tau: \alpha_i - \alpha_j \to \alpha_j - \alpha_i = -(\alpha_i - \alpha_j)$. Thus the total effect of τ on Δ_f is: $\tau(\Delta_f) = -\Delta_f$. The rest of the statement then follows from the fundamental theorem of Galois theory.

Hence if we want to determine whether the Galois group is a subgroup of the alternating group we may compute the discriminant. It will be an element of F. If its square root Δ_f is also in F then it will be fixed by every automorphism and consequently no automorphism can be an odd permutation of the zeros. Conversely, if Δ_f is not in F then some odd permutation of the zeros will have to belong to the Galois group, although not necessarily a transposition.

Remark.   We can now tell the origin of the term "alternating group." Let x and y be two indeterminates over a commutative ring R. A polynomial $h(x, y) \in R[x, y]$ is called alternating if $h(y, x) = -h(x, y)$. More generally, if $x_1, \ldots, x_n$ are n indeterminates over R, a polynomial $h(x_1, \ldots, x_n) \in R[x_1, \ldots, x_n]$ is said to be alternating in the n indeterminates if it is alternating in each pair. Thus

$$\Delta(x_1, \ldots, x_n) = \prod_{1 \leq i < j \leq n} (x_i - x_j)$$

is alternating in the n indeterminates. The subgroup of permutations on the symbols $x_1, \ldots, x_n$ that fix every alternating form is called the alternating group. And $\Delta(x_1, \ldots, x_n)$ is the litmus test for the alternating group.

The computation of the discriminant is always a little tedious. The following result helps.

13.3.3 Theorem.   Computation of the Discriminant from the Derivative

If $f(x)$ is a monic polynomial, $f(x) = \prod_{i=1}^{n}(x - \alpha_i)$ then

$$\Delta_f^2 = (-1)^{\frac{n(n-1)}{2}} \prod_{j=1}^{n} f'(\alpha_j)$$

Proof.   Using the properties of the formal derivative of $f(x)$ we calculate:

$$f'(x) = \sum_j \left(\prod_{i \neq j} (x - \alpha_i) \right)$$

and so

$$f'(\alpha_j) = \prod_{i \neq j} (\alpha_j - \alpha_i)$$

thus

$$\prod_{j=1}^{n} f'(\alpha_j) = \prod_{j=1}^{n} \left(\prod_{i \neq j} (\alpha_j - \alpha_i) \right) \tag{1}$$

Thus we need only to check up on the signs since

$$\Delta_f^2 = \prod_{i < j} (\alpha_i - \alpha_j)^2$$

while in (1), for fixed $i < j$, both $\alpha_i - \alpha_j$ and $\alpha_j - \alpha_i$ occur. Thus for each ordered pair, (i, j), $i < j$, if each term $(\alpha_j - \alpha_i)$ is replaced by $(-1)(\alpha_i - \alpha_j)$, the number of factors (-1) introduced is the number of ordered pairs (i, j) where $1 \le i < j \le n$. This number is $n(n - 1)/2$.

13.3.4 Example.   The Discriminant of a Quadratic and a Cubic

1. The discriminant of the quadratic $x^2 + bx + c$ is $b^2 - 4c$. Let

$$x^2 + bx + c = (x - \alpha_1)(x - \alpha_2) = x^2 - (\alpha_1 + \alpha_2)x + \alpha_1\alpha_2$$

so that $\alpha_1 + \alpha_2 = -b$ and $\alpha_1\alpha_2 = c$. Now $f'(x) = 2x + b$ so

$$f'(\alpha_1)f'(\alpha_2) = (2\alpha_1 + b)(2\alpha_2 + b) = 4\alpha_1\alpha_2 + 2b(\alpha_1 + \alpha_2) + b^2$$
$$= 4c - 2b^2 + b^2$$

and so the discriminant of $x^2 + bx + c$ is

$$\Delta_2^2 = (-1)^{\frac{2 \cdot 1}{2}}(4c - b^2) = b^2 - 4c$$

as is familiar from high school algebra!

2. The discriminant of the cubic $f(x) = x^3 + px + q$ is $-(27q^2 + 4p^3)$. Let

$$f(x) = (x - \alpha_1)(x - \alpha_2)(x - \alpha_3)$$
$$= x^3 - (\alpha_1 + \alpha_2 + \alpha_3)x^2 + (\alpha_1\alpha_2 + \alpha_2\alpha_3 + \alpha_1\alpha_3)x - \alpha_1\alpha_2\alpha_3$$

so that

$$\alpha_1 + \alpha_2 + \alpha_3 = 0$$
$$\alpha_1\alpha_2 + \alpha_2\alpha_3 + \alpha_1\alpha_3 = p$$
$$\alpha_1\alpha_2\alpha_3 = -q$$

Then

$$f'(\alpha_1)f'(\alpha_2)f'(\alpha_3)$$
$$= (3\alpha_1^2 + p)(3\alpha_2^2 + p)(3\alpha_3^2 + p)$$
$$= 27(\alpha_1\alpha_2\alpha_3)^2 + 9p(\alpha_1^2\alpha_2^2 + \alpha_2^2\alpha_3^2 + \alpha_1^2\alpha_3^2) + 3p^2(\alpha_1^2 + \alpha_2^2 + \alpha_3^2) + p^3$$

Now $\alpha_1\alpha_2\alpha_3 = -q$ and

$$(\alpha_1^2 + \alpha_2^2 + \alpha_3^2) = (\alpha_1 + \alpha_2 + \alpha_3)^2 - 2(\alpha_1\alpha_2 + \alpha_2\alpha_3 + \alpha_1\alpha_3) = -2p$$

Finally

$$(\alpha_1^2\alpha_2^2 + \alpha_2^2\alpha_3^2 + \alpha_1^2\alpha_3^2) = (\alpha_1\alpha_2 + \alpha_2\alpha_3 + \alpha_1\alpha_3)^2$$
$$- 2\alpha_1\alpha_2\alpha_3(\alpha_1 + \alpha_2 + \alpha_3)$$
$$= p^2$$

Hence by substituting these values we obtain

$$\Delta^2 = (-1)^{\frac{3 \cdot 2}{2}}(27q^2 + 9p(p^2) + 3p^2(-p) + p^3) = -(27q^2 + 4p^3)$$

EXERCISE SET 13.3

13.3.1 Determine the possibilities of $[F(\Delta):F]$ where Δ is a square root of the discriminant of $f(x) \in F[x]$.

13.3.2 Show that the discriminants of $f(x)$ and $t(x) = f(x + c)$ are equal.

13.3.3 If E is the splitting fields of $f(x)$ and $g(x)$ over a field F and Δ_f and Δ_g are square roots of their respective discriminants, what if any is the relation between Δ_f and Δ_g? What if any is the relation between $F(\Delta_f)$ and $F(\Delta_g)$?

13.3.4 Let $f(x) \in F[x]$ be the quartic $x^4 + bx^2 + cx + d$. Find the discriminant of $f(x)$.

13.3.5 Let $f(x) \in F[x]$ have degree 4 and zeros α_1, α_2, α_3, and α_4 in some extension of F. Define

$$\rho_1 = \alpha_1\alpha_2 + \alpha_3\alpha_4$$
$$\rho_2 = \alpha_1\alpha_3 + \alpha_2\alpha_4$$
$$\rho_3 = \alpha_1\alpha_4 + \alpha_2\alpha_3$$

Show that $g(x) = (x - \rho_1)(x - \rho_2)(x - \rho_3)$ lies in $F[x]$ and that f and g have the same discriminant. (This is sometimes an easier way to compute the discriminant of a quartic than the straightforward way of the previous exercise.)

Chapter **14**

Solutions in Radicals

Now we take a first look at the following classical problem:

Given a polynomial $f(x) \in F[x]$, when is it possible to express the zeros of $f(x)$ in terms of elements of the form $\sqrt[n]{a}$?

The question is vaguely put. Of course, F is a field and in this text we assume that $\text{char}(F) = 0$. Without loss of generality we always assume that f is monic.

We do have some examples of what we mean:

The zeros of a quadratic polynomial

$$x^2 + bx + c \tag{1}$$

have the form

$$\frac{1}{2}\left(-b \pm \sqrt{b^2 - 4c}\right) \tag{2}$$

The number $\sqrt{b^2 - 4c}$ denotes, of course, a zero of the polynomial

$$x^2 - (b^2 - 4c)$$

The expression (2) represents a formula for the zeros of the quadratic (1) in the following sense. Regard b and c as indeterminates over a field K and let $F = K(b, c)$. Then upon substitution for any elements of K for b and c, say, β and ψ, we obtain the expression for the zeros of

$$x^2 + \beta x + \psi$$

by substituting β for b and ψ for c in (2).

Similar but more complicated expressions exist for the zeros of cubics and quartics. However, there are no such formulas for general polynomials whose degrees exceed 4.

Of course, some polynomials of degree higher than 4 can be expressed in terms of radicals. For example, in Exercise 12.1.6 the zeros of

$$x^4 + x^3 + x^2 + x + 1$$

were expressed in the form

$$\frac{1}{2}\left(-c \pm \sqrt{c^2 - 4}\right)$$

where

$$c = \frac{1}{2}\left(-1 \pm \sqrt{5}\right)$$

Thus the fifth roots of unity over $\mathbb{Q}$ can be expressed in terms of radicals or expressions that involve other radicals.

In this chapter we make these notions precise and present the criterion due to Galois for a polynomial to have its zeros expressible in terms of radicals. The prerequisites are Chapters 1 to 10, Sections 12.2 and 12.4, and Chapter 13.

14.1 SOLUTIONS IN RADICALS: A NECESSARY CONDITION

First we introduce a notion that contains the essence of the phrase "expressible in radicals."

14.1.1 Definition.   Root Tower

A *root tower* over a field F is a finite chain of extensions

$$F = F_0 \subseteq F_1 \subseteq \cdots \subseteq F_k$$

where for each $i, 1 \le i \le k$,

$$F_i = F_{i-1}(\alpha_i)$$

and there is an integer n_i such that

$$\alpha_i^{n_i} = \beta_i \in F_i$$

The top of the root tower is the field F_k. A field E over F is said to be contained in a root tower if E is the subfield of the top of a root tower over F.

Remark.   The sense of this definition is that any element that is contained in a root tower can be expressed as a combination of basis elements that involve only zeros of polynomials of the form $x^m - b$.

Informally, we say that

$$\alpha_i \text{ is an } n_i\text{th root of } \beta_i$$

and write

$$\alpha_i = \sqrt[n_i]{\beta_i}$$

although this notation does not specify which root is designated.

Note that each element of F_i can be expressed in the form

$$\sum_{k=0}^{n_i-1} c_k \alpha_i^k$$

where $c_k \in F_{i-1}$. Note that α_i simply denotes one (unspecified) n_ith root of β_i. There is, in general, no canonical way of denoting a particular root. True, in the case that $F = \mathbb{Q}(\alpha)$ where $\alpha^2 \in Q$ and is positive, we do designate

$$\sqrt{\alpha^2}$$

as the *positive* square root. Such a luxury does not exist more generally. The laws governing algebraic operations with roots that are valid for roots of positive real numbers simply do not generalize. Watch your symbol manipulation like a hawk. What's wrong with the following?

$$-1 = \left(\sqrt{-1}\right)^2 = \sqrt{-1}\sqrt{-1} = \sqrt{(-1)(-1)} = \sqrt{1} = 1$$

14.1.2 Definition.   Solution in Terms of Radicals

Let F be a field and let $f(x) \in F[x]$. The equation

$$f(x) = 0$$

is said to be *solvable in terms of radicals* if its splitting field over F is contained in a root tower over F. We say the zeros of $f(x)$ may be expressed in terms of radicals if the splitting field is contained in a root tower.

The classical result in this subject is due to Galois. He found a condition on the group of $f(x)$ over F, that is, the group of the automorphisms of the splitting field of $f(x)$ over F that fix each element of F, which is both necessary and sufficient for $f(x)$ to be solvable in terms of radicals. Naturally, a group having this condition is called "solvable." In this section we introduce this condition and show that it is necessary.

The first lemma is a special case of the necessity. Two things make it special. One is that we assume the top of the root tower to be a normal extension of F. The other is that we assume that F contains the roots of unity of all the orders present in the root tower. Of the two special conditions it is the presence of the roots of unity that will be the most tedious to remove.

14.1.3 Lemma.   The Necessary Condition

Suppose that the splitting field E of $f(x)$ over F is the top of a root tower

$$F_0 = F \subseteq F_1 \subseteq \cdots \subseteq F_k = E$$

and suppose further that F contains the n_ith roots of unity whenever we have

$$F_i = F_{i-1}(\alpha_i) \quad \text{with} \quad \alpha_i^{n_i} \in F_{i-1}$$

in the root tower.

Under these conditions the Galois group $\mathbb{G}(E/F)$ contains a sequence of subgroups

$$\mathbb{G}(E/F) = \mathbb{G}_0 \supseteq \mathbb{G}_1 \supseteq \cdots \supseteq \mathbb{G}_k = \text{identity}$$

such that $\mathbb{G}_{j+1}$ is a normal subgroup of $\mathbb{G}_j$ and

$$\mathbb{G}_j/\mathbb{G}_{j+1} \text{ is cyclic}$$

(A group with this property will be called solvable.)

Proof. Since $F_i = F_{i-1}(\alpha_i)$ with $\alpha_i^{n_i} \in F_{i-1}$ and since F_{i-1} contains all the n_ith roots of unity, we see that F_i is the splitting field of

$$x^{n_i} - \alpha_i^{n_i}$$

over F_{i-1} and that

$$\mathbb{G}(F_i/F_{i-1})$$

is cyclic (Theorem 12.2.11).

But according to the fundamental theorem of Galois theory,

$$\mathbb{G}(F_i/F_{i-1}) \cong \mathbb{G}(F_k/F_{i-1})/\mathbb{G}(F_k/F_i)$$

Thus the Galois group of $E = F_k$ over $F = F_0$ has this chain of subgroups:

$$\mathbb{G}(F_k/F_0) = \mathbb{G}_0 \supseteq \mathbb{G}(F_k/F_1) = \mathbb{G}_1 \supseteq \cdots \supseteq \mathbb{G}(F_k/F_k) = \mathbb{G}_k = \text{identity}$$

in which each member is a normal subgroup of the subgroup above it in the chain and the factor group is *cyclic*.

Note that this does not imply that the groups in the chain are normal subgroups of $\mathbb{G}_k$.

14.1.4 Definition. Solvable Group

A group G is called *solvable* if there is a chain of subgroups

$$G = G_0 \supseteq G_1 \supseteq \cdots \supseteq G_k = \text{identity}$$

such that for each i, G_{i+1} is a normal subgroup of G_i and

$$G_i/G_{i+1} \text{ is cyclic}$$

We shall have more to say about solvable groups from the group point of view. Here are some examples of solvable groups.

14.1.5 Examples. Some Solvable Groups

First, any cyclic group is solvable. We can choose $k = 1$ in the definition !
Second, any dihedral group is solvable. Let $G = \langle a, b \rangle$ where $a^m = 1$ and

$b^2 = 1$ and $ba = a^{-1}b$. The following sequence satisfies the definition:

$$G \supset \langle a \rangle \supset \text{identity}$$

Third, any finite abelian group is solvable. Recall that a finite abelian group is the direct product of cyclic groups, say,

$$A = C_1 \times C_2 \times \cdots \times C_k$$

Now a sequence satisfying the definition is

$$A \supset C_2 \times \cdots \times C_k \supset C_3 \times \cdots \times C_k \supset \cdots \supset C_k \supset \text{identity}$$

Finally, here is a class of nonsolvable groups! For $n \geq 5$ both the symmetric groups $\mathbb{S}_n$ and the alternating group $\mathscr{A}_n$ on n symbols are *not* solvable. This is because $\mathscr{A}_n$ has no normal subgroups except itself and the identity; hence no chain satisfying the definition can even begin. The group $\mathbb{S}_n$ has only itself, $\mathscr{A}_n$, and the identity as normal subgroups, and hence a chain satisfying the definition would have to begin with $\mathscr{A}_n$ as its first term, but then there would be no candidates for the next subgroup of the chain.

We shall prove in Section 14.4 that if a polynomial $f(x) \in F[x]$ is solvable in terms of radicals over F then the Galois group of $f(x)$ over F is solvable in the sense we've just defined. The full details of the proof of the necessity of this solvability condition on the group require a bit more about solvable groups and a rather more complicated argument than is pleasant, so we shall postpone them.

Instead we look at the sufficiency of the condition. Here the results are constructible and lead, among other things, to the formulas for the zeros of quadratic, cubic, and quartic polynomials.

14.2 CONSTRUCTING SOLUTIONS IN TERMS OF RADICALS

To prove that if the group of a polynomial is solvable then its zeros may be expressed in terms of radicals we must find a way to construct a root tower. This process is made possible by the presence of certain roots of unity in the underlying field.

We first study the case of a normal extension whose group is cyclic over a field containing the roots of unity corresponding to the order of the group. This is the simplest situation. The general case follows when we adjoin the necessary roots of unity and then show how to use the simple case construction over and over again to build the root tower. The clever device that does the job is called the *Lagrange resolvent*.

14.2.1 Definition.   Lagrange Resolvent

Let F be a field containing ζ, an nth root of unity. Let E be an extension of F. Let $a \in E$ and let σ be an automorphism of E fixing each element of F.

The Lagrange resolvent $\lambda(\zeta, a, \sigma)$ is defined:

$$\lambda(\zeta, a, \sigma) = a + \zeta \cdot \sigma(a) + \zeta^2 \cdot \sigma^2(a) + \cdots + \zeta^{n-1} \cdot \sigma^{n-1}(a)$$

Here a power k on the automorphism σ denotes the composition of k iterations of the automorphism σ.

The utility of the resolvent comes from the next lemma.

14.2.2 Lemma.   An Element Fixed by σ

Let σ be an automorphism of a field E. Let ζ be an nth root of unity fixed by σ. If the order of σ divides n then the nth power of the Lagrange resolvent is fixed by σ and

$$\sigma(\lambda(\zeta, a, \sigma)) = \zeta^{-1}\lambda(\zeta, a, \sigma)$$

Proof.   Let $a \in E$, and to simplify notation, let

$$v = \lambda(\zeta, a, \sigma)$$

We compute the effect of the automorphism σ on v and show that $\sigma(v^n) = v^n$.

Since $\zeta^n = 1$ and σ fixes ζ and has order n,

$$\sigma(v) = \sigma(a) + \zeta \cdot \sigma^2(a) + \zeta^2 \cdot \sigma^3(a) + \cdots + \zeta^{n-1} \cdot \sigma^n(a)$$
$$= \zeta^{n-1}a + \sigma(a) + \zeta \cdot \sigma^2(a) + \cdots + \zeta^{n-2} \cdot \sigma^{n-1}(a)$$
$$= \zeta^{-1}v$$

Hence

$$\sigma(v) = \zeta^{-1}v$$

Now raise both sides to the nth power.

$$(\sigma(v))^n = (\zeta^{-1})^n v^n = v^n$$

On the other hand, since σ is an automorphism of E we have

$$(\sigma(v))^n = \sigma(v^n)$$

so

$$\sigma(v^n) = v^n$$

Thus v^n is fixed by σ and so belongs to the fixed field of $\langle \sigma \rangle$.

Now we can prove the essential step in constructing zeros in terms of radicals.

14.2.3 Theorem.   Cyclic Extensions

Let F be a field of characteristic zero and let E be a normal extension of F such that $\mathbb{G}(E/F)$ is cyclic of order n. Let F contain the nth roots of unity. Under these assumptions E is the splitting field of a polynomial

$$(x^n - b_1) \ldots (x^n - b_r)$$

where each $b_k \in F$.

In particular E is the top of a root tower from F.

$$F \subseteq F\left(\sqrt[n]{b_1}\right) \subseteq F\left(\sqrt[n]{b_1}, \sqrt[n]{b_2}\right) \subseteq \cdots \subseteq F\left(\sqrt[n]{b_1}, \ldots, \sqrt[n]{b_r}\right) = E$$

Moreover if n is a prime p then E is the splitting field of a polynomial $x^p - b$ which is *irreducible* in $F[x]$.

Proof. Since E is a finite extension of a field of characteristic zero we know that $E = F(a)$ by the simple extension theorem 12.4.2.

Let σ generate the Galois group $\hat{G}(E/F)$ which we've assumed to be cyclic.

Let ζ be a primitive nth root of unity which we're assuming to be in F.

Now we write down the Lagrange resolvents for the powers of ζ

$$v_k = \lambda\left(\zeta^k, a, \sigma\right)$$

for $k = 1, \ldots, n$. Explicitly

$$v_1 = a + \zeta \cdot \sigma(a) + \cdots + \zeta^k \cdot \sigma^k(a) + \cdots + \zeta^{n-1} \cdot \sigma^{n-1}(a)$$

$$v_2 = a + \zeta^2 \cdot \sigma(a) + \cdots + \zeta^{2k} \cdot \sigma^k(a) + \cdots + \zeta^{2(n-1)} \cdot \sigma^{n-1}(a)$$

$$\vdots$$

$$v_n = a + 1 \cdot \sigma(a) + \cdots + 1 \cdot \sigma^k(a) + \cdots + 1 \cdot \sigma^{n-1}(a)$$

Now add the columns:

Every column on the right-hand side except the first sums to zero! This is so because, if ω is any nth root of unity except 1, the sum

$$1 + \omega + \omega^2 + \cdots + \omega^{n-1} = \frac{\omega^n - 1}{\omega - 1} = 0$$

Thus we have

$$v_1 + v_2 + \cdots + v_n = n \cdot a$$

In particular this equation shows that

$$F(v_1, v_2, \ldots, v_n) = F(a) = E$$

Moreover Lemma 14.2.2 tells us that

$$b_k = v_k^n \in F$$

Thus E becomes the splitting field of the product

$$(x^n - b_1)(x^n - b_2) \ldots (x^n - b_n)$$

and is the top of the root tower obtained by successive adjunction of $v_1, \ldots, v_n$.

Finally, suppose that $n = p$ is a prime, then the pth roots of unity are all primitive so that for any power, say k, the roots of unity

$$\zeta^k, \zeta^{2k}, \ldots, \zeta^{k(n-1)}$$

are all distinct. Since for some k

$$v_k = \lambda(\zeta^k, a, \sigma) \neq 0$$

and by Lemma 14.2.2

$$\sigma^h(v_k) = \zeta^{-hk}v_k$$

it follows that the conjugates of v_k are the elements

$$v_k, \sigma(v_k), \sigma^2(v_k), \ldots, \sigma^{p-1}(v_k)$$

These conjugates are all distinct and they are, in some order,

$$v_k, \zeta v_k, \ldots, \zeta^{p-1}v_k$$

Thus v_k has p distinct conjugates that are zeros of $x^p - b_k$. This means that this polynomial is irreducible over F because zeros of different irreducible factors over F cannot be conjugate over F. And because $p = [E : F]$ it must be that $E = F(v_k)$ and so is the splitting field of $x^p - v_k^p$.

14.2.4 Example. Solving the Cubic

Before beginning the proof of the sufficiency of Galois' criterion for the solvability in radicals of a polynomial equation we pause to develop the solutions of a cubic polynomial equation. The methods we employ here give a clear indication of what must be done in general.

Let F be a field whose characteristic is not 2 or 3. Let

$$f(x) = x^3 + ax^2 + bx + c$$

To simplify the computation we translate to

$$g(x) = f\left(x - \frac{a}{3}\right) = x^3 + px + q$$

Note that $f(x)$ is reducible if and only if $g(x)$ is reducible and that the zeros of $f(x)$ are $\alpha - a/3$ where α is a zero of $g(x)$.

Let the other zeros of $g(x)$ be β and γ so that a splitting field of $g(x)$ over F is $F(\alpha, \beta, \gamma)$.

Although it is not important for the work here, to avoid always catching your thoughts if some of the fields we mention are not distinct, we assume that $g(x)$ is irreducible over F. Moreover we assume that Δ, a square root of the discriminant of $g(x)$, which according to Example 13.3.4 is a square root of

$$-(27q^2 + 4p^3)$$

is not in F. (Recall that Exercise 13.3.2 shows that the discriminants of $g(x)$ and $f(x)$ are the same.)

These two facts mean that the group of $g(x)$, as represented by permutations on the zeros of $g(x)$, is S_3. This is so because the group is a subgroup of S_3 and since $g(x)$ is irreducible we know that 3 divides the degree of the splitting field E of $g(x)$ over F; hence 3 divides the order of the Galois group. Second, since the square root of the discriminant is not in F, we see that 2 also divides the degree of E over F. Hence 6 divides $[E : F]$, and so 6 divides the order of the Galois group. Thus the group is $\mathbb{S}_3$.

We know that

$$\mathbb{S}_3 \supset \mathcal{A}_3 \supset I$$

is a sequence of subgroups that shows that $\mathbb{S}_3$ is solvable. The sequence of subfields of $F(\alpha, \beta, \gamma)$ under the Galois correspondence is

$$F \subset F(\Delta) \subset F(\alpha, \beta, \gamma)$$

In order to use the Lagrange resolvent technique we need to have the cube roots of unity in $F(\Delta)$. They may of course belong to $F(\Delta)$ (where Δ is a square root of the discriminant). But if not we begin by adjoining them to $F(\Delta)$.

Let us assume that $x^2 + x + 1$ is irreducible over $F(\Delta)$ and that the zeros, the cube roots of unity, are

$$\omega = (-1 + d)/2 \quad \text{and} \quad \omega^2 = (-1 - d)/2 \tag{1}$$

where d is a square root of the discriminant of $x^2 + x + 1$ which is -3. If F is a subfield of the complex numbers, it would be quite correct to write

$$d = i\sqrt{3}$$

The resulting fields and subfields of interest are shown in Figure 14.1.

The important fact for our construction is that $F(\Delta, \omega, \alpha)$ is a normal, cyclic extension of $F(\Delta, \omega)$. This is so because it has degree 3 and because $F(\Delta, \omega, \alpha)$ is the splitting field of $g(x)$ over $F(\Delta, \omega)$. Thus the only choice for the Galois group of $F(\Delta, \omega, \alpha)$ over $F(\Delta, \omega)$ is the cyclic group of order 3. We don't yet know the elements that make the extension $F(\Delta, \omega, \alpha)$ over $F(\Delta, \omega)$ a root extension. We'll use the Lagrange resolvent to do this.

What we should learn from this example for the later proof of the sufficiency is that we have adjoined the necessary roots of unity and then built a root tower around and containing the splitting field $F(\alpha, \beta, \gamma)$. The se-

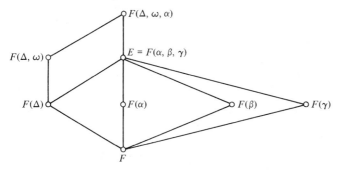

Figure 14.1 The subfields of an irreducible cubic whose zeros are α, β, γ over a field F. The illustration assumes that neither Δ, a square root of the discriminant, is in F nor is a primitive cube root of unity, ω, in $F(\Delta)$. Not all subfields are shown, but the required root tower is $F \subset F(\Delta) \subset F(\Delta, \omega) \subset F(\Delta, \omega, \alpha)$.

quence of subfields of $F(\alpha, \beta, \gamma)$ over F corresponding to the sequence of subgroups of S_3 has been extended to this root tower.

First our notation. We have

$$g(x) = x^3 + px + q = (x - \alpha)(x - \beta)(x - \gamma)$$

so that

$$0 = \alpha + \beta + \gamma$$
$$p = \alpha\beta + \alpha\gamma + \beta\gamma$$
$$q = -\alpha\beta\gamma$$

To construct the roots that we need, we employ the Lagrange resolvents in the case $n = 3$. Here they are:

$$r = \lambda(\omega, \alpha, \sigma) = \alpha + \omega\beta + \omega^2\gamma$$
$$s = \lambda(\omega^2, \alpha, \sigma) = \alpha + \omega^2\beta + \omega\gamma$$
$$t = \lambda(1, \alpha, \sigma) = \alpha + \beta + \gamma = 0.$$

And so by addition

$$r + s + t = 3\alpha$$
$$\omega^2 r + \omega s + t = 3\beta$$
$$\omega r + \omega^2 s + t = 3\gamma$$

Our Lagrange theory tells us that r^3, s^3, and t^3 are in F, so we proceed to calculate these powers. However, because the calculations are identities, they do not depend upon the cyclic automorphism (α, β, γ) for their validity; thus they hold whether or not $F(\Delta, \omega, \alpha)$ is a proper extension of $F(\Delta, \omega)$.

$$r^3 = (\alpha + \omega\beta + \omega^2\gamma)^3$$

and expanding we have

$$r^3 = (\alpha^3 + \beta^3 + \gamma^3) + 3\omega(\alpha^2\beta + \beta^2\gamma + \alpha\gamma^2)$$
$$+ 3\omega^2(\alpha\beta^2 + \beta\gamma^2 + \alpha^2\gamma) + 6\alpha\beta\gamma$$

Now using the basis $\{1, d\}$ of $F(\Delta, \omega)$ over $F(\Delta)$, and substituting for ω and ω^2 in this basis, as expressed in equations (1), we have

$$r^3 = (\alpha^3 + \beta^3 + \gamma^3)$$
$$- \frac{3}{2}(\alpha^2\beta + \beta^2\gamma + \alpha\gamma^2 + \alpha\beta^2 + \beta\gamma^2 + \alpha^2\gamma)$$
$$+ \frac{3}{2}d(\alpha^2\beta + \beta^2\gamma + \alpha\gamma^2 - \alpha\beta^2 - \beta\gamma^2 - \alpha^2\gamma)$$
$$+ 6\alpha\beta\gamma$$

The first and second terms are fixed by all automorphisms of E and so belong to F. We compute these expressions below. The third term can be recognized as $3d\Delta/2$,

$$\Delta = (\alpha - \beta)(\alpha - \gamma)(\beta - \gamma)$$

a square root of the discriminant. The last term we know to be $-6q$. Here are the calculations for the first and second terms:

$$\alpha^2\beta + \beta^2\gamma + \alpha\gamma^2 + \alpha\beta^2 + \beta\gamma^2 + \alpha^2\gamma$$
$$= (\alpha + \beta + \gamma)(\alpha\beta + \beta\gamma + \alpha\gamma) - 3\alpha\beta\gamma = 3q$$

$$\alpha^3 + \beta^3 + \gamma^3 = (\alpha + \beta + \gamma)^3 - 3(\alpha^2\beta + \beta^2\gamma + \alpha\gamma^2 + \alpha\beta^2 + \beta\gamma^2 + \alpha^2\gamma)$$
$$- 6\alpha\beta\gamma = -3q$$

Substituting these computations we determine

$$r^3 = -\frac{27}{2}q + \frac{3}{2}d\,\Delta$$

To get s^3 simply interchange ω and ω^2 to find

$$s^3 = -\frac{27}{2}q - \frac{3}{2}d\,\Delta$$

Hence we have

$$r = \sqrt[3]{-\frac{27}{2}q + \frac{3}{2}d\,\Delta} \quad \text{and} \quad s = \sqrt[3]{-\frac{27}{2}q - \frac{3}{2}d\,\Delta} \tag{2}$$

and so

$$\alpha = \frac{1}{3}\left(\sqrt[3]{-\frac{27}{2}q + \frac{3}{2}d\,\Delta} + \sqrt[3]{-\frac{27}{2}q - \frac{3}{2}d\,\Delta} \right) \tag{3}$$

Similarly, we obtain

$$\beta = (\omega^2 r + \omega s)/3 \quad \text{and} \quad \gamma = (\omega r + \omega^2 s)/3$$

There is one thing yet to be settled. There are three cube roots of any element. The notation used in (2) is not uniquely determined. Which cube roots are to be used? There is a simple requirement. We find that the product of the Lagrange resolvents

$$rs = (\alpha + \omega\beta + \omega^2\gamma)(\alpha + \omega^2\beta + \omega\gamma)$$
$$= \alpha^2 + \beta^2 + \gamma^2 - (\alpha\beta + \beta\gamma + \alpha\gamma) = -3p$$

Thus the cube roots chosen for r and s in equations (2) must satisfy the condition

$$rs = -3p \tag{4}$$

Conversely if this condition holds then formula (3) gives a solution we can test by substituting

$$\alpha = (r + s)/3$$

and assuming that $rs = -3p$:

$$\alpha^3 + p\alpha + q = \frac{1}{27}(r^3 + 3r^2s + 3rs^2 + s^3) + \frac{p}{3}(r + s) + q$$

$$= \frac{1}{27}(r^3 + s^3) + q = -q + q = 0$$

Equations (2) and (3) are called "Cardano's formulas" for the solution of the cubic. Usually we work within the field of complex numbers. Note, however, even if the zeros are all real we have had to use complex numbers ω and ω^2 to express the solution. Does it seem surprising to have to use complex numbers to express in radicals the zeros of polynomial that are all real? Before calculators and computers it seemed important to have formulas for determining these zeros "exactly." The formulas have lost a great deal of their appeal with the advent of numerical methods for approximating the real solutions. By the way, rapidly converging iterative schemes such as Newton's method work just as well to locate complex zeros as to find real zeros. Start with a complex first guess!

EXERCISE SET 14.2

14.2.1 Consider the splitting field E of $x^7 - 1$ over Q.

    1.1 Show that E is a cyclic extension of Q.

    1.2 Show that there is no root tower from Q to E.

    1.3 Find all the subfields of E.

    1.4 Show that if ω is a primitive cube root of unity then $(x^7 - 1)/(x - 1)$ remains irreducible over $Q(\omega)$.

    1.5 Find two root towers from $Q(\omega)$ to $E(\omega)$.

    *1.6 Find the "roots" that generate each step of the root towers.

14.2.2 Find the Galois group of $x^3 - 3x + 1$ over the rationals. Express the zeros of the polynomial in terms of radicals.

14.2.3 Use one of the root towers developed in Problem 14.2.1 above to express the seventh roots of unity in terms of radicals.

14.2.4 This exercise gives a method for determining the group of an irreducible fourth-degree equation over a field F of characteristic zero. Let

$$f(x) = x^4 + c_1x^3 + c_2x^2 + c_3x + c_4$$

have *distinct zeros* $\alpha, \beta, \delta, \gamma$. Let G be the Galois group of $f(x)$ over F represented as a permutation group on the zeros of $f(x)$. Let E be the splitting field of $f(x)$ over F. Recall (reprove) that G is a transitive permutation group on the zeros $\alpha, \beta, \delta, \gamma$. Let

$$r = \alpha\beta + \delta\gamma$$

$$s = \alpha\delta + \beta\gamma$$

$$t = \alpha\gamma + \beta\delta$$

We know from Exercise 13.3.5 that the discriminants of $f(x)$ and

$$g(x) = (x - r)(x - s)(x - t)$$

are the same.

1. Let V be the four-group of $\mathbb{S}_4$ generated by the permutations $(\alpha, \beta)(\delta, \gamma)$, $(\alpha, \beta)(\beta, \gamma)$, $(\alpha, \gamma)(\beta, \delta)$. Show that the fixed field of $G \cap V$ is $F(r, s, t)$.
2. Determine the transitive subgroups of $\mathbb{S}_4$.
3. Let H be the Galois group of $g(x)$ over F. Show that H is isomorphic to $G/(G \cap V)$.
4. Show that if $G = \mathbb{S}_4$ then H is $\mathbb{S}_3$.
5. Show that if $G = \mathscr{A}_4$ then H has order 3.
6. Show that if $G = V$ then H is the identity.
7. Show that if G is cyclic of order 4 then H has order 2.
8. Show that if G is dihedral of order 4 then H has order 2.
9. Show that if H has order 2 then cases 7 and 8 can be distinguished on the basis of the reducibility (case 7) or the irreducibility (case 8) of $f(x)$ over $F(\Delta)$.
10. Determine the Galois group of $x^4 + 3x^3 - 3x - 2$ over $\mathbb{Q}$.

14.3 A SUFFICIENT CONDITION FOR SOLUTION IN RADICALS

Now we prove the first half of the classic theorem of Galois on the solvability of a polynomial equation in terms of radicals. We begin by adjoining certain roots of unity to the given field F. The first lemma shows that we can do this without affecting seriously the Galois group of the polynomial.

14.3.1 Lemma.   Join of Normal Extensions

Let E and K be normal extensions of F contained in an extension L. Then $E \vee K$ is a normal extension of F and

$$G(E \vee K/K) \text{ is isomorphic to a subgroup of } G(E/F)$$

In particular:

1. $[E \vee K : K] = |G(E \vee K/K)|$ which divides $|G(E/F)| = [E : F]$.
2. If $G(E/F)$ is cyclic so is $G(E \vee K/K)$.

Proof.   Figure 14.2 shows the relationships of the field and their corresponding Galois groups. First note that E and K are splitting fields of polynomials over F. The field $E \vee K$ is the splitting field of the product of these polynomials over F.

Thus we have by the diamond lemma on subgroups that

$$G(E \vee K/K) \text{ is isomorphic to } G(E \vee K/K \cap E)/G(E \vee K/E)$$

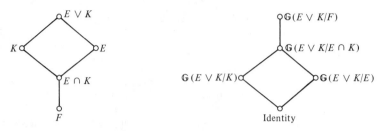

Figure 14.2 The subfields and the corresponding subgroups for Lemma 14.3.1. The field $E \vee K$ is a normal extension of F and the subfields E, K, and $E \cap K$ are normal extensions of F.

and by the fundamental theorem of Galois theory the latter factor group is isomorphic to $G(E/K \cap E)$ which is a subgroup of $G(E/F)$.

Now we are ready for the proof of the sufficiency.

14.3.2 Theorem. Solvability of the Galois Group Implies Solvability in Terms of Radicals

Let F be a field of characteristic zero and let E be the splitting field of $f(x) \in F[x]$. If $G(E/F)$ is solvable then $f(x) = 0$ is solvable in terms of radicals.

Proof. Let

$$G(E/F) = G_0 \supset G_1 \cdots \supset G_r = \text{identity}$$

be a sequence of subgroups such that each G_{i+1} is normal in G_1 and G_i/G_{i+1} is cyclic. Let

$$F = H_0 \subset H_1 \subset \cdots \subset H_r = E$$

be the corresponding sequence of subfields under the Galois correspondence of the fundamental theorem of Galois theory. This theorem shows that

$$G(E/H_i) = G_i$$

and that H_{i+1} is a normal extension of H_i. Thus

$$G(H_{i+1}/H_i) \text{ is isomorphic to } G_i/G_{i+1}$$

and so H_{i+1} is a cyclic extension of H_i. However, since we cannot be sure that the proper roots of unity lie in H_i we cannot yet apply the Lagrange resolvent technique. So we fix this up first. Let $[E : F] = n$. Adjoin to F the nth roots of unity, and call the enlarged field $\bar{F}$. Thus

$$\bar{F} = F(\omega)$$

where ω is a primitive nth root of unity. This is both a normal extension and, by definition, an acceptable first step of a root tower over F. (This seems almost like cheating, and we will prove a much better theorem on the adjunction of roots of unity in terms of radicals. See Theorem 14.4.6.)

Let

$$\overline{\mathbb{H}}_i = \overline{F} \vee \mathbb{H}_i = \mathbb{H}_i(\omega)$$

Now $\overline{\mathbb{H}}_i$ is a normal extension of $\mathbb{H}_i$ and

$$\overline{\mathbb{H}}_{i+1} = \overline{\mathbb{H}}_i \vee \mathbb{H}_{i+1}$$

Apply Lemma 14.3.1 with $F = \mathbb{H}_i$, $E = \mathbb{H}_{i+1}$, and $K = \overline{\mathbb{H}}_i$. We see that $\mathbb{G}(\overline{\mathbb{H}}_{i+1}/\overline{\mathbb{H}}_i)$ is isomorphic to a subgroup of $\mathbb{G}(\mathbb{H}_{i+1}/\mathbb{H}_i)$ and so is cyclic.

Since $\mathbb{H}_i$ contains the nth roots of unity and since $[\mathbb{H}_{i+1}:\mathbb{H}_i]$ divides n we know that $\overline{\mathbb{H}}_i$ contains the $[\mathbb{H}_{i+1}:\mathbb{H}_i]$th roots of unity.

For these reasons the theorem on cyclic extensions applies (Theorem 14.2.3) and so we find that $\overline{\mathbb{H}}_{i+1}$ is the top of a root tower from $\overline{\mathbb{H}}_i$.

Concatenating these root towers builds a root tower from F to $\overline{E} = E(\omega)$ and thus guarantees that the zeros of $f(x)$ may be expressed in terms of radicals.

14.4 SOLVABLE GROUPS

In this section we collect the important theorems about solvable groups. First we recall Definition 14.1.4: A group G is called solvable if there is a chain of subgroups

$$G = G_0 \supset G_1 \supset \cdots \supset G_k = \text{identity}$$

such that for each i, G_{i+1} is a normal subgroup of G_i and

$$G_i/G_{i+1} \text{ is cyclic}$$

When we made this definition we also pointed out that every abelian group and every dihedral group were solvable. Now we obtain some further important and useful properties of solvability.

14.4.1 Theorem.   Solvability Is Hereditary

1. A subgroup of a solvable group is solvable.
2. A homomorphic image of a solvable group is solvable. Conversely,
3. If a group has a solvable homomorphic image whose kernel is solvable, then the group is solvable.

Proof of 1.   Let G be a solvable group and let

$$G = G_0 \supset G_1 \supset \cdots \supset G_k = \text{identity}$$

be a chain of subgroups of G so that G_{i+1} is normal in G_i and G_i/G_{i+1} is cyclic. Let S be any subgroup of G.

Consider the chain (see Figure 14.3):

$$S \cap G_0 = S = S_0 \supset S_1 = S \cap G_1 \supset \cdots \quad S_k = S \cap G_k = \text{identity}$$

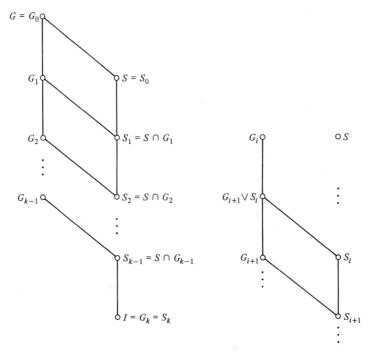

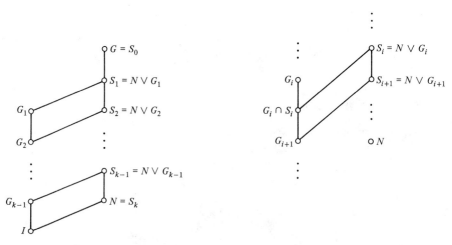

Figure 14.3 These two diagrams show that a subgroup S of a solvable group G is solvable. The diagram on the left shows how a chain of subgroups from S to the identity is constructed from a chain of subgroups from G to the identity. The lattice diagram on the right shows that each factor group S_i/S_{i+1} is isomorphic to a subgroup of G_i/G_{i+1}.

Figure 14.4 These two diagrams show that the homomorphic image of a solvable group is solvable. The diagram on the left shows how a chain of subgroups from G to a normal subgroup may be constructed from chain of subgroups from G to the identity. The diagram on the right shows that the factor group S_i/S_{i+1} is isomorphic to a subgroup of the factor group G_i/G_{i+1}.

We know from the diamond isomorphism theorem that because G_{i+1} is normal in G_i it follows that $S \cap G_{i+1}$ is normal in $S \cap G_i$. Moreover

$$(G_{i+1} \vee S_i)/G_{i+1} \cong S_i/S_{i+1}$$

and so S_i/S_{i+1} is isomorphic to a subgroup of G_i/G_{i+1} and so is cyclic. Thus the sequence

$$S = S_0 \supset \cdots \supset S_i \supset \cdots \supset S_k = \text{identity}$$

satisfies the definition to show that S is solvable.

Proof of 2. Let N be a normal subgroup of G such that G/N is isomorphic to the homomorphic image of G. As shown in Figure 14.4 define

$$S_i = G_i \vee N$$

Consider the sequence

$$G = S_0 \supset \cdots \supset S_i \supset \cdots \supset S_k = N$$

We first show that S_{i+1} is normal in S_i. Because N is normal in G we have

$$S_i = G_i \vee N = \{ xn : x \in G_i \text{ and } n \in N \}$$

Thus to establish normality of S_{i+1} in S_i we must show that if $x \in S_{i+1}$, $y \in S_i$, and $n, m \in N$ then

$$(ym)^{-1}(xn)(ym) = m^{-1}(y^{-1}xy)(y^{-1}ny)m \in S_{i+1} \vee N$$

which is so because $y^{-1}xy \in S_{i+1}$ and $y^{-1}ny \in N$. Then by the diamond lemma we have that

$$S_i/S_{i+1} \cong G_i/(G_i \cap S_{i+1})$$

which is a factor group of a cyclic group and so is cyclic. Thus the sequence

$$G = S_0 \supset \cdots \supset S_i \supset \cdots \supset S_k = N$$

satisfies the condition of solvability for the factor group G/N.

Proof of 3. Let G/N be the homomorphic image of G that is solvable and we suppose that N is solvable. Thus there is a chain of subgroups of G from G to N so that each is normal in the preceding one and the factor groups are cyclic. Also there is a similar chain from N to the identity. By concatenating these chains we have a chain from G to the identity satisfying the condition for solvability.

While we need no further theorems about solvable groups for the proof of the necessity of the solvability of the Galois group of a polynomial for the solvability of the zeros of the polynomial in terms of radicals, there are other important and interesting properties of solvability and chains of subgroups.

14.4.2 Definition. Subnormal Chains and Composition Series

Let G be a group. A chain of subgroups

$$G = H_0 \supset H_1 \supset \cdots \supset H_r = \text{identity}$$

is called a *subnormal* chain if each H_{i+1} is a normal subgroup of H_i. The chain is called a *composition series* if each H_{i+1} is a maximal normal subgroup of H_i. Thus, the definition of a solvable group may be restated to say that G has a subnormal chain in which every factor group is cyclic. In fact

14.4.3 Lemma. Equivalent Conditions for Solvability

The following are equivalent:

1. G has a subnormal chain in which every index $[H_i : H_{i+1}]$ is a prime.
2. G has a subnormal chain in which every factor group H_i/H_{i+1} is cyclic (i.e., G is solvable).
3. G has a subnormal chain in which each factor group is abelian.

Proof. Since groups of prime order are cyclic and since cyclic groups are abelian it follows immediately that 1 implies 2 and 2 implies 3.

We work backward to show that 3 implies 2. In fact that is just the argument we gave in Examples 14.1.5 to show that an abelian group is solvable. Here are the details.

Let $\bar{G} = H_i/H_{i+1}$ be abelian. Then by the argument of Examples 14.1.5, $\bar{G}$ is solvable and has a subnormal chain in which the factor groups are cyclic. The preimage of this chain in the homomorphism yields a chain from H_i down to H_{i+1} in which each member is normal in the preceding one and in which the factor groups are cyclic. Concatenating these chains gives a subnormal sequence for G that satisfies the condition.

This same argument shows that 2 implies 1 since we can insert subgroups in a cyclic group so that the consecutive indices are primes.

Now we want to make the important connection between solvability and the so-called derived series.

Recall the commutator subgroup of a group G. A commutator in a group G is an element of the form $xyx^{-1}y^{-1}$.

The commutator subgroup G', also called the derived subgroup of G, is the subgroup generated by the set of all commutators in G:

$$G' = \langle \{xyx^{-1}y^{-1} : x, y \in G\} \rangle$$

Remember that the set of commutators themselves do not always form a group. Remember too that G' is a normal subgroup of G because the conjugate of a commutator is a commutator:

$$z(xyx^{-1}y^{-1})z^{-1} = (zxz^{-1})(zyz^{-1})(zxz^{-1})(zyz^{-1})^{-1}$$

The other important fact is that if N is a normal subgroup of G then

$$G/N \text{ is abelian if and only if } N \supset G'$$

Thus G/G' is the largest abelian factor group of G. This characterization is often helpful in determining the commutator subgroup. For example, the commutator subgroup of the symmetric group $\mathbb{S}_n$ is $\mathscr{A}_n$, the alternating group, if $n \geq 5$ since $\mathscr{A}_n$ has no normal subgroup except itself and the identity.

14.4.4 Definition.   Derived Series

The *derived series* is the chain of subgroups in which each successive subgroup is the commutator subgroup of the preceding:

$$G = \supseteq G' \supseteq G'' = (G')' \supseteq \cdots \supseteq (G^{(k)})' = G^{(k+1)}$$

This is not a subnormal chain because it may not end at the identity. Indeed, the derived series for $\mathbb{S}_5$ is just

$$\mathbb{S}_5 \supset \mathscr{A}_5$$

since the commutator subgroup of C_5 is itself. The next theorem shows that a group is solvable if and only if the derived chain ends at the identity.

14.4.5 Theorem.   Solvability and the Derived Series

A finite group G is solvable if and only if G has a subnormal chain in which every member is the commutator subgroup of the preceding subgroup; that is, G is solvable if and only if the derived series of G ends at the identity.

Proof.   If the derived series does end at the identity then each factor group has the form

$$H/H'$$

and so is abelian. Hence by Lemma 14.4.3, G is solvable.
   Conversely suppose that

$$G = H_0 \supset \cdots \supset H_i \supset \cdots \supset H_r = I$$

is a subnormal chain in which the successive factor groups are cyclic.
   For $i = 1, \ldots, r$, let

$$G^{(i)} = (G^{(i-1)})'$$

be the derived subgroup of $G^{(i-1)}$. We prove, by induction on the index, that

$$H_i \supseteq G^{(i)} \tag{1}$$

This will do it, because the H-chain ends at the identity and the containment (1) then forces the derived series to end at the identity.
   To prove (1) we first verify that $H_1 \supseteq G'$, which is true because G/H_1 is abelian. Now for the inductive step. If we know that $H_i \supset G^{(i)}$ then it follows that

$$(H_i)' \supseteq (G^{(i)})' = G^{(i+1)}$$

since every commutator formed from the elements of $G^{(i)}$ is already a commutator of H_i. And we also have the $H_{i+1} \supseteq (H_i)'$ since H_i/H_{i+1} is cyclic. Thus the induction step holds and the proof is complete.

Now that we know more about solvable groups we can fill the aesthetic gap concerning the roots of unity in the discussion of the sufficiency of solvability for solutions in radicals in Section 14.3.

14.4.6 Theorem.   Adjunction of Roots of Unity by Radicals

Let F be a field of characteristic 0. The nth roots of unity belong to a root tower over F

$$F = K_0 \subset K_1 \subset \cdots \subset K_m$$

where each $K_{i+1} = K_i(a_i)$ and the degree $[K_{i+1} : K_i]$ is a prime p_i and

$$x^{p_i} - a_i^{p_i}$$

is irreducible over K_i.

Proof.   The proof is by induction on n. We have it for $n = 1$ and 2 trivially. Let ζ be a primitive nth root of unity over F. We know that $F(\zeta)$ is a normal extension and that $\mathbf{G}(F(\zeta)/F)$ is abelian, hence solvable, and hence that there is a subnormal chain of subgroups in which each successive index is a prime p_i.

Moreover, corresponding to this chain of subgroups there is a chain of subfields from F to $F(\zeta)$ in which each is a normal extension of prime degree over its predecessor. Moreover, since

$$[F(\zeta): F] \le \varphi(n) < n$$

it follows that each of the primes p_i that divide the index $[F(\zeta): F]$ are less than n. Let these primes be $p_1, \ldots, p_t$.

Let $E_0 = F$. By the induction hypothesis, we may build, successively, root towers from fields E_i to E_{i+1} to contain the (p_{i+1})th roots of unity for $i = 0, 1, \ldots, t - 1$. In this way we obtain, by concatenation, a root tower from F to a field $F^* = E_t$. Since this field will be a field obtained by adjunction of all the roots of unity of various (prime) orders it will be a normal extension of F. Moreover, by Theorem 14.2.3 on cyclic extensions, the degrees of the root towers will be primes and the adjunction will be irreducible polynomials of the form

$$x^q - a^q$$

where q is a prime and where a^q belongs to the preceding field.

Now apply Lemma 14.3.1 with $F = F$, $E = F(\zeta)$, and $K = F^*$. We have that the group of $F^* \vee F(\zeta)$ over F^* is isomorphic to a subgroup of $\mathbf{G}(F(\zeta)/F)$ and from the proof of part 1 of Theorem 14.4.1 the chain of subgroups of $\mathbf{G}(F(\zeta)/F)$ leads to a chain of subgroups of $\mathbf{G}((F^* \vee F(\zeta)/F^*)$ in which the successive factor groups are subgroups of the chain of $\mathbf{G}(F(\zeta)/F)$ and hence are cyclic of prime order. This chain of subgroups corresponds to a chain of subfields from F^* to $F^* \vee F(\zeta)$ in which successive extensions are cyclic of prime order. But now for each of these primes, the roots of unity of that order belong to the field F^*. Hence, by Theorem 14.2.3 (cyclic extensions) each one of these extensions is the splitting field of a polynomial

$$x^q - a^q$$

over the preceding field. Here q is a prime and so, by Theorem 14.2.3, the polynomial is irreducible over the preceding field.

Thus, by concatenating the roots towers from F to F^* and from F^* to $F^* \vee F(\zeta)$ we obtain a root tower of the type desired in the theorem. Since $F(\zeta)$ is a subfield of the top of the root tower we have shown that the nth roots of unity can be expressed in terms of radicals of irreducible polynomials of degrees less than n.

EXERCISE SET 14.4

14.4.1 Prove that every group of order pq where p and q are primes is solvable.

14.4.2 Show that the quaternion group is solvable.

14.4.3 Show that every group of order 12 is solvable.

14.4.4 Show that a group of order 24 is solvable. (Use the fact that every group of order 24 has a normal subgroup.)

14.4.5 Show that the direct product of solvable groups is solvable.

14.4.6 Prove or disprove: If S and T are subgroups of a group G then $S' \supset T'$ if and only if $S \supset T$.

14.4.7 Express the eleventh roots of unity in terms of radicals.

14.4.8 Express the thirteenth roots of unity in terms of radicals.

14.5 PROOF OF NECESSITY OF SOLVABILITY FOR SOLUTIONS IN RADICALS

Now we return to the problem of showing that solvability of the Galois group of a polynomial is a necessary condition for the zeros of the polynomial to be expressible in terms of radicals. We are ready for the proof of the full theorem.

14.5.1 Theorem.   Necessary Condition for Solutions in Radicals

Let F be a field of characteristic 0 and let $f(x) \in F[x]$. If the equation

$$f(x) = 0$$

is solvable in radicals then the Galois group of $f(x)$ over F is solvable.

Proof Strategy.   Let E be the splitting field of $f(x)$ over F. We are given that E is contained in a root tower

$$F = F_0 \subset F_1 \subset \cdots \subset F_k$$

and

$$F \subset E \subset F_k$$

We want to show that the Galois group $\mathbb{G}(E/F)$ is solvable. We want to reduce the theorem to that of the special case, Lemma 14.1.3. To do this we must first adjoin certain roots of unity. Then we must build a root tower culminating in a normal extension. In this multistage process we repeatedly use the following lemma.

14.5.2 Lemma.   Tower Adjunction

Let L be a field of characteristic 0. Let fields K and M be such that $L \subseteq K \subseteq M$. Suppose that K is a normal extension of L. Suppose further that $\alpha \in M$ is such that $\alpha^n \in K$ and that K contains the nth roots of unity. Under these assumptions there is a root tower

$$K = T_0 \subset \cdots \subset T_u$$

such that $L(\alpha) \subset T_u$ and T_u is a normal extension of L.

Proof.   Let $u = [K : L]$ and let the Galois group of K over L be

$$\mathbb{G}(K/L) = \{I = \sigma_1, \ldots, \sigma_u\}$$

Consider

$$h(x) = \prod_{i=1}^{i=u} (x^n - \sigma_i(\alpha^n)) = h_i(x) \cdots h_u(x)$$

Now the coefficients of $h(x)$ are fixed by each element of $\mathbb{G}(K/L)$ and so, by the fundamental theorem of Galois theory, they belong to L. Consider the root tower defined by

$$T_i = T_{i-1}(\beta_i)$$

where β_i is a zero of

$$h_i(x) = x^n - \sigma_i(\alpha^n)$$

for $i = 1, \ldots, u$. Since K contains the nth roots of unity, T_i contains all the zeros of $h_i(x)$; in particular, T_1 contains all the zeros of $x^n - \alpha^n$ (and we may take $\beta_1 = \alpha$).

Finally, note that T_u contains all the zeros of $h(x)$; in fact

$$T_u = K(\beta_i, \ldots, \beta_u)$$

and T_u is the splitting field of $h(x)$ over L and so in a normal extension of L. Since T_1 contains α, it follows that T_1 contains $L(\alpha)$; hence $T_u \supseteq L(\alpha)$.

Proof of Theorem 14.5.1.   We are given $f(x) \in F[x]$ and that its splitting field over F is contained in a root tower, say,

$$F = K_0 \subset K_1 = F(\alpha_1) \subset \cdots \subset K_i = K_{i+1}(\alpha_i) \subset \cdots \subset K_r$$

where for $1 \le i \le r$, there is a positive integer n_i such that

$$\alpha_i^{n_i} \in K_{i+1}$$

Since each α_i is algebraic over F there is some polynomial in $F[x]$ whose zeros include each α_i and the zeros of $f(x)$. Let M be the splitting field of this polynomial over F. All the action takes place in this larger field M.

First let m be the least common multiple,

$$m = \text{lcm}[n_1, \ldots, n_r]$$

and construct $F^* = F(\zeta)$ where ζ is a primitive mth root of unity. Thus F^* contains all the n_ith roots of unity. Of course, F^* is a normal extension of F.

Now apply the tower adjunction Lemma 14.5.1 with M as M, F^* as K, F as L, and α_1 as α. Thus there is a root tower

$$F^* = T_{10} \subseteq T_{11} \subseteq \cdots \subseteq T_{1n_1}$$

where $T_{11} \supset K_1$ and T_{1n_1} is a normal extension of F.

Now apply the tower adjunction lemma again with M as M, T_{1n_1} as K, F as L, and α_2 as α. Thus there is a root tower

$$T_{1n_1} = T_{20} \subseteq T_{21} \subseteq \cdots \subseteq T_{2n_2}$$

where $T_{21} \supset K_2$ and T_{2n_2} is a normal extension of F.

After at most r applications we have finally the root tower

$$F \subset F^* \subseteq \cdots \subseteq T_{ij} \subseteq \cdots \subseteq T_{rn_r}$$

where the top field is a normal extension of F and contains K_r and thus also E.

Then by Lemma 14.1.3 we have that $\mathbb{G}(T_{tn_r}/F^*)$ is solvable.

We know that $\mathbb{G}(F^*/F)$ is abelian and hence is solvable. It follows from the hereditary properties of solvable groups (Theorem 14.4.1) that $\mathbb{G}(T_{rn_r}/F)$ is solvable. Finally then, since $\mathbb{G}(E/F)$ is a factor group of this group, it follows that it is solvable, as was to be proved!

(It always seems to be such a complicated mess to get all the details of this theorem after the relatively easy steps of Lemma 14.1.3.)

We complete this section by constructing an equation of fifth degree with integer coefficients whose zeros cannot be expressed in radicals. To do this we need an interesting result on the symmetric group.

14.5.3 Theorem.   Generators for the Symmetric Group

Let p be a prime. The full symmetric group $\mathbb{S}_p$ is generated by a transposition and a p-cycle.

Proof.   Let the *p*-cycle be

$$\pi = (a_1, \ldots, a_p)$$

Let the transposition be (a_i, a_j) where we may assume that $i < j$. Now recall (Theorem 6.2.3) that π^k sends a_i into a_{i+k} and, since p is a prime π^k is a *p*-cycle. Hence we can assume, by replacing π by a power if necessary, that the *p*-cycle sends a_i to a_j. By renumbering we may assume for simplicity that

$$\pi = (1, 2, \ldots, p) \quad \text{and} \quad \tau = (1, 2)$$

Now we shall show that every transposition is in the subgroup generated by π and τ and hence that

$$\mathbb{S}_p = \langle \pi, \tau \rangle$$

First we have the conjugates

$$\pi^k \tau \pi^{-k} = (k + 1, k + 2)$$

for $k = 0, 1, \ldots, p - 1$, taking integers mod p.
  Next we have

$$(2, 3)(1, 2)(2, 3) = (1, 3)$$

and inductively get all transpositions of the form $(1, k)$. Indeed if we have $(1, k)$ then

$$(k, k + 1)(1, k)(k, k + 1) = (1, k + 1)$$

and thus all transpositions can be generated from π and τ.
  Now we prove a sufficient condition on the zeros of an irreducible polynomial in $\mathbb{Q}[x]$ to have $\mathbb{S}_p$ as its Galois group.

14.5.4 Theorem.   Some Polynomials with Symmetric Groups

Let $f(x)$ be an irreducible polynomial in $\mathbb{Q}[x]$ of prime degree p having exactly two complex zeros and $p - 2$ real zeros. Then the group of $f(x)$ over $\mathbb{Q}$ is the full symmetric group $\mathbb{S}_p$.

Proof.   By hypothesis we are assuming that the splitting field E of $f(x)$ over $\mathbb{Q}$ is (isomorphic to) a subfield of $\mathbb{C}$, the field of complex numbers.
  Let α be a zero of $f(x)$. Since $f(x)$ is irreducible we know the degree $[\mathbb{Q}(\alpha) : \mathbb{Q}] = p$ so that the index of $\mathbb{G}(E / \mathbb{Q}(\alpha))$ in $\mathbb{G}(E / \mathbb{Q})$ is p. Thus p divides the order of the Galois group of $f(x)$.
  Since the Galois group may be represented as a group of permutations on the p zeros of $f(x)$ we see that $\mathbb{G}(E / \mathbb{Q})$ contains a *p*-cycle. Next we argue

that $G(E/Q)$ contains a transposition. Indeed let γ be a complex zero of $f(x)$ and let $\bar{\gamma}$ be the complex conjugate of γ. Under the automorphism of the complex numbers that maps every complex number into its conjugate we see that every real number is fixed; in particular the real zeros of $f(x)$ are fixed, and γ and $\bar{\gamma}$ are interchanged. The restriction of this conjugation map of the complex numbers to E becomes an automorphism of E which is represented by the transposition $(\gamma, \bar{\gamma})$.

It follows from the previous result that

$$G(E/Q) = S_p$$

We can use this result to construct a fifth-degree polynomial in $Q[x]$ whose Galois group over Q is S_5 and hence is not solvable in terms of radicals.

14.5.5 Example.   A Polynomial Not Solvable in Radicals

The Galois group over Q of the polynomial

$$p(x) = x^5 - 10x + 5$$

is S_5. First observe that by Eisenstein's criterion $p(x)$ is irreducible. Next note that its derivative

$$p'(x) = 5(x^4 - 2)$$

has two real zeros; hence $p(x)$ has just two critical points and thus can have at most three real zeros.

These critical points are a and $-a$ where

$$a = \sqrt[4]{2}$$

Moreover we find that $p(x) = x(x^4 - 10) + 5$ so that

$$p(a) = a(-8) + 5 < 0 \quad \text{and} \quad p(-a) = -a(-8) + 5 > 0$$

Hence $p(x)$ has a negative zero less than $-a$, another zero between $-a$ and a, and one greater than a. Hence $p(x)$ has exactly three real zeros and must therefore have two complex zeros. Thus Theorem 14.5.4 applies and the Galois group of $p(x)$ over Q is S_5.

This construction may be generalized to construct a polynomial of degree p whose group is S_p for each prime p. The general question of whether for each finite group there is a polynomial with rational coefficients whose group over Q is G is unsolved. Much progress has been made, and we now know, thanks to the work of the Russian mathematician I. R. Shafarevich in the 1950s, that if G is a solvable group then there is a polynomial with rational coefficients whose group is solvable. (For further information see N. Jacobson, *Basic Algebra*, volume 1, W. H. Freeman, 1985.)

EXERCISE SET 14.5

14.5.1 Find an equation whose group over $\mathbb{Q}$ is $\mathscr{S}_7$.

14.5.2 Show that the Galois group of $x^8 - 72x^6 + 180x^4 - 144x^2 + 36$ over $\mathbb{Q}$ is the quaternion group.

14.5.3 Let $y_1, \ldots, y_n$ be n indeterminates over a field F of characteristic 0. Let

$$s_1 = y_1 + y_2 + \cdots + y_n$$
$$s_2 = y_1 y_2 + \cdots + y_1 y_n + y_2 y_3 + \cdots + y_{n-1} y_n$$

$$\vdots$$

$$s_k = \sum_{1 \le i_1 \le i_1 \le \cdots \le j_k \le n} y_{i_1} y_{i_2} \cdots y_{i_k}$$

$$\vdots$$

$$s_n = y_1 \ldots y_n$$

Show that $f(x) = (x - y_1)(x - y_2)\ldots(x - y_n) \in F(s_1, \ldots, s_n)$ and that the Galois group of $f(x)$ over $F(s_1, \ldots, s_n)$ is $\mathbb{S}_n$.

14.5.4 If p is an odd prime determine the Galois group of $x^p - 2px + p$ over $\mathbb{Q}$.

---

*See R. Dean, "A Rational Polynomial Whose Group Is the Quaternions," *American Mathematical Monthly*, vol. 88, pages 42–45 (1981).

Our Lagrange theory tells us that r^3, s^3, and t^3 are in F, so we proceed to calculate these powers. However, because the calculations are identities, they do not depend upon the cyclic automorphism (α, β, γ) for their validity; thus they hold whether or not $F(\Delta, \omega, \alpha)$ is a proper extension of $F(\Delta, \omega)$.

Chapter 15

Linear Transformations on Vector Spaces

15.1 LINEAR TRANSFORMATIONS

There is of course a theory of homomorphisms from one vector space to another. Since a vector space is an abelian group together with a scalar multiplication a vector space homomorphism must also be a group homomorphism. The homomorphism should also preserve scalar multiplication, and so there is a question about the underlying field of scalars. To simplify our treatment here we suppose that the underlying field of scalars is the same for both fields. The prerequisites are Chapters 1 to 10.

From time to time we also assume that the vector spaces we discuss are finite-dimensional. In some cases this simplifies proofs; in others it simplifies the statement of the result. So please read the statements carefully to know whether finite dimension is an important part of the hypothesis.

The algebraic term given to a vector space homomorphism is "linear transformation." The term reflects a geometric property. Suppose that the vector space is the real plane or space. If a linear transformation is one-to-one then lines and planes are transformed into lines and planes, respectively.

Notation. In this section we use the usual function notation; the image of a vector v under the linear transformation L is denoted by $L(v)$.

15.1.1 Definition. Linear Transformations

Let V and W be vector spaces over a field F. A mapping L from V into W is called a *linear transformation* if for all v, $w \in V$ and all $a \in F$:

1. $L(v + w) = L(v) + L(w)$
2. $L(av) = aL(v)$

Condition 1 simply says that a linear transformation is a group homomorphism and so vector sums are preserved. Condition 2 guarantees that scalar products are preserved.

The kernel of L,

$$K = \{ v \in V : L(v) = 0 \}$$

is a subspace of V and determines whether or not L is one-to-one. From the group homomorphism property this means that L is one-to-one if and only if $K = \{0\}$. If L is not one-to-one we say that L is *singular*; otherwise we say that L is *nonsingular*.

If $V = W$ then linear transformations may be composed, and as we shall prove, if the dimension of V is finite the set of all nonsingular linear transformations from V into V form a group. This group is called the general linear group. If the dimension of V is finite, say, n, then this group depends only on n and on the underlying field.

In the finite-dimensional case calculations with linear transformations are most easily done with matrices. Because of properties 1 and 2 of the definition if the images of a linear transformation are known for any basis of V then the images of all vectors in V can be calculated. Indeed, let $v \in V$. Express v in terms of a basis $\{e_1, \ldots e_n\}$, say,

$$v = \sum_{1 \leq i \leq n} v_i e_i$$

then calculate

$$L(v) = L\left(\sum_{1 \leq i \leq n} v_i e_i \right) = \sum_{1 \leq i \leq n} v_i L(e_i)$$

We say that to "know" a linear transformation it is enough to "know" it on a basis. One important consequence of this is that if we prescribe the images of a basis for V then there is a unique linear transformation whose images of the basis are the prescribed vectors. We exploit this property in Theorem 15.1.4.

The introduction of matrices to describe a linear transformation, as well as to calculate images with the matrix, now comes about as follows:

15.1.2 Definition.   The Matrix of a Linear Transformation

Let V and W be finite-dimensional vector spaces over a field F. Let L be a linear transformation from V into W. Let bases

$$\{e_1, \ldots e_n\} \qquad \text{for } V \quad \text{and} \quad \{f_1, \ldots f_m\} \qquad \text{for } W$$

be chosen. Express each image $L(e_j)$ in terms of the f-basis for W: For each $1 \leq j \leq n$,

$$L(e_j) = \sum_{1 \leq i \leq m} a_{ij} f_i$$

The *matrix of L with respect to the chosen bases* is the array of elements from F:

$$\begin{bmatrix} a_{11} & a_{12} & \cdots & a_{1j} & \cdots & a_{1n} \\ a_{21} & a_{22} & \cdots & a_{2j} & \cdots & a_{2n} \\ \vdots & \vdots & & \vdots & & \vdots \\ a_{i1} & a_{i2} & \cdots & a_{ij} & \cdots & a_{in} \\ \vdots & \vdots & & \vdots & & \vdots \\ a_{m1} & a_{m2} & \cdots & a_{mj} & \cdots & a_{mn} \end{bmatrix}$$

Thus the jth column is the column vector made from the sequence of coordinates of $L(e_j)$ in the f-basis of W. The element a_{ij} in the ith row and jth column is the ith coordinate of $L(e_j)$ in the f-basis.

As you see, the number of columns of the matrix is the dimension of the domain V, and the number of rows is the dimension of W.

It is very important to understand the dependence of the matrix associated with the linear transformation upon the bases chosen. Here are two examples. For simplicity we take the field of scalars to be $\mathbb{Q}$.

15.1.3 Examples. Linear Transformations and Matrices in V_2 and V_3

1. Let $V = V_2(\mathbb{Q})$ and $W = V_3(\mathbb{Q})$. Use the standard bases for V and W

$$e_1 = \begin{bmatrix} 1 \\ 0 \end{bmatrix} \quad e_2 = \begin{bmatrix} 0 \\ 1 \end{bmatrix} \quad \text{and} \quad f_1 = \begin{bmatrix} 1 \\ 0 \\ 0 \end{bmatrix} \quad f_2 = \begin{bmatrix} 0 \\ 1 \\ 0 \end{bmatrix} \quad f_3 = \begin{bmatrix} 0 \\ 0 \\ 1 \end{bmatrix}$$

Let L be the transformation that sends e_1 and e_2 into

$$L(e_1) = f_1 + 2f_2 + 3f_3 \quad \text{and} \quad L(e_2) = 4f_1 + 5f_2 - 6f_3$$

respectively. Then the matrix associated with L is

$$\begin{bmatrix} 1 & 4 \\ 2 & 5 \\ 3 & -6 \end{bmatrix}$$

2. For our next example we choose $V = W = V_3(\mathbb{Q})$ and choose L to be the identity transformation. In this and Example 15.1.9 we denote this identity transformation by J. Thus $J(v) = v$ for all $v \in V$.
Choose, following the notation above, bases:

$$e_1 = \begin{bmatrix} 1 \\ 0 \\ 0 \end{bmatrix} \quad e_2 = \begin{bmatrix} 0 \\ 1 \\ 0 \end{bmatrix} \quad e_3 = \begin{bmatrix} 0 \\ 0 \\ 1 \end{bmatrix}$$

$$f_1 = \begin{bmatrix} 1 \\ 0 \\ 0 \end{bmatrix} \quad f_2 = \begin{bmatrix} 1 \\ 1 \\ 0 \end{bmatrix} \quad f_3 = \begin{bmatrix} 1 \\ 1 \\ 1 \end{bmatrix}$$

To find the matrix associated with the identity transformation J and these bases we must express $J(e_i)$ in terms of the basis $\{f_1, f_2, f_3\}$. We compute

$$J(e_1) = e_1 = f_1 + 0f_2 + 0f_3$$

$$J(e_2) = e_2 = -f_1 + f_2 + 0f_3$$

$$J(e_3) = e_3 = 0f_1 - f_2 + f_3$$

The matrix associated with J we denote by $\hat{J}$. We find

$$\hat{J} = \begin{bmatrix} 1 & -1 & 0 \\ 0 & 1 & -1 \\ 0 & 0 & 1 \end{bmatrix}$$

15.1.4 Theorem. The Correspondence between Linear Transformations and Matrices

Let V and W be finite-dimensional vector spaces over a field F. There is a one-to-one correspondence between linear transformations from V to W and matrices of size $\dim(W) \times \dim(V)$ with elements in F.

Proof. For the remainder of this proof fix a basis $\{e_1, \ldots, e_n\}$ for V and a basis $\{f_1, \ldots, f_m\}$ for W.

Definition 15.1.2 shows how a matrix can be associated with a linear transformation. The association defines a mapping of transformations into matrices. It is a one-to-one mapping since transformations will differ if and only if they differ at the image of some vector in the chosen basis for V. But then the differing images for a basis vector yield differing entries for the matrix. Thus the mapping is one-to-one.

The mapping is onto since given a matrix

$$\begin{bmatrix} a_{11} & a_{12} & \cdots & a_{1j} & \cdots & a_{1n} \\ a_{21} & a_{22} & \cdots & a_{2j} & \cdots & a_{2n} \\ \vdots & \vdots & & \vdots & & \vdots \\ a_{i1} & a_{i2} & \cdots & a_{ij} & \cdots & a_{in} \\ \vdots & \vdots & & \vdots & & \vdots \\ a_{m1} & a_{m2} & \cdots & a_{mj} & \cdots & a_{mn} \end{bmatrix}$$

of the proper size we may define the linear transformation on the given basis for V in terms of the coefficients in the matrix. Then we may extend this definition from the basis to all the vectors in V in the natural way: If

$$v = \sum_{1 \le j \le n} v_j e_j$$

define

$$L(v) = \sum_{1 \le j \le n} v_j L(e_j) = \sum_{1 \le j \le n} v_j \left(\sum_{1 \le i \le m} a_{ij} f_i \right)$$

$$= \sum_{1 \le i \le m} \left(\sum_{1 \le j \le n} a_{ij} v_j \right) f_i$$

It is routine to show that L satisfies Properties 1 and 2 of Definition 15.1.1.

In a moment we deal with the question of how the matrix changes when a change in the basis for V or for W is made. But first we want to introduce a multiplication of a matrix and a vector.

Let $v \in V$. Expressing v in the e-basis for V:

$$v = \sum_h v_h e_h$$

we can calculate $L(v)$ as follows:

$$L(v) = L\left(\sum_h v_h e_h \right) = \sum_h v_h L(e_h)$$

$$= \sum_h v_h \left(\sum_i a_{ih} f_i \right) = \sum_i \left(\sum_h a_{ih} v_h \right) f_i$$

Thus the coordinates of $L(v)$ in the f-basis for W are

$$\sum_{h=1}^{n} a_{1h} v_h, \quad \sum_{h=1}^{n} a_{2h} v_h, \ldots, \sum_{h=1}^{n} a_{mh} v_h$$

And if we write v as a column vector of its coordinate in the e-basis for V as

$$\begin{bmatrix} v_1 \\ v_2 \\ \vdots \\ v_h \\ \vdots \\ v_n \end{bmatrix}$$

then the image $L(v)$ can be written in the f-basis as a column

$$\begin{bmatrix} \sum_h a_{1h} v_h \\ \sum_h a_{2h} v_h \\ \vdots \\ \sum_h a_{ih} v_h \\ \vdots \\ \sum_h a_{mh} v_h \end{bmatrix}$$

We now define the matrix-column vector product to reflect this outcome.

15.1.5 Definition.   Matrix-Vector Multiplication

The product of the $(m \times n)$ matrix $A = [a_{ih}]$ and a column vector $v = [v_h]$ of length n will be

$$A(v) = \begin{bmatrix} a_{11} & a_{12} & \cdots & a_{1h} & \cdots & a_{1n} \\ a_{21} & a_{22} & \cdots & a_{2h} & \cdots & a_{2n} \\ \vdots & \vdots & & \vdots & & \vdots \\ a_{i1} & a_{i2} & \cdots & a_{ih} & \cdots & a_{in} \\ \vdots & \vdots & & \vdots & & \vdots \\ a_{m1} & a_{m2} & \cdots & a_{mh} & \cdots & a_{mn} \end{bmatrix} \begin{bmatrix} v_1 \\ v_2 \\ \vdots \\ v_h \\ \vdots \\ v_n \end{bmatrix} = \begin{bmatrix} \sum_h a_{1h}v_h \\ \sum_h a_{2h}v_h \\ \vdots \\ \sum_h a_{ih}v_h \\ \vdots \\ \sum_h a_{mh}v_h \end{bmatrix} \quad (1)$$

This is commonly referred to as "row-column" multiplication of matrices.

Remark.   It is very useful to note that if the columns of the matrix A are denoted by $A_1, A_2, \ldots, A_n$ then we may express

$$A(v) = \sum_k v_k A_k$$

Thus $A(v)$ is a *linear combination of the columns of A*. The coefficients used are the entries in the (column) vector v.

15.1.6 Theorem.   Range Space and Column Space

Let V and W be finite dimensional vector spaces over F and let L be a linear transformation from V into W. If A is a matrix corresponding to L then $L(V)$, the range of L, is isomorphic to the column space of A. In particular, the dimension of the range of L is the rank of A.

Proof.   We use the isomorphism of W to $V_m(F)$, the space of m-tuples of elements in F. These vectors are written here as columns to make the proof transparent.

Suppose that we have chosen bases $\{e_1, \ldots, e_n\}$ for V and $\{f_1, \ldots, f_m\}$ for W. Then W is isomorphic to $V_m(F)$ through the correspondence

$$f_i \leftrightarrow \begin{bmatrix} 0 \\ \vdots \\ 1 \\ \vdots \\ 0 \end{bmatrix}$$

where the 1 is in the ith row. Now we follow the notation of Theorem 15.1.4 and Definition 15.1.5. We know that if $v = \sum_j v_j e_j$ then

$$L(v) = L\left(\sum_j v_j e_j\right) = \sum_i \left(\sum_h a_{ih}v_h\right) f_i$$

corresponds to the column vector

$$\bar{v}_m = \begin{bmatrix} \sum_h a_{1h}v_h \\ \sum_h a_{2h}v_h \\ \vdots \\ \sum_h a_{ih}v_h \\ \vdots \\ \sum_h a_{mh}v_h \end{bmatrix}$$

The mapping Φ: $L(v) \to \bar{v}_m$ from the range of L to the column space satisfies the homomorphism condition. Further $L(v) = 0$ if and only if for each i, $\sum_h a_{ih}v_h = 0$. From this it follows that Φ is one-to-one. Thus we may determine the dimension of the range space by computing the dimension of the column space of A. But this is the rank of A by definition and Theorem 10.1.20.

It is frequently useful to create new linear transformations from old ones. One way of doing this is by addition. Another is by composition. In the next two theorems we determine the corresponding matrix computations.

Notation. To simplify eye-boggling displays like the ones above we denote a matrix by giving the notation for the entry in the ith row and jth column. Thus we denote

$$\begin{bmatrix} a_{11} & a_{12} & \cdots & a_{1j} & \cdots \\ a_{21} & a_{22} & \cdots & a_{2j} & \cdots \\ \vdots & \vdots & & \vdots & \\ a_{i1} & a_{i2} & \cdots & a_{ij} & \cdots \\ \vdots & \vdots & & \vdots & \end{bmatrix} = [a_{ij}]$$

15.1.7 Theorem. Addition of Linear Transformations and Matrices

Let V and W be vector spaces over a field F. If R and S are two linear transformations from V into W then their sum, defined by

$$(R + S)(v) = R(v) + S(v)$$

is a linear transformation from V to W.

If V and W are finite-dimensional and $\hat{R}$ and $\hat{S}$ are the corresponding matrices with respect to bases chosen and fixed for V and W, say,

$$\hat{R} = [r_{ij}] \quad \text{and} \quad \hat{S} = [s_{ij}]$$

then the matrix corresponding to the linear transformation $R + S$ with respect to the same bases is

$$[r_{ij} + s_{ij}]$$

in which every entry is the sum of the corresponding entries in R and S.

Correspondingly we define the *sum* of two matrices of the same size

$$[r_{ij}] + [s_{ij}] = [r_{ij} + s_{ij}] \tag{2}$$

Proof.   The details of the proof are routine, and we omit them. Definition (2) of addition of matrices is given so that the correspondence between linear transformations and matrices established in Theorem 15.1.4 now *preserves* addition if the same underlying bases are used.

15.1.8 Theorem.   Composition of Linear Transformations and Multiplication of Matrices

Let U, V, and W be vector spaces over a field F.

1. If R is a linear transformation from U into V and S is a linear transformation from V into W then the composition SR defined by

$$SR(u) = S(R(u)) \qquad \text{for all } u \in U$$

   is a linear transformation from U into W.
2. If R and S are nonsingular then so is SR.
3. If U, V, and W are finite-dimensional, say, $\dim(U) = n$, $\dim(V) = m$ and $\dim(W) = p$, and bases have been chosen for U, V, and W so that the corresponding $m \times n$ matrix for R is

$$\hat{R} = [r_{ij}]$$

   and the $p \times m$ matrix for S is

$$\hat{S} = [s_{ij}]$$

   then the matrix corresponding to SR has size $p \times m$ and is

$$\left[\sum_h s_{ih} r_{hj} \right]$$

   in which the entry in the (i, j) position is the "row-column" product of the ith row of $\hat{S}$ and the jth column of $\hat{R}$.

   Correspondingly we define the *product* of an $m \times n$ and an $n \times p$ matrix as

$$[r_{ih}][s_{hj}] = \left[\sum_{1 \le h \le n} r_{ih} s_{hj} \right] \tag{3}$$

4. The columns of the product $[r_{ih}][s_{hj}]$ are linear combinations of the columns of $[r_{ih}]$, the matrix on the left.
5. The rows of the product $[r_{ih}][s_{hj}]$ are linear combinations of the rows of $[s_{hj}]$, the matrix on the right.

6. The matrix transpose of a product is the product of the transposed matrices in the reverse order:

$$(\hat{S}\hat{R})^T = \hat{R}^T\hat{S}^T$$

Proof of 1 and 2. We already know that the composition of group homomorphisms is a group homomorphism and so the only thing we need to verify is that scalar multiplication is preserved over compositions. That is routine, and so we conclude that SR is a linear transformation.

It is also routine to show that if both R and S are one-to-one, then so is the composition SR.

Proof of 3. To determine the jth column of the matrix corresponding to SR we simply need to make the computation. Begin by choosing bases $\{e_1, \ldots, e_n\}$ for U, $\{f_1, \ldots, f_m\}$ for V, and $\{g_1, \ldots, g_p\}$ for W. Then

$$SR(e_j) = S(R(e_j)) = S(\Sigma_i r_{ij} f_i)$$

Now note that $R(e_j)$ is a vector in V and we have expressed it in its coordinates in the f-basis for V. These coordinates are the jth column of R. We calculate the image of $R(e_j)$ under S using equation (1) above. The matrix A is replaced by the matrix $\hat{S}$ since it corresponds to the linear transformation S from V to W. Thus

$$
\begin{bmatrix}
s_{11} & s_{12} & \cdots & s_{1h} & \cdots \\
s_{21} & s_{22} & \cdots & s_{2h} & \cdots \\
\vdots & \vdots & & \vdots & \\
s_{i1} & s_{i2} & \cdots & s_{ih} & \cdots \\
\vdots & \vdots & & \vdots &
\end{bmatrix}
\begin{bmatrix}
v_1 \\
v_2 \\
\vdots \\
v_h \\
\vdots
\end{bmatrix}
=
\begin{bmatrix}
\Sigma_h s_{1h} v_h \\
\Sigma_h s_{2h} v_h \\
\vdots \\
\Sigma_h s_{ih} v_h \\
\vdots
\end{bmatrix}
$$

Now we substitute for v_h the hth coordinate of $R(e_j)$ which is r_{hj}. Hence the entry in the ith row of column j is

$$\sum_h s_{ih} v_h = \sum_h s_{ih} r_{hj}$$

Thus

$$\hat{S}\hat{R} = [\Sigma_h s_{ih} r_{hj}]$$

In view of this theorem we define multiplication of matrices by (3).

Note that to add matrices they must have the same size. To multiply two matrices the number of columns of the matrix on the left must be the same as the number of rows of the matrix on the right.

Proof of 4, 5 and 6. The rule for determining the transpose of a product as the product of the transposed matrices in permuted order is simply the result of a careful computation. The fact that the columns of the product of two matrices $\hat{S}\hat{R}$ is a linear combination of the columns of the matrix on the right $[s_{ih}]$ is an immediate result of the remark following Definition 15.1.5 of the matrix-vector product. The corresponding fact that the rows of the product are linear combinations of the rows of the matrix on the right follows from the previous observation on the transposed matrices $\hat{R}^T\hat{S}^T$.

As result of these observations and the fact that the rank of a matrix is either its column rank or its row rank we may conclude that the rank of a product of two matrices is at most the rank of either factor.

15.1.9 Theorem.   The Ring of Linear Transformations

Let V and W be a vector spaces over a field F.

1. The set of linear transformations of V into W form a vector space over F under the definitions of addition and scalar multiplication introduced in Theorems 15.1.7 and 15.1.8. If $\dim(V) = n$ and $\dim(W) = m$ then the dimension of the space of linear transformations is nm. If bases are chosen for V and W the set of matrices corresponding to these linear transformations with respect to these bases form a vector space of dimension nm.

2. In addition, if $V = W$ the set of linear transformations of V into V form a ring under the definitions of addition and multiplication of Theorems 15.1.6 and 15.1.7. In particular, if $\dim(V) = n$ is finite and a basis $\{e_1, \ldots, e_n\}$ is chosen and fixed for V, the set of $n \times n$ matrices over F corresponding to the linear transformations with respect to the e-basis form both a vector space and a ring.

3. If $\dim(V) = n$ then the units in the ring of linear transformations from V into V are the nonsingular transformations. Each nonsingular transformation corresponds to a nonsingular matrix A for which there is an inverse matrix A^{-1} such that

$$A^{-1}A = AA^{-1}$$

Proof of 1.   The vector addition is to be the addition of linear transformations as given in Theorem 15.1.7. The multiplication of a linear transformation L by an element $k \in F$ is defined to be kL and

$$(kL)(v) = k(L(v))$$

It is routine to verify the vector space axioms. For matrices this means that the multiplication of a matrix by an element of $k \in F$ is given by multiplying every element of the matrix by k:

$$k[a_{ij}] = [ka_{ij}]$$

If $\dim(V) = n$ and $\dim(W) = m$, a basis for this vector space over F is given by the following set of linear transformations E_{ij}.

Let $\{e_1, \ldots, e_n\}$ be a basis for V over F and $\{f_1, \ldots, f_m\}$ be a basis for W over F. For each ordered pair of basis elements (e_i, f_j) define transformations E_{ij} by

$$E_{ij}(e_h) = \begin{cases} 0 \text{ if } h \neq i \\ f_j \text{ if } h = i \end{cases}$$

Thus the entries for the matrix E_{ij} are all 0 except for the entry in row i and column j, which is 1. It is not difficult to show that these $\dim(V) \times \dim(W)$ transformations are a basis for the space of all linear transformations of V into W.

Proof of 2.   If $V = W$ the vector space can be made into a ring by defining multiplication as a composition of linear transformation. For the corresponding matrices to be consistent under composition we require that the same basis be used for the domain and the range spaces. Then the corresponding operations coincide with the matrix operations as defined in Theorems 15.1.7 and 15.1.8.

Remark.   The symbol 0 gets a lot of action. We know it as an element of the field F, the additive identity. It also denotes a vector, the additive identity of the vector space. As a third use it denotes the zero linear transformation. A linear transformation L is the *zero transformation* if and only if

$$L(v) = 0 \text{ (vector)}$$

for all $v \in V$. It is sufficient to know that $L(v) = 0$ for all members of a basis for V.

And then there is another use for the symbol 0. It is used to denote the matrix, of all zeros, of course, that corresponds to the zero transformation. You will need to keep these uses in mind. The context is supposed to be clear in each case.

Now we continue with the details of the proof of part 2 of Theorem 15.1.9.

As we have just noted, the zero of this ring is the linear transformation that maps all vectors into the zero vector of V. The multiplicative identity is just the identity transformation.

The identity transformation corresponds to the identity $n \times n$ matrix

$$I = \begin{bmatrix} 1 & 0 & \cdots & 0 \\ 0 & 1 & \cdots & 0 \\ & & \ddots & \\ 0 & 0 & \cdots & 1 \end{bmatrix}$$

with 1's down the main diagonal and 0's elsewhere. This is so because we are using the same basis to express vectors in V as a domain and range space.

The associative law holds because composition of functions is associative. Therefore, because of the one-to-one correspondence of transformations to matrices which, by the definition of multiplication, is preserved under multiplication, it follows that multiplication of matrices is also associative. By the way, this argument, which appeals to the associative law for composition of functions and the correspondence of matrices to functions (linear transformations), avoids a grungy calculation involving triple sums that would have been necessary to prove that the associative law for multiplication of matrices from definition (3) of Theorem 15.1.8.

To prove that multiplication distributes over addition we again appeal to linear transformations. Using the definition of addition and composition of functions alone we can prove that $(S + T)R = SR + TR$.

$$(S + T)R(v) = (S + T)(R(v)) = S(R(v)) + T(R(v))$$

while on the other hand

$$(SR + TR)(v) = SR(v) + TR(v) = S(R(v)) + T(R(v))$$

The proof of the other distributive law, $R(S + T) = RS + RT$, uses the linearity of R in the equality below marked by an asterisk.

$$R(S + T)(v) = R((S + T)(v)) = R(S(v) + T(v))$$
$$\overset{*}{=} R(S(v)) + R(T(v))$$

and on the other hand

$$(RS + RT)(v) = RS(v) + RT(v) = R(S(v)) + R(T(v))$$

Again the correspondence of linear transformations to matrices gives a proof that the distributive law holds for matrices.

Proof of 3. Let L be a nonsingular transformation from a finite-dimensional vector V into itself. Let $\{e_1, \ldots, e_n\}$ be a basis for V and let $f_i = L(e_i)$. We first show that the set of images $\{f_1, \ldots, f_n\}$ are independent. Since there are n vectors in this set it follows that $\{f_1, \ldots, f_n\}$ is a basis for V and so the transformation L is onto. It is at this point that we use the finiteness of the dimension of V in a crucial way.

To show independence, suppose there are scalars $x_1, \ldots, x_n$ such that

$$\sum_i x_i f_i = 0$$

We will show that all the x's must be zero. In any event we may write

$$0 = \sum_i x_i f_i = \sum_i x_i L(e_i) = L\left(\sum_i x_i e_i\right)$$

Now since L is nonsingular this implies that $\sum_i x_i e_i = 0$. But then the independence of the e's implies that all the coefficients $x_i = 0$.

Since the f's are a basis $L(V)$ has dimension n and so A has rank n by Theorem 15.1.6. This means that A is a nonsingular matrix.

Next we define a linear transformation L^{-1} into V by

$$L^{-1}(f_i) = e_i$$

Clearly we have

$$LL^{-1} = L^{-1}L = \text{identity}$$

Now suppose that we have chosen a basis for V to use in its role as an image space and another to use in its role as the range space. Let A be the matrix corresponding to L using these bases. Let A^{-1} correspond to L^{-1} using the bases in reverse roles. Then our rules of composition and the corresponding rules for matrix multiplication imply that

$$AA^{-1} = A^{-1}A = \text{identity}$$

Conversely suppose that A is a nonsingular matrix and so its column rank is n. From Theorem 15.1.6 the range space of L has dimension n. Suppose that $\{L(w_1), \ldots, L(w_n)\}$ is a basis for the range space. It follows that $\{w_1, \ldots, w_n\}$ must be independent in V and is thus a basis for V. It follows that $L(v) = 0$ only if $v = 0$ and thus L is nonsingular.

15.1.10 Theorem.   Change of Basis and Similar Matrices

Let V be a finite-dimensional vector space over a field F and let L be a linear transformation from V into V.

Let $\{e_1, \ldots, e_n\}$ be a basis for V. Let A be the matrix corresponding to L using the e-basis for both the domain and the image spaces.

Let $\{f_1, \ldots, f_n\}$ be a basis for V and let B be the matrix corresponding to L using the f-basis for both the domain and image spaces.

Under these conditions

$$B = S^{-1}AS$$

where S is the matrix that expresses the f-basis vectors in terms of the e-basis.

Terminology.   Two matrices with elements in F are said to be *similar* over F if and only if they correspond to the same linear transformation of a vector space V over F. In view of Theorem 15.1.10 we can say that two matrices A and B are similar over F if and only if there is a nonsingular matrix S with elements in F such that

$$B = S^{-1}AS$$

The dependence on the field F is not crucial. The test for similarity given in Theorem 16.5.5 shows that two matrices with elements in F are similar over an extension field of F if and only if they are similar over F.

Proof.   For this proof we distinguish three different notations for each vector in V.

w, a vector in V.

$[w]_e$, a column vector of the coordinates of w in the e-basis.

$[w]_f$, a column vector of the coordinates of w in the f-basis.

Let A be the matrix corresponding to L in the e-basis. Thus for each $v \in V$

$$[L(v)]_e = A[v]_e$$

Similarly, let B denote the matrix corresponding to L in the f-basis. Thus

$$[L(v)]_f = B[v]_f$$

The e-basis and the f-basis are related by a matrix S corresponding to the identity transformation from V into V with respect to the f- and e-bases. Thus to repeat the computations of Theorem 15.1.4 in this case, let

$$f_j = \sum_{1 \le i \le n} s_{ij} e_i$$

then the matrix $S = [s_{ij}]$ effects the change of basis. For each $w \in V$

$$[w]_e = S[w]_f$$

and

$$S^{-1}[w]_e = [w]_f$$

Now all this can be put together. In the next display the first line shows the composition of transformations, and the second line shows the corresponding matrix calculations reflecting the changing bases.

$$
\begin{array}{ccccccc}
 & \xrightarrow{\text{identity}} & & \xrightarrow{\;L\;} & & \xrightarrow{\text{identity}} & \\
v & & v & & L(v) & & L(v) \\
[v]_f & \longrightarrow & [v]_e & \longrightarrow & [L(v)]_e & \longrightarrow & [L(v)]_f \\
 & = S[v]_f & & = A[v]_e & & = S^{-1}[L(v)]_e \\
 & & & = AS[v]_f & & = S^{-1}AS[v]_f
\end{array}
$$

Thus, reading from extreme left to extreme right, we read off that the matrix corresponding to L using the f-basis is $S^{-1}AS$ and hence

$$B = S^{-1}AS$$

15.1.11 Example.   A Change of Basis in $V_3(\mathbb{Q})$

Again consider $V = W = V_3(\mathbb{Q})$ and choose the e-basis and the f-basis to be those used in part 2 of Example 15.1.3.

Let L be the linear transformation such that

$$L(e_1) = e_1$$
$$L(e_2) = e_1 + 2e_2$$
$$L(e_3) = e_1 + 2e_2 + 3e_3$$

Using the e-basis for L we find the matrix A corresponding to L is

$$A = \begin{bmatrix} 1 & 1 & 1 \\ 0 & 2 & 2 \\ 0 & 0 & 3 \end{bmatrix}$$

According to Theorem 15.1.10 we shall find the matrix corresponding to L using the f-basis by the matrix computation

$$B = S^{-1}AS$$

where S is the matrix of the change of basis from the f-basis to the e-basis. In Example 15.1.3, we found the matrix $\hat{J}$ that gave the e-basis elements in terms of the f-basis. So $S^{-1} = \hat{J}$ in this case.

To complete this computation we need the matrix $S = \hat{J}^{-1}$. In this case it is easy to determine S, for it corresponds to the identity transformation using the e-basis to the f-basis and so we need only express the f's in terms of the e's. We have

$$f_1 = e_1$$
$$f_2 = e_1 + e_2$$
$$f_3 = e_1 + e_2 + e_3$$

and so

$$S = \begin{bmatrix} 1 & 1 & 1 \\ 0 & 1 & 1 \\ 0 & 0 & 1 \end{bmatrix}$$

We compute

$$B = \begin{bmatrix} 1 & -1 & 0 \\ 0 & 1 & -1 \\ 0 & 0 & 1 \end{bmatrix} \begin{bmatrix} 1 & 1 & 1 \\ 0 & 2 & 2 \\ 0 & 0 & 3 \end{bmatrix} \begin{bmatrix} 1 & 1 & 1 \\ 0 & 1 & 1 \\ 0 & 0 & 1 \end{bmatrix} = \begin{bmatrix} 1 & 0 & -1 \\ 0 & 2 & 1 \\ 0 & 0 & 3 \end{bmatrix}$$

Of course, we could have found B just by computing the effect of L on the f-basis in terms of the f-basis. This makes a nice check of our work. We find

$$L(f_1) = L(e_1) = e_1 = f_1$$
$$L(f_2) = L(e_1 + e_2) = L(e_1) + L(e_2)$$
$$\quad = e_1 + e_1 + 2e_2 = 2(e_1 + e_2) = 2f_2$$
$$L(f_3) = L(e_1 + e_2 + e_3) = L(e_1) + L(e_2) + L(e_3)$$
$$\quad = e_1 + (e_1 + 2e_2) + (e_1 + 2e_2 + 3e_3)$$
$$\quad = e_2 + (3e_1 + 3e_2 + 3e_3) = f_2 - f_1 + 3f_3 = -f_1 + f_2 + 3f_3$$

and so, by definition,

$$B = \begin{bmatrix} 1 & 0 & -1 \\ 0 & 2 & 1 \\ 0 & 0 & 3 \end{bmatrix}$$

Remark.   There is an important consequence of the fact that the dimension of the space of all linear transformations from V into V is finite if $\dim(V)$ is finite. We find immediately that there is a polynomial $f(x) \in F[x]$ such that

$$f(L) = 0 \text{ transformation}$$

Indeed, as members of this vector space

$$I, L, L^2, \ldots, L^{(n^2)}$$

are $n^2 + 1$ elements. The dimension is n^2 and so these transformations must be linearly dependent. Hence there are elements of F, c_i such that

$$\Sigma_i c_i L^i = 0 \text{ transformation}$$

Thus if we take

$$f(x) = \Sigma_i c_i x^i$$

we shall have the desired polynomial. Actually more is true; we show in Exercises 15.1.2–15.1.8 that there is a polynomial $m(x)$ of degree at most the dimension of V such that $m(L) = 0$.

EXERCISE SET 15.1

15.1.1 Let L be a linear transformation from V into W. Show that $\dim(V) = \dim(\text{kernel}) + \dim(\text{range})$.

The purpose of Problems 15.1.2 to 15.1.8 is to give a determinant free proof that if V is a finite-dimensional vector space over a field F and $L: V \to V$ is a linear transformation from V into V then there exists a polynomial $m(x) \in F[x]$ of degree at most $\dim(V)$ such that $m(L) = 0$ transformation. Equivalently, to prove that for every $n \times n$ matrix with entries in F there is a polynomial $m(x)$ of degree at most n such that $m(A) = 0$ matrix. The notation is continued from problem to problem. Problems 15.1.9 and 15.1.10 give applications of this result.

15.1.2 Consider

$$\mathbf{A}_V(L) = \{ f(x) \in F[x] : f(L) = 0 \}$$

Prove that $\mathbf{A}_V(L)$ is a nonzero ideal of $F[x]$ and hence

$$\mathbf{A}_V(L) = \langle m(x) \rangle$$

When $m(x)$ is chosen to be monic, as it may, it is called the *minimal polynomial* for L. Show that if $h(x)$ has degree smaller than the degree of $m(x)$ then there is a vector $v \in V$ such that $h(L)v \neq 0$.

15.1.3 Show that if S is a subspace of V then

$$\mathbf{A}_S(L) = \{ f(x) \in F[x] : f(L)s = 0 \text{ for all } s \in S \}$$

is an ideal containing $\mathbf{A}_V(L)$ and thus

$$\mathbf{A}_S(L) = \langle s(x) \rangle \quad \text{and} \quad s(x) | m(x)$$

15.1.4 Show that if $m(x) = r(x)s(x)$ and $(r(x), s(x)) = 1$ then V is the direct sum of subspaces S and T such that $\mathbf{A}_S(L) = \langle s(x) \rangle$ and $\mathbf{A}_T(L) = \langle r(x) \rangle$. *Hint:* Write $1 = r(x)u(x) + s(x)v(x)$ and hence $I = r(L)u(L) + s(L)w(L)$. Let $S = \{ r(L)u(L)v : v \in V \}$. Similarly for T.

15.1.5 Extend the result of Exercise 15.1.4 to show that if

$$m(x) = p_1^{e_1}(x) \ldots p_k^{e_k}(x)$$

then V is the direct sum of k subspaces S_i such that

$$\mathbf{A}_{S_i}(L) = \langle p_i^{e_i}(x) \rangle$$

15.1.6 Let $v \in V$. Show that $\{v, L(v), \ldots, L^n(v)\}$ are linearly dependent. Let S be the subspace they span and let

$$\mathbf{A}_S(L) = \langle s(x) \rangle$$

Show that $\deg(s(x)) \leq \dim(S)$.

15.1.7 If $m(x) = p^e(x)$ where $p(x)$ is an irreducible polynomial, show that

$$\deg(m(x)) \leq \dim(V)$$

Hint. Consider a basis and apply Problem 15.1.6 to each vector in turn.

15.1.8 Show that $\deg(m(x)) \leq \dim(V)$

15.1.9 A nonzero vector v is called an *eigenvector* for L if there is a $\lambda \in F$ such that $L(v) = \lambda v$. If v is an eigenvector, the corresponding λ is called the *eigenvalue* corresponding to v. An element $\mu \in F$ is called an *eigenvalue* of L if there is an eigenvector for L whose corresponding eigenvalue is μ.

Show that if v is an eigenvector for L then v has only one corresponding eigenvalue.

Show that if μ is an eigenvalue of L then the set of eigenvectors whose eigenvalue is μ is a subspace.

Show that if the minimal polynomial $m(x)$ for L has a zero in F then that zero is an eigenvalue of L.

Suppose that $A = [a_{ij}]$ is the $n \times n$ matrix corresponding to L. Show that μ is an eigenvalue of L if and only if the following homogeneous system of linear equations has a nontrivial solution in F:

$$(a_{11} - \mu)x_1 + \quad a_{12}x_2 + \cdots + \quad a_{1n}x_n = 0$$
$$a_{21}x_1 + (a_{22} - \mu)x_2 + \cdots + \quad a_{2n}x_n = 0$$
$$\vdots$$
$$a_{n1}x_1 + \quad a_{n2}x_2 + \cdots + (a_{nn} - \mu)x_n = 0$$

15.1.10 Prove that an $n \times n$ matrix S is similar to a diagonal matrix over F if and only if its minimal polynomial is the product of distinct linear factors in $F[x]$.

Exercises 15.1.11 to 15.1.13 introduce the *dual* vector space and its properties. Let V be a vector space of dimension n over a field F. Classically, a function λ from V into F is called a *linear functional* if $\lambda(av + bw) = a\lambda(v) + b\lambda(w)$ for all $a, b \in F$ and all $v, w \in V$. Thus a linear functional is a linear transformation from V to F.

15.1.11 Show that the set of linear functionals from V to F is a vector space over F if addition of functionals and scalar multiplication are defined by

$$(\lambda + \varphi)(v) = \lambda(v) + \varphi(v)$$
$$(a\lambda)(v) = \lambda(av)$$

15.1.12 Show that V^* is isomorphic to V as follows: Pick a basis $\{v_1, \ldots, v_n\}$ for V. For each $w \in V$ let $w = a_1v_1 + \cdots + a_nw_n$. Now define

$$\lambda_w(v) = \sum_{i=1}^{n} a_ib_i$$

where $v = b_1v_1 + \cdots + b_nv_n$.
Show that for each $w \in V$, λ_w is a linear functional and that the mapping of $w \rightarrow \lambda_w$ is an isomorphism of V onto V^*.

15.1.13 Show that for every basis $\{w_1, \ldots, w_n\}$ of V there is a basis $(\omega_1, \ldots, \omega_n)$ for V^* such that

$$\omega_i(w_j) = 1 \quad \text{if } i = j, \quad \text{and } 0 \text{ otherwise}$$

These bases are said to be dual to each other.

15.2 DETERMINANTS

15.2.1 Definition.   Determinants over a Commutative Ring

Let R be a commutative ring with identity. Let $M = [m_{ij}]$ be an $n \times n$ matrix whose entries belong to R. The determinant of M is an element of R defined by

$$\det(M) = \sum_{\pi \in S_n} \operatorname{sgn}(\pi) m_{1(1)\pi} m_{2(2)\pi} \cdots m_{n(n)\pi} \tag{1}$$

where

$$\operatorname{sgn}(\pi) = \left(\begin{array}{ll} 1 & \text{if } \pi \text{ is an even permutation} \\ -1 & \text{if } \pi \text{ is an odd permutation} \end{array} \right.$$

and where the sum extends over all $n!$ permutations in the symmetric group S_n.

Thus we evaluate $\det(A)$ by choosing an entry from each row and column of A, no two elements in the same row or column, taking the product of these n elements, attaching a $+$ or $-$ sign, repeating this for every permutation, and then summing over all of them!

The next theorem gives the elementary properties for determinants. The proofs are straightforward, and we shall leave most of them to the exercises.

15.2.2 Theorem.   Elementary Properties of Determinants

Let the columns of M be $C_1, \ldots, C_n$. As a function of the columns

1. Homogeneity: For all $a \in R$, $\det[C_1, \ldots, aC_k, \ldots, C_n] = a \det[C_1, \ldots, C_k, \ldots, C_n]$.
2. Additivity in each column: If, for a fixed index k, $C_k = A_k + B_k$ then $\det[C_1, \ldots, (A_k + B_k), \ldots, C_n] = \det[C_1, \ldots, A_k, \ldots, C_n] + \det[C_1, \ldots, B_k, \ldots, C_n]$.
3. Zero condition: If two columns of M are equal then $\det(M) = 0$.
4. Normalization: $\det(\text{identity}) = 1$.
5. The determinant is unchanged if rows and columns are interchanged: $\det(M) = \det(M^T)$, where M^T is the transpose of M.
6. Expansion by cofactors: For a fixed pair of indices i, j, denote by M_{ij} the $((n-1) \times (n-1))$ matrix obtained from M by deleting the ith row and jth column from M. The cofactor c_{ij} is defined

$$c_{ij} = (-1)^{i+j} \det(M_{ij})$$

Then

$$\det(M) = \sum_{1 \le h \le n} m_{ih}c_{ih} = \sum_{1 \le h \le n} m_{hj}c_{hj}$$

Proof. As a sample of the arguments we shall prove that $\det(M) = 0$ if two (or more) columns are equal. Suppose that column i and column j are equal; that is, for all k

$$m_{ki} = m_{kj}$$

Now consider an even permutation π and its product with the transposition (i, j). Let

$$\sigma = \pi(i, j)$$

Note that

$$(h)\pi = i \qquad \text{if and only if } (h)\sigma = j$$

and

$$(h)\pi = j \qquad \text{if and only if } (h)\pi = i$$

Otherwise

$$(h)\pi = (h)\sigma$$

Now we claim

$$m_{1(1)\pi} \cdots m_{n(n)\pi} = m_{1(1)\sigma} \cdots m_{n(n)\sigma}$$

This is so because most factors are unchanged in switching from π to σ. Only two factors change, but remember

$$m_{hi} = m_{hj}$$

so, if $(h)\pi = i$ then

$$m_{h(h)\pi} = m_{hi} = m_{hj} = m_{h(h)\sigma}$$

and if $(h)\pi = j$ then

$$m_{h(h)\pi} = m_{hj} = m_{hi} = m_{h(h)\sigma}$$

Now we observe that $\mathrm{sgn}(\pi) = -\mathrm{sgn}(\sigma)$; hence these two equal products cancel as summands in (1). Since $\mathbb{S}_n = \mathscr{A}_n \cup \mathscr{A}_n(i, j)$ all terms in (1) will cancel. Hence $\det(M) = 0$.

15.2.3 Theorem.   Product of Matrices and the Product of Determinants

1. The determinant of the product of two matrices is the product of their determinants: If M and N are two $n \times n$ matrices

$$\det(MN) = \det(M)\det(N)$$

2. Define the *adjoint* of $M = [a_{ij}]$ to be the $(n \times n)$ matrix formed by taking the transpose of the matrix in which every element of M has

been replaced by its cofactor. Thus

$$\text{adj}(M) = [d_{ij}] \qquad \text{where } d_{ij} = c_{ji}$$

is the cofactor of a_{ji}.

$$M \text{ adj}(M) = \text{adj}(M)\, M = \text{diag}(\det(M), \ldots, \det(M))$$

$$= \begin{pmatrix} \det(M) & 0 & \cdots & 0 \\ 0 & \det(M) & \cdots & 0 \\ \cdot & & \cdots & \cdot \\ 0 & 0 & \cdots & \det(M) \end{pmatrix}$$

In particular,

3. If $\det(M)$ is a unit in R then $M^{-1} = (1/\det(M))\text{adj}(M)$.

4. A matrix is nonsingular if and only if its determinant is nonzero.

Proof of 1. Let the columns of M be $C_1, \ldots, C_n$. From Theorem 15.1.8 we know that the columns of MN are linear combinations of these columns. Specifically, if the jth column of N is

$$D_j = \begin{bmatrix} d_{1j} \\ d_{2j} \\ \vdots \\ d_{nj} \end{bmatrix}$$

then the jth column of MN is

$$\sum_{k=1}^{k=n} C_k d_{kj}$$

Thus expressing the determinant of MN as a function of its columns we obtain

$$\det MN = \det\left[\sum_{k=1}^{k=n} C_k d_{k1}, \quad \sum_{k=1}^{k=n} C_k d_{k2}, \ldots, \quad \sum_{k=1}^{k=n} C_k d_{kn} \right] \qquad (2)$$

Now we use the fact that the determinant is a linear function of its columns. We use this linearity to its fullest extent, expanding the right-hand side of (2) completely. Begin by working only on the first column. We use h as the summation index in the first column. We obtain

$$\det MN = \sum_{h=1}^{h=n} \det\left[C_h d_{h1}, \quad \sum_{k=1}^{k=n} C_k d_{k2}, \ldots, \quad \sum_{k=1}^{k=n} C_k d_{kn} \right]$$

At this point $\det MN$ has been expressed as the sum of n determinants. Each term in this sum can be expanded according to the linear combination of its second column. Moreover note that this second column is the same for all these n determinants and is the second column of MN. After this expansion we shall have expressed $\det MN$ as the sum of n^2 determinants. Continue to expand each of these n^2 determinants according to the linear combination of the third column. The process stops when all possible expansions have been made; at this point det MN will be expressed as the sum of n^n determinants. A typical term in the sum looks like

$$\det\left[C_{\pi(1)}d_{\pi(1)1}, C_{\pi(2)}d_{\pi(2)2}, \ldots, C_{\pi(n)}d_{\pi(n)n}\right] \tag{3}$$

where the function $\pi(j)$ simply tells which choice has been made from the expression

$$\sum_{k=1}^{k=n} C_k d_{kj}$$

in the jth column of det MN. Thus π is a function from the set of indices $\{1, \ldots, n\}$ into itself. As we have just seen, each term in the ultimate expression determines such a function, and conversely, each such function determines a term in the expansion.

We may of course factor out the scalar coefficients in the determinant (3) and write

$$\det\left[C_{\pi(1)}, C_{\pi(2)}, \ldots, C_{\pi(n)}\right] d_{\pi(1)1} d_{\pi(2)2} \cdots d_{\pi(n)n}$$

Now, because a determinant is zero if two columns are equal, we may delete all terms for which the function π is not a permutation. Thus we may express

$$\det MN = \sum_{\pi \in \mathbb{S}_n} \det\left[C_{\pi(1)}, C_{\pi(2)}, \ldots, C_{\pi(n)}\right] d_{\pi(1)1} d_{\pi(2)2} \cdots d_{\pi(n)n}$$

as the sum over the permutations in the full symmetric group $\mathbb{S}_n$.

Since the determinants in this sum are merely a permutation of the columns of M, we may replace each of these by det M sgn(π). Recall that sgn π is 1 if π is an even permutation; -1 otherwise. Now factor out det M. Thus

$$\det MN = \det M\left(\sum_{\pi \in \mathbb{S}_n} \text{sgn}(\pi)\, d_{\pi(1)1}\, d_{\pi(2)2} \cdots d_{\pi(n)n}\right)$$

Now the expression inside the parentheses is equivalent to the definition of the determinant of N. Just replace an index j by $i = \pi^{-1}(j)$ and sum over π^{-1} instead of π. Since sgn $\pi = $ sgn π^{-1} there is no change in the sum. Thus

$$\det MN = \det M \det N$$

Proof of 2. Using the notation introduced in part 2, compute

$$M(\text{adj } M) = \left[\Sigma_h a_{ih} d_{hj}\right] = \left[\Sigma_h a_{ih} c_{jh}\right]$$

Consider the sum

$$\Sigma_h a_{ih} c_{jh} \tag{4}$$

If $i = j$ then (4) is just the expansion of det M by the cofactors of the ith row. Thus each diagonal entry is det M. However, if $i \neq j$ then we may view the sum (4) as the same expansion of the determinant of the matrix made from M by replacing the jth row by the ith row. The determinant of such a matrix is zero since it has two equal rows.

The result for the product (adj M)M comes from the same sort of argument, except that the expansions are column expansions instead of row expansions.

Proof of 4. From part 3 we see that if det M is a unit in R then M^{-1} exists. This means that if linear transformations are associated with M and M^{-1} with respect to the same basis, their composition is the identity transformation. Thus M (and M^{-1}) is nonsingular.

Conversely, if M is nonsingular, then a transformation associated with M is nonsingular and thus has an inverse transformation and if N is the matrix associated with that transformation it must be that $MN = I$. But then

$$\det(MN) = \det(M)\det(N) = \det I = 1$$

so that det $M \neq 0$.

EXERCISE SET 15.2

15.2.1 Prove the properties of determinants in Theorem 15.2.2.

15.2.2 Let

$$M = \begin{bmatrix} c_{11} & c_{12} \\ c_{21} & c_{22} \end{bmatrix} \quad \text{and} \quad N = \begin{bmatrix} d_{11} & d_{12} \\ d_{21} & d_{22} \end{bmatrix}$$

Write out all the steps involved in the proof of Theorem 15.2.3, including expressing det(MN) as the sum of 2^2 terms.

15.2.3 Let $x_1, \ldots, x_n$ be indeterminates over $\mathbb{Z}$. Prove that the Van der Monde identity holds in $\mathbb{Z}[x_1, x_2, \ldots, x_n]$.

$$\det \begin{bmatrix} 1 & 1 & 1 & \cdots & 1 \\ x_1 & x_2 & x_3 & \cdots & x_n \\ x_1^2 & x_2^2 & x_3^2 & \cdots & x_n^2 \\ x_1^3 & x_2^3 & x_3^3 & \cdots & x_n^3 \\ & & \cdots & & \\ x_1^{n-1} & x_2^{n-1} & x_3^{n-1} & \cdots & x_n^{n-1} \end{bmatrix} = \prod_{i<j}(x_j - x_i)$$

Hint: Remember a determinant is zero if $x_i = x_j$ and that $\mathbb{Z}[x_1, \ldots, x_n]$ is a unique factorization domain.

Chapter **16**

Modules and Canonical Forms

In this chapter another important class of algebraic systems is introduced. These systems are called modules. The concept of a module generalizes the concept of a vector space. We use the theory of modules to derive the basic structure of finitely generated abelian groups and the standard canonical forms for linear transformations of a vector space. The prerequisites are Chapters 1 to 10 and Chapter 15. For the applications you should know the results of Sections 11.3 and 11.4.

16.1 INTRODUCTION TO MODULES

Speaking loosely, just to give the idea of what we shall soon be treating with considerable precision, a module is a "vector space" over a ring instead of a field. Indeed, as you will see in the definition, the modules we consider in this text have axioms that are identical with those of a vector space over a field except that scalars for a module are permitted to come from a ring whereas the scalars for a vector space must come from a field. Since a field is a ring every vector space is a module.

In general the ring may be quite "wild." Rings may or may not be commutative; they may or may not have a multiplicative identity. As a result one must proceed with care. In this text we assume that the ring is always a principal ideal ring (PID). Remember, a PID is a commutative ring that is an integral domain in which every ideal is principal. Recall that a field is a PID; indeed a field has but two ideals, the zero ideal $\{0\} = \langle 0 \rangle$ and the field itself $\langle 1 \rangle$.

We shall constantly watch two classes of examples of modules. One is the class of abelian groups with the ring of scalars chosen to be the integers. The other is the class of vector spaces over a field. If the field is F, the ring of scalars for the module is chosen to be $F[x]$, the ring of polynomials over F. The action of $F[x]$ on the vector space will be determined by a given linear transformation of the vector space. Thus there is great flexibility in this class of examples, and therein lies the power of the module concept. See Example 16.1.4. Important concepts from a vector space that do not carry over to modules are the existence of a basis and a dimension of a vector space.

We begin with the definition.

16.1.1 Definition.   Module over a Ring

A *module M* over a ring R (briefly, an R-module) is an abelian group $\langle M, + \rangle$ over which R acts as a ring of scalars; that is

1. For each $a \in R$ and each $m \in M$, $am \in M$.
2. For each $a, b \in R$ and each $m \in M$, $(a + b)m = am + bm$.
3. For each $a \in R$ and each $m, n \in M$, $a(n + m) = an + am$.
4. For each $a, b \in R$ and each $m \in M$, $(ab)m = a(bm)$.
5. If 1 is the identity of R then $1m = m$ for all $m \in M$.

Remark.   Axiom 5 is not always required of a module. When it is, and we always make that requirement in this text, the module is called *unital*.

As with vector spaces, Axiom 2 implies that the scalar product of $0 \in R$ and any $m \in M$ equals 0, the identity of the abelian group $\langle M, + \rangle$.

$$0m = 0$$

Note that the symbol 0 appears in different roles on each side of the preceding equality. And, using the same conventions used with vector spaces, we use $+$ to denote addition, both in the module M and in the ring R. While this practice may be confusing it is more so to insist on separate symbols for addition and multiplication in R and in $\langle M, + \rangle$.

Also it is true, as in vector space, that

$$(-1)m = -m$$

where $-m$ is the additive inverse of m in the abelian group $\langle M, + \rangle$.

16.1.2 Examples of Modules

1. A vector space over a field F is a module over F.
2. A ring R is itself a module over R. Strictly speaking, we should say that if R is a ring then $\langle R, + \rangle$ is a module over R.
3. If R is a ring and A is an ideal of R then A is an R-module. The swallow-up property of ideals shows that Axiom 1 holds. The remaining axioms are ring axioms that hold in R.

4. If R is a ring and A is an ideal then the factor ring R/A is an R-module under the usual addition of residue classes. The natural definition for the action of R on the residue classes is defined as follows: For any residue class $r + A$ define

$$a(r + A) = ar + A$$

for any $a \in R$. It is easy to see that this definition is independent of the residue class representative chosen.

A specific instance of this category of examples is that

$$\langle \mathbb{Z}_n, + \rangle$$

is a $\mathbb{Z}$-module; that is, the integers mod n form a module over the integers. Thus we've been working with modules since Chapter 1, even if we didn't know it.

5. *Module of n-Tuples* Let n be a positive integer. Let R be any commutative ring with identity. Consider the set of n-tuples

$$M_n = \{(r_1, \ldots, r_n) : r_i \in R\}$$

This set forms an R-module under componentwise addition and multiplication. That is,

$$(r_1, \ldots, r_n) + (s_1, \ldots, s_n) = (r_1 + s_1, \ldots, r_n + s_n)$$

and

$$s(r_1, \ldots, r_n) = (sr_1, \ldots, sr_n)$$

for all n-tuples and elements $s \in R$. In the next section we learn to call this a free R-module on n generators.

16.1.3 Example.   An Abelian Group Is a $\mathbb{Z}$-module

Let $\langle M, + \rangle$ be an abelian group. For each positive integer a define

$$am = m + \cdots + m \qquad \text{(a summands)}$$

and if a is negative, $am = -((-a)m)$.

The usual rules for arithmetic in an abelian group verify the module axioms. Notice that the group is unchanged by its consideration as a $\mathbb{Z}$-module. However, we shall gain additional insight about abelian groups by studying them as $\mathbb{Z}$-modules.

It is instructive to note in Example 16.1.2, 5 (where the elements of the module are n-tuples of ring elements) that if $R = \mathbb{Z}$, then the abelian group of the module is just the direct sum of n copies of the integers.

16.1.4 Example.   A Vector Space over F Is an $F[x]$-module

Let V be a vector space over a field F. Let T be a linear transformation of V. (If you wish you may think of V as being finite-dimensional and T as a matrix with coefficients in F. For a more specific example see Example 16.1.10.)

Addition of module elements is defined to be the underlying addition of the vectors in V.

Scalar multiplication is defined as follows. For each $f(x) \in F[x]$ and each $v \in V$

$$f(x)v = f(T)(v)$$

Thus if $f(x) = x$ then $xv = T(v)$ and if $f(x)$ is a constant polynomial, say, $f(x) = a$ then $f(x)v = aI(v) = av$.

The action of the ring $F[x]$ on V through scalar multiplication depends upon the linear transformation T. Indeed, as we see in later sections, the structure of V as an $F[x]$-module depends heavily on T. This dependence will permit us to obtain a lot of information about T itself and even to develop an efficient canonical form for the matrix representing T.

Just as cyclic groups provide easily understood examples and form the fundamental building blocks for constructing abelian groups, cyclic modules play the same role in the theory of modules. In the next definition note how the notion of a "generator" of a module is an extension of a generator of a group.

16.1.5 Definition.   Cyclic Module

A *cyclic R-module* is one in which there is an element u such that

$$M = \langle u \rangle = \{ ru: r \in R \}$$

The element u is called a generator for the module.

Note that if a cyclic group has u as a generator in the group sense then u is also a generator when the group is regarded as a $\mathbb{Z}$-module in the sense of Example 16.1.3.

Another example of a cyclic module is a vector space of dimension 1 over a field F. Indeed, if u is any nonzero vector then the span of u,

$$\langle u \rangle = \{ au: a \in F \}$$

is a subspace of dimension 1 and is a cyclic F-module.

Cyclic modules will form the basic building blocks for the modules we study. But before getting into that we first introduce the notions of submodule, homomorphism, and direct sum as extensions of the corresponding concepts for abelian groups.

16.1.6 Definition.   Submodule

A *submodule* N of an R-module M is a subset of M that is an R-module under the operations of addition and scalar multiplication given for M.

As with the other algebraic systems we have studied the verification that a subset is a submodule depends only on the closure properties of the subset under module addition, additive inverses, and scalar multiplication. In particular note that N is a submodule of M if N is a subgroup of M and if, for all $a \in R$,

$$an \in N \quad \text{whenever} \quad n \in N$$

16.1.7 Theorem. Submodules of Cyclic Modules

If R is a PID then every submodule of a cyclic R-module is cyclic.

Proof.   The proof is very much like that for subgroups of a cyclic group. You may want to review the proof of Theorem 3.2.5.

To fix notation let $M = \langle u \rangle = \{ru: r \in R\}$ be the cyclic R-module. Let N be an R-submodule. Define

$$A = \{a \in R: au \in N\}$$

Verify that A is an ideal of R. Because R is a PID, A is principal. Let $A = \langle \hat{a} \rangle$. We shall prove that

$$N = \langle \hat{a}u \rangle = \{(r\hat{a})u: r \in R\}$$

In any event since $\hat{a}u \in N$ by definition it follows that

$$N \supseteq \langle \hat{a}u \rangle$$

On the other hand, if $su \in N$ for some $s \in R$ then $s \in A$ by definition so that

$$s = r\hat{a}$$

for some $r \in R$ and hence

$$su = r\hat{a}u \in \langle \hat{a}u \rangle \quad \text{and so} \quad N \subseteq \langle \hat{a}u \rangle$$

Hence $N = \langle \hat{a}u \rangle$.

The submodules form a complete lattice under set intersection and a join that is defined as the join of subgroups in an abelian group. Exercise 16.1.4 calls for a verification of this.

16.1.8 Definition. Homomorphisms and Direct Sums of Modules

If M and M^* are R-modules then a mapping ϕ,

$$\phi: m \mapsto m^*$$

is an R-module *homomorphism* from M into M^* if

  1. ϕ is a group homomorphism from $\langle M, + \rangle$ to $\langle M^*, + \rangle$.
  2. $a\phi(m) = \phi(am)$ for all $m \in M$ and $a \in R$.

If ϕ is one-to-one, ϕ is called an injection and if in addition the range of ϕ is M^* then ϕ is called an isomorphism.

It follows, just as in the theory of groups, that the image of a module homomorphism is a submodule of M^*, that the kernel N is a submodule of M, and that the range of ϕ is module isomorphic to a factor (now called *difference*) module and denoted $M - N$. As with groups ϕ is an injection if and only if the kernel is just $\{0\}$. The action of R on the cosets is the natural one:

$$a(m + N) = am + N$$

Here again we use the additive notation for cosets.

An R-module is the *direct sum* of two submodules N and N' if

$$M = N \vee N' \quad \text{and} \quad N \wedge N' = \{0\}$$

In this case we write

$$M = N \oplus N'$$

and it is easy to see that

$$M \cong \{(n, n'): n \in N \quad \text{and} \quad n' \in N'\}$$

where, of course, the set of ordered pairs is made into an R-module by defining addition and scalar multiplication componentwise.

As with groups we may extend this definition to direct sums of more than two submodules. An alternate characterization of a direct sum that proves useful later is outlined in Problem 16.1.7. Use it to show that a vector space of finite dimension regarded as a module is the direct sum of cyclic submodules, the ones generated by a basis.

We shall prove that a wide class of modules are direct sums of cyclic modules. For finite-dimensional vector spaces the cyclic subspaces (submodules) in a direct sum decomposition are not unique, their number is. That number is the dimension of the vector space. That proof depended heavily on field properties, and as the next example shows, the same result does not hold for modules in general.

However, we shall be able to determine a canonical form for R-modules when R is a PID. This canonical form will depend on the notion of an *order* of a submodule. Yes, you guessed it, "order" in a module will be a generalization of "order" in a group. The tricky part is to see how to replace the cardinal (size) aspect of order as it is used in groups with an algebraic notion.

But first an example.

16.1.9 Example. Module Decompositions of the Cyclic Group of Order 6 as a Z-module

Let M be the $\mathbb{Z}$-module that is the cyclic group of order 6. Thus $M = \mathbb{Z}_6$. This module is a cyclic module generated by the residue class [1]. It is also the direct sum of a submodule of order 2, generated by [3], and a submodule of order 3, generated by [2]. Thus

$$\mathbb{Z}_6 = \mathbb{Z}_2 \oplus \mathbb{Z}_3$$

since [3] is isomorphic to $\mathbb{Z}_2$ and [2] is isomorphic to $\mathbb{Z}_3$.

This example shows that it is possible to express a module as the direct sum of cyclic submodules in several ways. In particular there is nothing unique about the number of submodules present in a direct sum decomposition.

16.1.10 Example. The Vector Space $V_2(\mathbb{Q})$ as a Cyclic Module over $\mathbb{Q}[x]$

Let T be a linear transformation on $V = V_2(\mathbb{Q})$, the vector space of ordered pairs over $\mathbb{Q}$ where the vector and scalar operations are done componentwise. It is also good to think of V as euclidean 2-space, except that we shall not make use of the metric properties of that space. To define the transformation T we need to give the image of a basis for V. Let us use the standard basis

$$\left\{ \begin{bmatrix} 1 \\ 0 \end{bmatrix}, \begin{bmatrix} 0 \\ 1 \end{bmatrix} \right\}$$

The linear transformation T can be represented by a matrix. For this example let it be

$$A = \begin{bmatrix} 1 & 1 \\ 1 & 2 \end{bmatrix}$$

Thus for vectors

$$v = \begin{bmatrix} h \\ k \end{bmatrix} \qquad \text{compute } T(v) = Av = \begin{bmatrix} 1 & 1 \\ 1 & 2 \end{bmatrix} \begin{bmatrix} h \\ k \end{bmatrix} = \begin{bmatrix} h+k \\ h+2k \end{bmatrix}$$

Now let R be the well-known PID $\mathbb{Q}[x]$. Define the action of $\mathbb{Q}[x]$ on the vector space, described in Example 16.1.4, as follows:

$$\text{For } f(x) \in \mathbb{Q}[x] \quad \text{let} \quad f(x)v = f(T)v = f(A)v$$

Thus, in particular,

$$xv = Av$$

and for each $q \in \mathbb{Q}$,

$$qv = qIv = \begin{bmatrix} qh \\ qk \end{bmatrix}$$

Similarly,

$$(1+x)v = (I+A)v = \left(\begin{bmatrix} 1 & 0 \\ 0 & 1 \end{bmatrix} + \begin{bmatrix} 1 & 1 \\ 1 & 2 \end{bmatrix} \right) v = \begin{bmatrix} 2 & 1 \\ 1 & 3 \end{bmatrix} \begin{bmatrix} h \\ k \end{bmatrix} = \begin{bmatrix} 2h+k \\ h+3k \end{bmatrix}$$

We want to show that V is a cyclic $\mathbb{Q}[x]$-module. We claim that

$$u = \begin{bmatrix} 1 \\ 0 \end{bmatrix}$$

is a generator for $V_2(\mathbb{Q})$ as a $\mathbb{Q}[x]$-module.

To verify this, note that

$$xu = \begin{bmatrix} 1 \\ 1 \end{bmatrix} \quad \text{and} \quad (1+x)u = \begin{bmatrix} 2 \\ 1 \end{bmatrix}$$

and since these vectors are linearly independent, they span the space; for any vector v there are numbers s and $t \in \mathbb{Q}$ such that

$$v = s\begin{bmatrix} 1 \\ 1 \end{bmatrix} + t\begin{bmatrix} 2 \\ 1 \end{bmatrix} = s(xu) + t(1+x)u = [t + (s+t)x]u$$

16.1.11 Definition. Order of a Module

Let M be an R-module. The *order* of M is the set of elements

$$o(M) = \{r \in R: ru = 0 \text{ for all } u \in M\}$$

If R is a PID, as it will be for all our applications, then $o(M)$ is a principal ideal and thus

$$o(M) = \langle a \rangle \qquad \text{for some } a \in R$$

We leave the verification that $o(M)$ is an ideal of R as an exercise, but please do note that the commutativity of R plays a role in the verification.

If $o(M) = \langle a \rangle$ we also refer to the order of M as simply a. There is no real danger in this abuse of language since two generators of the same ideal differ by a unit of R because R is an integral domain.

It is important to observe that if M is a finite cyclic group of order n, then when we regard M as a $\mathbb{Z}$-module the module order of M is $\langle n \rangle$. It is in this way that we have replaced the concept of order as a cardinal number with an algebraic property. The order of a module is an ideal.

More generally, if M is an abelian group, not necessarily finite, the corresponding order of the $\mathbb{Z}$-module is $\langle n \rangle$, where n is the *exponent* of the abelian group, if n is not zero. In this case the order of an ideal does not reflect a cardinal property of the abelian group.

Now let us return to Example 16.1.10 and determine the order of the $\mathbb{Q}[x]$-module considered there.

16.1.12 Example. The Order of the Cyclic $\mathbb{Q}[x]$-module of Example 16.1.10

Again let $V = V_2(\mathbb{Q})$ be regarded as a $\mathbb{Q}[x]$-module where the action of the scalar multiplication is determined by the matrix

$$A = \begin{bmatrix} 1 & 1 \\ 1 & 2 \end{bmatrix}$$

A polynomial $q(x)$ is in $o(V)$ if and only if $q(A)v = 0$ for all $v \in V$. We want to find $p(x)$ such that $o(V) = \langle p(x) \rangle$. There are a couple of ways to find a candidate for $p(x)$.

One way is to verify that A satisfies its *characteristic* polynomial which is defined to be

$$c(x) = \det(A - Ix) = \det \begin{bmatrix} 1 - x & 1 \\ 1 & 2 - x \end{bmatrix}$$

and so $c(x) = x^2 - 3x + 1$. Compute that $c(A)$ is the zero matrix. This is a special case of the Hamilton-Cayley theorem which we will prove in Corollary 16.5.4. These calculations show that $c(x) \in o(V)$. Because $c(x)$ is irreducible in this case it follows that $o(V) = \langle c(x) \rangle$.

A second and more constructive way is to take advantage of the fact that the module V is cyclic and is generated by

$$u = \begin{bmatrix} 1 \\ 0 \end{bmatrix}$$

so that $q(x)$ is in the order of V if and only if

$$q(x)u = q(A)u = 0$$

We computed $u = Iu$ and Au in Example 16.1.10. Now compute A^2u.

$$A^2u = A(Au) = \begin{bmatrix} 1 & 1 \\ 1 & 2 \end{bmatrix}\begin{bmatrix} 1 \\ 1 \end{bmatrix} = \begin{bmatrix} 2 \\ 3 \end{bmatrix} = 3\begin{bmatrix} 1 \\ 1 \end{bmatrix} - \begin{bmatrix} 1 \\ 0 \end{bmatrix}$$

$$= 3A(u) - u$$

so that

$$(A^2 - 3A + 1)u = 0$$

Thus $x^2 - 3x + 1$ is in $o(V)$. As before let

$$c(x) = x^2 - 3x + 1$$

On the other hand, $o(V) = \langle p(x) \rangle$ so that $p(x)$ must divide $c(x)$. But we observe that $c(x)$ is irreducible in $\mathbb{Q}[x]$; thus it must be that

$$o(V) = \langle x^2 - 3x + 1 \rangle$$

We conclude this section by determining the submodules and their orders of a cyclic module over a PID. Please note the similarity of Lemma 16.1.13 and Theorem 3.2.6; in fact, that theorem is a special case of this lemma.

16.1.13 Lemma. Submodules of a Cyclic Module

Let R be a PID and let $\langle u \rangle$ be a cyclic R-module whose order is $\langle r \rangle$.

1. If $r \neq 0$ then each nonzero submodule of $\langle u \rangle$ has the form $\langle du \rangle$ where $d \mid r$.
2. If $r \neq 0$ then the order of $\langle tu \rangle$ is $\langle r/\gcd(r, t) \rangle$.
3. If $r = 0$ then the order of every nonzero submodule is also 0.

Remark.   The import of this lemma is that if the order of a cyclic module is not zero there is one-to-one correspondence between submodules and divisors of the order.

Proof.   We know from Lemma 16.1.7 that every submodule of $\langle u \rangle$ has the form $\langle tu \rangle$. If $t = 0$ then the submodule is $\{0\}$ and its order is 1. If $t \neq 0$ and if for some nonzero $s \in R$ it is true that $s(tu) = (st)u = 0$ then the order of $\langle u \rangle$ is not zero since $st \neq 0$. Thus if the order of $\langle u \rangle$ were zero it would follow that the order of every nonzero submodule is zero also. This proves part 3.

Now suppose that the order of $\langle u \rangle$ is $r \neq 0$. We shall in fact prove that $\langle tu \rangle = \langle du \rangle$ where $d = \gcd(t, r)$.

Since R is a PID it is true for suitable $x, y \in R$ that

$$d = xr + yt$$

and so, since $ru = 0$,

$$du = xru + ytu = y(tu) \in \langle tu \rangle$$

On the other hand, since $d \mid t$, $tu \in \langle du \rangle$. This proves part 1.

To prove part 2 first observe that in view of part 1 we may assume that $\langle tu \rangle = \langle du \rangle$ where $d \mid r$. Then certainly r/d is in the order of $\langle du \rangle$ since

$$\frac{r}{d} du = ru = 0$$

Conversely, suppose that $w \in o(\langle du \rangle)$; thus $(wd)u = 0$ so that $wd \in o(\langle u \rangle)$ and so $r \mid wd$ and hence r/d divides w. Thus

$$o(\langle du \rangle) = \left\langle \frac{r}{d} \right\rangle$$

EXERCISE SET 16.1

16.1.1   Show that in any R-module, $0v = 0$ for all module elements u. Show that $(-r)v = -(rv)$ for all ring elements r and module elements v.

16.1.2   Let R be a commutative ring with identity and let M be an ideal of R. Show that M is an R-module.

16.1.3   Let $\langle A, + \rangle$ be an abelian group. Show that there is only one way to define a "scalar multiplication" so that A becomes a $\mathbb{Z}$-module.

16.1.4   Prove that the submodules of an R-module form a complete lattice under set inclusion. Prove that

$$H \vee K = \{ h + k : h \in H \text{ and } k \in K \}$$

Prove a diamond lemma: $(H \vee K) - K \cong H - (H \cap K)$ as R-modules.

16.1.5   Prove that the modular law:

$$\text{If } A \subseteq B \quad \text{then} \quad A \vee (B \cap C) = B \cap (A \vee B)$$

holds in the lattice of submodules of a module.

16.1.6   Consider M to be the integers mod n, $M = \mathbb{Z}_n = \mathbb{Z}/\langle n \rangle$ for $n > 0$, as a $\mathbb{Z}$-module. Show that M is never the direct sum of two nonzero submodules if n is the power of a prime. What is the case for other integers n? Is $\mathbb{Z}$ itself the direct sum of two submodules?

16.1.7   Extend the notion of the direct sum of two modules to finitely many by induction: M is the direct sum of R-modules $N_1, N_2, \ldots, N_s$ if $M = N^* \oplus N_k$ where N^* is the direct sum of $N_1, \ldots, N_{s-1}$. Show that M is the direct sum of s submodules, $N_1, \ldots, N_s$, if and only if
1. $M = N_1 \vee \cdots \vee N_s$ and
2. For all i, $1 \le i \le s$, $N_i \cap (N_1 \vee \cdots \vee N_{i-1} \vee N_{i+1} \vee \cdots \vee N_s) = 0$.

16.1.8 Show that M is the direct sum of n submodules $N_1, N_2, \ldots, N_s$ if and only if for each element $u \in M$ there is a unique representation for u,

$$u = n_1 + n_2 + \cdots + n_s$$

where each $n_i \in N_i$.

16.1.9 Let M be the $\mathbb{Q}[x]$-module defined as in Example 16.1.10 except that the transformation is given by the matrix B,

$$B = \begin{bmatrix} 1 & 1 \\ 1 & 1 \end{bmatrix}$$

1. Show that $V_2(\mathbb{Q})$ is cyclic and find its order.
2. Find the order of the cyclic submodules generated by $\begin{bmatrix} 1 \\ 1 \end{bmatrix}$ and $\begin{bmatrix} 1 \\ -1 \end{bmatrix}$.

16.1.10 For what transformations T will the $\mathbb{Q}[x]$-module $V_2(\mathbb{Q})$ not be cyclic?

16.1.11 Let R be a PID and let M be an R-module. Let $N = \langle n \rangle$ be a cyclic submodule generated by $n \in M$. Let the order of N be the ideal $\langle a \rangle$. Show that N is an R^*-module where R^* is the factor ring $R/\langle a \rangle$. Show that N is isomorphic to R^*.

Interpret this result in the case that $R = \mathbb{Z}$ and M is an abelian group. Be sure to include the extreme cases in which the order ideal is $\{0\}$ and $\mathbb{Z}$.

16.1.12 Show that a cyclic module over a PID whose order is $\{0\}$ is isomorphic over R to R as an R-module. More generally show that if $M = \langle u \rangle$ is a cyclic R-module whose order is $\langle r \rangle$ then $\langle u \rangle \cong R/\langle r \rangle$ as R-modules.

16.2 FINITELY GENERATED MODULES

In this section we present a wide class of modules. Fortunately, this class includes the abelian groups and the vector spaces we find so useful in all kinds of mathematics. It is fortunate because for this class of modules we can obtain some very nice structure theorems. These theorems enable us to classify, to construct, and to compute in these algebraic structures.

16.2.1 Definition.   Finitely Generated Modules

A module M over a ring R is said to be *finitely generated* if there exist finitely many elements $v_1, \ldots, v_n$ such that each element $m \in M$ can be expressed:

$$m = r_1 v_1 + \cdots + r_n v_n \tag{1}$$

where the r's belong to R. In this case we write

$$M = \langle v_1, \ldots, v_n \rangle$$

This definition extends to modules the notion of "span" in a vector space. We might very well say that M is spanned by the elements $\{v_1, \ldots, v_n\}$.

In a vector space the notion of a spanning set was refined to that of a basis. To do this we used the notion of independence. These concepts do not extend to modules. The problem is that in a vector space the order of any nonzero cyclic subspace $\langle u \rangle$ is $\{0\}$ because in a vector space, $au = 0$ if and only if $a = 0$. This in turn is the direct result of field properties: If $a \neq 0$ and $au = 0$ then

$$u = a^{-1}(au) = 0$$

In PIDs, not every nonzero element has to have an inverse, and so we cannot expect to have this property.

Note too, that in (1) there is no requirement that there be a *unique* representation for $m \in M$. But we can extend the inheritance property of finite dimension of vector spaces to finitely generated modules.

16.2.2 Theorem. Finite Generation Is Inherited

Let R be a PID and let M be a finitely generated R-module. If M can be generated by n elements then so can every homomorphic image and every submodule of M.

Proof. Let M be generated by a set of n elements. Clearly every homomorphic image can be generated by the images of the v's and hence can be generated by at most n elements.

Now we shall prove by induction on n, the number of generators of M, that every submodule of M can be generated by n or fewer elements. Please note that the proof given here is not constructive. No algorithm is given to obtain the generators of the submodule.

Begin the induction at $n = 1$. Then M is cyclic, and as we know, every submodule is also cyclic, that is, generated by one element.

Now to fix notation, suppose that $M = \langle v_1, \ldots, v_n \rangle$ and let S be a submodule of M. Since every element of M, and in particular of S, can be expressed in terms of the v's we may consider the set of elements of R that occur as coefficients of v_1 in expressing the elements of S. Define A_1:

$$A_1 = \{ a_1 \in R \colon \text{There exists } s \in S \text{ such that } s = a_1 v_1 + \cdots + a_n v_n \}$$

Note that since $0 \in S$ and $0 = 0v_1 + \cdots + 0v_n$ it follows that $0 \in A_1$; in particular A_1 is not empty. It is routine to show that A_1 is an ideal of R and since R is a PID,

$$A_1 = \langle e_1 \rangle$$

Hence there is some $g_1 \in S$ such that

$$g_1 = e_1 v_1 + b_2 v_2 + \cdots + b_n v_n$$

This means that if $g = a_1 v_1 + \cdots + a_n v_n \in S$ then e_1 divides a_1 and thus

$$g^* = g - \frac{a_1}{e_1} g_1 = c_2 v_2 + \cdots + c_n v_n \tag{2}$$

is an element of S. This suggests that we consider the elements of S that can be expressed in terms of $v_2, \ldots, v_n$. These will be elements of a submodule of M, call it P, generated by the $n - 1$ elements $v_2, \ldots, v_n$. Thus an inductive hypothesis can be invoked. To proceed:

Consider the submodule $P = \langle v_2, \ldots, v_n \rangle$ of M and its submodule $T = P \cap S$. By induction we know that T is generated by $n - 1$ or fewer elements, say,

$$T = \langle g_2, \ldots, g_{n-1} \rangle$$

The proof is completed by showing that

$$S = \langle g_1, g_2, \ldots, g_n \rangle$$

The computation has in effect already been done in (2). Just note that $g*$ is in $P \cap S = T$. Now we investigate a very important class of finitely generated modules, free modules.

16.2.3 Definition. Free Modules

An R-module F is said to be *free* on n generators $\{x_1, \ldots, x_n\}$ over R if whenever

$$r_1 x_1 + r_2 x_2 + \cdots + r_n x_n = 0$$

then all coefficients $r_i = 0$ in R.

As an example verify that the module of n-tuples M_n defined in Example 16.1.2 is free on the generators x_i, $i = 1, \ldots, n$, where

$$x_i = (0, \ldots, 1, \ldots, 0)$$

where the single 1 occurs in the ith coordinate. As it turns out, any free module is isomorphic to M_n for some n.

16.2.4 Lemma.   Representation of Free Modules

If F is a *free* R-module on n generators then F is isomorphic to M_n.

Proof.   The proof is a routine verification that the map

$$r_1 x_1 + \cdots + r_n x_n \mapsto (r_1, \ldots, r_n)$$

is an isomorphism.

Of more utility, but with an equally easy proof, is another characterization of free modules. By the way, it is this property that is used more generally in algebra to extend the notion of free systems.

16.2.5 Theorem. Homomorphism Property of Free Modules

Every finitely generated R-module is the homomorphic image of a free R-module. Specifically:

If M is an R-module that is finitely generated by $\{v_1, \ldots, v_n\}$ and if F is free on n generators, $\{x_1, \ldots, x_n\}$ then the map

$$r_1 x_1 + \cdots + x_n r_n \mapsto r_1 v_1 + \cdots + r_n v_n \tag{3}$$

is a homomorphism of F onto M.

Conversely if M is finitely generated by $\{v_1,\ldots,v_n\}$ and if for any R-module S and any set of n elements $\{y_1,\ldots,y_n\}\subseteq S$ the map

$$r_1v_1 + \cdots + r_nv_n \mapsto r_1y_1 + \cdots + r_ny_n \tag{4}$$

is a homomorphism of M into S, then M is free on $\{v_1,\ldots,v_n\}$.

Proof. For the proof of the first part, in view of the preceding lemma, it can be assumed that $F = M_n$ and that

$$x_i = (0,\ldots,1,\ldots,0)$$

And as in the proof of the lemma it is routine to show that the map (3) is a homomorphism of M_n onto M.

To prove the converse, choose S to be M_n and

$$y_i = (0,\ldots,1,\ldots,0)$$

then verify that the map (4) is an isomorphism.

There are other nice properties of free R-modules. One result requires the following lemma, a lemma that is interesting in its own right.

16.2.6 Lemma.   A Property of Matrices over a Commutative Ring

If A and B are $n \times n$ matrices with elements in a commutative ring R with identity then

$$AB = \text{identity} \quad \text{if and only if} \quad BA = \text{identity}$$

Proof. Recall Theorem 15.2.3, which states among other things that

$$\det(AB) = \det(A)\det(B)$$

so that if $AB = $ identity then $\det(A)$ is a unit of the ring. Thus we are led to consider the set of $n \times n$ matrices whose determinant is a unit of R. Let U be this set.

$$U = \{A : \det(A) \text{ is a unit in } R\} \tag{5}$$

We claim that U is a group under the matrix multiplication given in Theorem 15.1.8. The set U is closed under multiplication because of (5) and the fact that the product of units in a ring is again a unit of the ring (Lemma 7.1.8). Theorem 15.1.8 shows that the multiplication is associative and the identity is the $n \times n$ matrix with 1's down the main diagonal and 0's elsewhere. Theorem 15.2.3 shows that if $\det(A)$ is a unit then an inverse matrix exists.

Finally then we know that in any group an element and its inverse commute.

16.2.7 Theorem.   Exchange Property for Free Modules

Let F be a free R-module. If F is free on $\{x_1,\ldots,x_n\}$ and if

$$F = \langle y_1,\ldots,y_n \rangle$$

then F is also free on $\{y_1,\ldots,y_n\}$.

Proof. Since the x's and the y's each generate F, each x can be expressed in terms of the y's and conversely, each y can be expressed in terms of the x's.

Thus for each i there are coefficients $a_{ij} \in R$ such that

$$x_i = \sum_{j=1}^{n} a_{ij} y_j \qquad i = 1, \ldots, n \tag{6}$$

Let A be the $n \times n$ matrix of coefficients of this system of equations:

$$A = [a_{ij}]$$

Similarly, for each j there are coefficients b_{jk} in R such that

$$y_j = \sum_{k=1}^{n} b_{jk} x_k \qquad j = 1, \ldots, n \tag{7}$$

Let B be the matrix of coefficients of this system of equations.

$$B = [b_{jk}]$$

We shall prove that $AB =$ identity matrix. In (6) substitute for each y_j the expression in (7). Thus

$$x_i = \sum_{j=1}^{n} a_{ij} \left(\sum_{k=1}^{n} b_{jk} x_k \right)$$

$$= \sum_{k=1}^{n} \left(\sum_{j=1}^{n} a_{ij} b_{jk} \right) x_k$$

Since F is free on $\{x_i, \ldots, x_n)$, we can conclude that each coefficient of x_k is zero unless $k = i$, in which case it is 1. That is,

$$\sum_{j=1}^{n} a_{ij} b_{jk} = \begin{cases} 1 & \text{if } k = i \\ 0 & \text{if } k \neq i \end{cases}$$

But this means in matrix terms that $AB = I$. From Lemma 16.2.6 it follows that $BA = I$ as well.

Now we prove that F is free on $\{y_1, \ldots, y_n\}$. Suppose, therefore, that there were a relation, say,

$$r_1 y_1 + \cdots + r_n y_n = 0$$

Substitute here for the y's in terms of the x's:

$$0 = \sum_{j=1}^{n} r_j y_j = \sum_{j=1}^{n} r_j \left(\sum_{k=1}^{n} b_{jk} x_k \right) = \sum_{k=1}^{n} \left(\sum_{j=1}^{n} r_j b_{jk} \right) x_k$$

Since F is free on the x's it follows that

$$\sum_{j=1}^{n} r_j b_{jk} = 0 \qquad \text{for all } k$$

In matrix terms, using row and column multiplication,

$$(r_1, \ldots, r_n)B = (0, \ldots, 0)$$

Hence

$$(r_1, \ldots, r_n)BA = (0, \ldots, 0)A = 0$$

But $BA = I$; thus

$$(r_1, \ldots, r_n) = (0, \ldots, 0)$$

and so each $r_i = 0$; F is free on $\{ y_1, \ldots, y_n \}$.

The main objective of this chapter is to show that any finitely generated R-module is the direct sum of cyclic modules, moreover that their cyclic summands may be chosen so that their orders form a descending chain in R. With this proviso, the decomposition into the direct sum of cyclic modules is unique up to isomorphism. As we shall see, the proof that this decomposition exists is algorithmic and constructive. We carry out this algorithm not only in the proof of the theorems but also in the class of abelian groups. In Section 16.5 we repeat the algorithm for vector spaces and a linear transformation. In this text we assume that R is a euclidean domain. The theorem is true in the case that R is a PID. There is an algorithm in that case as well. (For the details of this more complicated result, see the treatment by Nathan Jacobson in *Basic Algebra*, vol. 1, W. H. Freeman, 1985, pp. 176–180.)

The main algorithm theorem tells how, given a set of generators for a module and the generators for a submodule, we can construct generators for the module and the submodule so that the generators of the submodule are given in a very simple way from the generators of the module. This is done in the next theorem for modules over euclidean domains.

16.2.8 Theorem.   The Main Algorithm for Choosing Generators for a Submodule

Let R be a euclidean ring and let M be an R-module that is finitely generated over R by n elements. For each submodule K generated by a finite number of elements, there exist

1. Generators $\{ y_1, \ldots, y_n \}$ for M.
2. Elements $\{ r_1, \ldots, r_n \}$ in R such that
3. K is generated by $\{ r_1 y_1, \ldots, r_n y_n \}$ and
4. As ideals in R, $\langle r_1 \rangle \supseteq \langle r_2 \rangle \supseteq \cdots \supseteq \langle r_n \rangle$.

Remark.   Notice that although we know from Theorem 16.2.2 that K can be generated by n or fewer elements, this theorem does not require that information. Theorem 16.2.2 was an existence theorem; Theorem 16.2.8 is a constructive result. Once generators for M and K are given, this theorem constructs the y's and the r's.

Also note that (4) is equivalent to

$$\text{In } R, \ r_1 \mid r_2, r_2 \mid r_3, \ldots, r_{i-1} \mid r_i \quad \text{and} \quad r_i = r_{i+1} = \cdots = r_n = 0 \quad (4')$$

(It is possible that all the r's $= 0$ (this happens if $K = \{0\}$) or that none are zero.)

Proof of Theorem 16.2.8.   First, if $K = \{0\}$ then we can choose all the r's to be zero and the theorem holds. So from now on we assume that K is not the zero submodule.

Second, we need a bit of notation. The ring R is euclidean; denote its size function by δ. Let

$$M = \langle x_1, \ldots, x_n \rangle$$

and let

$$K = \langle g_1, \ldots, g_m \rangle$$

Now it must be that the g's can be expressed in terms of the x's:

$$g_1 = a_{11}x_1 + a_{12}x_2 + \cdots + a_{1n}x_n$$
$$g_2 = a_{21}x_1 + a_{22}x_2 + \cdots + a_{2n}x_n$$
$$\cdots \quad\quad (8)$$
$$g_m = a_{m1}x_1 + a_{m2}x_2 + \cdots + a_{mn}x_n$$

We can summarize the information in (8) in a matrix of m rows and n columns

$$A = [a_{ij}]$$

whose (i, j)-entry is the coefficient $a_{ij} \in R$ in the system of equations (8).

We shall systematically alter the generators for M and K. These alterations provide the algorithm. At the conclusion of all the steps we will have constructed the generators for M and the elements r_i required in the theorem.

We use three types of alterations of the generators of M and three others on the generators of the submodule K. By iterating these operations we obtain the desired result. The order of these operations forms the algorithm of the theorem. After an appropriate sequence of operations the matrix A is transformed into a diagonal matrix

$$\text{diag}[r_1, r_2, \ldots, r_n]$$

after possibly having added or deleted superfluous rows of zeros to make the result of size $n \times n$.

Operations on the Generators of *M*

1. Permute the generators of M. If π is a permutation of $1, \ldots, n$, replace each generator x_i by generator $x_{(i)\pi}$.
2. Replace generator x_j of M by $y_j = ex_j$ where e is a unit of R.
3. Replace the generator x_j by $y_j = x_j + qx_s$ where $q \in R$ and $s \neq j$.

The thing to observe is that if the set of generators $\{x_1, \ldots, x_n\}$ is changed, by either a permutation or a replacement of type 2 or 3, the new set of elements again generates M because the original x's can be recovered from the modified set. If π is a permutation and $y_i = x_{(i)\pi}$ then $x_i = y_{(i)\pi^{-1}}$; if e is a unit of R and $y_j = ex_j$ then $x_j = e^{-1}y_j$; or if $y_j = x_j + qx_s$ then $x_j = y_j - qx_s$.

If these alterations are made and the generators of K are expressed in terms of the new generators, the matrix of coefficients will be altered in the following way:

The columns of A will be permuted.

The jth column of A will have been multiplied by the *inverse* of e.

The sth column of A will have had q times the jth column of A *subtracted* from it.

Thus we may alter the matrix A by the corresponding *column* operations:

1′. Replace each column i by column $(i)\pi^{-1}$.
2′. Multiply the jth column of the *inverse* of the unit used in 2.
3′. *Subtract q times the jth column of A from the sth column.*

Now we describe the analogous operations on the generators of the submodule K.

Operations on the Generators of the Submodule *K*

4. Permute the generators of K; if σ is a permutation of $1, \ldots, m$, replace each generator g_i by generator $g_{(i)\sigma}$.
5. Replace generator g_i by $h_i = eg_i$ where e is a unit of R.
6. Replace generator g_j by $h_j = g_j + qg_s$ where $q \in R$ and $s \neq j$.

As before, it follows that if the generators g_j are altered by any of these operations, the new elements generate the submodule K. And the module A has been altered by the corresponding *row* operations:

4′. Replace each row i by row $\sigma(i)$.
5′. Multiply the ith row of A by the unit used in 5.
6′. Add q times the sth row of A to the jth row of A.

Now we prove by an induction on n, the number of generators of M, that there is a finite sequence of operations 1 to 6 to modify the generators of M and K, ending with generators $\{y_1, \ldots, y_n\}$ for M and elements $\{r_1, \ldots, r_n\}$ in R such that the conditions of the theorem are satisfied.

Equivalently, we prove that there is a sequence of operations 1′ to 6′ on the matrix A ending with a diagonal matrix

$$\text{diag}[r_1, \ldots, r_n]$$

for which the r's satisfy the theorem. It is easier to describe the operations in terms of the matrix A, and we use this equivalent form.

· The proof is in several stages. In the first stage we prove that the entries of A can be altered so that all those in the first row and column are zero except for a_{11} and that a_{11} divides all the other entries in the matrix A. The proof of this stage is by induction on μ, the minimum value of δ over the nonzero entries in A. Let

$$\mu = \min_{i,\,j} \delta(a_{ij})$$

Then the proof is completed by applying induction to the submatrix consisting of the last $m-1$ rows and $n-1$ columns.

Stage 1.   The induction begins with the value of μ equal to the least possible value it can have in R. Recall Theorem 9.3.2, if $a \in R$ and $\delta(a)$ is minimal in R then a is a unit of R. Suppose then that some entry of A is a unit of R.

By permuting rows and columns of A if necessary we may suppose that a_{11} is a unit and thus divides all other entries, particularly those in the first row and column. And so by subtracting an appropriate multiple of the first row from each of the other rows we may replace the other entries in the first column by zero. Similarly we may obtain zeros in the first row. The entry a_{11} has not been changed; it is still a unit and so divides all elements. Thus stage 1 is complete in case an entry of A is a unit.

Now we assume that stage 1 may be completed if the minimum δ-value of the entries in A is less than μ and that some entry of A has μ as its δ-value. Again by permuting rows and columns if necessary we may suppose that $\delta(a_{11}) = \mu$.

Suppose there is some entry in the first column that a_{11} does not divide, say a_{i1}. Divide a_{i1} by a_{11} and get a remainder whose δ-value is less than $\mu = \delta(a_{11})$:

$$a_{i1} = a_{11}q + b_{i1} \quad \text{with} \quad \delta(b_{i1}) < \delta(a_{11}) = \mu$$

Now alter A by subtracting q times the first row from the ith row. The new matrix will have b_{i1} in position $(i,1)$ and by the induction hypothesis, it follows that this matrix, and hence A, can be altered so that stage 1 has been completed. Thus we may suppose that a_{11} divides all entries in the first column. Now all the other entries in the first column may be replaced by zero by subtracting off appropriate multiples of the first row from rows 2 through m.

Next we work on the elements of the first row. If a_{11} does not divide a_{1j} then, as above, we may divide a_{j1} by a_{11} to obtain a remainder with a δ-value less than μ. Then by subtracting a multiple of the first column from the jth we obtain a new matrix whose $(1, j)$-entry has a δ-value less than μ. Note that since the entries in the columns are all zero except for a_{11} these operations only alter entries in the first row. Moreover the entries of the first column are unchanged. By induction stage 1 may be completed on this matrix and hence on A.

At this point we may assume that the matrix A looks like

$$\begin{bmatrix} a_{11} & 0 & \cdots & 0 \\ 0 & a_{22} & \cdots & a_{2n} \\ & & \ddots & \\ 0 & a_{m2} & \cdots & a_{mn} \end{bmatrix} \tag{9}$$

Now we can easily prove the theorem if $n = 1$. Indeed, if $n = 1$ then (9) is but a column of zeros except for the first entry, and the theorem is proved. The generator for M is just x_1 and has never been modified, for we have done no column operation. The generator for K is $a_{11}x$; the other entries are $0 = 0x$ and may be deleted from the list of generators for K.

Now we have the main induction on n, the number of columns. The induction hypothesis is that if M is generated by fewer than n generators, a finite sequence of operations 1 to 6 will produce generators for M and elements of R satisfying the theorem.

However, if $n > 1$ we do not yet know that a_{11} divides all the entries in the matrix. Suppose, for example, that a_{11} does not divide a_{ij}. Add the ith row to the first, thus creating a matrix whose entry in the first row and jth column is not divisible by a_{11}. Now the procedure above may be performed again to reduce the minimal δ-value; induction may be applied and stage 1 completed.

Thus we may assume that A has the form (9) and that a_{11} divides all the entries in the submatrix consisting of the last $m - 1$ rows and $n - 1$ columns. This means that at the end of stage 1 we have found generators $\{x_1, \ldots, x_n\}$ for M and generators $\{g_1, \ldots, g_n\}$ for the submodule K such that

$$g_1 = a_{11}x_1$$
$$g_2 = 0x_1 + a_{22}x_2 + \cdots + a_{2n}x_n$$
$$\vdots$$
$$g_m = 0x_1 + a_{m2}x_2 + \cdots + a_{mn}x_n$$

and where a_{11} divides all the entries of the matrix A';

$$A' = \begin{bmatrix} a_{22} & \cdots & a_{2n} \\ \vdots & \ddots & \vdots \\ a_{m2} & \cdots & a_{mn} \end{bmatrix}$$

Now we can apply the induction hypothesis to the submodule M',

$$M' = \langle x_2, \ldots, x_n \rangle$$

and to its submodule K',

$$K' = \langle g_2, \ldots, g_m \rangle$$

to find that there are generators $\{y_2, \ldots, y_n\}$ for M' and elements $\{r_2, \ldots, r_n\}$ so that

$$K' = \langle r_2 y_2, \ldots, r_n y_n \rangle$$

and such that as ideals in R,

$$\langle r_2 \rangle \supseteq \langle r_3 \rangle \supseteq \cdots \supseteq \langle r_n \rangle$$

To complete the proof we define $r_1 = a_{11}$ and need only verify that r_1 divides r_2. This is easy for it requires only the check to see that the operations $1'$ to $6'$ on A' do not alter the divisibility of its elements by $a_{11} = r_1$. Thus the proof of this theorem is complete.

16.2.9 Example.   A Submodule of $\mathbb{Z} \oplus \mathbb{Z}$

In the free $\mathbb{Z}$-module, $\mathbb{Z}^2 = \mathbb{Z} \oplus \mathbb{Z}$ generated by $\{x, y\}$, let

$$K = \langle 3x, 4y, 6x + 2y \rangle$$

Apply the algorithm of Theorem 16.2.8 to find generators $\{u, v\}$ for $\mathbb{Z}^2$ and integers $\{r, s\}$ such that $K = \langle ru, sv \rangle$ where $r \mid s$.

The matrix A of the theorem is in this case

$$\begin{bmatrix} 3 & 0 \\ 0 & 4 \\ 6 & 2 \end{bmatrix}$$

Carry out the following operations:

$$\begin{bmatrix} 3 & 0 \\ 0 & 4 \\ 6 & 2 \end{bmatrix} \xrightarrow{\boxed{1}} \begin{bmatrix} 6 & 2 \\ 0 & 4 \\ 3 & 0 \end{bmatrix} \xrightarrow{\boxed{2}} \begin{bmatrix} 2 & 6 \\ 4 & 0 \\ 0 & 3 \end{bmatrix} \xrightarrow{\boxed{3}} \begin{bmatrix} 2 & 0 \\ 4 & -12 \\ 0 & 3 \end{bmatrix}$$

$$\xrightarrow{\boxed{4}} \begin{bmatrix} 2 & 0 \\ 0 & -12 \\ 0 & 3 \end{bmatrix} \xrightarrow{\boxed{5}} \begin{bmatrix} 2 & 3 \\ 0 & -12 \\ 0 & 3 \end{bmatrix} \xrightarrow{\boxed{6}} \begin{bmatrix} 2 & 1 \\ 0 & -12 \\ 0 & 3 \end{bmatrix} \xrightarrow{\boxed{7}} \begin{bmatrix} 1 & 2 \\ -12 & 0 \\ 3 & 0 \end{bmatrix}$$

$$\xrightarrow{\boxed{8}} \begin{bmatrix} 1 & 0 \\ -12 & 24 \\ 3 & -6 \end{bmatrix} \xrightarrow{\boxed{9}} \begin{bmatrix} 1 & 0 \\ 0 & 24 \\ 0 & -6 \end{bmatrix} \xrightarrow{\boxed{10}} \begin{bmatrix} 1 & 0 \\ 0 & 0 \\ 0 & -6 \end{bmatrix} \xrightarrow{\boxed{11}} \begin{bmatrix} 1 & 0 \\ 0 & 0 \\ 0 & 6 \end{bmatrix}$$

Where the operations are as follows:

$\boxed{1}$. Interchange rows 1 and 3.

$\boxed{2}$. Interchange columns 1 and 2.

$\boxed{3}$. Subtract 3 times column 1 from column 2.

$\boxed{4}$. Subtract 2 times row 1 from the row 2.

$\boxed{5}$. Add row 3 to row 1.

$\boxed{6}$. Subtract column 1 from column 2.

$\boxed{7}$. Interchange column 1 and column 2.

$\boxed{8}$. Subtract two times column 1 from column 2.

$\boxed{9}$. Add 12 times row 1 to row 2 and (-3) times row 1 to row 3.

$\boxed{10}$. Add 4 times row 3 and row 2.

$\boxed{11}$. Multiply column 2 by the unit -1.

At the end of these operations we know that the module is generated by two elements, say u and v, and the submodule is generated by $u, 0 = 0u + 0v$ and $6v$. The generator 0 is superfluous and is deleted from the list of generators.

One mystery remains: How can we find u and v from the original generators x and y? This mystery is not very deep, and we answer it in general in the next remark.

Remark. The column and row operations can also be described in terms of matrix operations. Each column operation corresponds to a multiplication of A on the right by an $n \times n$ matrix; each row operation corresponds to a multiplication of A on the left by an $m \times m$ matrix. Through these multiplications we can automatically keep track of the new generators for M and for the submodule K.

To see how all this comes about we begin with the observation that the initial system (8) of equations may be expressed in matrix notation as

$$
\begin{bmatrix} g_1 \\ g_2 \\ \vdots \\ g_m \end{bmatrix} = \begin{bmatrix} a_{11} & a_{12} & \cdots & a_{1n} \\ a_{21} & a_{22} & \cdots & a_{2n} \\ & & \ddots & \\ a_{m1} & a_{m2} & \cdots & a_{mn} \end{bmatrix} \begin{bmatrix} x_1 \\ x_2 \\ \vdots \\ x_m \end{bmatrix}
$$

Briefly, using an arrow to denote column vectors,

$$ \vec{g} = A\vec{x} $$

The next thing to note is that every column operation 1′, 2′, or 3′ can be accomplished by a multiplication of the matrix A on the *right* by a matrix J. We describe J in each of the three cases.

Denote by E_i the usual unit column vector of all 0's except for a single 1 in row i. The matrix J is given by listing its sequence of columns in each of the three cases.

Corresponding to operation 1′, if π^{-1} is the permutation of the columns,

$$ J = \left[E_{(1)\pi^{-1}}, \ldots, E_{(n)\pi^{-1}} \right] $$

Corresponding to the operation 2′, if column j is to be multiplied by the unit e^{-1},

$$ J = \left[E_1, \ldots, e^{-1}E_j, \ldots, E_n \right] $$

Corresponding to the operation 3′, if q times column j is to be subtracted from columns,

$$ J = \left[E_1, \ldots, E_j, \ldots, E_s - qE_j, \ldots, E_n \right] $$

Note that in all cases det J is a unit in R.

If J is any one of these matrices, the matrix equation $\vec{g} = A\vec{x}$ may then be rewritten

$$\vec{g} = AJ(J^{-1}\vec{x})$$

and so the change of generators is given by

$$\vec{y} = J^{-1}\vec{x}$$

A sequence of column operations $J_1, J_2, \ldots, J_s$ carried out on A can be handled in the same way by setting

$$J = J_1 J_2 \ldots J_s$$

Then the same matrix equations hold and in particular the new generators are obtained from the old by premultiplication by J^{-1}. For ease of computation the sequence of operations giving rise to $J_1, \ldots, J_s$ may be carried out on the columns of the $n \times n$ identity matrix. The result is the matrix J. Then J^{-1} can be computed.

If this is done in Example 16.2.9 using the sequence of column operations $\boxed{2}, \boxed{3}, \boxed{6}, \boxed{7}, \boxed{8}, \boxed{11}$ we obtain the matrix

$$J = \begin{bmatrix} 1 & 2 \\ -4 & -9 \end{bmatrix}$$

whose inverse is

$$J^{-1} = \begin{bmatrix} 9 & 2 \\ -4 & -1 \end{bmatrix}$$

so that the new generators for the module M are

$$\begin{bmatrix} u \\ v \end{bmatrix} = \begin{bmatrix} 9 & 2 \\ -4 & -1 \end{bmatrix}\begin{bmatrix} x \\ y \end{bmatrix}$$

so that $u = 9x + 2y$, $v = -4x - y$.

Similarly we can keep track of the row operations 4', 5', and 6' by altering the rows of an $m \times m$ identity matrix in the same sequence as the row operations are performed on A. If H is the result of this action, the corresponding matrix equations are

$$H\vec{g} = HA\vec{x}$$

and det H is a unit of R.

In Example 16.2.9 the row operations are $\boxed{1}, \boxed{4}, \boxed{5}, \boxed{9}, \boxed{10}$. The result of carrying them out on a 3×3 identity matrix is

$$H = \begin{bmatrix} 1 & 0 & 1 \\ 4 & 1 & -2 \\ -2 & 0 & -3 \end{bmatrix}$$

Combining both the sequence of row and column operations we obtain the matrix equations

$$H\vec{g} = HAJ(J^{-1}\vec{x})$$

The new matrix is HAJ. In Example 16.2.9 $HAJ =$

$$\begin{bmatrix} 1 & 0 & 1 \\ 4 & 1 & -2 \\ -2 & 0 & -3 \end{bmatrix} \begin{bmatrix} 3 & 0 \\ 0 & 4 \\ 6 & 2 \end{bmatrix} \begin{bmatrix} 1 & 2 \\ -4 & -9 \end{bmatrix} = \begin{bmatrix} 1 & 0 \\ 0 & 0 \\ 0 & 6 \end{bmatrix}$$

The new generators for the submodule are given in $H\vec{g}$,

$$h_1 = g_1 + g_3 = 9x + 2y = u$$
$$h_2 = 4g_1 + g_2 - 2g_3 = 12x + 4y - 2(6x + 2y) = 0$$
$$h_3 = -2g_1 - 3g_3 = -6x - 18x - 6y = 6v$$

as alleged. Finally we obtain a 2×2 diagonal matrix

$$\text{diag}[1, 6]$$

by deleting the superfluous rows of zeros from HAJ.

As a corollary of this proof of the main algorithm we may express the result in terms of matrices.

16.2.10 Corollary.   Diagonal Form for Matrices

If $[a_{ij}]$ is an $m \times n$ matrix with elements in a euclidean ring, then there exists a nonsingular $m \times m$ matrix H and a nonsingular $n \times n$ matrix J such that

$$H[a_{ij}]J = [r_1, \ldots, r_n]$$

where all terms are zero except the diagonal terms $a_{ii} = r_i$ and $r_i \mid r_{i+1}$ for $i = 1, \ldots, k$ and $r_h = 0$ if $h > k$.

The determinants of H and J are units in R and if $n = m$ then $\det[a_{ij}] = \Pi_i r_i$.

EXERCISE SET 16.2

16.2.1 Let R be a PID. Show that any submodule of a free R-module is again a free R-module. Show that not every submodule of a free $\mathbb{Z}_6$-module is free.

16.2.2 Let M be the free $\mathbb{Z}$-module on $\{x, y, z\}$. Let K be the submodule generated by $\{a, b, c, d\}$ where

$$a = x + 2y + 4z$$
$$b = 3x + 6y + 12z$$
$$c = -12x - 24y - 48z$$
$$d = -2x + y + 7z$$

Find integers r, s, and t for which there exist generators $\{u, v, w)$ for M such that K is generated by $\{ru, sv, tw\}$ *and* so that $r \mid s$ and $s \mid t$.

16.2.3 Let M be a free $\mathbb{Q}[t]$-module on $\{x_1, x_2, x_3\}$. Let S be the submodule generated by $\{g_1, g_2, g_3\}$ where

$$g_1 = (3 - t)x_1 + x_2$$

$$g_2 = -x_1 + (1 - t)x_2$$

$$g_3 = x_1 + x_2 + (2 - t)x_3$$

Find polynomials $r_1(t), r_2(t), r_3(t)$ such that $r \mid r_2 \mid r_3$ in $\mathbb{Q}[t]$ for which there exist generators y_1, y_2, y_3 for M such that S is generated by $\{r_1(t)y_1, r_2(t)y_2, r_3(t)y_3\}$. Express the y's in terms of the x's.

16.3 MODULE DECOMPOSITION

We are now ready to prove the existence of a direct sum decomposition of a finitely generated R-module into the direct sum of cyclic modules whose orders form a descending sequence. This theorem does hold if R is a PID, but we prove it only in the case that R is a euclidean domain.

16.3.1 Theorem.   The Existence of the Canonical Direct Sum Decomposition

If R is a principal ideal domain then every R-module is the direct sum of cyclic submodules:

$$M = \langle u_1 \rangle \oplus \langle u_2 \rangle \oplus \cdots \oplus \langle u_n \rangle$$

whose orders form a descending chain of ideals in R:

$$\langle r_1 \rangle \supseteq \langle r_2 \rangle \supseteq \cdots \supseteq \langle r_n \rangle$$

Moreover, $\langle u_i \rangle \cong R/\langle r_i \rangle$; that is, each cyclic summand is isomorphic to the ring R modulo the order of its generator.

Proof. Suppose that M is an R-module generated by n elements $\{g_1, \ldots, g_n\}$. Consider the R-module F free on n generators $\{x_1, \ldots, x_n\}$.

By Theorem 16.2.5, the homomorphism property of free modules, there is a module homomorphism ϕ from F to M such that

$$\phi: x_i \mapsto g_i = \phi(x_i)$$

for all i. Let K be the kernel of this homomorphism. By Theorem 16.2.2, K is finitely generated and from the main algorithm (Theorem 16.2.8) we know there exist generators $\{y_1, \ldots, y_n\}$ for F and elements $\{r_1, \ldots, r_n\}$ in R satisfying the divisibility criteria and such that the elements

$$h_i = r_i y_i$$

for $i = 1, \ldots, n$ generate K. Moreover, by Theorem 16.2.7, F is freely generated by $\{y_1, \ldots, y_n\}$.

Let u_i denote the image of y_i under ϕ; $u_i = \phi(y_i)$ for all i. Because $r_i y_i \in K$, the kernel of ϕ, it follows that for all i,

$$r_i u_i = \phi(r_i y_i) = 0$$

We shall prove that M is the direct sum of the u's:

$$M = \langle u_1 \rangle \oplus \cdots \oplus \langle u_n \rangle$$

and that the order of $\langle u_i \rangle$ is $\langle r_i \rangle$. It then follows from Exercise 16.1.12 that

$$\langle u_i \rangle \cong R/\langle r_i \rangle$$

This exercise is easily verified by using the homomorphism from R onto $\langle u_i \rangle$ given by $a \mapsto au_i$ for all $a \in R$ and confirming that the kernel is $\langle r_i \rangle$.

To prove the direct sum decomposition it is enough to prove that each element $m \in M$ has a unique representation in the form

$$m = v_1 + \cdots + v_n \tag{1}$$

where each $v_i \in \langle u_i \rangle$. Since the y's generate F and since ϕ is a homomorphism of F onto M it follows that the images of the y's, which are the u's, generate M. Hence each $m \in M$ can be expressed as in (1).

(By the way, since $v_i = c_i u_i$ for some $c_i \in R$, it is tempting to try to prove that the coefficients c_i are uniquely determined. However, this is not the case; each coefficient c_i is determined only up to congruence mod r_i.)

To prove the uniqueness of the v's it suffices to prove that if

$$0 = v_1 + \cdots + v_n$$

then each $v_i = 0$. Suppose that $v_i = c_i u_i$ for $c_i \in R$. Under ϕ,

$$\phi: c_1 y_1 + \cdots + c_n y_n \mapsto v_1 + \cdots + v_n = 0$$

and so $\Sigma c_i y_i$ belongs to the kernel K. Since K is generated by $(r_1 y_1, \ldots, r_n y_n\}$ it is true that for some elements d_i,

$$c_1 y_1 + \cdots + c_n y_n = d_1 r_1 y_1 + \cdots + d_n r_n y_n$$

in F. Since F is free on the y's it follows that for all i,

$$c_i = d_i r_i$$

in R and hence that

$$c_i u_i = d_i r_i u_i$$

in M. Since $r_i u_i = 0$ in M we conclude that $y_i = c_i u_i = 0$ in M. This shows that M is the direct sum of the cyclic submodules $\langle u_i \rangle$.

To complete the proof of the theorem we must determine the order of $\langle u_i \rangle$. In any event we know that $r_i u_i = 0$ so that r_i belongs to the order of $\langle u_i \rangle$. On the other hand, suppose that $c_i u_i = 0$ in M. Thus under ϕ, $c_i y_i \in K = \langle r_i y_i, \ldots, r_n y_n \rangle$ and so for some coefficients $d_j \in R$,

$$c_i y_i = \sum d_j r_j y_j$$

Since the y's are free this means that

$$d_j r_j = 0 \qquad \text{if } j \neq i \quad \text{and} \quad c_i = d_i r_i$$

thus $c_i \in \langle r_i \rangle$ and so the order of $\langle u_i \rangle$ is $\langle r_i \rangle$.

The descending chain condition $\langle r_i \rangle \supseteq \langle r_{i+1} \rangle$ of the theorem follows directly from the divisibility conditions provided by the main algorithm theorem.

Remark. If one of the r_i is a unit then $\langle r_i \rangle = R$; in particular 1 is in the order and hence $u_i = 0$. In this case the direct summand $\langle u_i \rangle$ is superfluous and can be deleted from the summands. At the other extreme, if one of the $r_i = 0$ then the homomorphism $y_i \mapsto u_i$ is an isomorphism and the direct summand $\langle u_i \rangle \cong R$.

In the next section we show that after eliminating the superfluous zeros and retaining the descending chain condition on the order of the cyclic summands, the representation is unique up to isomorphism. That is, it is the orders that uniquely determine the direct sum decomposition.

As an application of this theorem we can now establish the direct sum decomposition for abelian groups.

16.3.2 Theorem. Decomposition of Abelian Groups

Every finitely generated abelian group is the direct product of cyclic groups $C_1, C_2, \ldots, C_n$ and there is an integer h, $0 \leq h \leq n + 1$ such that

$$|C_i| \, | \, |C_{i+1}| \qquad \text{if } 1 \leq i \leq h$$

and

$$C_i \cong \langle \mathbb{Z}, + \rangle \qquad \text{if } h < i \leq n$$

The extreme case that $h = 0$ means that all summands are isomorphic to $\mathbb{Z}$. The extreme case $h = n + 1$ means that all summands have finite order.

16.3.3 Definition. Finitely Presented Modules

Let R be a PID. An R-module M is said to be *finitely presented* by a finite set of generators $\{w_1, \ldots, w_n\}$ and a finite set of relations

$$\sum_j a_{ij} w_j = 0$$

for $i = 1, \ldots, m$ where $a_{ij} \in R$ if M is the homomorphic image of the free R-module F on $\{x_1, \ldots, x_n\}$ defined by $x_i \mapsto w_i$ $(i = 1, \ldots, n)$ whose kernel is generated by

$$g_i = \sum_j a_{ij} x_j$$

16.3.4 Example.   Abelian Group Example

Let G be an abelian group finitely presented by generators a, b, c, and d and satisfying the following relations:

$$2a + 3b = 0$$
$$4a = 0$$
$$5c + 11d = 0$$

Our task is to express G as the direct product of cyclic groups.

Let F be the free $\mathbb{Z}$-module generated by $\{x_1, x_2, x_3, x_4\}$ and let K be the submodule generated by

$$\{2x_1 + 3x_2, 4x_1, 5x_3 + 11x_4\}$$

We know that G is isomorphic to the difference module $F - K$ (if we were using group terminology we would say the factor group F/K) and according to Theorem 16.3.1 we need only find generators $\{y_1, \ldots, y_4\}$ for F and integers r_1, r_2, r_3, r_4 so that K is generated by $r_i y_i$ $(i = 1, \ldots, 4)$. Then we shall find that

$$G \cong \mathbb{Z}/\langle r_1 \rangle \oplus \mathbb{Z}/\langle r_2 \rangle \oplus \mathbb{Z}/\langle r_3 \rangle \oplus \mathbb{Z}/\langle r_4 \rangle$$

It is a little tidier if we add the redundant generator 0 to the list of generators for K. Then we apply the main algorithm and work with the matrix

$$A = \begin{bmatrix} 2 & 3 & 0 & 0 \\ 4 & 0 & 0 & 0 \\ 0 & 0 & 5 & 11 \\ 0 & 0 & 0 & 0 \end{bmatrix}$$

The following sequence of operations:

Subtract column 2 from column 1.

Multiply row 1 by -1.

Multiply column 2 by -1.

produces

$$\begin{bmatrix} 1 & 3 & 0 & 0 \\ 4 & 0 & 0 & 0 \\ 0 & 0 & 5 & 11 \\ 0 & 0 & 0 & 0 \end{bmatrix}$$

The following sequence of operations:

Subtract 4 times row 1 from row 2.

Multiply row 2 by -1.

Subtract 3 times column 1 from column 2.

produces

$$\begin{bmatrix} 1 & 0 & 0 & 0 \\ 0 & 12 & 0 & 0 \\ 0 & 0 & 5 & 11 \\ 0 & 0 & 0 & 0 \end{bmatrix}$$

Now the following sequence of operations:

Subtract 2 times column 3 from column 4.

Subtract 5 times column 4 from column 3.

Permute the columns: Send column 2 to column 3, column 3 to column 4, and column 4 to column 2.

Interchange row 2 and row 3.

produces

$$\begin{bmatrix} 1 & 0 & 0 & 0 \\ 0 & 1 & 0 & 0 \\ 0 & 0 & 12 & 0 \\ 0 & 0 & 0 & 0 \end{bmatrix}$$

At this point we know that $r_1 = 1$, $r_2 = 1$, $r_3 = 12$, $r_4 = 0$ and that F is generated by four generators, say, y_i. The kernel K is generated by

$$r_1 y_1 = y_1 \qquad r_2 y_2 = y_2 \qquad r_3 y_3 = 12 y_3 \qquad r_4 y_4 = 0$$

Since

$$\mathbb{Z}/\langle 1 \rangle \simeq 0 \qquad \mathbb{Z}/\langle 12 \rangle = \mathbb{Z}_{12} \qquad \mathbb{Z}/\langle 0 \rangle \simeq \mathbb{Z}$$

it follows that

$$G \simeq \mathbb{Z}_{12} \oplus \mathbb{Z}$$

is the direct sum of a cyclic group of order 12 and an infinite cyclic group.

If you should want to express the generators for the summands in terms of the original generators $\{ a, b, c, d \}$, here is one way to do this. First, find the generators of the kernel K. Then find their images under the homomorphism from F to G. To do this, as discussed in Example 16.2.9, calculate the matrix, call it J, that results when the sequence of column operations is carried out on the identity matrix. Then determine J^{-1} and compute

$$J^{-1} \begin{bmatrix} x_1 \\ x_2 \\ x_3 \\ x_4 \end{bmatrix}$$

When we do this in this example we find

$$J = \begin{bmatrix} 1 & 0 & -3 & 0 \\ -1 & 0 & 2 & 0 \\ 0 & -2 & 0 & 11 \\ 0 & 1 & 0 & -5 \end{bmatrix}$$

and

$$J^{-1} = \begin{bmatrix} -2 & -3 & 0 & 0 \\ 0 & 0 & 5 & 11 \\ -1 & -1 & 0 & 0 \\ 0 & 0 & 1 & 2 \end{bmatrix}$$

so that the generators for K are

$$-2x_1 - 3x_2 \qquad 5x_3 + 11x_4 \qquad -x_1 - x_2 \qquad x_3 + 2x_4$$

which map into

$$0, 0, -a - b, c + 2d$$

Notice that $a + b$ has order 12 in G; it is less immediate that $c + 2d$ has infinite order, but that is what our calculations have shown!

16.3.5 Example. The Decomposition of the Cyclic Q[*t*]-module of Example 16.1.10

Let us take a second look at Example 16.1.10 to see how our theory and the main algorithm determine the cyclic nature of that module.

Recall that in this example the module M is the vector space of dimension 2 over $\mathbb{Q}$. The ring is $\mathbb{Q}[t]$ where the scalar multiplication is defined to reflect the action of a linear transformation. The linear transformation was described by a matrix A acting on the standard basis

$$\left\{ E_1 = \begin{bmatrix} 1 \\ 0 \end{bmatrix}, \qquad E_2 = \begin{bmatrix} 0 \\ 1 \end{bmatrix} \right\}$$

where

$$A = \begin{bmatrix} 1 & 1 \\ 1 & 2 \end{bmatrix}$$

The scalar multiplication then was as follows: If v is a vector and $p(t) \in \mathbb{Q}[t]$

$$p(t)v = p(A)v$$

To put the problem in our current framework, we must consider the free $\mathbb{Q}[t]$-module F on two generators x_1, x_2. We want to find generators for the kernel of the homomorphism ϕ from F to M whose action is

$$\phi: x_1 \mapsto E_1 \quad \text{and} \quad \phi: x_2 \mapsto E_2$$

Since every element of the ring is a polynomial in t it suffices to know the action of ϕ on tx_1 and tx_2. We calculate:

$$\phi: tx_1 \mapsto tE_1 = AE_1 = \begin{bmatrix} 1 & 1 \\ 1 & 2 \end{bmatrix}\begin{bmatrix} 1 \\ 0 \end{bmatrix} = E_1 + E_2$$

and

$$\phi: tx_2 \mapsto tE_2 = AE_2 = \begin{bmatrix} 1 & 1 \\ 1 & 2 \end{bmatrix}\begin{bmatrix} 0 \\ 1 \end{bmatrix} = E_1 + 2E_2$$

Thus there are these two relations in M:

$$0 = IE_1 - AE_1 + IE_2 = (I - A)E_1 + E_2$$
$$0 = IE_1 + 2IE_2 - AE_2 = IE_1 + (2I - A)E_2$$

This means that the kernel contains

$$g_1 = (1 - t)x_1 + x_2$$

and

$$g_2 = x_1 + (2 - t)x_2$$

As we shall prove, these two elements do in fact generate the kernel. This is the result of a general theorem (Theorem 16.5.3) that is proved later. In a moment we shall give an argument for this case. Right now we want to apply the main algorithm to determine the decomposition of M, assuming that the kernel is generated by g_1 and g_2.

We form the matrix of coefficients expressing g_1 and g_2 in terms of x_1 and x_2; we call this the characteristic matrix, char(A):

$$\text{char}(A) = \begin{bmatrix} 1 - t & 1 \\ 1 & 2 - t \end{bmatrix}$$

and carry out the main algorithm:

Permute the two rows. Subtract $(1 - t)$ times row 1 from row 2 to obtain

$$\begin{bmatrix} 1 & 2 - t \\ 0 & 1 - (2 - t)(1 - t) \end{bmatrix}$$

Now subtract $(2 - t)$ times the first column from the second and then multiply the second row by -1 to obtain

$$\begin{bmatrix} 1 & 0 \\ 0 & t^2 - 3t + 1 \end{bmatrix}$$

This means that free generators y_1 and y_2 for F can be found so that the submodule of F generated by g_1 and g_2 is also generated by y_1 and $(t^2 - 3t + 1)y_2$.

This tells us that the difference module is the direct sum of two modules. The first has order $\langle 1 \rangle = \mathbb{Q}(t)$ and hence the corresponding direct summand is $\langle 0 \rangle$. This is redundant in the decomposition; therefore, the second summand, which is the cyclic module whose order is $\langle t^2 - 3t + 1 \rangle$, is the difference module itself.

A generator of this difference module could be found from the algorithm, but we can give an ad hoc argument to determine a generator u as follows: Whatever u is, it must be that u and tu ($tu = Au$) are linearly independent in the vector space. Were this not the case we should have rational numbers a and b such that

$$au + bAu = 0$$

and hence that

$$(a + bt)u = 0$$

which would mean that $(a + bt)$ belongs to the order of u; but we know that the order of u is $\langle t^2 - 3t + 1 \rangle$, a contradiction. Therefore, u and Au form a basis for the vector space.

Now if we can find, and it is not difficult, a vector u so that u and Au are linearly independent this will show two things: first that u generates the difference module, and second that the difference module is all of M. In fact,

$u = E_1$ does the job. Looking back, this proves that the kernel is generated by g_1 and g_2.

Were we to follow the main algorithm, note that the only column operation was to subtract $(2 - t)$ times the first column from the second. This means that generator x_1 was replaced by $y_1 = x_1 + (2 - t)x_2$ while the second generator x_2 was unchanged. Apply the homomorphism to find the image of y_1 is zero; hence the image of the second generator x_2 $(= E_2)$ is a generator.

EXERCISE SET 16.3

16.3.1 Let A be the abelian group finitely presented by generators a, b, and c which satisfy these relations:

$$a + 2b + 4c = 0$$
$$3a + 6b + 12c = 0$$
$$-12a - 24b - 48c = 0$$
$$-2a + b + 7c = 0$$

Use your solution to Problem 16.2.2 to determine A as the direct product of cyclic groups.

16.3.2 Let A be the abelian group finitely presented by generators a, b, c, d, and e and satisfying

$$a - 7b \quad\quad +14d - 21e = 0$$
$$5a - 7b - 2c + 10d - 15e = 0$$
$$3a - 3b - 2c + 6d - 9e = 0$$
$$a - b \quad\quad + 2d - 3e = 0$$

Express A as the direct sum of cyclic groups whose orders form a descending chain.

16.3.3 Let V_3 be the vector space of dimension 3 over $\mathbb{Q}$. Let a linear transformation be defined by the action of the matrix A given below on the standard basis, E_1, E_2, E_3:

$$A = \begin{bmatrix} 2 & 2 & -1 \\ -1 & -1 & 1 \\ -2 & -4 & 3 \end{bmatrix}$$

Make V_3 into a $\mathbb{Q}[t]$-module as in the text. Determine V_3 as the direct sum of cyclic $\mathbb{Q}[t]$-modules whose orders form a descending chain.

16.4 UNIQUENESS OF MODULE DECOMPOSITION

In this section we wish to prove that the decomposition of a finitely generated module into the direct sum of cyclic submodules is determined only by the descending sequence of orders associated with the decomposition.

16.4.1 Theorem.   Uniqueness of the Direct Sum Decomposition

Let R be a PID and let M and M' be isomorphic R-modules. If

$$M = \langle x_1 \rangle \oplus \cdots \oplus \langle x_n \rangle$$

where each summand $\langle x_i \rangle$ has $\langle r_i \rangle$ as its order ideal, where no r_i is a unit, and where

$$\langle r_1 \rangle \supseteq \cdots \supseteq \langle r_n \rangle$$

and if

$$M' = \langle y_1 \rangle \oplus \cdots \oplus \langle y_m \rangle$$

where each summand $\langle y_j \rangle$ has $\langle s_j \rangle$ as its order ideal, where no s_j is a unit, and where

$$\langle s_1 \rangle \supseteq \cdots \supseteq \langle s_m \rangle$$

then $n = m$ and for all i, $\langle r_i \rangle = \langle s_j \rangle$ and so in particular, $\langle x_i \rangle \cong \langle y_i \rangle$.

Proof Strategy.   Our proof begins with the observation that because of the descending nature of the orders, the order of M is $\langle r_1 \rangle$ and the order of M' is $\langle s_1 \rangle$. Since M and M' are isomorphic, they must have the same order. Thus we have established that $\langle r_1 \rangle = \langle s_1 \rangle$.

Now what we should like to do is to cancel the isomorphic factors $\langle x_1 \rangle$ and $\langle y_1 \rangle$. Then we should proceed by an induction just as we did in the proof of the fundamental theorem of arithmetic. But it is not obvious that we can cancel isomorphic factors. We would like to have a cancellation law as follows: "If A, B, A', and B' are groups and if $A \times B \cong A' \times B'$ and $B \cong B'$ then $A \cong A'$."

Unfortunately such a general cancellation law is false. Here is an example in which it fails. Let $C_1, C_2, \ldots$ be an infinite number of groups each isomorphic to the cyclic group of order 2. Let

$$A = C_1$$
$$B = C_2 \times C_3 \times \ldots$$
$$A' = C_1 \times C_3 \times C_5 \ldots$$
$$B' = C_2 \times C_4 \times C_5 \ldots$$

Notice that the direct product of A and B and the direct product of A' and B' are each isomorphic to $C_1 \times C_2 \times \ldots$. Thus $A \times B \cong A' \times B'$. Similarly B and B' are isomorphic, but A and A' are certainly not isomorphic.

The ease with which this example has been constructed depends on the fact that the group $A \times B$ requires an infinite number of generators, the elements of the cyclic groups C_i. Another example due to W. R. Scott is given in Exercise 16.4.5 in which the groups, although not abelian, are finitely generated.

Fortunately there is a version of this cancellation law that is just the right stuff for the modules we are considering. The theorem is due to P. M. Cohn and E. A. Walker and answers a question of Irving Kaplansky.

16.4.2 Theorem (Cohn, Walker).   Cancellation of a Cyclic Summand

Let R be a PID. Let M and M' be isomorphic R-modules such that

$$M = N \oplus \langle a \rangle \quad \text{and} \quad M' = N' \oplus \langle a' \rangle$$

If the orders of $\langle a \rangle$ and $\langle a' \rangle$ are equal in R then N and N' are isomorphic R-modules.

Proof.   Since R is a PID the common order of $\langle a \rangle$ and $\langle a' \rangle$ has the form $\langle s \rangle$ where $s \in R$. The proof treats three cases: $s = 0$, s is a power of a prime, and s is the product of several different primes. Since the unique factorization theorem holds in R, this exhausts all possibilities.

It will ease the details of the proof if we assume that $M = M'$. We can do this because the isomorphism of M and M' takes N into a submodule N'' of M' and a into $a'' \in M'$. The isomorphism guarantees that $N \cong N''$, that the order of a'' is the same as the order of a, and that

$$M' \cong N' \oplus \langle a'' \rangle$$

So, if we prove that the submodules N' and N'' of M' are isomorphic it follows that N and N' are isomorphic as well.

Case 1.   The order of $\langle a \rangle$ and $\langle a' \rangle$ is $\{0\}$, that is, $s = 0$.

In this case both $\langle a \rangle$ and $\langle a' \rangle$ are free R-modules on one generator. The proof focuses on the role of N and N' in M. The lattice relationships are pictured in Figure 16.1. The elements u and u' are defined in the argument that follows.

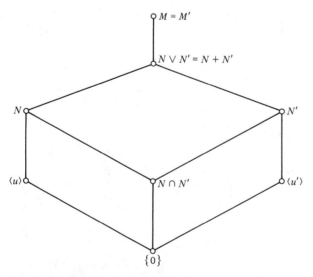

Figure 16.1 The lattice of submodules involved in the proof of case 1 of Theorem 16.4.2. The elements u and u' are constructed in the proof.

Recall (Exercise 16.1.4) that the lattice join $N \vee N'$ of two submodules is

$$N \vee N' = N + N' = \{n + n': n \in N \text{ and } n' \in N'\}$$

By the diamond lemma for modules (Exercise 16.1.4) the difference modules $(N \vee N') - N'$ and $N - (N \cap N')$ are isomorphic via the mapping of cosets in N' in $N \vee N'$ to cosets of $N \cap N'$ in N given by

$$x + N' \to x + N \cap N' \qquad \text{for all } x \in N$$

For brevity let

$$S' = (N \vee N') - N' \quad \text{and} \quad S = N - (N \cap N')$$

Note that S' is a submodule of $M - N' = \langle a' \rangle$ which is a free module on one generator. Since any submodule of a free module on one generator is also a free module on one generator unless it is $\{0\}$, there are now two subcases:

$$S' = \{0\} \quad \text{or} \quad S' = \langle \bar{u}' \rangle \text{ is a free } R\text{-module for some } \bar{u}' \in S'$$

Because S and S' are isomorphic it follows that

$$S' = S = \{0\} \quad \text{or} \quad S' \cong S = \langle \bar{u} \rangle \text{ is a free } R\text{-module for some } \bar{u} \in S$$

If $S = S' = \{0\}$ then $N = N \vee N' = N'$ and we are done. Suppose then that neither S nor S' is $\{0\}$ and that $S = \langle \bar{u} \rangle$ is a free R-module. Since $\bar{u}$ is a coset of $N \cap N'$ in N we may write $\bar{u} = u + N \cap N'$ for some $u \in N$. Note that $\langle u \rangle$ is also a free R-module since its homomorphic image $\langle \bar{u} \rangle$ is free.

Now we want to show that

$$N = (N \cap N') \oplus \langle u \rangle$$

In any event $N = (N \cap N') + \langle u \rangle$ since $N - (N \cap N') = \langle u + N \cap N' \rangle$.

Next we must verify that $\langle u \rangle \cap N' = \{0\}$. A typical element of $\langle u \rangle$ has the form ru where $r \in R$. If, in addition, $ru \in N'$ then $ru \in N \cap N'$ and so

$$ru + (N \cap N') = N \cap N'$$

and so

$$r\bar{u} = r(u + N \cap N') = ru + N \cap N' = N \cap N' = 0 + N \cap N'$$

Now since S is free it follows that $r = 0$ and hence that $\langle u \rangle \cap N' = \{0\}$. Thus $N = (N \cap N') \oplus \langle u \rangle$. Similarly,

$$N' = (N \cap N') \oplus \langle u' \rangle$$

for some $u' \in N'$. Now $\langle u \rangle$ and $\langle u' \rangle$ are isomorphic since each is a free R-module on a single generator. This isomorphism can be extended then to show that

$$(N \cap N') \oplus \langle u \rangle \cong (N \cap N') \oplus \langle u' \rangle$$

and thus to prove that $N \cong N'$.

Case 2. The order of $\langle a \rangle$ is generated by a prime power, that is, $s = p^e$, where p is a prime in R.

There are two stages to this case. In the first stage we show that there is an element $b \in M$ so that $\langle b \rangle$ has order $\langle p^e \rangle$ and such that

$$N \cap \langle b \rangle = \{0\} \quad \text{and} \quad N' \cap \langle b \rangle = \{0\}$$

In the second stage we show that for this element b,

$$M = N \oplus \langle b \rangle = N' \oplus \langle b \rangle$$

and hence that both N and N' are isomorphic to the difference module $M - \langle b \rangle$.

Now for the first stage. If it happens that

$$N \cap \langle a' \rangle = \{0\} \quad \text{or} \quad N' \cap \langle a \rangle = \{0\}$$

then we may choose $b = a'$ or $b = a$, respectively. If neither of the above alternatives holds we now show that the choice $b = a + a'$ satisfies the conditions. Here are the details.

In any event the intersection $N \cap \langle a' \rangle$ is a submodule of the cyclic module $\langle a' \rangle$ whose order is $\langle p^e \rangle$ and hence (Lemma 16.1.13)

$$N \cap \langle a' \rangle = \langle p^f a' \rangle \qquad \text{where } f < e$$

The strict inequality holds since $p^e a' = 0$ and we are assuming that $N \cap \langle a' \rangle \neq \{0\}$. In any event $p^{e-1} a' \in N$. By a similar argument, $p^{e-1} a \in N'$. All this means that

$$p^{e-1}(a' + a) \neq 0$$

for otherwise

$$p^{e-1} a = -p^{e-1} a' \in N$$

and this implies that $p^{e-1} a \in \langle a \rangle \cap N = \{0\}$, a contradiction of the order of $\langle a \rangle$.

More generally we can argue that

$$p^{e-1}(a + a') \notin N$$

since otherwise

$$p^{e-1} a = p^{e-1}(a + a') - p^{e-1} a' \in N \cap N'$$

a contradiction. Thus

$$\langle a + a' \rangle \cap N = \{0\}$$

Similarly

$$\langle a + a' \rangle \cap N' = \{0\}$$

and moreover the order of $\langle a + a' \rangle$ is $\langle p^e \rangle$ since $p^{e-1}(a + a') \neq 0$. Thus $b = a + a'$ satisfies the conditions demanded in the first stage.

Now for the second stage. Since $N \cap \langle b \rangle = \{0\}$ we may consider the submodule of M

$$N^* = N \vee \langle b \rangle \cong N \oplus \langle b \rangle$$

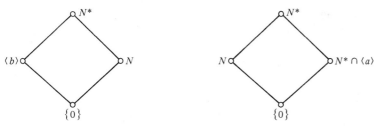

Figure 16.2 Two sublattices of submodules of M entering in the proof of case 2 of Theorem 16.4.2. Together they show that $\langle b \rangle \cong N^* \cap \langle a \rangle$.

We shall show that $a \in N^*$ and hence that

$$M \supseteq N^* \supseteq N \vee \langle a \rangle \supseteq M$$

and thus $M = N^* = N \oplus \langle b \rangle$.

Figure 16.2 shows the diagrams of two sublattices of the lattice of submodules of M. We shall apply the diamond lemma to each of the sublattices as soon as we have verified that they are sublattices.

The only join or meet that is not immediate from the definitions, hypotheses, and the preceding results is the join of N and $N^* \cap \langle a \rangle$. We apply the modular law (Exercise 16.1.5):

$$N \vee (N^* \cap \langle a \rangle) = N^* \cap (N \vee \langle a \rangle)$$

and since $N \vee \langle a \rangle = M$ and $M \supseteq N^*$ it follows that

$$N \vee (N^* \cap \langle a \rangle) = N^*$$

Now from the diamond lemma applied to the lattice of the left side of Figure 16.2 we obtain $N^* - N \cong \langle b \rangle$. Applying the diamond lemma to the lattice of the right side of Figure 16.2, we obtain $N^* - N \cong N^* \cap \langle a \rangle$ and hence we conclude that $\langle b \rangle \cong N^* \cap \langle a \rangle$. But this means that the orders of $\langle b \rangle$ and $N^* \cap \langle a \rangle$ are $\langle p^e \rangle$, which is also the order of $\langle a \rangle$. It now follows (Lemma 16.1.13) that in fact these two submodules are equal:

$$N^* \cap \langle a \rangle = \langle a \rangle$$

and that $a \in N^*$. Thus as we foretold:

$$M = N \oplus \langle b \rangle$$

A similar argument shows that $M = N' \oplus \langle b \rangle$, and the second stage is complete. Thus both N and N' are isomorphic to the difference module $M - \langle b \rangle$.

Case 3. The order ideal is generated by the product of several different primes. Then we may write $s = p^e t$ where p does not divide t.

The proof proceeds by induction on the number of distinct primes appearing in the factorization of s. If there is only one, then this is exactly case 2. If there is more than 1, then t is not a unit and $\gcd(p, t) = 1$. Since R is a

PID there exist u and v such that

$$1 = up^e + vt$$

and so

$$a = up^e a + vta$$

Now $\langle p^e a \rangle$ has order t and $\langle ta \rangle$ has order $\langle p^e \rangle$. Also, since $\gcd(p, t) = 1$, it follows that these two submodules have only 0 in common. Thus it follows that

$$\langle a \rangle = \langle p^e a \rangle \oplus \langle ta \rangle$$

A similar argument shows that

$$\langle a' \rangle = \langle p^e a' \rangle \oplus \langle ta' \rangle$$

Thus

$$M = N \oplus \langle p^e a \rangle \oplus \langle ta \rangle = N \oplus \langle p^e a' \rangle \oplus \langle ta' \rangle$$

By case 2 the last two cyclic modules may be canceled so that

$$N \oplus \langle p^e a \rangle \cong N \oplus \langle p^e a' \rangle$$

Thus by a routine induction on the number of primes appearing in the prime factorization of s the proof of this cancellation theorem may be completed.

16.4.3 Proof of the Uniqueness Theorem 16.4.1

Begin by refreshing your memory of the notation and the statement of the theorem at the beginning of this section!

   Our proof begins with the remark already made that the order of M is $\langle r_1 \rangle$ and the order of M' is $\langle s_1 \rangle$ and these orders must be equal since M and M' are isomorphic modules. Moreover, since R is a PID it follows that the elements r_1 and s_1 can differ by at most a unit multiple of R.

   The proof proceeds by an induction on k, the sum $n + m$ ($k = n + m$) of the number of summands appearing in the given decomposition of M and M'. The least possible value for this sum is 2 when $n = m = 1$. The induction starts there ($k = 2$) and the theorem holds by the remarks of the preceding paragraph.

   The induction hypothesis states that if isomorphic modules can be expressed as the direct sum of cyclic submodules whose orders satisfy the descending chain condition and if fewer than k summands are involved, then the two decompositions are isomorphic; that is, there are the same number of submodules in each decomposition and the order ideals are equal at each step.

   Now the induction step is handled by the cancellation theorem. Suppose that there is a decomposition for M and one for M' satisfying the hypothesis of the uniqueness theorem and suppose that $n + m = k$. By the cancellation theorem we may cancel $\langle x_1 \rangle$ and $\langle y_1 \rangle$ since we have seen that $\langle r_1 \rangle = \langle s_1 \rangle$. Thus

$$M_1 = \langle x_2 \rangle \oplus \cdots \oplus \langle x_m \rangle$$

and

$$M_1' = \langle y_2 \rangle \oplus \cdots \oplus \langle y_m \rangle$$

are isomorphic and the cyclic summands have the same order ideals as before. But now there are only $k - 2 = (n - 1) + (m - 1)$ summands involved. Hence by the induction hypothesis it follows that $n - 1 = m - 1$ and that the order ideals are equal at each step. This completes the proof.

As the first application of this general uniqueness theorem we formulate the standard uniqueness theorem for the decomposition of finitely generated abelian groups. Remember that we have already proved the existence of such a decomposition. And as a result of this uniqueness theorem we will be able to determine, up to isomorphism, all abelian groups of a given finite order n in terms of the prime factorization of n.

16.4.4 Theorem.   Uniqueness of Direct Sum Decomposition for Abelian Groups

If $C_1 \oplus \cdots \oplus C_r$ and $D_1 \oplus \cdots \oplus D_s$ are two decompositions of an abelian group into the direct sum of cyclic groups and if

$1 < |C_i|$ divides $|C_{i+1}|$ for $i = 1, \ldots, (k - 1)$ and $C_g \cong \mathbb{Z}$ for all $g > k$,

and similarly, if

$1 < |D_j|$ divides $|D_{j+1}|$ for $j = 1, \ldots, (h - 1)$ and $C_g \cong \mathbb{Z}$ for all $g > h$,

then $k = h$, $r = s$, and $C_i \cong D_i$ for all i.

The abelian group of the hypothesis is finitely generated by the generators of either direct sum representation. Thus we can recognize the abelian group as a $\mathbb{Z}$-module. And now translate the general theorem to this case. Details are called for in Exercise 16.4.1.

16.4.5 Determination of the Abelian Groups of Order n

We now know that each finite abelian group may be expressed as the direct sum of a finite number of nonidentity cyclic groups,

$$C_1 \oplus \cdots \oplus C_r$$

and that this decomposition is unique provided the orders of these summands satisfy the condition

$$|C_1| \,|\, |C_{i+1}| \qquad \text{for } i = 1, \ldots, r - 1$$

We also know that the order of the group n is the product of the orders of the summands:

$$n = \prod_{i=1}^{i=r} |C_i|$$

Thus to determine, up to isomorphism, all finite abelian groups of order n we need only see how to determine all ways of factoring n into a product $|C_1| \times \cdots \times |C_r|$ so that each factor divides the next.

Begin with the factorization of n into its distinct prime power factors. Let

$$n = p_1^{e_1} p_2^{e_2} \cdots p_s^{e_s}$$

We suppose that for each i,

$$|C_i| = p_1^{e_{i1}} p_2^{e_{i2}} \cdots p_s^{e_{is}}$$

where now $0 \le e_{ij} \le e_j$ for $j = 1, \ldots, s$. Let us focus on one prime, say, p_j. To satisfy the divisibility criterion,

$$e_{1j} \le e_{2j} \le \cdots \le e_{rj} \tag{1}$$

To satisfy the total order,

$$e_{1j} + e_{2j} + \cdots + e_{rj} = e_j \tag{2}$$

And these conditions hold for each $j = 1, \ldots, s$.

Conversely, if we start by selecting any set of nonnegative integers that satisfy (2), then the inequalities (1) may be obtained by defining the e_{ij} in nondecreasing order. If we make such a choice for each j, $j = 1, \ldots, s$, then we may define C_i to be the cyclic group of order $p_1^{e_{i1}} p_2^{e_{i2}} \cdots p_s^{e_{is}}$. An abelian group of order n is obtained forming the direct sum of these cyclic groups. The uniqueness theorem guarantees that if two constructions differ in the sum (2) for at least one index j then the two groups are nonisomorphic. Thus to determine all abelian groups of order n we need only know the ways of expressing e_j, the exponent on p_j in the prime factorization of n, as a sum of nonnegative integers (2). The total number of the abelian groups of order n is simply the product, over the j's, of the number of ways of doing this for each j.

Here is a specific example.

16.4.6 The Abelian Groups of Order 2250

Factor $2250 = 2 \times 3^2 \times 5^3$. For each prime factor we determine the possible sums of the form (2) for the exponent of each prime appearing in the factorization of 2250. The largest possible number of summands is 3, the largest exponent occurring in the factorization. This anticipates the choice of the partition $3 = 1 + 1 + 1$. Now we list all the possibilities.

For the prime divisor 2 the exponent is 1. The partition is:

$$1 = 0 + 0 + 1$$

For the prime divisor 3 the exponent is 2. There are two partitions:

$$2 = 0 + 1 + 1$$
$$2 = 0 + 0 + 2$$

For the prime divisor 5 the exponent is 3. There are three partitions:

$$3 = 1 + 1 + 1$$
$$3 = 0 + 1 + 2$$
$$3 = 0 + 0 + 3$$

To form an abelian group simply choose a partition for each prime factor.

Suppose we have chosen
$$1 = 0 + 0 + 1 \qquad 2 = 0 + 0 + 2 \quad \text{and} \quad 3 = 0 + 1 + 2$$
Now form the prescribed cyclic groups (denote by $\mathbb{Z}_m$ the additive group of integers mod m; in particular, $\mathbb{Z}_1$ is just the identity):
$$C_1 = \mathbb{Z}_{2^0} \oplus \mathbb{Z}_{3^0} \oplus \mathbb{Z}_{5^0} = \mathbb{Z}_1 \oplus \mathbb{Z}_1 \oplus \mathbb{Z}_1 \simeq \mathbb{Z}_1$$
$$C_2 = \mathbb{Z}_{2^0} \oplus \mathbb{Z}_{3^0} \oplus \mathbb{Z}_{5^1} = \mathbb{Z}_1 \oplus \mathbb{Z}_1 \oplus \mathbb{Z}_5 \simeq \mathbb{Z}_5$$
$$C_3 = \mathbb{Z}_{2^1} \oplus \mathbb{Z}_{3^2} \oplus \mathbb{Z}_{5^2} = \mathbb{Z}_2 \oplus \mathbb{Z}_9 \oplus \mathbb{Z}_{25} \simeq \mathbb{Z}_{450}$$
We can eliminate the identity $\mathbb{Z}_1$. Thus the group we have just constructed is isomorphic to $\mathbb{Z}_5 \oplus \mathbb{Z}_{450}$.

The total number of groups of order 2250 is the total number of choices of the partitions. In this case it is product $1 \times 2 \times 3 = 6$. If we exhaust these possibilities systematically we find these groups of order 2250:

THE ABELIAN GROUPS OF ORDER 2250

	Prime power factor	Partition of exponent	Cyclic factors
No. 1	2^1	$1 = 0 + 0 + 1$	$\mathbb{Z}_{2^0} \oplus \mathbb{Z}_{3^0} \oplus \mathbb{Z}_{5^1} \cong \mathbb{Z}_5$
	3^2	$2 = 0 + 1 + 1$	$\mathbb{Z}_{2^0} \oplus \mathbb{Z}_{3^1} \oplus \mathbb{Z}_{5^1} \cong \mathbb{Z}_{15}$
	5^3	$3 = 1 + 1 + 1$	$\mathbb{Z}_{2^1} \oplus \mathbb{Z}_{3^1} \oplus \mathbb{Z}_{5^1} \cong \mathbb{Z}_{30}$

The abelian group is $\mathbb{Z}_5 \oplus \mathbb{Z}_{15} \oplus \mathbb{Z}_{30}$.

	Prime power factor	Partition of exponent	Cyclic factors
No. 2	2	$1 = 0 + 0 + 1$	$\mathbb{Z}_{2^0} \oplus \mathbb{Z}_{3^0} \oplus \mathbb{Z}_{5^0} \cong \mathbb{Z}_1$
	3	$2 = 0 + 1 + 1$	$\mathbb{Z}_{2^0} \oplus \mathbb{Z}_{3^1} \oplus \mathbb{Z}_{5^1} \cong \mathbb{Z}_{15}$
	5	$3 = 0 + 1 + 2$	$\mathbb{Z}_{2^1} \oplus \mathbb{Z}_{3^1} \oplus \mathbb{Z}_{5^2} \cong \mathbb{Z}_{150}$

The abelian group is $\mathbb{Z}_{15} \oplus \mathbb{Z}_{150}$.

	Prime power factor	Partition of exponent	Cyclic factors
No. 3	2	$1 = 0 + 0 + 1$	$\mathbb{Z}_{2^0} \oplus \mathbb{Z}_{3^0} \oplus \mathbb{Z}_{5^0} \cong \mathbb{Z}_1$
	3	$2 = 0 + 1 + 1$	$\mathbb{Z}_{2^0} \oplus \mathbb{Z}_{3^1} \oplus \mathbb{Z}_{5^0} \cong \mathbb{Z}_3$
	5	$3 = 0 + 0 + 3$	$\mathbb{Z}_{2^1} \oplus \mathbb{Z}_{3^1} \oplus \mathbb{Z}_{5^3} \cong \mathbb{Z}_{750}$

The abelian group is $\mathbb{Z}_3 \oplus \mathbb{Z}_{750}$.

	Prime power factor	Partition of exponent	Cyclic factors
No. 4	2	$1 = 0 + 0 + 1$	$\mathbb{Z}_{2^0} \oplus \mathbb{Z}_{3^0} \oplus \mathbb{Z}_{5^1} \cong \mathbb{Z}_5$
	3	$2 = 0 + 0 + 2$	$\mathbb{Z}_{2^0} \oplus \mathbb{Z}_{3^0} \oplus \mathbb{Z}_{5^1} \cong \mathbb{Z}_5$
	5	$3 = 1 + 1 + 1$	$\mathbb{Z}_{2^1} \oplus \mathbb{Z}_{3^2} \oplus \mathbb{Z}_{5^1} \simeq \mathbb{Z}_{90}$

The abelian group is $\mathbb{Z}_5 \oplus \mathbb{Z}_5 \oplus \mathbb{Z}_{90}$.

	Prime power factor	Partition of exponent	Cyclic factors
No. 5	2	$1 = 0 + 0 + 1$	$\mathbb{Z}_{2^0} \oplus \mathbb{Z}_{3^0} \oplus \mathbb{Z}_{5^0} \cong \mathbb{Z}_1$
	3	$2 = 0 + 0 + 2$	$\mathbb{Z}_{2^0} \oplus \mathbb{Z}_{3^0} \oplus \mathbb{Z}_{5^1} \cong \mathbb{Z}_5$
	5	$3 = 0 + 1 + 2$	$\mathbb{Z}_{2^1} \oplus \mathbb{Z}_{3^2} \oplus \mathbb{Z}_{5^2} \cong \mathbb{Z}_{450}$

The abelian group is $\mathbb{Z}_5 \oplus \mathbb{Z}_{450}$.

	Prime power factor	Partition of exponent	Cyclic factors
No. 6	2	$1 = 0 + 0 + 1$	$\mathbb{Z}_{2^0} \oplus \mathbb{Z}_{3^0} \oplus \mathbb{Z}_{5^0} \cong \mathbb{Z}_1$
	3	$2 = 0 + 0 + 2$	$\mathbb{Z}_{2^0} \oplus \mathbb{Z}_{3^0} \oplus \mathbb{Z}_{5^0} \cong \mathbb{Z}_1$
	5	$3 = 0 + 0 + 3$	$\mathbb{Z}_{2^1} \oplus \mathbb{Z}_{3^2} \oplus \mathbb{Z}_{5^3} \cong \mathbb{Z}_{2250}$

The abelian group is $\mathbb{Z}_{2250}$.

Reference: P. M. Cohn, The Complement of a Finitely Generated Direct Summand of an Abelian Group, *Proceedings of the American Mathematical Society*, vol. 7, pp. 520–521, 1956. E. A. Walker, Cancellation in Direct Sums of Groups, *Proceedings of the American Mathematical Society*, vol. 7, pp. 898–902, 1956.

EXERCISE SET 16.4

16.4.1 Give the complete details to show that the uniqueness theorem for abelian groups is an instance of the uniqueness theorem for modules.

16.4.2 Find all abelian groups of order $2^5 5^3 7^4$. (Determine the number of nonisomorphic abelian groups of this order before you start an enumeration as a check on your list.)

16.4.3 Decompose the multiplicative group of units in $\mathbb{Z}_{91}$ as a direct product of cyclic groups satisfying the conditions for the uniqueness theorem. Then do the same for the multiplicative group of units in $\mathbb{Z}_{175}$.

16.4.4 Show that a finite abelian group is the direct product of its Sylow subgroups. Note that each Sylow subgroup is an abelian p-group and is thus the direct product of cyclic p-groups. State and prove a theorem classifying abelian groups as the direct product of cyclic p-groups for the various primes dividing the order of the group.

   Use your theorem to list all the abelian groups of order 2250.

16.4.5 (W.R. Scott) (The group operation for the groups in this exercise is denoted multiplicatively because the groups are not all abelian.) Let F be the free abelian group on two generators y and z. Thus

$$yz = zy, \langle y \rangle \cap \langle z \rangle = 1, \text{ and } y^h = 1 \text{ or } z^h = 1 \qquad \text{if and only if } h = 0$$

Let c be the cyclic group of order 11 generated by an element x; that is,

$$C = \langle x \rangle \quad \text{and} \quad x^{11} = 1$$

Let L be the semidirect product of C by F where

$$\phi_y(x) = x^2 \quad \text{and} \quad \phi_z(x) = x^8$$

(Alternatively L is a group generated by elements x, y, and z with defining relations: $x^{11} = 1$, $yz = zy$, $yxy^{-1} = x^8$, and $zxz^{-1} = x^2$.) Define subgroups of L by

$$A = \langle x, y \rangle \cong \langle x \rangle \times_{\phi_y} \langle y \rangle$$

$$A' = \langle x, z \rangle \cong \langle x \rangle \times_{\phi_z} \langle y \rangle$$

$$B = \langle y^7 z \rangle \quad \text{and} \quad B' = \langle yz^3 \rangle$$

1. Compute $y^{-1}xy$ and $z^{-1}xz$.
2. Show that A and A' are normal subgroups of L.
3. Show that B and B' are normal subgroups of L.
4. Show that L is isomorphic to both $A \times B$ and $A' \times B'$.
5. Show that B and B' are isomorphic but that A and A' are not isomorphic.

16.5 CANONICAL FORMS FOR A LINEAR TRANSFORMATION

In this section we apply the module decomposition theory to obtain the standard canonical forms for a linear transformation of a finite-dimensional vector space. Here is the setup: Let V be a vector space of dimension n over a field F. Because we need to distinguish between elements of F, $F[t]$, V, and linear transformations we shall place an arrow over the letters that denote vectors in V. Let L be a linear transformation of V into V. Recall from Section 15.1 that we may associate with L an $n \times n$ matrix with entries from F in the following fashion:

Choose a basis, say

$$\vec{b}_1, \vec{b}_2, \ldots, \vec{b}_n$$

and express $L(\vec{b}_j)$, the image of $\vec{b}_j$ under L, in this basis:

$$L\left(\vec{b}_j\right) = \sum_i a_{ij}\vec{b}_i \qquad (1)$$

From this system of equations (1) form the matrix

$$\begin{bmatrix} a_{11} & \cdots & a_{ij} & \cdots & a_{1n} \\ a_{21} & \cdots & a_{2j} & \cdots & a_{2n} \\ \vdots & & \vdots & & \vdots \\ a_{n1} & \cdots & a_{nj} & \cdots & a_{nn} \end{bmatrix}$$

whose jth column is the sequence of coefficients in $L(\vec{b}_j)$.

The reason for this convention is that if

$$\vec{v} = \sum_j v_j\vec{b}_j$$

then

$$L(\vec{v}) = \sum_j v_j L\left(\vec{b}_j\right) = \sum_j v_j\left(\sum_i a_{ij}\vec{b}_i\right) = \sum_i\left(\sum_j a_{ij}v_j\right)\vec{b}_i$$

and this can be computed from the usual matrix-vector product

$$\begin{bmatrix} a_{11} & \cdots & a_{ij} & \cdots & a_{1n} \\ a_{21} & \cdots & a_{2j} & \cdots & a_{2n} \\ \vdots & & \vdots & & \vdots \\ a_{n1} & \cdots & a_{nj} & \cdots & a_{nn} \end{bmatrix}\begin{bmatrix} v_1 \\ v_2 \\ \vdots \\ v_n \end{bmatrix}$$

One of the goals of this section is to show how we may choose a basis for V so that the matrix associated with L has as many zero entries as possible. Anyone who has spent even a little time doing matrix calculations can well appreciate the efficacy of using matrices with lots of zeros! Ideally we should like to obtain a basis for which the matrix is a diagonal matrix

$$\text{diag}[d_1, d_2, \ldots, d_n] = \begin{bmatrix} d_1 & \cdots & 0 & \cdots & 0 \\ 0 & d_2 & 0 & \cdots & 0 \\ \vdots & & \vdots & & \vdots \\ 0 & \cdots & 0 & \cdots & d_n \end{bmatrix}$$

This of course cannot be realized in all cases; we describe what can be done in this section. The essence of what can be done is localized in the study of a cyclic submodule. Begin by making V into an $F[t]$-module as we have described in earlier sections. The action of the polynomial ring $F[t]$ on the vectors is given by

$$f(t)\vec{v} = f(L)\vec{v}$$

in more detail, if

$$f(t) = \Sigma_i c_i t^i$$

then

$$f(t)\vec{v} = f(L)(\vec{v}) = c_0\vec{v} + c_1 L(\vec{v}) + \cdots + c_m L^m(\vec{v})$$

As we know, our general theory tells us that V, viewed as an $F[t]$-module, is the direct sum of cyclic submodules whose orders form a descending chain. We begin with

16.5.1 Lemma.   Cyclic Submodule as a Subspace

Let L be a linear transformation on V. Let $\langle \vec{u} \rangle$ be a cyclic submodule of V with order $\langle r(t) \rangle$. Then

1. $r(t)$ is a nonzero polynomial, of degree d, that we may take to be monic, say,

$$r(t) = c_0 + c_1 t + \cdots + c_{d-1} t^{d-1} + t^d$$

2. If $r(t)$ is a constant in F then $\langle \vec{u} \rangle = \{0\}$; otherwise
3. A vector basis for the cyclic submodule $\langle \vec{u} \rangle$ is

$$\{ \vec{u}, L(\vec{u}), \ldots, L^{d-1}(\vec{u}) \}$$

In particular the dimension of the subspace $\langle \vec{u} \rangle$ is the degree of the polynomial generating the order of $\langle \vec{u} \rangle$.

4. Regarding L as a linear transformation on $\langle \vec{u} \rangle$, and using the basis in (3), the matrix for L is

$$\begin{bmatrix} 0 & 0 & \cdots & 0 & -c_0 \\ 1 & 0 & \cdots & 0 & -c_1 \\ \vdots & & & \vdots & \vdots \\ 0 & 0 & \cdots & 1 & -c_{d-1} \end{bmatrix}$$

Remark. This is a $d \times d$ matrix with 1's just below the main diagonal, whose last column is the negative of the coefficients of the polynomial $r(t)$, and whose other entries are all zero. This matrix is called the *companion matrix* of the monic polynomial $r(t)$.

Proof. First notice that any submodule is a subspace of V and *invariant* under L; that is, $L(\vec{u})$ is in the submodule if $\vec{u}$ is. This is a consequence of the definition of a submodule; that is, if $\vec{u}$ is in the submodule so is

$$t\vec{u} = L(\vec{u})$$

Hence we are free to apply the general theory to the vector space which is the cyclic module $\langle \vec{u} \rangle$.

Proof of 1. If the order of $\langle \vec{u} \rangle$ were $\{0\}$, then $\langle \vec{u} \rangle$ would be a free $F[t]$-module; in particular the vectors in the following sequence

$$\vec{u}, \, t\vec{u} = L(\vec{u}), \ldots, t^n\vec{u} = L^n(\vec{u})$$

would be independent, but as $(n + 1)$ vectors in an n-dimensional space they must be linearly dependent. The polynomial $r(t)$ that generates the order may be taken to be monic since multiplying through by a unit of the ring, here an element of F, does not change the ideal.

Proof of 2. If $r(t)$ is a constant then $r(t) = c$. By (1), $c \neq 0$, so that $r(t)$ is a unit of the ring $F[t]$. But that means that

$$r(t)\vec{u} = \vec{0}$$

hence $\vec{u} = \vec{0}$.

Proof of 3. First we show that the vectors

$$\vec{u}, \, L(\vec{u}), \ldots, L^{d-1}(\vec{u})$$

are linearly independent. If

$$\sum_{h=1}^{d-1} b_h L^h(\vec{u}) = \vec{0} = \left(\sum_{h=1}^{d-1} b_h L^h \right)(\vec{u})$$

then

$$\sum_{h=1}^{d-1} b_h t^h \in o(\langle \vec{u} \rangle) = \langle r(t) \rangle$$

and hence is divisible by $r(t)$, a contradiction since the degree of $r(t)$ is d.

Second we show that these vectors span $\langle \vec{u} \rangle$ as a subspace. Suppose that $\vec{w}$ is any member of the cyclic module; then

$$\vec{w} = f(t)\vec{u}$$

for some polynomial $f(t) \in F[t]$. Divide $f(t)$ by $r(t)$ and get a remainder $s(t)$ of degree less than d:

$$\vec{w} = f(t)\vec{u} = [r(t)q(t) + s(t)]\vec{u} = s(t)(\vec{u})$$

Now $r(t)\vec{u} = \vec{0}$ since $r(t)$ is in the order of $\langle \vec{u} \rangle$. Hence

$$\vec{w} = s(t)(\vec{u}) = s(L)\vec{u}$$

is in the space spanned by

$$\vec{u}, \, L(\vec{u}), \ldots, L^{d-1}(\vec{u})$$

Proof of 4. Simply write out the action of L on the basis of choice and convert the coefficients in the system of equations to the matrix. Thus for

$i = 0, \ldots, d - 2$,

$$L(L^i(\vec{u})) = L^{i+1}(\vec{u}) = 0 + \cdots + L^{i+1}(\vec{u}) + \cdots + 0$$

and finally,

$$L(L^{d-1}(\vec{u})) = L^d(\vec{u}) = -c_0\vec{u} - c_1 L(\vec{u}) - \cdots - c_{d-1} L^{d-1}(\vec{u})$$

We can make an immediate application of this lemma to obtain a canonical form for a linear transformation or for a square matrix.

16.5.2 Theorem.   Rational Canonical Form and Invariant Factors

Let L be a linear transformation of a finite-dimensional vector space over a field F. Then a basis may be chosen for V so that L corresponds to a matrix of the form

$$\mathrm{diag}[C_1, C_2, \ldots, C_k] = \begin{bmatrix} \boxed{C_1} & \cdots & 0 & \cdots & 0 \\ 0 & \boxed{C_2} & 0 & \cdots & 0 \\ \vdots & & \vdots & & \vdots \\ 0 & \cdots & 0 & \cdots & \boxed{C_k} \end{bmatrix}$$

where each C_i is a companion matrix of a polynomial $r_i(t)$ and $r_i(t)$ divides $r_{i+1}(t)$. The polynomials $r_i(t)$ are called the *invariant factors* of L.

The last polynomial $r_k(t)$ is a minimal polynomial for L, that is, a polynomial of least degree such that

$$m(L) = 0 \text{ transformation}$$

or equivalently,

$$m(A) = 0 \text{ matrix}$$

Moreover two linear transformations are equal if and only if they have, to within unit factors, the same sequence of invariant factors.

Proof.   We know that V, regarded as an $F[t]$-module defined by the action of L, is the direct sum of cyclic submodules $\langle \vec{u}_i \rangle$ whose orders $\langle r_i(t) \rangle$ are generated by polynomials with the divisibility property.

Each submodule is a subspace. By Lemma 16.5.1 the action of L on the subspace can be associated with a companion matrix.

Now put this information together. Since V as a module is the direct sum of submodules, each is a subspace. It follows immediately that V as a space is the direct sum of the subspaces; in particular a basis for V is obtained by taking the set union of all the bases for each cyclic submodule. The matrix describing L on the whole space then is exactly as pictured above.

Since for $1 \le i \le k$ we have that $r_i(t) | r_k(t)$ it follows that

$$r_k(L)(\vec{u}_i) = \vec{0}$$

for all i and hence

$$r_k(L)(\vec{v}) = \vec{0}$$

for all $\vec{v} \in V$. Hence $r_k(L)$ is the zero transformation on V.

On the other hand, if $m(t)$ is a polynomial such that

$$m(L)(\vec{v}) = 0$$

for all $\vec{v} \in V$ then $m(t)$ belongs to the order of $\langle \vec{u}_k \rangle$ and in particular is divisible by $r_k(t)$.

The proof of the uniqueness of the sequence of invariant factors follows immediately from the uniqueness of the decomposition of a module into the direct sum of cyclic submodules (Theorem 16.4.1).

Now the question is to give an algorithm to determine the polynomials $r_i(t)$ and vectors $\vec{u}_i$. As we know, the polynomials will be unique if we insist on the divisibility criterion and choose the polynomials to be monic. The vectors $\vec{u}_i$, on the other hand, will not be uniquely determined and there seems to be no particular way of norming them.

16.5.3 Theorem. Determination of the Invariant Factors of a Linear Transformation

Let F be a field, V a vector space of dimension n over F, L a linear transformation of V into V, and A an $n \times n$ matrix representing L with respect to a basis. The *characteristic matrix* of A is defined

$$A - It = \begin{bmatrix} a_{11} - t & \cdots & a_{ij} & \cdots & a_{1n} \\ a_{21} & a_{22} - t & a_{2j} & \cdots & a_{2n} \\ \vdots & & \vdots & & \vdots \\ a_{n1} & \cdots & a_{nj} & \cdots & a_{nn} - t \end{bmatrix}$$

The polynomials $r_i(t)$ generating the orders in the canonical decomposition of V as an $F[t]$-module defined by L into cyclic submodules are determined by carrying out the main algorithm (Theorem 16.2.7) on the transpose of the characteristic matrix.

Proof. Let

$$\vec{b}_1, \vec{b}_2, \ldots, \vec{b}_n$$

be a basis and let

$$A = \begin{bmatrix} a_{11} & \cdots & a_{ij} & \cdots & a_{1n} \\ a_{21} & \cdots & a_{2j} & \cdots & a_{2n} \\ \vdots & & \vdots & & \vdots \\ a_{n1} & \cdots & a_{nj} & \cdots & a_{nn} \end{bmatrix}$$

be the matrix representing L with respect to this basis.

Now we know that V, as an $F[t]$-module is the image of a free $F[t]$-module on n generators

$$x_1, x_2, \ldots, x_n$$

under the mapping

$$x_j \to b_j$$

and we need only find the generators for the kernel. Since

$$tx_j \to t\vec{b}_j = L(\vec{b}_j) = a_{1j}\vec{b}_1 + a_{2j}\vec{b}_2 + \cdots + a_{jj}\vec{b}_j + \cdots + a_{nj}\vec{b}_n$$

we have

$$0 = a_{1j}\vec{b}_1 + a_{2j}\vec{b}_2 + \cdots + (a_{jj} - t)\vec{b}_{jj} + \cdots + a_{nj}\vec{b}_n$$

and hence

$$g_j = a_{ij}x_1 + a_{2j}x_2 + \cdots + (a_{jj} - t)x_j + \cdots + a_{nj}x_n \tag{2}$$

is in the kernel.

We shall now prove that

$$g_1, g_2, \ldots, g_n$$

generate the kernel. (It may be helpful to look back at Example 16.3.5.)

Let k belong to the kernel. As an element in the free module we have

$$k = c_1 x_1 + \cdots + c_n x_n \tag{3}$$

where the coefficient $c_j = c_j(t)$ is a polynomial in t which we now write as

$$c_j(t) = d_j(t)t + c_{j0}$$

and thus

$$c_j(t)x_j = d_j(t)(tx_j) + c_{j0}x_j$$

However, we may express each tx_j in terms of the elements g_j. We have from (2)

$$tx_j = -g_j + (a_{1j}x_1 + a_{2j}x_2 + \cdots + a_{jj}x_j + \cdots + a_{nj}x_n)$$

and so substituting in (3) and collecting all terms x_i with constant coefficients we have an expression for k in the form:

$$k = \left(\sum_j e_j(t)g_j\right) + f_1 x_1 + f_2 x_2 + \cdots + f_j x_j + \cdots + f_n x_n \tag{4}$$

where the f's are constants. We shall see that all are 0.

Now apply to both sides of (4) the homomorphism of the free module to V. We have

$$0 = 0 + f_1\vec{b}_1 + f_2\vec{b}_2 + \cdots + f_j\vec{b}_j + \cdots + f_n\vec{b}_n$$

But the $\vec{b}$'s are a basis for V and so are independent; hence each constant

$$f_j = 0$$

and so we have in (4)

$$k = \sum_j e_j(t)g_j$$

an expression of k in terms of the g_j's with coefficients from the ring $F[t]$.

Thus we need only apply the main algorithm to the matrix of coefficients obtained from the system of equations (2) expressing the generators of the kernel in terms of the generators of the free module. This matrix is

$$A^T - tI = (A - tI)^T$$

Note that it is the transpose A^T of the matrix A that represents the linear transformation L in terms of the basis of vectors $\vec{b}_j$ that is used because of our conventions in writing the matrix from the two systems of equations (1) and (2).

It is immaterial to the algorithm whether one works with $A - tI$ or with its transpose; however, if one wants to keep track of the generators of the free module, and thus find from the algorithm the generators of the cyclic module in terms of the $\vec{b}$'s, it is important.

In view of Corollary 16.2.10 we have the following corollary.

16.5.4 Corollary.   Characteristic Polynomial and Invariant Factors and the Hamilton-Cayley Theorem

Let A be an $n \times n$ matrix with elements in a field F. There exist matrices H and J with coefficients in $F[t]$ whose determinants are elements of F such that

$$H(A - tI)J = \text{diag}[r_1(t), \ldots, r_n(t)]$$

where $r_i(t)|r_{i+1}(t)$ for $i = 1, \ldots, n - 1$. Moreover $\det(A - tI) = u\prod_i r_i(t)$ where $u \in F$.

The polynomial $\det(A - tI)$ is called the *characteristic polynomial of A*. We have just proved that up to constant factors it is the product of its invariant factors. And because it is divisible by $r_n(t)$, which is the minimal polynomial for A, and so $r_n(A) = 0$, it follows that A is also a zero of its characteristic polynomial. This result is known as the Hamilton-Cayley theorem. (It is tempting, but oh so wrong, to substitute A for t in $\det(A - tI)$ to say $\det(A - AI) = 0$. What goes wrong with that argument?)

It is important to note that the characteristic polynomial depends only on the linear transformation L and not on the basis chosen for V or the matrix A representing L. If another basis were chosen and another matrix B represented L, then the two characteristic polynomials $\det(A - tI)$ and $\det(B - tI)$ are equal. This is so because A and B are similar since they represent the same linear transformation. And so there is a nonsingular matrix S such that $B = SAS^{-1}$ and so

$$\det(B - tI) = \det(SAS^{-1} - tI) = \det(S(A - tI)S^{-1})$$
$$= \det(S)\det(A - tI)\det(S^{-1}) = \det(A - tI)$$

The uniqueness of the invariant factors for L established in Theorem 16.5.2 can be applied to give a test for the similarity of two matrices. Note that all the calculations and all matrices in the next theorem have elements in F or in $F[t]$. It is not necessary to go to an extension of F even if the characteristic polynomial of L does not split in F.

16.5.5 Theorem.   Test for Similarity of Matrices

Two $n \times n$ matrices with entries in a field F are similar over F if and only if they have the same invariant factors; that is,

If A and B are $n \times n$ matrices over a field F there exists a nonsingular matrix C such that

$$B = C^{-1}AC$$

if and only if, after the completion of the main algorithm on the two matrices

$$A - tI \quad \text{and} \quad B - tI$$

the polynomials on the diagonals differ at most by a nonzero factor in F.

Proof.   We know from the rational canonical form theorem that an $n \times n$ matrix A is similar to a block diagonal matrix

$$D = \text{diag}[C_1, C_2, \ldots, C_k]$$

where the C's are companion matrices of the invariant factors. We may assume that these polynomials are monic. Thus if A and B have the same monic invariant factors we must have

$$A = SDS^{-1} \quad \text{and} \quad B = TDT^{-1}$$

so that

$$A = ST^{-1}BTS^{-1} = (TS^{-1})^{-1}B(TS^{-1})$$

so that A and B are similar.

On the other hand, if A and B are similar then they define the same linear transformation on V and hence must have the same set of invariant factors. Thus performing the main algorithm on $A - tI$ and $B - tI$ must give rise to the same diagonal matrix of companion matrices.

16.5.6 Example.   Determination of the Invariant Factors of a Matrix over $\mathbb{Q}$

Let $F = \mathbb{Q}$, the field of rational numbers. Let V be a four-dimensional space over V and let a linear transformation be defined by the matrix

$$A = \begin{bmatrix} 3 & 16 & -4 & -96 \\ 1 & 9 & -2 & -48 \\ 5 & 40 & -9 & -240 \\ 0 & 0 & 0 & 1 \end{bmatrix}$$

Now form the matrix $A^T - tI$:

$$A^T - tI = \begin{bmatrix} 3-t & 1 & 5 & 0 \\ 16 & 9-t & 40 & 0 \\ -4 & -2 & -9-t & 0 \\ -96 & -48 & -240 & 1-t \end{bmatrix}$$

Upon applying the main algorithm to this matrix we obtain

$$\begin{bmatrix} 1 & 0 & 0 & 0 \\ 0 & 1-t & 0 & 0 \\ 0 & 0 & 1-t & 0 \\ 0 & 0 & 0 & (1-t)^2 \end{bmatrix} \tag{5}$$

This means that if we use a basis from the cyclic submodules we may describe the linear transformation with the matrix B (see below) and hence conclude that A is similar to the matrix

$$B = \begin{bmatrix} \boxed{1} & 0 & 0 & 0 \\ 0 & \boxed{1} & 0 & 0 \\ 0 & 0 & \boxed{\begin{matrix} 0 & -1 \\ 1 & 2 \end{matrix}} \end{bmatrix}$$

If we keep track of the generators for the free module of which V is the image we find generators

$$y_1 = (3-t)x_1 + x_2 + 5x_3$$

$$y_2 = 2x_1 + x_3$$

$$y_3 = 48x_1 + x_4$$

$$y_4 = x_1$$

Similarly we find that the kernel is generated, as indicated in (5), by

$$h_1 = y_1$$

$$h_2 = (1-t)y_2$$

$$h_3 = (1-t)y_3$$

$$h_4 = (1-t)^2 y_4$$

Use as a basis for V

$$b_1 = \begin{bmatrix} 1 \\ 0 \\ 0 \\ 0 \end{bmatrix} \qquad b_2 = \begin{bmatrix} 0 \\ 1 \\ 0 \\ 0 \end{bmatrix} \qquad b_3 = \begin{bmatrix} 0 \\ 0 \\ 1 \\ 0 \end{bmatrix} \qquad b_4 = \begin{bmatrix} 0 \\ 0 \\ 0 \\ 1 \end{bmatrix}$$

Then we may take, as generators of the cyclic subspaces, the images of y_2, y_3, y_4 under the homomorphism $M_4 \to V$:

$$y_2 \to \vec{u}_2 = 2\vec{b}_1 + \vec{b}_3 = \begin{bmatrix} 2 \\ 0 \\ 1 \\ 0 \end{bmatrix}$$

$$y_3 \to \vec{u}_3 = 48\vec{b}_1 + \vec{b}_4 = \begin{bmatrix} 48 \\ 0 \\ 0 \\ 1 \end{bmatrix}$$

$$y_4 \to \vec{u}_4 = \vec{b}_4 = \begin{bmatrix} 0 \\ 0 \\ 0 \\ 1 \end{bmatrix}$$

The image of $\vec{y}_1$ is in the kernel and so we may expect that the computation

$$(3I - A)\vec{b}_1 + \vec{b}_2 + 5\vec{b}_3 = 3\vec{b}_1 + \vec{b}_2 + 5\vec{b}_3 - A\vec{b}_1 = \vec{0}$$

Indeed we get

$$\begin{bmatrix} 3 \\ 1 \\ 5 \\ 0 \end{bmatrix} - \begin{bmatrix} 3 & 16 & -4 & -96 \\ 1 & 9 & -2 & -48 \\ 5 & 40 & -9 & -240 \\ 0 & 0 & 0 & 1 \end{bmatrix} \begin{bmatrix} 1 \\ 0 \\ 0 \\ 0 \end{bmatrix} = \vec{0}$$

Sometimes it is not convenient to work with the decomposition in terms of the invariant factors, and is better to work with subspaces whose orders are prime powers. Next we see how to obtain another canonical form for V as the direct sum of subspaces whose orders are prime powers. Associated with this decomposition will be a form for the matrix into block diagonals which are the companion matrices of these polynomials that are powers of irreducibles.

16.5.7 Theorem. Decomposition into Cyclic Subspaces Whose Orders Are Prime Powers

Let R be a PID and let $\langle u \rangle$ be a cyclic R-module with order $\langle r \rangle$. If

$$r = p_1^{e_1} p_2^{e_2} \cdots p_k^{e_k}$$

is the prime power decomposition of r into the product of distinct prime powers in R then

$$\langle u \rangle = \langle u_1 \rangle \oplus \langle u_2 \rangle \oplus \cdots \oplus \langle u_k \rangle$$

is the direct sum of cyclic submodules $\langle u_i \rangle$ such that the order of $\langle u_i \rangle$ is $\langle p_i^{e_i} \rangle$.

Proof. The proof is by induction on the number k of prime power factors. If $k = 1$ there is nothing to prove. For the induction step let

$$r = p_1^{e_1} s$$

where s is not divisible by p_1 and so we have as ideals in R

$$R = \langle 1 \rangle = \langle p_1^{e_1}, s \rangle$$

and so we have as an equation in R

$$1 = p_1^{e_1}x + sy$$

where x and y belong to R. Hence as an equation in the module we have

$$u = 1 \cdot u = p_1^{e_1}xu + syu$$

and hence, as modules

$$\langle u \rangle = \langle p_1^{e_1}u \rangle + \langle su \rangle = \langle p_1^{e_1}u \rangle \vee \langle su \rangle$$

On the other hand, these two submodules have only 0 in common:

$$\langle p_1^{e_1}u \rangle \cap \langle su \rangle = \{0\}$$

since if tu is in both submodules, then t is a multiple of $p_1^{e_1}$ and also of s, and since these two elements of R are relatively prime, t must be a multiple of their product r. But $ru = 0$ in the module, so $tu = 0$.

Thus we have set up the conditions to show that

$$\langle u \rangle = \langle p_1^{e_1}u \rangle \oplus \langle su \rangle$$

Finally we see that the order of $\langle su \rangle$ is $\langle p_1^{e_1} \rangle$ since

$$zsu = 0 \text{ implies that } r|zs$$

and since $p_1^{e_1}$ and s are relatively prime in R, it follows that $p_1^{e_1}|z$. On the other hand, $p_1^{e_1}(su) = 0$. Similarly, you can prove that the order of $\langle p_1^{e_1}u \rangle$ is $\langle s \rangle$. Now s has fewer distinct prime factors than r so we may apply the induction hypothesis to the cyclic module $\langle p_1^{e_1}u \rangle$ whose order is $\langle s \rangle$.

16.5.8 Corollary.   Representation of a Matrix as a Direct Sum of Companion Matrices

Let F be a field and let L be a linear transformation of a vector space V of dimension n over F. We may choose a basis for V so that the matrix corresponding to L is

$$\text{diag}[C_1, \ldots, C_m]$$

where each C_i is the companion matrix of the polynomial

$$p_i^{e_i}(t)$$

where $p_i(t)$ is an irreducible polynomial in $F[t]$.

Proof.   From the rational canonical form theorem 16.5.2 we know that

$$V = \langle u_1 \rangle \oplus \cdots \oplus \langle u_k \rangle$$

where the order of the cyclic subspace $\langle u_i \rangle$ is the invariant factor $r_i(t)$. We now apply Theorem 16.5.7 to each cyclic subspace $\langle u_i \rangle$. Let

$$r_i(t) = p_{i1}^{e_{i1}}(t) \ldots p_{im(i)}^{e_{im(i)}}(t)$$

We conclude that

$$\langle u_i \rangle = \langle u_{i1} \rangle \oplus \cdots \oplus \langle u_{im(i)} \rangle$$

where the order of $\langle u_{ij} \rangle$ is $\langle p_{ij}^{e_{ij}}(t) \rangle$. We know that a basis for $\langle u_{ij} \rangle$ can be chosen so that the matrix associated with L on this cyclic subspace is the companion matrix to $p_{ij}^{e_{ij}}(t)$. Call this the matrix C_{ij}. Hence if we choose a basis for V consisting of the set union of these bases for all pairs (i, j) then the matrix associated with L on this basis is the desired form:

$$\text{diag}\left[C_{11}, \ldots, C_{1m(1)}, C_{21}, \ldots, C_{2m(2)}, \ldots, C_{k1}, \ldots, C_{km(k)}\right]$$

As a final application of Theorem 16.5.7 we present the canonical form for a matrix associated with L that is named after the mathematician Camille Jordan (1838–1922), who is famous for this and other contributions to algebra. In analysis he is known for the "Jordan Curve Theorem" which, speaking loosely, says that every continuous curve in the plane that starts and ends at the same point but has no other crossing has an inside and an outside!

16.5.9 Theorem.   Existence of the Jordan Canonical Form

Let V be a finite-dimensional vector space over a field F and let L be a linear transformation of V into V. If all the zeros of the minimal polynomial belong to F then a basis for V may be chosen so that the matrix representing L in this basis is a block diagonal matrix

$$\text{diag}[J_1, \ldots, J_m]$$

where each block matrix has the form

$$\begin{bmatrix} \lambda & 0 & 0 & \ldots & 0 & 0 \\ 1 & \lambda & 0 & \ldots & 0 & 0 \\ & & & \ldots & & \\ 0 & 0 & 0 & \ldots & \lambda & 0 \\ 0 & 0 & 0 & \ldots & 1 & \lambda \end{bmatrix} \tag{6}$$

where the diagonal element λ is a zero of the minimal polynomial of L and, if the size of the block is greater than 1, then 1's appear on the subdiagonal.

Terminology.   A matrix of the form (6) is called a Jordan matrix and the matrices $J_1, \ldots, J_m$ are called Jordan blocks of the linear transformation.

Proof.   Since the zeros of the minimal polynomial belong to F, each of the invariant factors of L is a product of linear polynomials in $F[t]$. Thus, when we apply the form guaranteed by Theorem 16.5.7, the prime power factors have the form

$$(t - \lambda)^e$$

for various zeros λ and exponents e. Thus to prove this theorem all that has to be done is to show how to choose a basis for the cyclic subspace $\langle u \rangle$ whose order is $\langle (t - \lambda)^e \rangle$. In Theorem 16.5.8 the companion matrix was used. Here we can make a different choice because of the simplicity of the form of the polynomial.

We claim that the basis that we seek is

$$\{\vec{u}, (L - \lambda I)\vec{u}, \ldots, (L - \lambda I)^{e-1}\vec{u}\} \tag{7}$$

First we show that these vectors do form a basis for the subspace $\langle \vec{u} \rangle$. Each member of the basis is clearly in the cyclic submodule. We already know that the dimension of the submodule is the degree of the polynomial generating the order of $\langle \vec{u} \rangle$; in this case it is e. Therefore, to show that they are a basis we need only show that they are linearly independent. But if there were constants

$$c_0, c_1, \ldots, c_{e-1}$$

such that

$$\sum_i c_i (L - \lambda I)^i (\vec{u}) = 0$$

then for the corresponding polynomial

$$g(t) = \sum c_i (t - \lambda)^i$$

we should have $g(L)\vec{u} = 0$ and so $g(t)$ would have to belong to the order of $\langle \vec{u} \rangle$ and be divisible by $(t - \lambda)^e$, a contradiction of the degrees.

Finally then we have to work out the form of the matrix corresponding to L using this basis. The key computation is that for any vector $\vec{w}$ we have

$$L(\vec{w}) = (L - \lambda I)(\vec{w}) + \lambda \vec{w} \tag{8}$$

Thus we can easily compute the effect of L on the basis (7) and verify that the corresponding matrix has the form guaranteed by the theorem. Substitute

$$\vec{w} = (L - \lambda I)^i (\vec{u})$$

in (8). Thus for each i in the range $1 \le i \le e - 1$:

$$L\left[(L - \lambda I)^i (\vec{u})\right] = (L - \lambda I)^{i+1}(\vec{u}) + \lambda (L - \lambda I)^i (\vec{u})$$

and since

$$(L - \lambda I)^e (\vec{u}) = \vec{0}$$

we have

$$L\left[(L - \lambda I)^{e-1}(\vec{u})\right] = \lambda (L - \lambda I)^{e-1}(\vec{u})$$

so that the matrix corresponding to L on $\langle \vec{u} \rangle$ is as stated.

16.5.10 Theorem. Uniqueness of the Jordan Canonical Form

Let V be a vector space over a field F. Let L_1 and L_2 be two linear transformations of V into V. If the characteristic polynomials for both L_1 and L_2 have all their zeros in F then L_1 and L_2 are equal if and only if their Jordan canonical forms differ at most by a permutation of the Jordan blocks.

Proof.   The uniqueness follows from the uniqueness of the invariant factors in the rational canonical form for L (Theorem 16.5.2). A $k \times k$ Jordan block corresponds to a particular prime power factor $(t - \lambda)^k$. The sequence of all Jordan blocks gives a sequence of prime power factors $(t - \lambda_i)^{k_i}$. To prove the uniqueness of the Jordan canoncial form, it is necessary only to see that from this sequence of prime powers the invariant factors of L can be uniquely determined. This is possible because of the divisibility criteria. For example, the last invariant factor must be the product of the all factors $(t - \lambda_i)^{m_i}$ where for each λ_i the exponent m_i is the highest power of the linear factor $(t - \lambda_i)$ occurring in the sequence. The next invariant factor is the product of the next highest powers of $(t - \lambda_i)$ occurring in the sequence of Jordan blocks. And so on until the invariant factors are determined.

EXERCISE SET 16.5

16.5.1   Using the rational canonical form theorem prove that an $n \times n$ matrix A with elements in a field F is similar, over F, to a diagonal matrix if and only if a minimal polynomial for A splits into linear factors in $F[t]$ and has no multiple zero.

16.5.2   Prove that the characteristic polynomial $\det[C - tI]$ of the companion matrix C to a monic polynomial $f(x)$ is $f(x)$.

16.5.3   Using the rational canonical form theorem prove that the monic minimal polynomial for a linear transformation divides the characteristic polynomial $\det[A - tI]$ where A is a matrix corresponding to L over a basis for V. Prove that every zero of the characteristic polynomial is a zero of a minimal polynomial.

16.5.4   Let $c(x) = (x^2 - 2)^3(x^2 + 1)$. List the possible rational canonical forms for linear transformations whose characteristic polynomials are $c(x)$. Classify these by their minimal polynomials. Do this considering $c(x)$ over the fields $\mathbb{Z}_5, \mathbb{Q}, \mathbb{R}$, and $\mathbb{C}$. For those fields for which $c(x)$ splits into linear factors give the corresponding Jordan canonical form.

16.5.5   The characteristic polynomial of each of the five matrices below is $(x - 2)^4$. Over $\mathbb{Q}$ determine the Jordan canonical form for A, B, C, and D. Determine which of A, B, C, or D is similar to E. Several methods are possible.

$$A = \begin{bmatrix} 1 & -3 & 3 & 8 \\ 1 & 3 & 0 & -3 \\ 0 & 1 & -1 & -7 \\ 0 & 0 & 1 & 5 \end{bmatrix} \qquad B = \begin{bmatrix} 2 & -2 & 5 & 11 \\ 0 & 2 & -2 & -6 \\ 0 & 1 & -1 & -7 \\ 0 & 0 & 1 & 5 \end{bmatrix}$$

$$C = \begin{bmatrix} 1 & -1 & 3 & 12 \\ 1 & 3 & 0 & -3 \\ 0 & 0 & -1 & -9 \\ 0 & 0 & 1 & 5 \end{bmatrix} \qquad D = \begin{bmatrix} 2 & 5 & 0 & 10 \\ 0 & 0 & 0 & -4 \\ 0 & -3 & 2 & -6 \\ 0 & 1 & 0 & 4 \end{bmatrix}$$

$$E = \begin{bmatrix} 177 & -5 & 30 & 10 \\ 945 & -25 & 162 & 54 \\ 1995 & -57 & 344 & 114 \\ -8575 & 245 & -1470 & -488 \end{bmatrix}$$

16.5.6 Note from Problem 16.5.5 that dissimilar matrices may have the same minimal polynomial. Prove that this is not the case if the dimension of the vector space is small. Determine the dimensions of V for which similarity can be determined from information about the minimal polynomial only.

16.5.7 Note from Problem 16.5.5 that if in addition to the minimal polynomial we know, for each eigenvalue (recall Exercise 15.1.9), the number of linearly independent eigenvectors then this information determines the similarity class of the matrix. For what dimensions for V does this hold true?

16.5.8 Prove that if the minimal polynomial of an $n \times n$ matrix over a field F is irreducible then the degree of the minimal polynomial is a divisor of n. State and prove an analog of this result for finite abelian groups.

16.5.9 Write a computer program to find the invariant factors of an $n \times n$ matrix with integer coefficients. (It is desirable that your program yields polynomials with integer coefficients.)

16.5.10 Let $r_1(t), \ldots, r_n(t)$ be the sequence of invariant factors of an $n \times n$ matrix over a field F. Prove that $r_1(t)$ is the greatest common divisor of all the 1×1 minors of $A - tI$. In general show that $r_i(t)$ is the quotient of the greatest common divisor of all the $(i) \times (i)$ minors divided by the greatest common divisor of all the $(i - 1) \times (i - 1)$ minors of $A - tI$.

Answers for Selected Problems and Hints

Here are some abbreviations we shall use in this section: WLOG means "without loss of generality." IFF means "if and only if." C_n denotes a cyclic group of order n. D_{2n} denotes a dihedral group of order $2n$. $\mathcal{A}_n$ denotes the alternating group on n symbols and $\mathbb{S}_n$ denotes the symmetric group on n symbols.

CHAPTER 1

Section 1.2

1.2.1 In the first triangle inequality put $a = x$, $b = y - x$, and conclude $|y - x| \geq |y| - |x|$. Interchange the roles of x and y and conclude $|x - y| \geq |x| - |y|$; hence $|x - y| \geq \max(|x| - |y|, |y| - |x|) = |(|x| - |y|)|$.

1.2.2 *Hint:* Show first that $|a| = |b| = 1$.

1.2.3 *Hint:* Show first that $|a| |b| > 0$.

1.2.4 WLOG $b \leq c$. Now distinguish three cases: $a \leq b \leq c$, $b \leq a \leq c$, and $b \leq c \leq a$.

1.2.5 From the division algorithm $b = ad + s$ where $0 \leq s < |a|$. Suppose $s > |a|/2$. If $a > 0$ write $b = a(d + 1) + (s - a)$ and let $r = s - a$. If $a < 0$ write $b = a(d - 1) + s + a$ and let $r = s + a$.

1.2.8 Use weak induction and the distributive law to prove the induction step:

$$(a_1 + b_1) \cdots (a_k + b_k)(a_{k+1} + b_{k+1})$$

$$= \left(\sum e_1 \cdots e_n\right) a_{n+1} + \left(\sum e_1 \cdots e_n\right) b_{n+1} = \sum e_1 \cdots e_n e_{n+1}$$

500

Section 1.3

1.3.2 *Hint:* $y = (x + y) - x$.

1.3.3 At least one of three consecutive integers is divisible by 2 and one, possibly the same integer, is divisible by 3. One of four consecutive integers is divisible by 4, and another will be even. At least one of four consecutive integers is divisible by 3. One possible generalization is that the product of n consecutive integers is divisible by $n!$.

1.3.4 Any common divisor of a and b divides $\gcd(a, b)$. Use part vi of Theorem 1.3.2.

1.3.7 If $n|m$, say $m = nt$, prove the identity $(x^t - 1) = (x - 1)(x^{t-1} + \cdots + 1)$, and show $(a^n - 1)|(a^m - 1)$. To prove the converse write $m = nt + r$ with $0 \le r < n$ and so $a^m - 1 = a^{nt+r} - 1 = a^{nt+r} - a^r + a^r - 1 = a^r(a^{nt} - 1) + (a^r - 1)$. Thus $a^n - 1$ divides $a^r - 1$, a contradiction if $0 < r < n$.

1.3.8 Note $d = ax + by = a(x - bm) + b(y + am)$. So if $|x|$ is not less than b an appropriate choice of m makes $u = x - bm$ satisfy that inequality. Complete the argument in the case that $d = 1$ by showing that $|y + am| < a$.

1.3.11 The condition is $\gcd(a, b, c) = 1$. Be sure your proof works in the case $a = 6$, $b = 10$, and $c = 15$.

1.3.12 Yes.

1.3.13 $\gcd(91, 195) = 13$

1.3.15 *Hint:* Let S be the set of positive integers t such that there is a rational number $r = c/t$ with the property that r cannot be expressed in lowest terms. Show that S must be empty.

1.3.19 The integer t is a least common multiple of m and n.

1.3.20 The integer t is a greatest common divisor of a and b.

1.3.21 If $(4k_1 - 1), \ldots, (4k_i - 1)$ are primes and N is their product, consider $M = 4N - 1$. Show that M cannot be divisible by any of the prime factors of N and cannot be divisible only by primes of the form $4k + 1$.

Section 1.4

1.4.4 To show that $(x \vee y) \wedge (x \vee z) = x \vee (y \wedge z)$, apply the given distributive law with $A = (x \vee y)$, $B = x$, and $C = z$ to find $(x \vee y) \wedge (x \vee z) = [(x \vee y) \wedge x] \vee [(x \vee y) \wedge z]$. Use the commutative and then the absorptive properties on the first term, and the commutative and then the given distributive property on the second term to find $[(x \vee y) \wedge x] \vee [(x \vee y) \wedge z] = x \vee [(z \wedge x) \vee (z \wedge y)]$. Now use associativity, commutativity twice, and then absorptivity to find $x \vee [(z \wedge x) \vee (z \wedge y)] = x \vee (y \wedge z)$. A similar and only slightly longer argument proves the other distributive property.

1.4.9 The symmetric difference operation on sets satisfies the commutative and associative properties and the distribute law: $A \cap (B \oplus C) = (A \cap B) \oplus (A \cap C)$. It does not satisfy the other distributive properties nor either absorptive property.

1.4.12 To prove $\binom{n}{k} = \binom{n}{n-k}$ set up a one-to-one correspondence between a subset and its complement.

1.4.15 Fix an element $a \in S$. Count the subsets of S having k elements by counting the subsets of $S - \{a\}$ having k elements and the subsets of S containing a and $k - 1$ elements of $S - \{a\}$.

1.4.16 Remember $0! = 1$. Use induction on n. Start with $n = 1$.

1.4.18 One example is $g(0) = 1$, $g(1) = 2$, and $g(2) = 0$.

1.4.19 Neither distributive law holds in general.

1.4.20 If A and B are ideals in $\mathbf{Z}$, then $A \cup B$ is an ideal IFF $A \subseteq B$ or $B \subseteq A$.

CHAPTER 2

Section 2.1

2.1.1 Division is not symmetric.

2.1.2 There are five of them.

2.1.3 There are 15 of them. Here is an abbreviated list which may be completed by permuting the elements of T. The relations are given by listing the equivalence classes:

$$Z: \quad \{1\}, \{2\}, \{3\}, \{4\}$$
$$W(1,2): \quad \{1,2\}, \{3\}, \{4\}$$
$$X(1,2,3): \quad \{1,2,3\}, \{4\}$$
$$Y(1,2)(1,3): \quad \{1,2\}, \{2,3\}$$
$$A: \quad \{1,2,3,4\}$$

There are six W's, four X's, and three Y's.

2.1.4 Try the argument on the set $\{0,1\}$ and the relation $\sim$ defined by $a \sim b$ IFF $a = b = 0$. Why isn't $1 \sim 1$?

2.1.8 $[7][7] = [9]$, $[8] + [5] - [9] = [4]$, $[7]([8] - [-2]) = [0]$

2.1.9 In $\mathbf{Z}_{18}$ $[-7] = [11]$ and $[-7][5] = [1]$.

2.1.10 $10 \equiv 1 \pmod{3}$ and so $10^k \equiv 1^k = 1 \pmod{3}$. If the decimal expansion of N is $N = a_n a_{n-1} \cdots a_0$, then $N = a_n 10^n + a_{n-1} 10^{n-1} + \cdots + a_1 10 + a_0 \equiv \Sigma a_k$.

2.1.11 Note $10 \equiv -1 \pmod{11}$, and so $10^k \equiv 1 \pmod{11}$ if k is even; $10^k \equiv -1 \pmod{11}$ if k is odd.

2.1.13 *Hint:* $2 \equiv -2 \pmod{n}$ means n divides 4.

2.1.14 A k satisfying the conditions of the exercise exists IFF n is even, in which case $k = n/2$.

2.1.15 The solutions are $x \equiv 1$ or $x \equiv -1$ modulo a prime p.

2.1.16 The solutions are $x \equiv 1$ or $x \equiv -1$ modulo $2p$ where p is an odd prime.

2.1.17 There are four solutions corresponding to the four cases derived from the possibilities $x \equiv \pm 1 \pmod{p}$ and $x \equiv \pm 1 \pmod{q}$.

2.1.18 The modulus n must be divisible by a square greater than 1 IFF there is a nonzero solution to $a^k \equiv 0 \pmod{n}$.

2.1.19 If $n = p_1^{e_1} \cdots p_k^{e_k}$ is the prime factorization of n, then

$$\varphi(n) = \left(p_1^{e_1} - p_1^{e_1-1} \right) \cdots \left(p_k^{e_k} - p_k^{e_k-1} \right) = n \prod_{i=1}^{k} (1 - 1/p_i)$$

2.1.20 $x = 61$.

Section 2.2

2.2.1

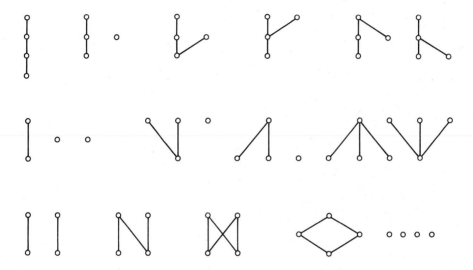

2.2.2 The divisors of 16 form a chain.

2.2.3 There is a one-to-one correspondence between subsets of $\{2, 3, 5\}$ and divisors of 30. One subset is contained in another IFF the integer corresponding to the first divides the integer corresponding to the second.

2.2.4 The divisors of 24 give a different lattice from the subsets of $\{3, 8\}$ because of the power of one of the divisors; in this case 2.

2.2.5 The divisors of n form a chain IFF n is a power of a prime. The divisors of n form a lattice identical with the lattice of subsets of the primes dividing n IFF no square divides the order of n.

2.2.7 No. There is no lower bound for the equivalence relations $(\{1, 2\}, \{3\}, \{4\}, \{5\})$ and $(\{1\}, \{2\}, \{3, 4, 5\})$.

2.2.11 Let $T = (a \cap b) \cup c$, RHS $= T \cap a$, and LHS $= [(T \cap a) \cup c] \cap a$. Note that RHS $\subseteq a$ and RHS $\subseteq$ RHS $\cup c$, and so RHS $\subseteq$ LHS. Similarly, LHS $\subseteq a$, and since $c \subseteq T$ it follows that $(T \cap a) \cup c \subseteq T$ and hence LHS $\subseteq T$; so LHS $\subseteq T \cap a =$ RHS.

2.2.13 No, there is no greatest lower bound to the set of rational numbers whose square is greater than 2. Even more immediate, there is no "top" element.

2.2.15 *Hint*: Begin by defining $f(a_1)$ as you please. Now suppose you have assigned the values of the first n elements in a one-to-one fashion obeying the condition. Show that $f(a_{n+1})$ can be chosen so that, given the previous choices, the condition is satisfied for the finite set $\{a_1, \ldots, a_n, a_{n+1}\}$.

2.2.16 *Hint*: Show that no sublattice of seven elements can be formed by deleting both a and b or by deleting any two of $\{r, s, t\}$.

2.2.18

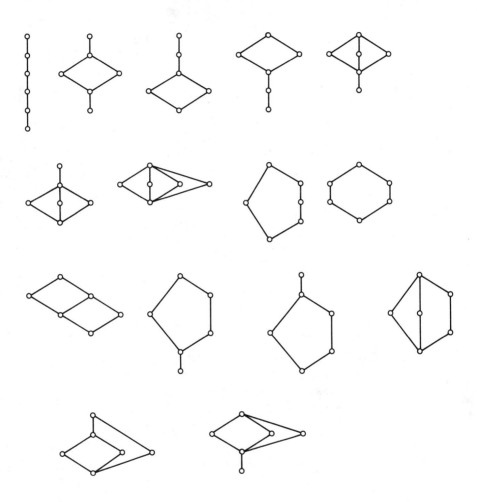

CHAPTER 3

Section 3.1

3.1.1   Yes, $\langle \mathbb{Z}, \odot \rangle$ is a group.

3.1.2   No, $\langle \mathbb{Q}^*, \div \rangle$ is not a group.

3.1.3   The group is not abelian.

3.1.4 The group has order 6, and the six functions may be written $I, f, g, g \circ g, f \circ g$, and $f \circ g \circ g$. You should write out a group table to show closure.

3.1.7 Notice that for all positive integers h, $(a^{-1}ba)^h = a^{-1}baa^{-1}ba \cdots a^{-1}ba = a^{-1}b^ha$. These two properties are used frequently in the work to come.

3.1.8 There is no left identity.

3.1.11 There are WLOG just three possible latin squares:

1	a	b	c
a	1	c	b
b	c	1	a
c	b	a	1

1	a	b	c
a	1	c	b
b	c	a	1
c	b	1	a

1	a	b	c
a	b	c	1
b	c	1	a
c	1	a	b

3.1.12 No. Associativity fails.

3.1.13 Yes.

3.1.14 Find the algebraic properties that separate these groups. For example, in (i) there is no solution to the equation $x + x = 3$. But all equations of the form $x + x = a$ have solutions in (ii). Do (i) and (ii) have more than one solution to $x + x = $ identity? Does (iii) have more than one solution to $x \cdot x = $ identity?

3.1.16 *Hint*: The elements that do not satisfy $x^2 = 1$ can be collected in pairs (x, x^{-1}). Now count elements.

3.1.18 Only the operation (iv) is guaranteed to give a group, and it will indeed be isomorphic to the original group.

Section 3.2

3.2.1 Yes, the groups are isomorphic. Number the vertices clockwise. Let the element a of Figure 3.2 be a clockwise rotation of $120°$, and let b be the reflection across the perpendicular bisector through the vertex numbered 1. Then the group table of the symmetries of the triangle is the table of Figure 3.2.

3.2.2 Every proper subgroup is cyclic. Let $A = \langle a \rangle$, $B = \langle b \rangle$, $C = \langle ab \rangle$, and $D = \langle a^2b \rangle$.

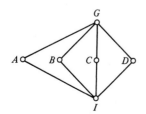

3.2.3 Number the vertices of the rectangle $1, 2, 3, 4$ in a clockwise fashion. The symmetries are the identity,

$$S = \begin{pmatrix} 1 & 2 & 3 & 4 \\ 2 & 1 & 4 & 3 \end{pmatrix}, \qquad T = \begin{pmatrix} 1 & 2 & 3 & 4 \\ 4 & 3 & 2 & 1 \end{pmatrix},$$

$$R = \begin{pmatrix} 1 & 2 & 3 & 4 \\ 3 & 4 & 1 & 2 \end{pmatrix}$$

The table is

I	R	S	T
R	I	T	S
S	T	I	R
T	S	R	I

3.2.5

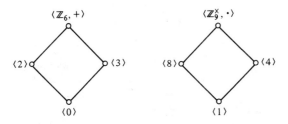

3.2.6 Example: Let G be the group $\langle \mathbb{Z}, + \rangle$ and let S be the set of positive integers.

3.2.8 Let a and b be positive integers and assume that $\gcd(a, b) = 1$.

$$\langle \mathbb{Z}, + \rangle \vee \langle a/b \rangle = \langle \mathbb{Z}, + \rangle \vee \langle 1/b \rangle = \{ n + (m/b) : n, m \in \mathbb{Z} \}$$

$$\langle 1/a \rangle \vee \langle 1/b \rangle = \langle 1/ab \rangle, \quad \langle 1/a \rangle \cap \langle 1/b \rangle = \langle 1 \rangle$$

3.2.10 The elements of $\langle \mathbb{Z}_{16}^{\times}, \cdot \rangle$ and their orders are

Elements: [1] [3] [5] [7] [−7] [−5] [−3] [−1]

Orders: 1   4   4   2   2   4   4   2

The lattice of subgroups is

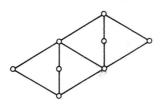

You provide the labeling!

3.2.11 $\langle \mathbb{Z}_{27}^{\times}, \cdot \rangle$ is a cyclic group of order 18 generated by [2]. Its lattice of subgroups is

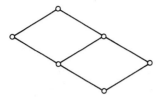

You provide the labeling!

3.2.12 If p is a prime the lattice of subgroups of a cyclic group of order p^n is a chain with $n + 1$ elements.

3.2.13 The lattice of subgroups of a cyclic group of order pq where p and q are distinct primes is

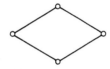

3.2.14 $\begin{pmatrix} 1 & 2 & 3 & 4 \\ 3 & 4 & 1 & 2 \end{pmatrix}$ and $\begin{pmatrix} 1 & 2 & 3 & 4 \\ 1 & 4 & 3 & 2 \end{pmatrix}$ will do.

3.2.15 *Hint*: Consider the subgroup generated by an element not in the (unique) subgroup covered by G.

3.2.17 *Hint*: Count the elements in a cyclic group of order n by counting the number of elements of each possible order.

3.2.18 The subgroup join of two subgroups is a subgroup IFF one subgroup contains the other.

3.2.19 *Hint*: An element in $B \vee C$ has the form bc where $b \in B$, $c \in C$. Since $A \supseteq B$, $b \in A$ also.

Section 3.3

3.3.2 The dihedral group D_{2n} has n elements of order 2 and $\varphi(d)$ elements of order d for every d dividing n.

3.3.3

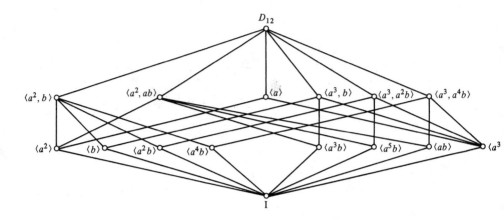

3.3.5 The number of subgroups of D_{2n} isomorphic to D_{2m} is n/m for each divisor m of n.

Section 3.4

3.4.1 The equivalence classes are the left cosets of S in G.

3.4.3 The number of elements in each equivalence class is the order of $A \cap B$.

3.4.9 *Hint*: If $xy = a$, then $yx = a$ in an abelian group, and so a will appear an even number of times off the diagonal.

3.4.10 *Hint*: Show that if $x(A \cap B)$ and $y(A \cap B)$ are distinct cosets of $A \cap B$ in B, then xA and yA are distinct cosets of A in $A \vee B$.

3.4.12 *Hint*: Remember that $4 \nmid 2s$ if s is odd.

3.4.13 There are two nonisomorphic groups of order 10, a cyclic group and a dihedral group.

3.4.14 There are two nonisomorphic groups of order 14, a cyclic group and a dihedral group.

3.4.15 There is only the cyclic group of order 85.

3.4.17 See the hint for Exercise 3.4.18.

3.4.18 *Hint*: The group $\mathbf{Z}_n^\times$ has order $\varphi(n)$.

3.4.19 How many elements of order 2 does $\mathbf{Z}_p^\times$ have? Then mimic Exercise 3.4.12.

3.4.20 Show $p \in \mathbf{Z}_{p^n-1}^\times$ and compute its order in this group.

CHAPTER 4

Section 4.1

4.1.1 *Hint:* Look for combinations of a, a^2, b, and b^2 that fix the vertices fixed by c and d. For example, $c = a^2 b^2$.

4.1.2 $b = cac^{-1}$

4.1.3 There are four elements in $\mathscr{A}_4$ having the form xax^{-1}.

4.1.4 Show that the complex $\langle a \rangle V$ is closed under multiplication and is a group. Now let u be any one of the elements of order 2. Determine b^2 by solving $u = ab^2$ and hence find a suitable b. Now, since $\mathscr{A}_4 = \langle a, b \rangle$ you can name all the elements in $G = \langle a \rangle \vee V$ according to the names in $\mathscr{A}_4$ and then verify that the group table of G is identical with that of the group table of $\mathscr{A}_4$.

Section 4.2

4.2.2 The order of the preimage of A^* is the product of the orders of A^* and the kernel.

4.2.9 *Hint:* Show that there can be but one subgroup of order p.

4.2.10 If $a \neq 0$, then $d = (1 + bc)a^{-1}$ and so the equation $ad - bc = 1$ has $3 \times 3 \times 2$ solutions. While if $a = 0$ the equation $ad - bc = 1$ has 3×2 solutions.

Section 4.3

4.3.2 The normal subgroups of the dihedral group D_{2n} are the subgroups of the cyclic subgroup of order n and, if they exist, the subgroups of index 2 in D_{2n} (each is isomorphic to $D_{2(n/2)}$). Hint for proof: Let $A = \langle a \rangle$ be the cyclic subgroup of order n. If S is a normal subgroup not in A, then $A \cap S = \langle a^k \rangle$ where $k|n$. Let $c \in S$ but not in A. Then $S = \langle c, a^k \rangle$ and $ca^k = a^{-k}c$. Then, if S is normal, $aca^{-1} = ca^{ku}$. Show $a^2 \in S$.

4.3.7 Remember that the commutator subgroup is normal. For these groups it is easy to examine the normal subgroups and decide which could be the commutator subgroup from the properties of the preceding exercise. Thus, the commutator subgroup of $\mathscr{A}_4$ is the subgroup of elements of order 2; the commutator subgroup of the quaternions is the subgroup of order 2; and the commutator subgroup of D_{2n} is the cyclic subgroup of order n if n is odd, of order $n/2$ if n is even.

4.3.10 *Hint:* Use the fact that $A \vee C = A \cdot C$ so that if $x \in A \vee C$, then $x = ac$. If x is also in B then $c = a^{-1}x \in B$ if $A \subseteq B$.

4.3.14 The automorphisms of the four-group are a group of order 6. The automorphisms of a cyclic group of order n are isomorphic to $\langle \mathbb{Z}_n^{\times}, \cdot \rangle$.

CHAPTER 5

Section 5.1

5.1.1   The groups of order 26 are the cyclic group and the dihedral group.

5.1.2   There are two abelian groups of order 12 and one abelian group of order 21.

5.1.7   Here is the lattice. You complete the labeling!

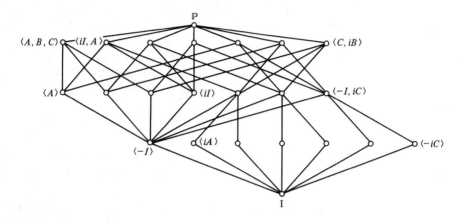

Section 5.2

5.2.3 $C_6 \times C_2 \cong V \times C_3$.

5.2.4   The group of automorphisms of $C_4 \times C_2$ is a four-group.

5.2.5 D_{2n} is a direct product of two proper subgroups IFF n is even.

Section 5.3

5.3.4   Here's the lattice of $C_4 \times C_4 = \langle a \rangle \vee \langle b \rangle$. You do the labeling.

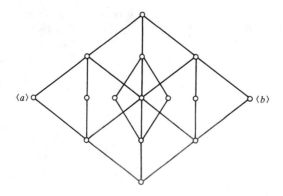

CHAPTER 6

Section 6.1

6.1.2 In $\mathscr{A}_5$ the identity, the 15 elements of order 2, and the 20 3-cycles can be written as the product of two transpositions.

6.1.3 There are 24 elements of order 5 in $\mathscr{A}_5$.

6.1.6 For $n = 3$ here is a possible sequence:

$$I, (12), (12)(13) = (123), (123)(23) = (13), (13)(12) = (132),$$
$$(132)(13) = (23), (23)(23) = I.$$

Section 6.2

6.2.3 There is the identity, 15 elements which are the product of two disjoint transpositions, 20 3-cycles, and 24 5-cycles.

6.2.4 The alternating group $\mathscr{A}_5$ has ten subgroups isomorphic to D_6 and six isomorphic to D_{10}.

6.2.5 The number of cyclic subgroups in $\mathscr{A}_5$ and $\mathbb{S}_5$ is given in this table:

Orders:	2	3	4	5	6
$\mathscr{A}_5$	15	10	0	6	0
$\mathbb{S}_5$	25	10	15	6	10

6.2.7 The elements that commute with $(1, 2, 3, 4, 5)$ are just its powers.

6.2.8 If π is an n-cycle, then π^d is the product of $\gcd(d, n)$ cycles each having length $n/\gcd(d, n)$.

Section 6.3

6.3.8 *Hint*: Show that a subgroup of order p_t is normal in G. If S_t is that group, consider G/S_t.

6.3.9 There are four groups of order 30. There is the cyclic group, the dihedral group, and two others, one of which may be generated by the 15-cycle $a = (1, 2, 3, 4, 5, 6, 7, 8, 9, 10, 11, 12, 13, 14, 15)$ and the element of order 2, $b = (2, 5)(3, 9)(4, 13)(7, 10)(8, 14)(12, 15)$.

CHAPTER 7

Section 7.1

7.1.4 *Hints*: Show that $2r = 0$ and hence that R has 2^k elements for some k. Show that the definition $a \leq b$ IFF $a = ab$ makes R into a partially ordered set in which $xy = \text{glb}(x, y)$ and $x + y + xy = \text{lub}(x, y)$. Show that the top element of this lattice is a multiplicative identity for R. Show that additive group $\langle R, + \rangle$ is the group direct sum of the subgroups of elements that cover 0. Now consider the correspondence $a \leftrightarrow \{x: x \text{ covers } 0 \text{ and } x \leq a\}$.

Section 7.2

7.2.9 There is one proper ideal in T. It consists of the matrices in T whose diagonal entries are 0.

7.2.13 Follow the notation of Exercise 7.2.11 and show that the factor rings $(A \vee B)/A$ and $B/(A \cap B)$ are isomorphic.

7.2.15 *Hint*: Show that $A \cap \mathbb{Z}$ and $\{y: x + y\sqrt{2} \in A\}$ are ideals of $\mathbb{Z}$.

CHAPTER 8

Section 8.1

8.1.6 *Hint*: Show that if R is a subring of $\mathbb{Q}$ containing 1, then if $a/b \in R$ and $\gcd(a,b) = 1$, then $1/b \in R$ and hence $1/p \in R$ for any p dividing b.

Section 8.2

8.2.9 If A is an ideal not R, then $1 \notin A$. Consider the set of ideals that contain A and are not equal to R.

Section 8.3

8.3.3 *Hint*: Compute the square of

$$\begin{bmatrix} 0 & 1 \\ 2 & 0 \end{bmatrix}$$

8.3.8 The smallest subfield of $\mathbb{R}$ containing $\sqrt{2}$, and $\sqrt{3}$ has as itself, $\mathbb{Q}$, $\mathbb{Q}(\sqrt{2})$, $\mathbb{Q}(\sqrt{3})$, and $\mathbb{Q}(\sqrt{6})$ as its only subfields.

8.3.11 A correct but trivial example is to choose ϕ to be the mapping of all elements to 0. A more interesting example would be the ring of diagonal matrices with rational entires and the mapping

$$\phi\left(\begin{bmatrix} a & 0 \\ 0 & b \end{bmatrix}\right) = \begin{bmatrix} a & 0 \\ 0 & 0 \end{bmatrix}$$

Does this example generalize?

CHAPTER 9

Section 9.1

9.1.1 One function from $\mathbb{Z}_6$ to $\mathbb{Z}_6$ that is not a polynomial function sends 0 to 1 and $1, 2, 3, 4, 5$ to 0.

9.1.3 There is one residue class for each polynomial function from R into R.

9.1.5 The ideal $\langle 2 \rangle$ in $\mathbb{Z}[x]$ is contained in the proper ideal $\langle 2, x \rangle$.

9.1.11 The distributive law fails.

9.1.12 If a is a unit of R, then the mapping $f(x) \to f(ax + b)$ is an automorphism of $R[x]$.

Section 9.2

9.2.2 $x + 1$ and $x + 1 = (x^4 + x^3 - 2x^2 - x + 1)(x^2 + 2)/4 - (x^4 + x^3 + 2x^2 + 3x + 1)(x^2 - 2)/4$.

9.2.3 $f(x) = -(x^2 + x - 1)/2$

9.2.4 $f(x) = x$ is a solution.

9.2.5 $u^{-1} = u + 1$.

9.2.7 *Hint*: Factor $g(x) = p_1(x) \cdots p_t(x)$ into powers of distinct prime factors and use induction on t. Strip off each prime in turn. Let $g_2(x) = p_2(x) \cdots p_t(x)$. Find $1 = p_1(x)u_2(x) + f_1(x)g_2(x)$. So

$$\frac{1}{g(x)} = \frac{f_1(x)}{p_1(x)} + \frac{u_2(x)}{g_2(x)}$$

Continue. Check the requirements on the degrees of f_1 and u_2.

9.2.10 Only the two cubics.

9.2.12 For irreducible $q(x)$ to have multiple zeros over a field of characteristic p, $q(x)$ must be a polynomial in x^p.

9.2.15 $2x + 1$ is a unit over Z_4. $(2x + 1)^2 = 4x^2 + 4x + 1 = 1$.

Section 9.3

9.3.2 For $a, b \in D$ consider the ideal $\langle a \rangle \cap \langle b \rangle$.

9.3.3 For $0 \neq x \in D_S$ write $x = a/b$ where $\gcd(a, b) = 1$. Define $\sigma(x) = |a|$.

Section 9.4

9.4.1 Degree 1: x and $x + 1$
Degree 2: $x^2 + x + 1$
Degree 3: $x^3 + x + 1$
           $x^3 + x^2 + 1$
Degree 4: $x^4 + x + 1$
           $x^4 + x^3 + 1$
           $x^4 + x^3 + x^2 + x + 1$

9.4.2 One way of constructing the irreducible monic polynomials of degree d is to have lists of all irreducible monic polynomials of degree less than d and use this to construct all *reducible* monic polynomials of degree d and eliminate them from a list of all monic polynomials of degree d.

9.4.4 If d is the degree of $p(x)$, then $f[x]/\langle p(x) \rangle$ has q^d elements.

9.4.5 The residue class $u = x + \langle p(x) \rangle$ has order 15 if $p(x) = x^4 + x + 1$ or $x^4 + x^3 + 1$ and order 5 if $p(x) = x^4 + x^3 + x^2 + x + 1$.

CHAPTER 10

Section 10.1

10.1.5 There are $(1/n!)\prod_{k=0}^{n-1}(p^n - p^k)$ bases in $V_n(F)$ and $\prod_{k=0}^{n-1}(p^n - p^k)$ $n \times n$ nonsingular matrices with entries in F.

10.1.6 The column space changes, but its dimension stays the same.

10.1.8 The solution vectors $(x, y, z) = (a, -1, 3 - a)$ where a is any element in the field of coefficients. Or one may say the set of solutions is the coset $(0, -1, 3)$ $+ \langle 1, 0, -1 \rangle$.

10.1.9 The necessary and sufficient conditions for a solution in integers are that $3|(a + b)$ and that $a + 2b - c = 0$.

10.1.10 The number of complementary spaces T is $3^{(n-k)k}$. Fix a particular basis for S. Let W be the number of sets of $n - k$ vectors that extend that basis of S to a basis for all of V_n. Each such extension spans a complementary space T. T has dimension $n - k$. Now divide W by the number of bases for T.

Section 10.2

10.2.5 The subfields of $\mathbf{Q}(\sqrt[3]{2}, \omega)$ are itself, $\mathbf{Q}$, $\mathbf{Q}(\sqrt[3]{2})$, $\mathbf{Q}(\omega \cdot \sqrt[3]{2})$, $\mathbf{Q}(\omega^2 \cdot \sqrt[3]{2})$, and $\mathbf{Q}(\omega)$.

10.2.6 $(1 + \theta)^{-1} = (2\theta^2 - 2\theta + 11)/5$

10.2.12 If neither u nor v is 0, then the quotient u/v must be a square in F.

Section 10.3

10.3.1 The splitting field of a quadratic polynomial has the form $F(\theta)$ where $\theta^2 \in F$ unless $\text{char}(F) = 2$. If $\text{char}(F) = 2$ about all that can be said is that θ is a zero of the quadratic polynomial.

10.3.5 Here is a table giving an irreducible factor of $x^n - 1$.

n	Polynomial factor
2	$x + 1$
3	$x^2 + x + 1$
4	$x^2 + 1$
6	$x^2 - x + 1$
8	$x^4 + 1$
9	$x^6 + x^3 + 1$
10	$x^4 - x^3 + x^2 - x + 1$
12	$x^4 - x^2 + 1$
p prime	$x^{p-1} + \cdots + x + 1$

10.3.6 *Hints:* α^2 is a zero of $x^2 - 2x - 2$; say $\alpha^2 = 1 + \sqrt{3}$. Moreover, $x^4 - 2x^2 - 2 = (x^2 - \alpha^2)(x^2 - \beta^2)$. An automorphism of E sending $\sqrt{c}$ into $-\sqrt{c}$ must send α to β or $-\beta$. If $\theta \in E$ and $\theta^2 \in \mathbf{Q}$ write $\theta = w + y\beta$ where $w, y \in \mathbf{Q}(\alpha)$ and show that either w or $y = 0$. Treat each case separately.

Show $\theta \in \mathbb{Q}(\alpha)$ implies $\theta \in \mathbb{Q}(\sqrt{3})$, while $\theta = y\beta$ implies $\theta \in \mathbb{Q}(\sqrt{-2})$ or $\theta \in \mathbb{Q}(\sqrt{-6})$. The group of automorphisms of E is the dihedral group of order 8.

10.3.8 The lattice of subfields of an extension of degree 6 over $\mathbb{Q}$ is either a chain of two or three elements or one of the following two lattices:

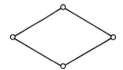 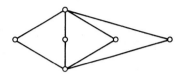

E will have either no subfields, one subfield of degree 2 over $\mathbb{Q}$, one subfield of degree 3 over Q, one subfield of degree 2 and one of degree 3 over $\mathbb{Q}$, or one subfield of degree 2 and three subfields of degree 3 over $\mathbb{Q}$.

Section 10.4

10.4.4 The zeros of $x^4 + x + 1$ and it reverse $x^4 + x^3 + 1$ are generators of the multiplicative group of GF(16). If α is one of these zeros, α^3 is a zero of $x^4 + x^3 + x^2 + x + 1$.

10.4.5 *Hint*: If α is a zero of $x^3 + 2x + 1$ in GF(3)/$\langle x^3 + 2x + 1 \rangle$, find a zero of $x^3 + 2x + 1$ in GF(3)/$\langle x^3 + 2x + 2 \rangle$ and map α to this element. Now extend this mapping.

INDEX